Introduction to Electrical Circuit Analysis

Introduction to Electrical Circuit Analysis

Özgür Ergül
Middle East Technical University, Ankara, Turkey

WILEY

This edition first published 2017

Registered Offices
John Wiley & Sons, Inc., 111 River Street, Hoboken, NJ 07030, USA
John Wiley & Sons Ltd, The Atrium, Southern Gate, Chichester, West Sussex, PO19 8SQ, UK

Editorial Office
The Atrium, Southern Gate, Chichester, West Sussex, PO19 8SQ, UK

For details of our global editorial offices, customer services, and more information about Wiley products visit us at www.wiley.com.

Wiley also publishes its books in a variety of electronic formats and by print-on-demand. Some content that appears in standard print versions of this book may not be available in other formats.

Library of Congress Cataloging-in-Publication data applied for

ISBN: 9781119284932

Cover design by Wiley
Cover image: (Circuit Board) © ratmaner/Gettyimages; (Electronic Components) © DonNichols/Gettyimages; (Formulas) © Bim/Gettyimages

Set in 10/12pt WarnockPro by SPi Global, Chennai, India

10 9 8 7 6 5 4 3 2 1

For my wife Ayça, three cats (Boncuk, Pepe, and Misket), and the dog, who all suffered during the writing of this book

Contents

Important Units

- Ampere (A)
- Coulomb (C)
- Farad (F): C/V
- Henry (H): weber/A
- Hertz (Hz): 1/s
- Joule (J): N m = kg m^2/s^2
- kilo (k…): ×1000
- meter (m)
- micro (μ …): $\times 10^{-6}$
- milli (m…): $\times 10^{-3}$
- Newton (N): kg m/s^2
- Second (s)
- Siemens (S): A/V
- Volt (V): J/C
- Volt-ampere (V A): J/s
- Watt (W): J/s

Conventions with Examples

- Fractions: $14/5\ \mathrm{A} = \dfrac{14}{5}\ \mathrm{A} = 2.8\ \mathrm{A}$
- Irrational numbers: $13/3\ \mathrm{A} = 4.33333333\ldots\ \mathrm{A} \neq 4.33\ \mathrm{A}$
- Approximation: $13/3\ \mathrm{A} \approx 4.33\ \mathrm{A}$
- Scientific notation: $3.4 \times 10^{3} = 3400$ and $3.4 \times 10^{-3} = 0.0034$
- Multiplication without sign: $v_a i_b = v_a \times i_b$
- Number ranges: $[9, 10]$ = all x that satisfy $9 \leq x \leq 10$
- Limit of a number from left and right: $10^{-} < 10 < 10^{+}$

Preface

Since the first known electricity experiments more than 25 centuries ago by Thales of Miletus, who believed that there should be better ways than mythology to explain physical phenomena, humankind has worked hard to understand and use electricity in many beneficial ways. The last three centuries have seen rapid developments in understanding electricity and related concepts, leading to constantly accelerating technology advancements in the last several decades. Today, most of us simply cannot live without electricity, and it is almost ubiquitous in daily life. We are so attached to and dependent on electricity that there are even post-apocalyptic fiction movies and film series based on sudden electrical power blackouts. And they are terrifying.

Electricity is one of a few subjects with which we have a strange relationship. The more we use it, less we know about it. Electrical and electronic devices, where electricity is somehow used to produce beneficial outputs, are a closed book to most of us, until we open them (not a suggested activity!) and see that they contain incredibly small but highly intelligent parts. These parts, some of which once had huge dimensions and even filled entire rooms, are now so tiny that we are able to place literally billions of them (at the time of writing) in a smartphone microprocessor. One billion is a huge number; at a rate of one a second, it takes 31 years to count. And we are able to put these uncountable (OK, countable, but not feasibly so) numbers of components together and make them work in harmony for our enjoyment. Yet most of us know little about how they actually work.

The topic of circuit analysis has naturally developed in parallel with electrical circuits and devices starting from centuries ago. To provide some intuition, Ohm's law has been known since 1827, while Kirchhoff's laws were described in 1845. Nodal and mesh analysis methods have been developed and used for systematically applying Kirchhoff's laws. Phasor notation is borrowed from mathematics to deal with time-harmonic circuits. These fundamental laws have not changed, and they will most probably remain the same in the coming years. In general, basic laws describe everything when they are wisely used. Hence, more and more sophisticated circuits in future technologies will also benefit from them, independent of their complexity.

Circuit analysis is naturally linked to all other technologies involving electricity, including medical, automotive, computer, energy, and aerospace industries, as well as all subcategories of electrical and electronic engineering. Interestingly, with the rapid development of technology, we tend to learn fundamental laws more superficially. One can identify two major factors, among many:

- As circuits become more complicated and specialized, we are attracted and guided to focus on higher-level representations, such as inputs and outputs of microchips with well-defined functions, without spending time on fundamental laws.
- Great advancements in circuit-solver software "eliminate" the need to fully understand fundamental laws and appreciate their importance in everyday life, reducing circuit analysis to numbers.

Unfortunately, without absorbing fundamental laws, we tend to make major conceptual mistakes. Most instructors have had a student who offers infinite energy by rotating something (usually a car wheel if s/he is a mechanical engineering student), disregarding the conservation of energy. It is often a confusing issue for a biomedical student to appreciate the necessity of grounding for medical safety. And it is probably a computational mistake but not a new technology if a circuit analyzer program provides a negative resistor value. The aim of this book is to gradually construct the basics of circuit analysis, even though they are not new material, while accelerating our understanding of electrical circuits and all technologies using electricity.

This is intended as an introductory book, mainly designed for college and university students who may have different backgrounds and, for whatever reason, need to learn about circuits for the first time. It mainly focuses on a few essential components of electrical components, namely,

- resistors,
- independent voltage and current sources,
- dependent sources (as closed components, not details),
- capacitors, and
- inductors.

On the other hand, transistors, diodes, OP-AMPs, and similar popular and inevitable components of modern circuits, which are fixed topics (and even starting points) in many circuit books, are not detailed. The aim of this book is not to teach electrical circuits, but rather to teach how to analyze them. From this perspective, the components listed above provide the required combinations and possibilities to cover the fundamental techniques, namely,

- Ohm's and Kirchhoff's laws,
- nodal analysis,
- mesh analysis,
- the black-box approach and Thévenin/Norton equivalent circuits.

This book also covers the analysis methods for both DC and AC cases in transient and steady states.

To sum up, the technology that is covered in this book is well established. The analysis methods and techniques, as well as components, listed above have been known for decades. However, the fundamental methods and components need to be known in sufficient depth in order to understand how electrical circuits work, including state-of-the-art devices and their ingredients. Many books in this area are dominated by an increasing number of new electrical and electronic components and their special working principles, while the fundamental techniques are squeezed into short descriptions and limited to a few examples. Therefore, the purpose of this book is to provide sufficient basic discussion and hands-on exercises (with solutions at the back of the book) before diving into modern circuits with higher-level properties.

Enjoy!

About the Companion Website

This book is accompanied by a companion website:

www.wiley.com/go/ergul4412

The website includes:

- Exercise sums and solutions
- Videos

1

Introduction

We start with the iconic figure (Figure 1.1), which depicts a bulb connected to a battery. Whenever the loop is closed and a full connection is established, the bulb comes on and starts to consume energy provided by the battery. The process is often described as the conversion of the chemical energy stored in the battery into electrical energy that is further released as heat and light by the bulb. The connection between the bulb and battery consists of two wires between the positive and negative terminals of the bulb and battery. These wires are shown as simple straight lines, whereas in real life they are usually coaxial or paired cables that are isolated from the environment.

The purpose of this first chapter is to introduce basic concepts of electrical circuits. In order to understand circuits, such as the one above, we first need to understand electric charge, potential, and current. These concepts provide a basis for recognizing the interactions between electrical components. We further discuss electric energy and power as fundamental variables in circuit analysis. The time and frequency in circuits, as well as related limitations, are briefly considered. Finally, we study conductivity and resistance, as well as resistors, independent sources, and dependent sources as common components of basic circuits.

1.1 Circuits and Important Quantities

An electrical circuit is a collection of components connected via metal wires. Electrical components include but are not limited to resistors, inductors, capacitors, generators (sources), transformers, diodes, and transistors. In circuit analysis, wire shapes and geometric arrangements are not important and they can be changed, provided that the connections between the components remain the same with fixed geometric topology. Wires often meet at intersection points; a connection of two or more wires at a point is called a node. Before discussing how circuits can be represented and analyzed, we first need to focus on important quantities, namely, electric charge, electric potential, and current, as well as energy and power.

1.1.1 Electric Charge

Electric charge is a fundamental property of matter to describe force interactions among particles. According to Coulomb's law, there is an attractive (negative) force between a

Introduction to Electrical Circuit Analysis, First Edition. Özgür Ergül.

Companion Website: www.wiley.com/go/ergul4412

Figure 1.1 A simple circuit involving a bulb connected to a battery. The connection between the bulb and battery is shown via simple lines.

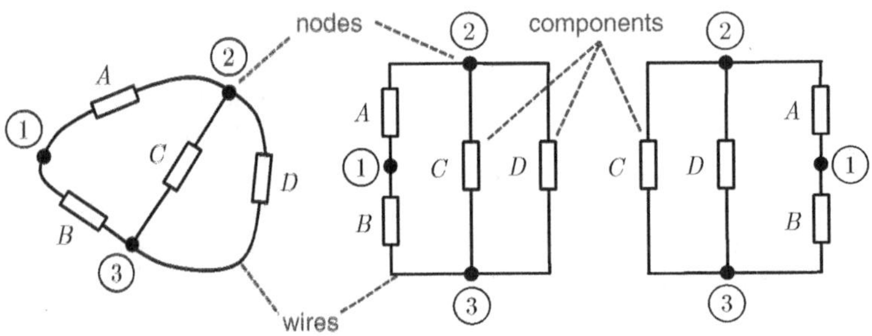

Figure 1.2 A circuit involving connections of four components labeled from *A* to *D*. From the circuit-analysis perspective, connection shapes are not important, and these three representations are equivalent.

proton and an electron given by

$$F_{pe} \approx -\frac{2.3071 \times 10^{-28}}{d^2} \quad \text{(newton (N))},$$

which is significantly larger than (around 1.2×10^{36} times) the gravity between these particles. In the above, d is the distance between the proton and electron, given in meters (m). This law can be rewritten by using Coulomb's constant

$$k \approx 8.9876 \times 10^9 \ \ (\text{N m}^2/\text{C}^2)$$

as

$$F_{pe} \approx k\frac{q_e q_p}{d^2} \ \ (\text{N}),$$

where

$$q_p \approx +1.6022 \times 10^{-19} \ \ (\text{C})$$
$$q_e = -q_p \approx -1.6022 \times 10^{-19} \ \ (\text{C})$$

are the electrical charges of the proton and electron, respectively, in units of coulombs (C). Coulomb's constant enables the generalization of the electric force between any arbitrary charges q_1 and q_2 as

$$F_{12} \approx k\frac{q_1 q_2}{d^2} \ \ (\text{N}),$$

where q_1 and q_2 are assumed to be point charges (theoretically squeezed into zero volumes), which are naturally formed of collections of protons and electrons.

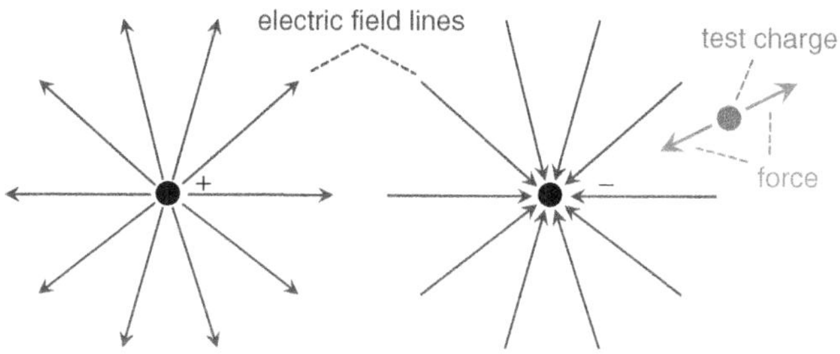

Figure 1.3 Electric field lines created individually by a positive charge and a negative charge. An electric field is assumed to be created whether there is a second test charge or not. If a test charge is located in the field, repulsive or attractive force is applied on it.

The definition of the electric force above requires at least two charges. On the other hand, it is common to extend the physical interpretation to a single charge. Specifically, a stationary charge q_1 is assumed to create an electric field (intensity) that can be represented as

$$E_1 \approx k\frac{q_1}{d^2} \text{ (N/C)},$$

where d is now the distance measured from the location of the charge. This electric field is in the radial direction, either outward (positive) or inward (negative), depending on the type (sign) of the charge. Therefore, we assume that an electric field is always formed whether there is a second test charge or not. If there is q_2 at a distance d, the electric force is now measured as

$$F_{12} = E_1 q_2 \text{ (N)},$$

either as repulsive (if q_1 and q_2 have the same sign) or attractive (if q_1 and q_2 have different signs).

The definition of the electric field is so useful that, in many cases, even the sources of the field are discarded. Consider a test charge q exposed to some electric field E. The force on q can be calculated as

$$F = qE \text{ (N)},$$

without even knowing the sources creating the field. This flexibility further allows us to define the electric potential concept, as discussed below.

1.1.2 Electric Potential (Voltage)

Consider a charge q in some electric field created by external sources. Moving the charge from a position b to another position a may require energy if the movement is opposite to the force due to the electric field. This energy can be considered to be absorbed by the charge. If the movement and force are aligned, however, energy is extracted from the charge. In general, the path from b to a may involve absorption and release of energy, depending on the alignment of the movement and electric force from position to position. In any case, the net energy absorbed/released depends on the start and end points, since the electric field is conservative and its line integral is path-independent.

Electric potential (voltage) is nothing but the energy considered for a unit charge (1 C) such that it is defined independent of the testing scheme. Specifically, the work done in

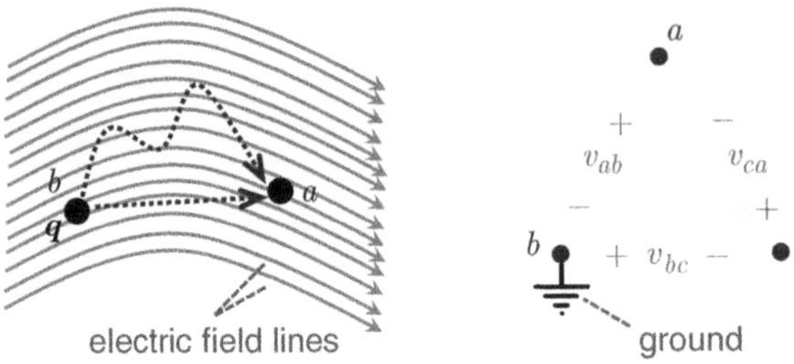

Figure 1.4 Movement of a charge in an electric field created by external sources. The energy absorbed or released by the charge does not depend on the path but depends on the potentials at the start and end points. The electric potential (voltage) is always defined between two points, while selecting a reference point as a ground enables unique voltage definitions at all points.

moving a unit charge from a point b to another point a is called the voltage between a and b. Conventionally, we have

$$v_{ab} = v_a - v_b$$

as the voltage between a and b, corresponding to the work done in moving the charge from b to a. If $v_a > v_b$, then work must be done to move the charge (the energy of the charge increases). On the other hand, if $v_b > v_a$, then the work done is negative, indicating that energy is actually released due to the movement of the charge. The unit of voltage is the volt (V), and 1 volt is 1 joule per coulomb (J/C).

A proper voltage definition always needs two locations and a polarity definition. Considering three separate points a, b, and c, we have

$$v_{ab} = v_a - v_b,$$
$$v_{bc} = v_b - v_c,$$
$$v_{ca} = v_c - v_a,$$

and

$$v_{ab} + v_{bc} + v_{ca} = 0.$$

The equality above is a result of the conservation of the electric energy (conservative electric field). On the other hand, v_a, v_b, and v_c are not yet uniquely defined. In order to simplify the analysis in many cases, a location can be selected as a reference with zero potential. In circuit analysis, such a location that corresponds to a node is called ground, and it allows us to define voltages at all other points uniquely. For example, if $v_b = 0$ in the above, we have $v_a = v_{ab} + v_b = v_{ab}$.

1.1.3 Electric Current

A continuous movement of electric charges is called electric current. Conventionally, the direction of a current flow is selected as the direction of movement of positive charges. The unit of current is the ampere (A), and 1 ampere is 1 coulomb per second (C/s). Formally, we have

$$i(t) = \frac{dq}{dt} \text{ (A)},$$

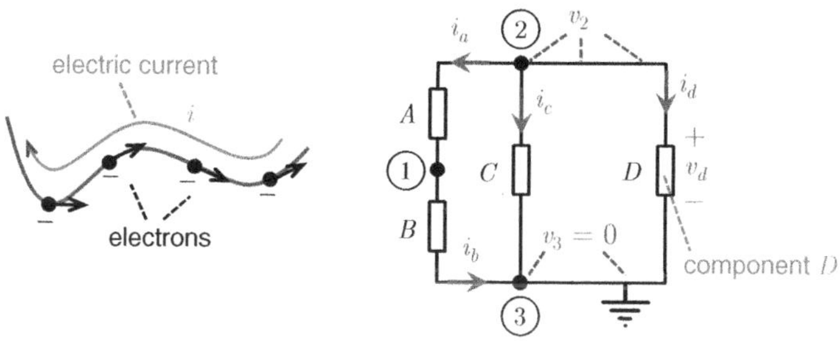

Figure 1.5 On a metal wire, the conventional current direction, which is defined as positive charge flow, is the opposite of the actual electron movements. In a circuit, voltages are defined at the nodes, as well as across components, using the sign convention.

where q and t represent charge and time, respectively. The current itself may depend on time, as indicated in this equation. But, in some cases, we only have steady currents, $i(t) = i$, where i does not depend on time.

Different types of current exist, as discussed in Section 1.2.1. In circuit analysis, however, we are restricted to conduction currents, where free electrons of metals (e.g., wires) are responsible for current flows. Since electrons have negative charges and an electric current is conventionally defined as the flow of positive charges, electron movements and the current direction on a wire are opposite to each other. Indeed, when dealing with electrical circuits, using positive current directions is so common that the actual movement of charges (electrons) is often omitted.

When charges move, they interact with each other differently such that they cannot be modeled only with an electric field. For example, two parallel wires carrying currents in opposite directions attract each other, even though they do not possess any net charges considering both electrons and protons. Similar to the interpretation that electric field leads to electric force, this attraction can be modeled as a magnetic field created by a current, which acts as a magnetic force on a test wire. Electric and magnetic fields, as well as their coupling as electromagnetic waves, are described completely by Maxwell's equations and are studied extensively in electromagnetics.

1.1.4 Electric Voltage and Current in Electrical Circuits

In electrical circuit analysis, charges, fields, and forces are often neglected, while electric voltage and electric current are used to describe all phenomena. This is completely safe in the majority of circuits, where individual behaviors of electrons are insignificant (because the circuit dimensions are large enough with respect to particles), while the force interactions among wires and components are negligible (because the circuit is small enough with respect to signal wavelength). The behavior of components is also reduced to simple voltage–current relationships in order to facilitate the analysis of complex circuits. The limitations of circuit analysis using solely voltages and currents are discussed in Section 1.6.

In an electrical circuit, voltages are commonly defined at nodes, while currents flow through wires and components. A wire is assumed to be perfectly conducting (see Section 1.2.4) such that no voltage difference occurs along it, that is, the voltage is the same on the entire wire. This is the reason why their shapes are not critical.

On the other hand, a voltage difference may occur across a component, depending the type of the component and the overall circuit. For unique representation of a node voltage, a reference node should be selected as a ground. However, the voltage across a component can always be defined uniquely since it is based on two or more (if the component has multiple terminals) points.

In circuit analysis, voltages and currents are usually unknowns to be found. Since they are not known, in most cases, their direction can be arbitrarily selected. When the solution gives a negative value for a current or a voltage, it is understood that the initial assumption is incorrect. This is never a problem at all. For consistency, however, it is useful to follow a sign convention by fixing the voltage polarity and current direction for any given component. In the rest of this book, the current through a component is always selected to flow from the positive to the negative terminal of the voltage.

1.1.5 Electric Energy and Power of a Component

Consider a component d with a current i_d and voltage v_d, defined in accordance with the sign convention. If $i_d > 0$, one can assume that positive charges flow from the positive to the negative terminal of the component. In addition, if $v_d > 0$, these positive charges encounter a drop in their potential values, that is, they release energy. This energy must be somehow used (consumed or stored) by the component. Formally, we define the energy of the component as

$$w_d(t) = \int_0^t v_d(t')i_d(t')dt' \text{ (J)},$$

where the time integral is used to account for all charges passing during $0 \leq t' \leq t$, assuming that the component is used from time $t' = 0$. If $w_d(t) > 0$, it is understood that the component consumes net energy during the time interval $[0, t]$. On the other hand, if $w_d(t) < 0$, the component produces net energy in the same time interval. We note that the unit of energy is the joule, as usual.

Energy as defined above provides information in selected time intervals. In many cases, however, it is required to know the behavior (change of the energy) of the component at a particular time. For a device d with a current i_d and voltage v_d, this corresponds to the time derivative of the energy, namely the power of the device, defined as

$$p_d(t) = \frac{dw}{dt} = v_d(t)i_d(t) \text{ (W)}.$$

Specifically, for a given component, its power is defined as the product of its voltage and current. The unit of power is the watt (W), where 1 watt is 1 volt ampere (V A) or 1 joule per second (J/s). If $p(t) > 0$, the component absorbs energy at that specific time. Otherwise (i.e., if $p(t) < 0$), the component produces energy.

Example 1: Electric power and energy are often underestimated. Consider an 80 W bulb, which is on for 24 hours. Using the energy spent by the bulb, how many meters can a 1000 kg object be lifted?

Solution: The energy spent by the bulb is

$$w_b = 24 \times 60 \times 60 \times 80 = 6.912 \times 10^6 \text{ J}.$$

Then, assuming $g = 10\,\text{m/s}^2$, and using $w_p = mgh$ for the potential energy, we have

$$1000 \times 10 \times h = 6.912 \times 10^6 \longrightarrow h = 691.2 \text{ m}.$$

Example 2: There are approximately 12×10^9 bulbs on earth. Assuming an average on period of 6 hours and 50 W average power, find the amount of coal required to produce the same amount of energy for 1 day. Assume that the thermal energy of coal is 3×10^4 J/kg and the efficiency of the conversion of the energy is 100%.

Solution: The required energy for the bulbs per day is

$$w_b = 12 \times 10^9 \times 50 \times 6 \times 60 \times 60 = 1.296 \times 10^{16} \text{ J}.$$

The corresponding amount of coal can be found as

$$13 \times 10^4 \times m_c = 1.296 \times 10^{16} \longrightarrow m_c = 432 \times 10^9 \text{ kg}.$$

Example 3: The voltage and current of a device are given by $v(t) = 100\exp(-3t)$ V and $i(t) = 2[1 - \exp(-3t)]$ A, respectively, as functions of time. Find the maximum power of the device.

Solution: We have

$$\begin{aligned} p(t) &= v(t)i(t) = 200\exp(-3t)[1 - \exp(-3t)] \\ &= 200[\exp(-3t) - \exp(-6t)] \text{ W} \end{aligned}$$

as the power of the device. We note that

$$\begin{aligned} p(0) &= 0, \\ p(\infty) &= 0. \end{aligned}$$

In order to find the maximum point for the power, we use

$$\frac{dp(t)}{dt} = 200[-3\exp(-3t) + 6\exp(-6t)] = 0,$$

leading to

$$16\exp(-6t) = 3\exp(-3t) \longrightarrow \exp(3t) = 2 \longrightarrow t = \ln(2)/3 \text{ s}.$$

Then the maximum power is

$$\begin{aligned} p(\ln(2)/3) &= 200[\exp(-\ln(2)) - \exp(-2\ln(2))] \\ &= 200(1/2 - 1/4) = 50 \text{ W}. \end{aligned}$$

Exercise 1: A device has a power of 60 W when it is active and 10 W when it is on standby. An engineer measures that it spends a total of 2664 kJ energy in 24 hours. How many hours was the device actively used?

1.1.6 DC and AC Signals

Until now, we have considered the time concept in circuits for studying energy (which needs to be defined in time intervals) and power (which may depend on time). In fact, the time dependence of the power of a component corresponds to the time dependence

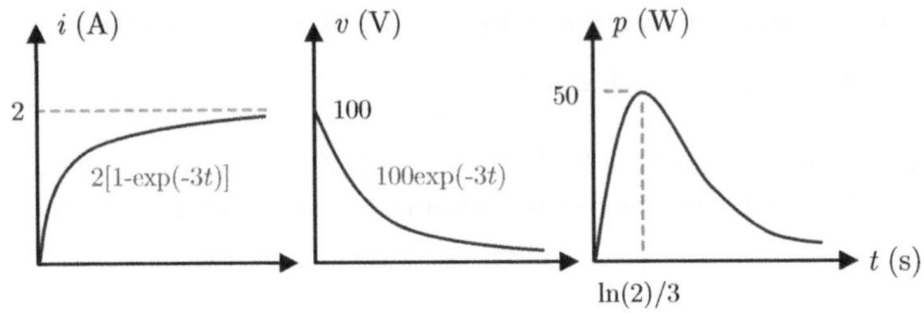

Figure 1.6 Power of a device for given current and voltage across it.

of its voltage and/or current. This brings us to the definition of direct current (DC) and alternating current (AC), which are important terms in describing and categorizing circuits and their components.

DC means a unidirectional flow of electric charges, leading to a current only in a single direction. However, the term 'DC signal' is commonly used to describe voltages and other quantities that do not change polarity. DC signals are produced by DC sources, whose voltages or currents are assumed to be fixed in terms of direction and amplitude. Examples of DC sources are batteries and dynamos. Voltage and current values of these sources may have very slight variations with respect to time, which are often neglected in circuit analysis.

AC describes electric currents and voltages that periodically change direction and polarity. This periodicity is generally imposed by AC sources, which may provide voltage and currents in sinusoidal, triangular, square, or other periodic forms. AC is commonly used in all electricity networks, including homes. The reason for its common usage is its well-known advantage when transmitting AC signals over long distances. Specifically, the electric power can be transmitted with less ohmic losses in the AC form in comparison to the DC form. In addition, AC signals can be amplified or reduced easily via transformers, making it possible to use different voltage and current values in different lines of electricity networks and electrical devices. In general, AC circuits have a fixed periodicity and frequency, which is set to 50–60 hertz (Hz = 1/s) in domestic usage. AC signals are also associated with electromagnetic waves (e.g., radiation from electrical components).

AC and DC signals can be converted into each other. The conversion from AC to DC is achieved by rectifiers, while inverters are used to convert DC signals into AC signals. DC to DC and AC to AC converters are also common when the properties but not the types of the signals need to modified.

1.1.7 Transient State and Steady State

Whether DC or AC, any circuit in real life has a time dependency, at least when switching the circuit on and off. The short-time state, in which variations in voltage and current values are encountered due to outer effects (e.g., switching), is called the transient state. Whether it is a DC or AC circuit, any circuit can be in a transient state before it reaches an equilibrium. A transient state is usually an unwanted state, where voltage, current, and power values involve fluctuations that are not designed on purpose.

In the long term, circuits that are not disturbed by outer effects enter into equilibrium, namely, the steady state. Theoretically, an infinite time is required to pass from transient

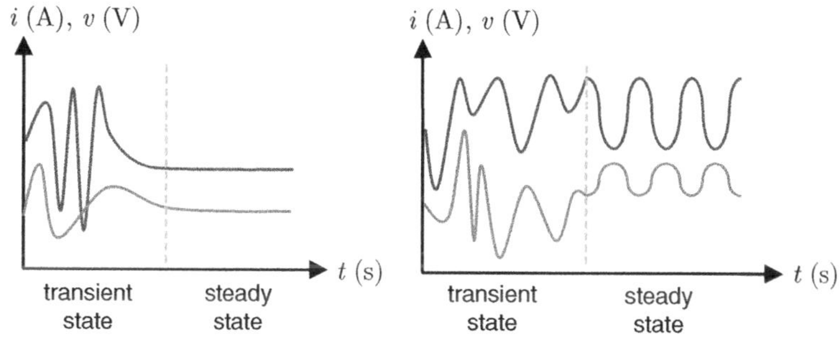

Figure 1.7 Transient state and steady state in DC and AC signals.

state to equilibrium, while most circuits are assumed to reach steady state after a sufficient period (i.e., when fluctuations become negligible). For DC circuits in steady state, voltage and current values are assumed to be constant. In the first few chapters, a steady state is automatically assumed when only resistors and DC sources are considered. In fact, the time needed to pass from transient state to steady state depends on a time constant, which is a contribution of both resistors and energy-storage elements (capacitors and inductors). Hence, circuits with only resistors and DC sources have zero transient time, that is, they can be assumed to be in steady state without any transient analysis. For AC circuits, voltages and currents in steady state oscillate with the time period dictated by the sources. Therefore, we emphasize that the steady state does not indicate constant properties for all circuits.

1.1.8 Frequency in Circuits

When AC sources are involved in a circuit, voltage and current values oscillate with respect to time. In most cases, the periodicity and frequency are fixed, that is, all voltages and currents change at the same rate, while there can be phase differences (delays) between them. The behavior of some components does not rely on the frequency, unless they are exposed to extreme conditions. As an example, resistors behave almost the same in a wide range of frequencies. On the other hand, many components, such as capacitors and inductors, strongly depend on the frequency. With DC sources, corresponding to zero frequency, capacitors/inductors act like open/short circuits, while they become almost the opposite at very high frequencies. Therefore, the behavior of an AC circuit directly depends on the frequency, as discussed extensively in time-harmonic analysis.

1.2 Resistance and Resistors

Resistors (Figure 11.1) are fundamental components in electrical circuits. They are basically energy-consuming elements that are used to control voltage and current values in circuits. In addition, the energy conversion ability of resistors can be useful in various applications, where these elements are directly used for heating and lighting (conventional bulbs). Specifically, the energy consumed by a resistor is usually released as heat, and sometimes as useful light. Resistance is a common property of all metals, and even very conductive wires have resistances, which may need to be included in circuit analysis.

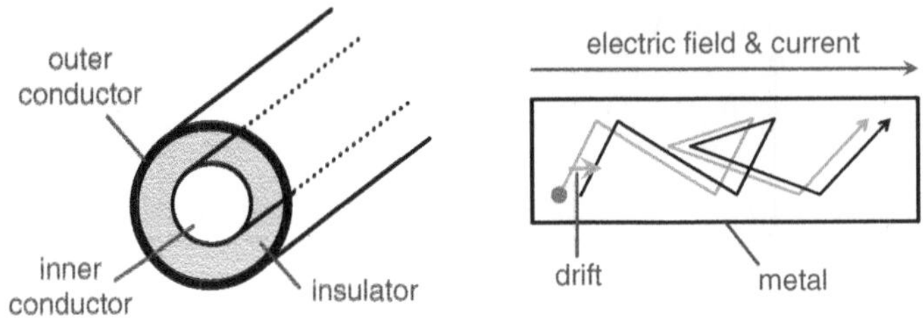

Figure 1.8 Structure of a general coaxial cable and a representation of the drift velocity of an electron under an electric field.

1.2.1 Current Types, Conductance, and Ohm's Law

In order to understand resistance and resistors, first we need to define the conduction current. As described in Section 1.1.3, current is a continuous flow of charges. In electrolytes, gases, and plasmas, currents may be formed by ions, and even by moving protons. In some applications, electrons can be injected from special devices, leading to a current flow in a vacuum. In circuits, however, currents are mostly formed by the conduction of metals.

In good conductors, one or more electrons from each atom is weakly bound to the atom. These electrons can move freely in the metal (especially on the surfaces), while these movements are random if the metal is not exposed to an electric field and potential. Therefore, without any excitation, there is no net flow of charges. When an electric field is applied, however, electrons collectively drift in the opposite direction, leading to a net measurable current. We note that the conventional current direction is also opposite to the movement of electrons, aligning it with the electric field. A simple relation between the current density and electric field intensity can be written by using Ohm's law as

$$J = \sigma E \ (\text{A/m}^2),$$

where σ is defined as the conductivity, given in siemens per meter (S/m). In the above equation, J represents the current density, whose surface integral (on the cross-section) gives the overall current flowing through the metal. All materials can be categorized in terms of their conductivity values, as discussed below.

1.2.2 Good Conductors and Insulators

Most metals are good conductors, with conductivity values in the 10^6–10^8 S/m range for a wide band of frequencies. For example, copper has a conductivity of approximately 6×10^7 S/m at room temperature. For all materials, conductivity values depend on temperature and other environmental conditions, as well as the frequency. Sea water is known to be conductive (with around 4–5 S/m conductivity), while its conduction mechanism is based on ions, not free electrons as in metals. Carbon has interesting properties, demonstrating extremely varying conductivity characteristics depending on the arrangement of its atoms. For example, diamond has a very low conductivity (around 10^{-13} S/m), while graphite is as conductive as some metals (greater than 10^5 S/m). A recently popular form of carbon called graphene may have conductivity values as large as 10^8 S/m.

There is often confusion between the velocity of electricity, velocity of electrons, and the drift velocity of electrons. In a typical metal without any excitation, electrons move randomly with a high (Fermi) velocity. These movements are of high speed (e.g., 1.57×10^6 m/s for copper). However, due to their random nature, no net current flows along the metal. When the metal is exposed to a voltage difference, leading to an electric field, electrons continue their random movements, while they tend to drift in the opposite direction to the electric field. The corresponding drift velocity is usually very low (e.g., only 10^{-5} m/s for a typical copper wire). On the other hand, the current measured along a wire is due to this drift velocity. Obviously, when AC sources are involved, electrons do not drift only in a single direction, but oscillate back and forth (in addition to high-velocity random movements) with the frequency of the signal. Since circuits are usually small with respect to wavelength, drift movements of electrons are almost synchronized through the entire circuit. Finally, the velocity of the electricity along a wire is not related to any actual movement of electrons. It is related to the speed of the electromagnetic wave through the wire (similar to sea waves that are not movements of water molecules). This speed is comparable to the speed of light in a vacuum, but it is reduced by a velocity factor depending on the properties of the material.

In general, materials with low conductivity values are called insulators. Wood, glass, rubber, air, and Teflon are well-known insulators in real-life applications. Insulators are also natural parts of all circuits, for example for isolating components and wires from each other, as well as the parts of electrical components. Since they are not electrically active, however, they are not considered directly in circuit analysis. For example, when considering wires in circuits, we assume perfectly conducting metals without any insulator, while in real life, electrical wires have shielded or coaxial structures with layers of conducting metals and insulating materials separating them.

1.2.3 Semiconductors

As their name suggests, semiconductors conduct electricity better than insulators and worse than good conductors. In addition, the conductivity of semiconductors can be altered by externally modifying their material properties permanently (via chemical processes) and temporarily (via electrical bias), making them suitable for controlling electricity. Silicon is the best-known semiconductor, and has been used in producing diverse components of integrated circuits. The key chemical operation is called doping, that is, modifying the conductivity of semiconductors by introducing impurities into their crystal lattice structures. This way, different (e.g., n-type, p-type) kinds of semiconductors can be produced and used to form junctions that enable control over electric current and voltage. Engineers use many different types of semiconducting devices, such as diodes and transistors, to construct modern circuits. These special components are discussed in Chapter 8.

1.2.4 Superconductivity and Perfect Conductivity

Perfect conductivity is a theoretical limit when the conductivity of a metal becomes infinite, that is, $\sigma \to \infty$. In this case, if a current J exists along the metal, $E \to 0$ and there is no potential difference over it. Therefore, a perfect conductor does not dissipate power while conducting electricity. Perfect conductivity is an idealized property as all metals actually have finite conductivity, while some metals can be assumed to be perfect conductors to simplify their modeling. In circuit analysis, all wires are assumed to be perfect

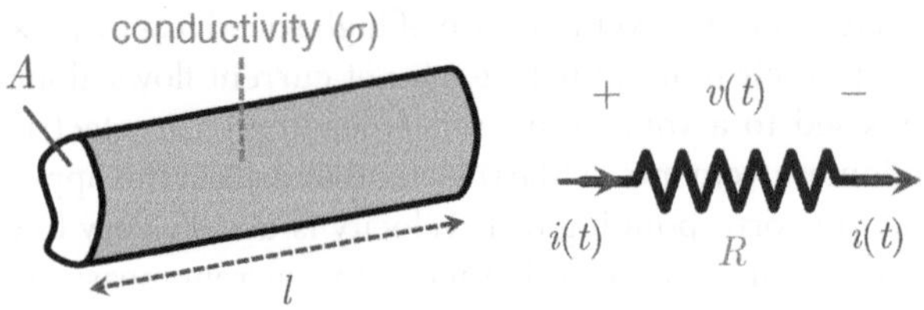

Figure 1.9 The resistance of a rod with conductivity σ is often approximated as $R = l/(\sigma A)$, where l and A are the length and cross-section area, respectively, of the rod. In circuit analysis, a resistor is a two-terminal device that usually consumes energy.

conductors (with no voltage drop across them), while any resistance due to imperfect conductivity can be modeled as a resistor component.

Under the perfect conductivity assumption, the electric field is zero anywhere on a metal. This also means that all charges are distributed on the surface of the metal. For electromagnetic fields, where electric and magnetic currents are coupled, a zero electric field leads to a zero magnetic field. On the other hand, perfect conductivity does not enforce any assumption on a static magnetic field. Specifically, a static magnetic field inside a perfect conductor does not violate Maxwell's equations.

Recently, superconductors have become popular due to their potential applications. Similar to perfect conductivity, superconductivity can be described as a limit case when the electrical conductivity goes to infinity. On the other hand, this infinite conductivity cannot be explained simply by electron movements, and quantum effects need to be considered to understand how a metal can become a superconductor. In a superconductor, magnetic fields are expelled toward its surface, making it different from theoretical perfect conductivity. Superconductivity is achieved in real life by cooling down special materials below critical temperatures.

1.2.5 Resistors as Circuit Components

As mentioned above, resistors are fundamental components of circuits. Given a resistor, the voltage–current relationship (obeying the sign convention) can be written as

$$v(t) = Ri(t),$$

where $R \geq 0$ is called the resistance. In general, the resistance of a structure depends on its dimensions and is inversely proportional to the conductivity of the material. The simple relationship above for the definition of the resistance is also commonly called Ohm's law. The unit of resistance is the ohm (Ω), and 1 ohm is 1 volt per ampere (V/A). The power of a resistor can be found from

$$p(t) = v(t)i(t) = Ri(t)i(t) = R[i(t)]^2 \geq 0,$$

which is always nonnegative. Therefore, resistors cannot produce energy themselves. In some cases, it is useful to use conductance, defined as

$$G = \frac{1}{R},$$

$$i(t) = Gv(t).$$

The unit of the conductance is the siemens (S), and 1 siemens is 1 ampere per volt (A/V).

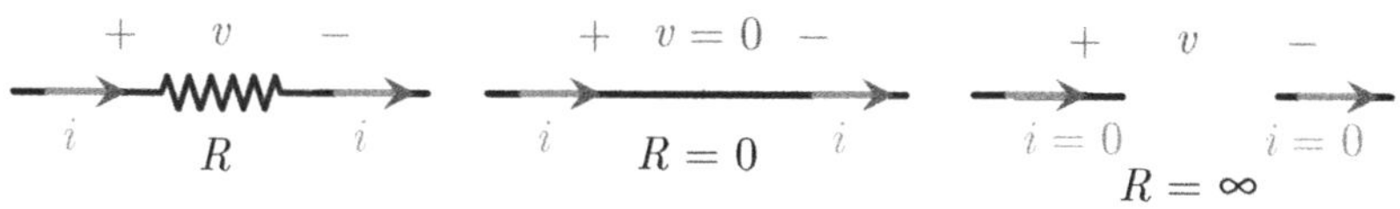

Figure 1.10 Short circuit and open circuit can be interpreted as special cases of resistors, with zero and infinite resistance values, respectively.

Resistors in real life are made of different materials, including carbon. In addition to standard resistors with fixed resistance values, there are also adjustable resistors, such as rheostats, which can be useful in different applications. The resistance of a fixed resistor also demonstrates nonlinear behaviors, that is, it may change with temperature, which may rise during the use of the resistor, leading to a complicated relationship between the voltage and current. In circuit analysis, however, these nonlinear behaviors are often discarded, and a fixed resistor has always the same resistance value.

Two limit cases of resistors are of particular interest in circuit analysis. When $R \to 0$, indicating a lack of resistance, we have a short circuit. Specifically, in a short circuit, we have

$$R = 0,$$
$$v(t) = 0,$$

while $i(t)$ can be anything (may not be zero). While all wires with zero resistances can be categorized as short circuits, this definition is often used to indicate a direct connection between two points that are not supposed to be connected. For many components and devices, having a short circuit means a failure or breakdown. At the other extreme case, two points without a direct connection between them can be interpreted as a resistor with infinite resistance. Such a case is called an open circuit, which can be defined as

$$R = \infty,$$
$$i(t) = 0,$$

while $v(t)$ can be anything (may not be zero). Any two points without a direct connection in a circuit can be interpreted as an open circuit, while this definition is again used to indicate a special case, particularly a breakdown of a connection.

1.3 Independent Sources

All circuits are excited with AC and DC sources. Among these, independent sources are defined as energy-delivering devices whose voltage or current values are fixed at a given value, independent of the rest of the circuit. Two types of independent sources are used in circuit analysis: voltage and current sources.

An ideal voltage source is defined as

$$v(t) = v_o(t),$$

where $v_o(t)$ is given and independent of other parts of the circuit. If the voltage source is DC, we further have $v(t) = v_o$ as a constant. We note that the current through a voltage

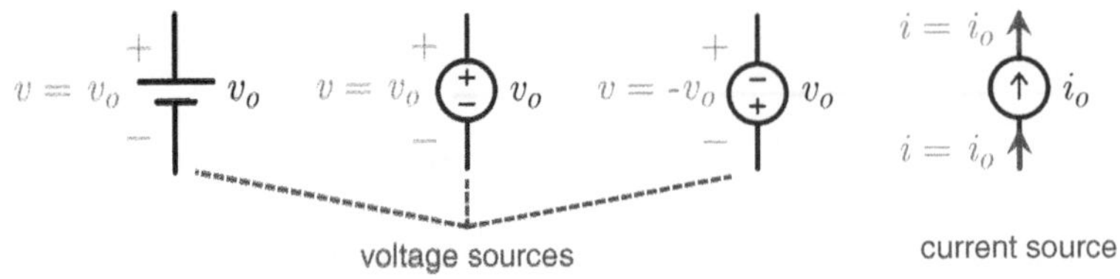

Figure 1.11 There are alternative symbols to show voltage and current sources; circular representations are used in this book. For any source, the polarity should be clearly indicated. In addition to sources with constant values (DC sources), the source values v_o and i_o may depend on time (AC sources).

source, $i(t)$, can be anything (not necessarily zero). In fact, if a voltage source is delivering energy, $i(t)$ must be nonzero.

An ideal current source is defined as

$$i(t) = i_o(t),$$

where $i_o(t)$ is given and independent of other parts of the circuit. Once again, if the current source is DC, we further have $i(t) = i_o$ as a constant. We note that the voltage across a current source, $v(t)$, can be anything (not necessarily zero).

In real life, batteries and dynamos can be considered as independent voltage sources. On the other hand, an independent current source, which has a predetermined current value no matter what the rest of the circuit does, is usually designed using a voltage source and some other components (e.g., diodes, transistors, and/or OP-AMPs). In this book, we always show an independent source as a single and ideal device, without detailed structures and any internal resistances. If a source has a nonideal resistance (e.g., nonzero resistance for a voltage source or finite resistance for a current source) it can be shown as a separate component in addition to the ideal part of the source.

Under normal circumstances, voltage and current sources provide energy to their circuits. However, depending on the rest of the circuit, a voltage or current source may consume energy, which is a perfectly valid scenario. A source that consumes energy indicates that there is at least one other source that delivers energy. For a given isolated circuit, the sum of powers of all components must be zero due to the conservation of energy.

1.4 Dependent Sources

Dependent sources are also energy-delivering devices, where, unlike independent sources, the voltage or current provided depends on another voltage or current in the circuit. While dependent sources are not frequently used practice, they are very common in circuit analysis for modeling a component, (e.g., transistors and OP-AMPs). Therefore, we assume that dependent sources exist as individual components of circuits, while the actual circuit structure may not be exactly the same. There are four types of dependent sources, which can be listed as follows.

- Voltage-controlled voltage source (VCVS): A voltage source whose voltage depends on another voltage in the circuit, i.e., $v_{ds} = A_{ds}v_o$, where A_{ds} is a unitless quantity.

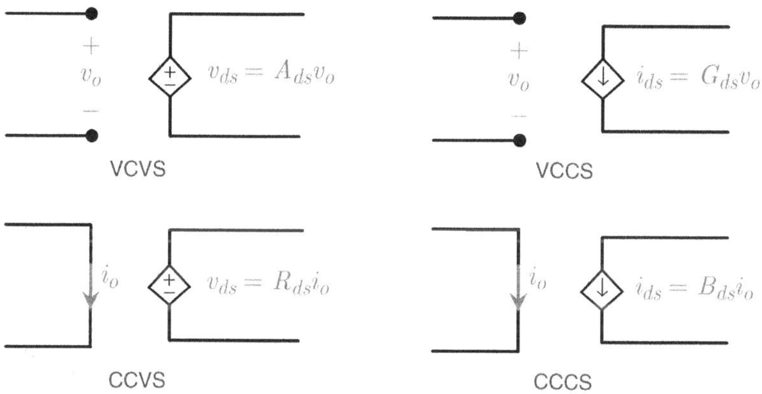

Figure 1.12 Dependent sources have fixed voltage/current values, depending on some other voltage/current values in the circuit.

- Voltage-controlled current source (VCCS): A current source whose current depends on a voltage in the circuit, i.e., $i_{ds} = G_{ds}v_o$, where G_{ds} is measured in siemens.
- Current-controlled voltage source (CCVS): A voltage source whose voltage depends on a current in the circuit, i.e., $v_{ds} = R_{ds}i_o$, where R_{ds} is measured in ohms.
- Current-controlled current source (CCCS): A current source whose current depends on another current in the circuit, i.e., $i_{ds} = B_{ds}i_o$, where B_{ds} is unitless.

The polarization of the voltage/current of a dependent source, as well as the reference voltage/current and the linkage constant (A_{ds}, B_{ds}, G_{ds}, R_{ds}), are given with the definition of the source. Similarly to independent sources, dependent sources can be DC or AC, depending on the reference voltage/current, v_o and i_o.

1.5 Basic Connections of Components

In any given circuit, components are connected via wires that intersect at nodes. Considering multiple components, two basic types of connection may occur: series and parallel.

If a common current flows through the components, they are connected in series. Hence, such components share the same current. If a common voltage is applied on the components, they are connected in parallel. Hence, such components share the same voltage. In general, series and parallel connections occur together, also with other types of connections, leading to a complex network.

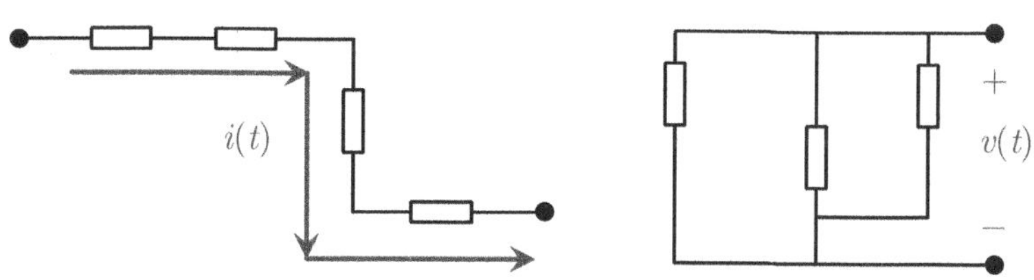

Figure 1.13 Series and parallel connections, where the current and voltage are common values, respectively, for the components.

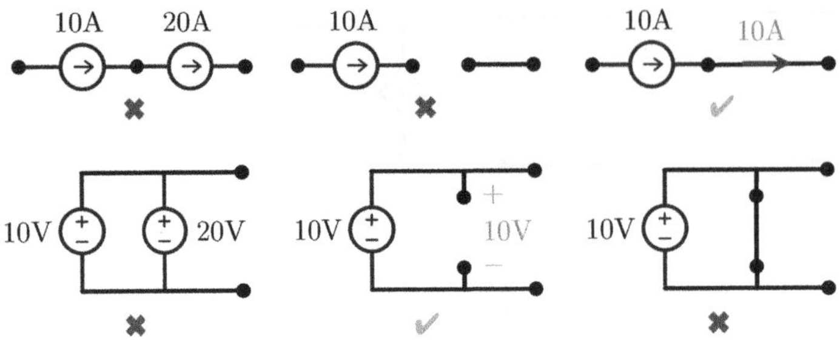

Figure 1.14 Some possible and impossible configurations using ideal components.

Considering ideal components, some of the connections are impossible. Some basic examples of possible and impossible scenarios are as follows.

- A 10 A current source and a 20 A current source cannot be connected in series.
- A 10 A current source and an open circuit cannot be connected in series.
- A 10 A current source and a short circuit can be connected in series. If these are the only components of the circuit (with a full connection on both terminals), no voltage occurs across the current source; hence, it does not produce any power.
- A 10 V voltage source and a 20 V voltage source cannot be connected in parallel.
- A 10 V voltage source and an open circuit can be connected in parallel. If these are the only components of the circuit, no current flows through the voltage source; hence, it does not produce any power.
- A 10 V voltage source and a short circuit cannot be connected in parallel.

In order to understand why a connection may not be possible, one can directly use the definition of the components. For example, consider a series connection of 10 A and 20 A current sources. The 10 A source indicates that 10 A is passing through the line. On the other hand, the 20 A source, by definition, needs 20 A current to flow in the same line. Therefore, there is an inconsistency, since a wire cannot have different current values at the same time. Similar inconsistencies can be found for all impossible cases. Such impossible configurations are not due to a modeling incapability in circuit analysis; they actually correspond to physically impossible practices in real life. Consider another example involving a parallel connection of two voltage sources with different values. In real life, this configuration never exists since voltage sources have internal resistances, while the wires between them are also not perfectly conducting, leading to a voltage drop. Therefore, a more realistic model of the physical scenario would require a resistor between the voltage sources, leading to a perfectly valid circuit that can be analyzed. All impossible configurations described above have similar missing components, which can be added to convert them into possible scenarios.

Impossible scenarios rarely occur, even when we consider ideal components in circuit analysis. In general, many circuits have multiple components and connections, where the voltage and current values become consistent. In order to find relations between voltage and current values, we use basic rules, namely Kirchhoff's laws, as described in the next chapter. These rules, which are based on the conservation of energy and charge, provide the necessary equations to relate different voltage and current values.

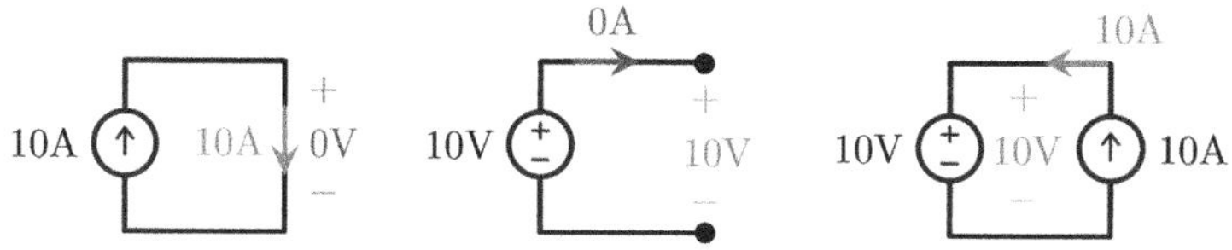

Figure 1.15 Some possible circuits involving only one or two components that are connected consistently. In the first and second circuits, where there is a current and a voltage source, the sources do not produce any power. However, they still provide the current and voltage values, in accordance with their definitions. In the third circuit, the voltage source absorbs power (100 W), while the current source delivers power (100 W).

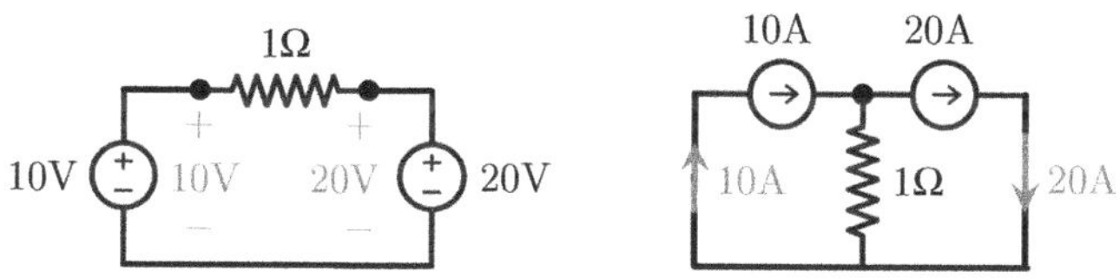

Figure 1.16 Two simple circuits that can be interpreted incorrectly as impossible. In fact, both two circuits are possible and they involve consistent voltage and current values. In the first circuit, a voltage drop (by an amount of 20 − 10 = 10 V) exists across the resistor. In the second circuit, a nonzero current (20 − 10 = 10 A) flows through the resistor. These values can easily be found by applying Kirchhoff's laws, as described in the next chapter.

At this stage, we can start analyzing some simple circuits, just by considering the definitions of the components.

Example 4: Consider a circuit involving a 10 V voltage source connected to a 5 Ω resistor.

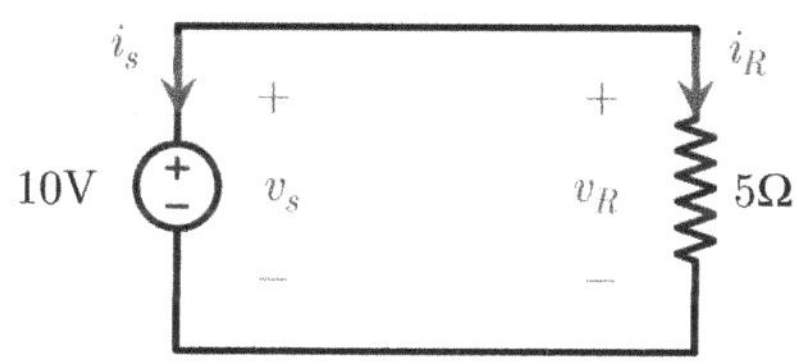

Note that voltage and current directions are defined arbitrarily, but they must obey the sign convention. We can analyze the circuit as follows.

- Using the definition of the voltage source: $v_s = 10$ V.
- Using the definition of the voltage between two points and considering that the voltage is fixed along a wire: $v_R = v_s = 10$ V.
- Using the definition of the resistor (Ohm's law): $i_R = v_R/5 = 2$ V.
- Using the definition of the current: $i_s = -i_R = -2$ A.
- Using the definition of the power: $p_s = v_s i_s = -20$ W.
- Using the definition of the power: $p_R = v_R i_R = 20$ W.

The signs of power values indicate that the voltage source delivers energy, while the resistor consumes the same amount of energy. As expected, we have $p_s + p_R = 0$ due to the conservation of energy.

Example 5: Consider a circuit involving a 10 A current source connected to a 5 Ω resistor.

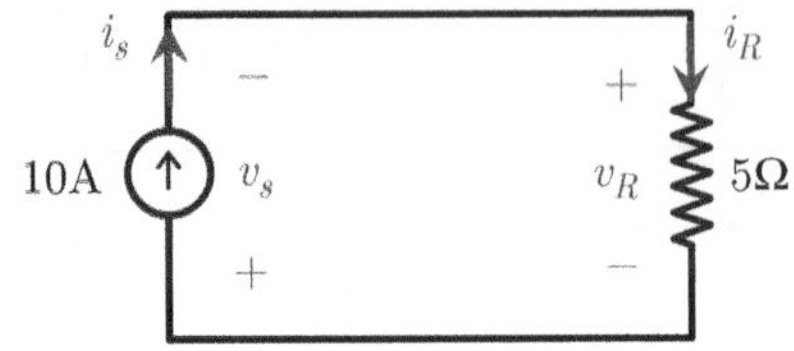

In this case, the current through the circuit is determined via the current source, while the voltage value is found by applying Ohm's law. We can analyze the circuit as follows.

- Using the definition of the current source: $i_s = 10$ A.
- Using the definition of the current: $i_R = i_s = 10$ A.
- Using the definition of the resistor (Ohm's law): $v_R = 5i_R = 50$ V.
- Using the definition of the voltage between two points and considering that the voltage is fixed along a wire: $v_s = -v_R = -50$ V.
- Using the definition of the power: $p_s = v_s i_s = -500$ W.
- Using the definition of the power: $p_R = v_R i_R = 500$ W.

Similarly to the previous example, the current source delivers energy, while the resistor consumes the same amount of energy.

Example 6: Consider a circuit involving a 10 V voltage source connected to a 10 A current source.

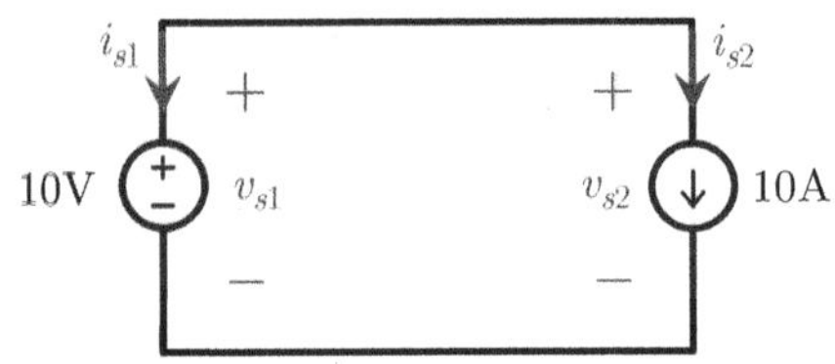

We can analyze the circuit as follows.

- Using the definition of the voltage source and voltage: $v_{s2} = v_{s1} = 10$ V.
- Using the definition of the current source and current: $i_{s1} = -i_{s2} = -10$ A.
- Using the definition of the power: $p_{s1} = v_{s1} i_{s1} = -100$ W and $p_{s2} = v_{s2} i_{s2} = 100$ W.

In this circuit, the voltage source produces energy, while the current source (despite also being a source) consumes energy. Once again the total power is zero due to the conservation of energy.

Example 7: Consider the following circuit involving a 10 V voltage source, a current-dependent voltage source, and two 10 Ω resistors.

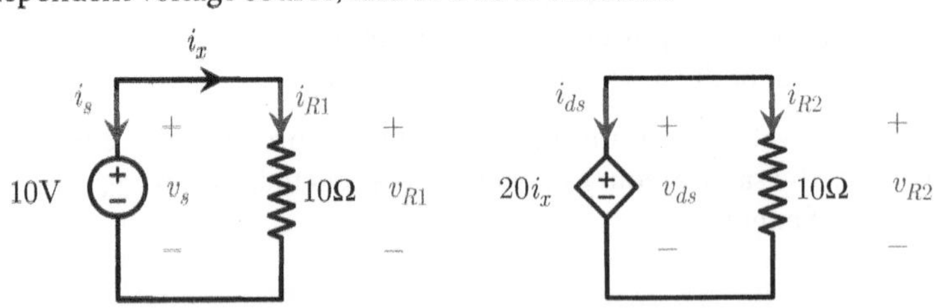

We can analyze the circuit as follows. First, using the definition of voltage, we have

$$v_{R1} = v_s = 10 \text{ V}.$$

Then, using Ohm's law, we get

$$i_{R1} = v_{R1}/10 = 10/10 = 1 \text{ A},$$
$$i_s = -i_{R1} = -1 \text{ A},$$

and

$$i_x = i_{R1} = 1 \text{ A}.$$

The definition of the current-dependent voltage source leads to

$$v_{ds} = 20i_x = 20 \text{ V}.$$

Therefore, using the definition of voltage again, we derive

$$v_{R2} = v_{ds} = 20 \text{ V}.$$

Using Ohm's law once again, we have

$$i_{R2} = v_{R2}/10 = 2 \text{ A}$$

and

$$i_{ds} = -i_{R2} = -2 \text{ A}.$$

Finally, the powers of all components can be found:

$$p_{R1} = v_{R1}i_{R1} = 10 \times 1 = 10 \text{ W},$$
$$p_{R2} = v_{R2}i_{R2} = 20 \times 2 = 40 \text{ W},$$
$$p_s = v_s i_s = 10 \times (-1) = -10 \text{ W},$$
$$p_{ds} = v_{ds}i_{ds} = 20 \times (-2) = -40 \text{ W}.$$

We note that both the independent and dependent source provides energy to the circuit, while both resistors consume.

1.6 Limitations in Circuit Analysis

A type of circuit analysis, which is used throughout this book, is based on lumped-element models. Specifically, we assume that the behavior of components and their interactions with each other can be described by voltage–current relations given by the descriptions of the components. In addition, we assume that all elements demonstrate their ideal characteristics, being independent of outer conditions. All these assumptions rely on two constraints on the sizes of the components and circuits.

- The circuit components are large enough to omit individual behaviors of protons and electrons. Hence, without dealing with individual particles, their bulk behaviors (i.e., voltage and current) are used directly to model the components.
- The circuit components and circuits are much smaller than the wavelength of the signals. For example, oscillations in the current and voltage values through a wire are synchronized and no phase accumulation occurs. In addition, voltage/current phase differences are well defined in all components.

Obviously, lumped-element models fail when the size constraints are not satisfied. For example, in circuits larger than the wavelength, connections may need to be modeled as transmission lines, where wave equations are solved. Some circuits may need the full application of Maxwell's equations to precisely describe the electromagnetic interactions of components with each other.

Depending on the complexity of the circuit model, further assumptions are often made to simplify the analysis of circuits. In this book, we accept the following assumptions that are also common in the circuit analysis literature.

- The voltage–current relationship defined for a component does not depend on outer conditions (temperature, pressure, light, etc.). This also means that the circuit behaves always the same (e.g., change in resistance due to rising temperature as the circuit is used is omitted).
- The voltage–current relationship defined for a component does not depend on other components. For example, a resistor of 10 Ω always satisfies Ohm's law as $v_R = 10i_R$, independent of other elements used in the same circuit. We also ignore cross-talk of circuits and their parts, other than the linkage through well-defined dependent sources.
- All components are ideal and we omit secondary effects, such as the resistance of a voltage source, inductance of a capacitor, or capacitance of a resistor. If these effects cannot be neglected, they can be represented as individual components. For example, the leakage of a capacitor can be represented by a resistor connected in parallel to the capacitor.

Despite all these limitations and assumptions, circuit analysis methods presented in this book are widely accepted and used to analyze diverse circuits and electrical devices. In many cases, lumped elements are used as starting models before more complicated analysis techniques are applied.

1.7 What You Need to Know before You Continue

Before proceeding to the next chapter, we summarize a few key points that need to be known to understand the higher-level topics.

- Sign convention: In this book, the current through a component is selected to flow from the positive to the negative side of the voltage.
- Steady state: For DC circuits in steady state, voltage and current values are assumed to be constant. In the first few chapters, steady state is automatically assumed when only resistors and DC sources are considered.
- Short circuit and open circuit: Short circuit and open circuit can be interpreted as special cases of resistors, with zero and infinite resistance values, respectively.
- Sources: There are alternative symbols to show voltage and current sources; circular representations are used in this book. DC/AC types are indicated in the context of source values.
- Energy conservation: For a given isolated circuit, the sum of powers of all components must be zero due to the conservation of energy.

- Series connection: Components that share the same current are connected in series.
- Parallel connection: Components that share the same voltage are connected in parallel.
- Impossible configurations: Some connections of ideal components are not allowed due to inconsistency of voltage and current values enforced by the definitions of components.

In the next chapter, we start with the most basic tools, namely Kirchhoff's laws, to analyze circuits.

2

Basic Tools: Kirchhoff's Laws

Circuits with a few components can be analyzed by using only the definitions of the components. On the other hand, as the circuits become more complicated, involving connections of many components, well-defined solution tools are required to derive the necessary equations. This chapter presents the most basic laws of circuit analysis based on the conservation of charges (Kirchhoff's current law) and conservation of energy (Kirchhoff's voltage law). These fundamental rules, collectively named Kirchhoff's laws, can be used to derive useful equations at nodes and in loops, in order to relate the voltages and currents of components to each other. Solutions of the resulting equations lead to numerical values of these variables, hence the analysis of the given circuit.

In general, Kirchhoff's laws should be sufficient to solve any type of circuit. On the other hand, when a circuit involves many resistors, a direct application of Kirchhoff's laws may lead to too many equations that can be difficult to solve. For scenarios of this kind, where multiple resistors are connected in series and parallel, one can derive shortcuts (again using Kirchhoff's laws) to simplify and represent the overall circuit using a few resistors. This chapter also presents such shortcuts for the analysis of large resistive circuits.

2.1 Kirchhoff's Current Law

According to Kirchhoff's current law (KCL), the sum of currents entering (+) and leaving (−) a node should be zero, that is,

$$\sum_{k=1}^{n} i_k = 0,$$

where n is the number of wires connected at the node.

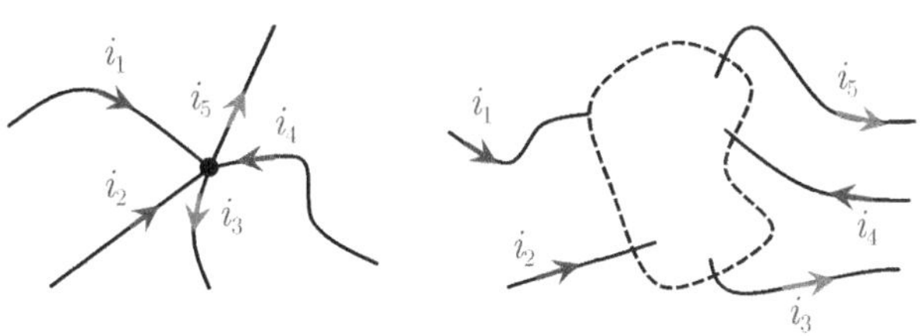

Introduction to Electrical Circuit Analysis, First Edition. Özgür Ergül

Companion Website: www.wiley.com/go/ergul4412

In the examples above, we have

$$i_1 + i_2 - i_3 + i_4 - i_5 = 0.$$

The conservation of charge is the background to KCL, which can be generalized to currents entering and leaving closed surfaces. While the selection could be arbitrary, we always select the entering and leaving currents as positive and negative, respectively. Obviously, KCL assumes that no charge accumulation occurs at the node or closed surface considered.

In the analysis of circuits, we usually apply KCL at nodes. The format that we adopt is

- KCL(x): equation derived by applying KCL,

where x represents the node.

2.2 Kirchhoff's Voltage Law

Kirchhoff's voltage law (KVL) states that the sum of voltages in a closed loop is zero, that is,

$$\sum_{k=1}^{n} v_k = 0,$$

where n is the number of segments (e.g., components) along the loop.

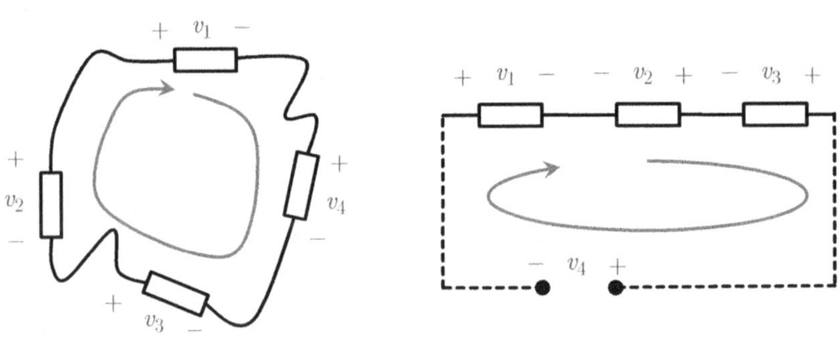

In the examples above, we have

$$v_1 - v_2 - v_3 + v_4 = 0.$$

KVL is based on the conservation of energy. When applying KVL, an added voltage can be a voltage across a component or any voltage between two consecutive nodes in the loop. While the direction can be selected arbitrarily, we always use the clockwise direction when adding the voltages. A plus or minus sign is used depending on the polarization of the added voltage.

In the analysis of circuits, we usually apply KVL in small closed loops, each of which is called a mesh. Similar to KCL, we use a format such as

- KVL($x \rightarrow y \rightarrow z \rightarrow x$): equation derived by applying KVL,

where x, y, and z represent the nodes forming the mesh.

Example 8: Consider the following circuit involving a voltage source, a current source, and two resistors.

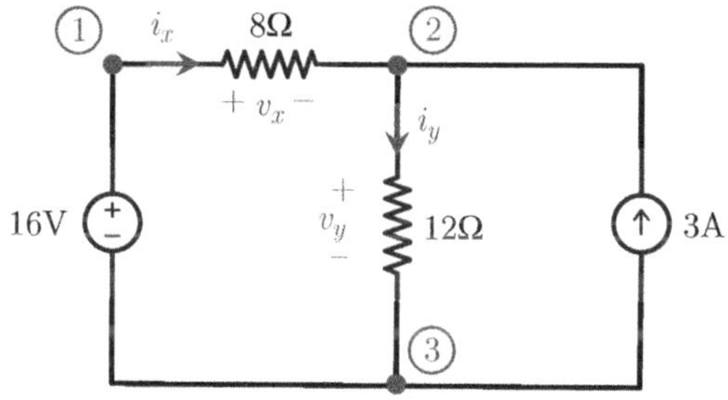

Find the voltage across the 8 Ω resistor.

Solution: With the given definitions of the voltages and currents on the circuit, we need to apply KCL and KVL systematically in order to solve the problem. One obtains

- KVL($1 \to 2 \to 3 \to 1$): $v_x + v_y - 16 = 0 \longrightarrow v_x + v_y = 16\,\text{V}$,
- KCL(2): $i_x + 3 - i_y = 0 \longrightarrow i_x - i_y = -3\,\text{A}$.

Then, using Ohm's law, $v_x = 8i_x$ and $v_y = 12i_y$, leading to $2i_x + 3i_y = 4$. Finally, solving for i_x, we obtain $i_x = -1\,\text{A}$, $i_y = 2\,\text{A}$, $v_y = 24\,\text{V}$, and

$$v_x = -8 \text{ V}$$

as the voltage across the 8Ω resistor. As one verification of the solution, one can calculate the powers of all components:

$$p_{8\Omega} = v_x i_x = 8 \text{ W},$$
$$p_{12\Omega} = v_y i_y = 48 \text{ W},$$
$$p_{16\text{V}} = 16 \times (-i_x) = 16 \text{ W},$$
$$p_{3\text{A}} = (-v_y) \times 3 = -72 \text{ W},$$

and check that

$$p_{8\Omega} + p_{12\Omega} + p_{16\text{V}} + p_{3\text{A}} = 8 + 48 + 16 - 72 = 0.$$

As far as the power values are concerned, we can conclude that the current source delivers power, while the resistors, as well as the voltage source, consume power.

Example 9: Consider the following circuit involving a voltage source, which is connected to two resistors and a bulb B.

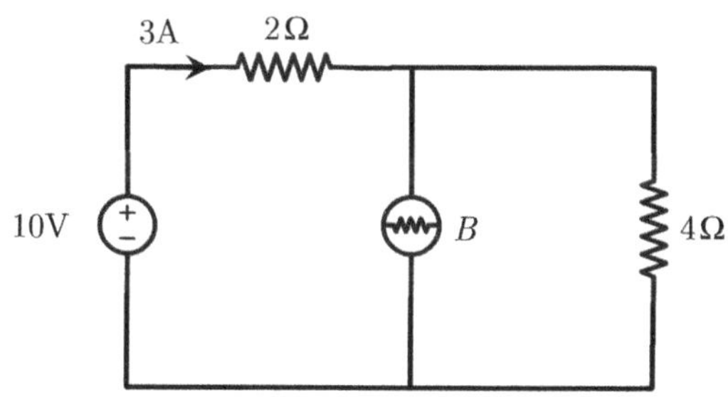

Find the power of the bulb.

Solution: In most circuits, the voltage and current directions are not defined a priori. Therefore, when analyzing such a circuit, we define the directions arbitrarily, while enforcing the sign convention for all components. When a current/voltage value is found to be negative, we understand that the initial assumption is not correct. However, this is not a problem at all, provided that we are consistent with the directions throughout the solution.

For the circuit above, we label the nodes, define the directions of the currents, and define the voltages in accordance with the sign convention, as follows.

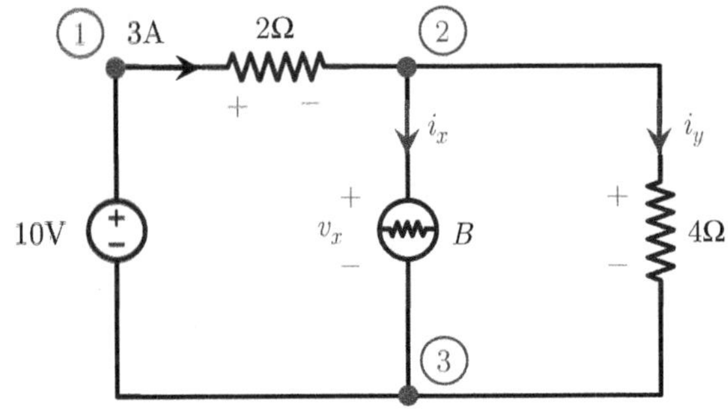

Using KVL, one obtains

- KVL($1 \rightarrow 2 \rightarrow 3 \rightarrow 1$): $-10 + 3 \times 2 + v_x = 0 \longrightarrow v_x = 4$ V.

Then, using Ohm's law, $i_y = v_x/4 = 1$ A, and we further have

- KCL(2): $3 - i_x - i_y = 0 \longrightarrow i_x = 3 - 1 = 2$ A.

Finally, the power of the device is found to be

$$p_B = v_x i_x = 4 \times 2 = 8 \text{ W}.$$

Example 10: Consider the following circuit involving two voltage sources connected to three resistors.

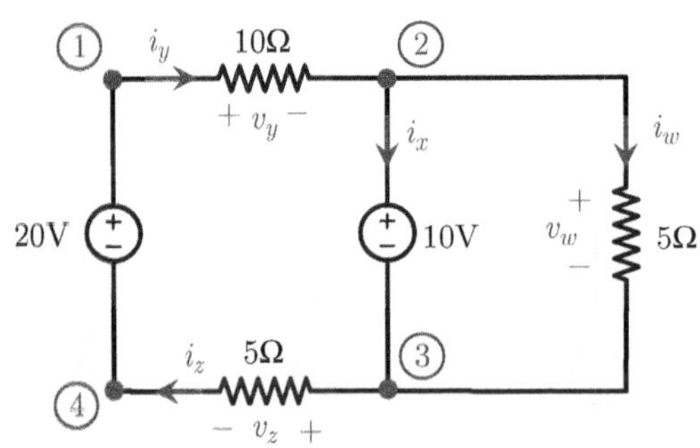

Find the value of i_x that passes through the 10 V source.

Solution: We again start with KVL to obtain

- KVL($1 \rightarrow 2 \rightarrow 3 \rightarrow 4 \rightarrow 1$): $-20 + v_y + 10 + v_z = 0 \longrightarrow v_y + v_z = 10\,\text{V}$.

We note that $i_z = i_y$, $v_y = 10i_y$, and $v_z = 5i_z = 5i_y$, leading to

$$10i_y + 5i_y = 10 \longrightarrow i_y = i_z = 2/3 \text{ A}.$$

Furthermore, $v_w = 10\,\text{V}$ and $i_w = v_w/5 = 2\,\text{A}$. Finally, applying KCL at node 2, we get

- KCL(2): $i_y - i_x - i_w = 0 \longrightarrow i_x = i_y - i_w = 2/3 - 2 = -4/3\,\text{A}$.

Example 11: Consider the following circuit, where a voltage source and a current source are connected to three resistors.

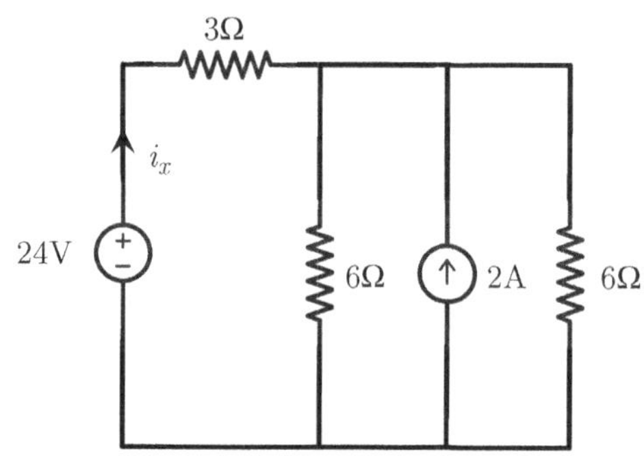

Find the value i_x, that is, the current across the 3 Ω resistor.

Solution: First, we label the nodes, define the directions of the currents, and define the voltages in accordance with the sign convention, as follows.

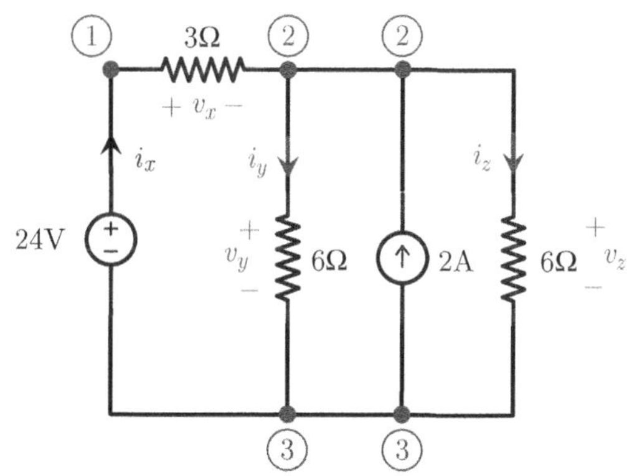

Then, using KVL and Ohm's law, we have $v_x = 3i_x$, $v_y = 6i_y$, and

- KVL($1 \rightarrow 2 \rightarrow 3 \rightarrow 1$): $-24 + v_x + v_y = 0 \longrightarrow v_x + v_y = 24\,\text{V}$.

Therefore, we have

$$3i_x + 6i_y = 24 \longrightarrow i_x + 2i_y = 8 \text{ A}.$$

Furthermore, using KCL (see below for some details), we derive

- KCL(2): $i_x - i_y - i_z = -2$ A.

Using $v_y = v_z$ and $i_y = v_y/6 = v_z/6 = i_z$, we obtain

$$i_x - 2i_y = -2 \text{ A}.$$

Finally, solving two equations with two unknowns, we get

$$i_x = 3 \text{ A}.$$

In the above, we note that node 2 (as well as node 3) is defined simultaneously at two intersections, and KCL is written accordingly by considering all entering and leaving currents, as follows.

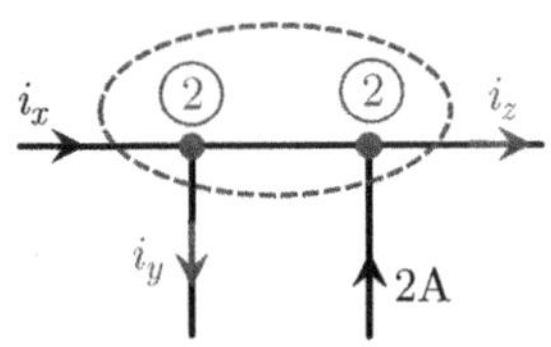

This is a common practice in circuit analysis in order to reduce the number of equations. Specifically, intersections without a component between them can be considered as a single node to avoid writing redundant equations with redundant unknowns. On the other hand, this not mandatory. For example, one can consider each intersection as a node, as follows.

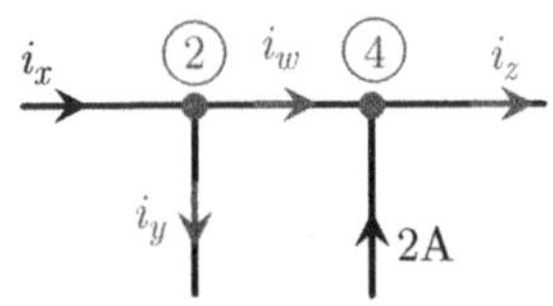

In this case, we need to define a current i_w between nodes 2 and 4. Writing KCL at the nodes, we now have

- KCL(2): $i_x - i_y - i_w = 0$,
- KCL(4): $i_w + 2 - i_z = 0$.

Obviously, when these equations are combined (directly added), we arrive at the same equation in the original solution,

$$i_x - i_y - i_z = -2 \text{ A}.$$

Example 12: Consider the following circuit.

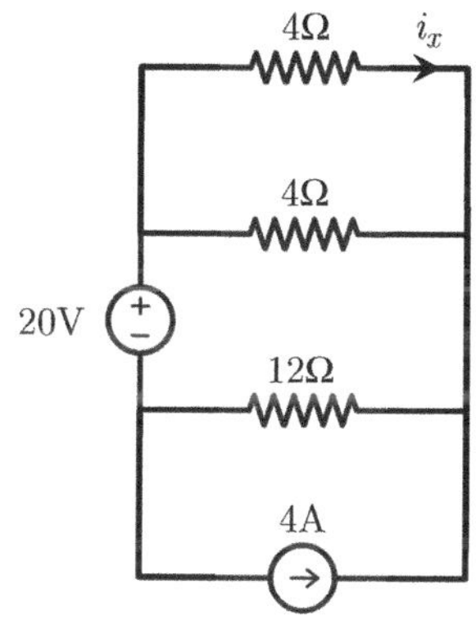

Find the value of i_x.

Solution: First, we label the nodes, define the directions of the currents, and define the voltages in accordance with the sign convention, as follows.

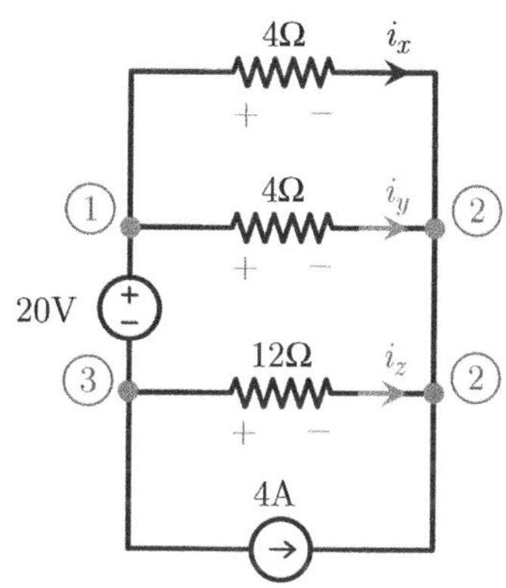

Note that we again define node 2 as the combination of two intersection points. Besides, in order to simplify the solution, we do not define voltage variables separately as they are already related to the currents via Ohm's law. Using KVL, we derive

- KVL($1 \rightarrow 2 \rightarrow 1$): $4i_x - 4i_y = 0 \longrightarrow i_x = i_y$,
- KVL($1 \rightarrow 2 \rightarrow 3 \rightarrow 1$): $4i_x - 12i_z - 20 = 0 \longrightarrow i_z = i_x/3 - 5/3$.

Then, using KCL, we obtain

- KCL(2): $i_x + i_y + i_z + 4 = 0$.

Finally, we have

$$i_x + i_x + i_x/3 - 5/3 + 4 = 0$$

$$7i_x/3 = -7/3 \longrightarrow i_x = -1 \text{ A}.$$

In solving this example, we further note the following.

- The voltage across the current source is unknown. Hence, it is not useful to write KVL for 3 → 2 → 3.
- The current across the voltage source is unknown. Hence, it is not useful to write KCL at 1 or 3.

In general, we avoid writing KCL at a node, to which a voltage source is connected, unless it is mandatory to find the current through the voltage source. In addition, applying KVL in a mesh containing a current source is usually not useful, unless the voltage across the current source must be found.

Example 13: Consider the following circuit involving a voltage source and a current source connected to four different resistors.

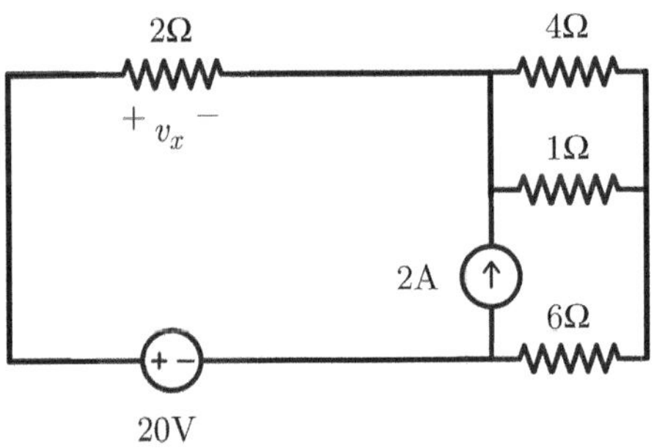

Find the value of v_x, that is, the voltage across the 2 Ω resistor.

Solution: As in the previous examples, we label the nodes, define the directions of the currents, and define the voltages in accordance with the sign convention.

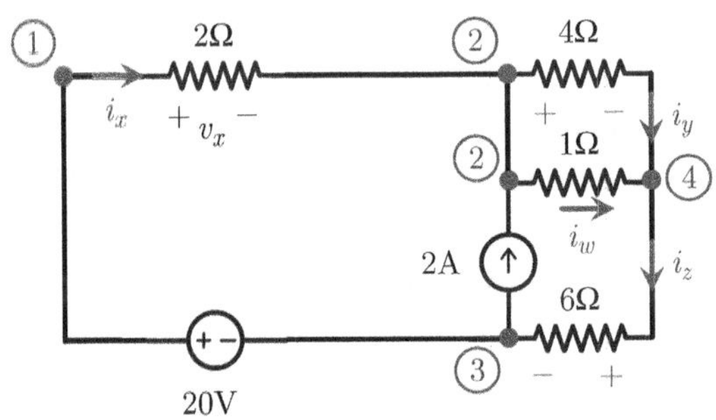

In order to simplify the solution, we again do not define voltage variables separately and write all equations in terms of currents. Applying KVL and KCL consecutively, we obtain

- KVL(2 → 4 → 2): $4i_y - 1i_w = 0 \longrightarrow i_w = 4i_y$,
- KCL(4): $i_y + i_w - i_z = 0 \longrightarrow i_z = 5i_y$,

- KCL(3): $i_z - i_x - 2 = 0 \longrightarrow i_x = 5i_y - 2$,
- KVL($1 \to 2 \to 4 \to 3 \to 1$): $2i_x + 4i_y + 6i_z - 20 = 0 \longrightarrow i_y = 6/11$ A.

Finally, we have $i_x = 30/11 - 2 = 8/11$ A and $v_x = 16/11$ V.

Example 14: Consider the following circuit involving a total of eight components.

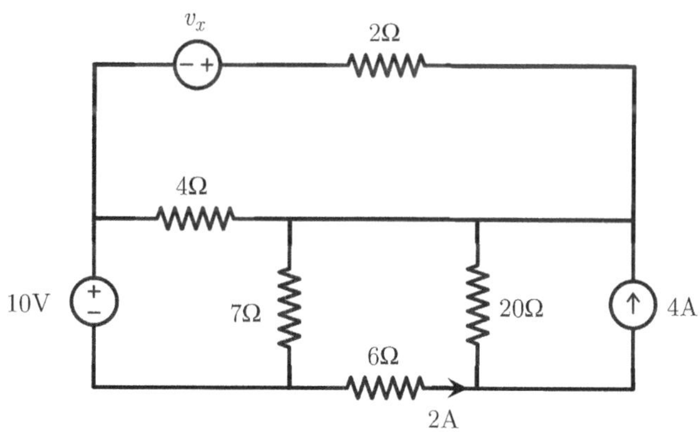

Find v_x.

Solution: We again label the nodes, from 1 to 4, and define the current directions. Node 2 is defined as the combination of three intersection points.

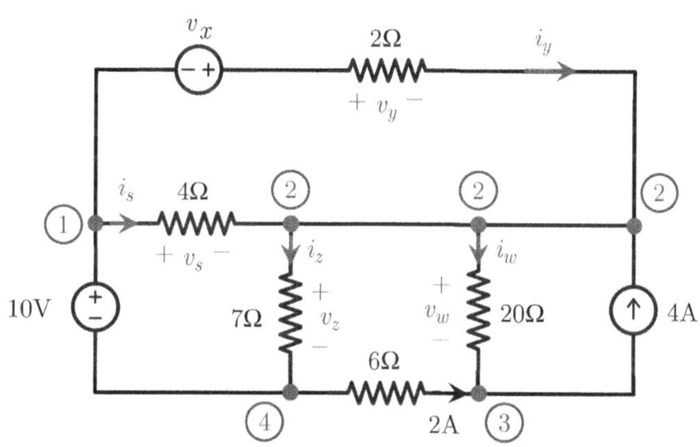

Applying KCL at nodes 3 and 2, we obtain

- KCL(3): $2 + i_w - 4 = 0 \longrightarrow i_w = 2$,
- KCL(2): $i_s - i_z - i_w + 4 + i_y = 0 \longrightarrow i_s + i_y - i_z = -2$.

Then, using KVL, we derive

- KVL($1 \to 2 \to 4 \to 1$): $4i_s + 7i_z = 10$,
- KVL($2 \to 3 \to 4 \to 2$): $20i_w - 2 \times 6 - 7i_z = 0 \longrightarrow i_z = 4$ A.

Using the updated information, we can also find i_s, as well as i_y as

$$4i_s = 10 - 7i_z = -18 \longrightarrow i_s = -9/2 \text{ A},$$
$$i_y = -2 - i_s + i_z = -2 + 9/2 + 4 = 13/2 \text{ A}.$$

Finally, we apply KVL($1 \to 2 \to 1$) to find v_x as

$$-v_x + 2i_y - 4i_s = 0 \longrightarrow v_x = 2i_y - 4i_s = 13 + 18 = 31 \text{ V}.$$

Example 15: Consider the following circuit involving six components.

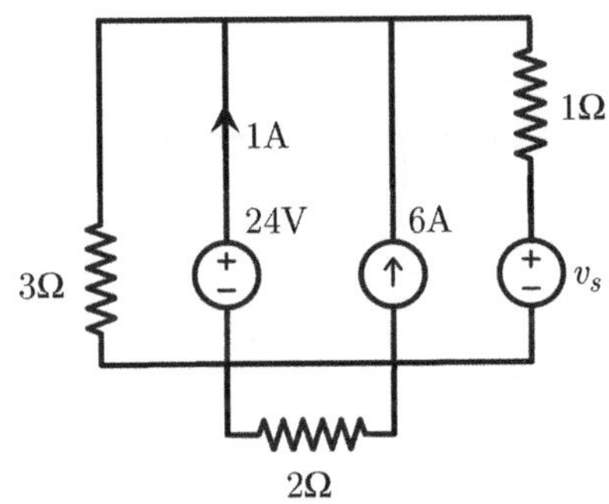

Find the value of v_s.

Solution: As in the previous examples, we label the nodes, define the directions of the currents, and define the voltages using the sign convention.

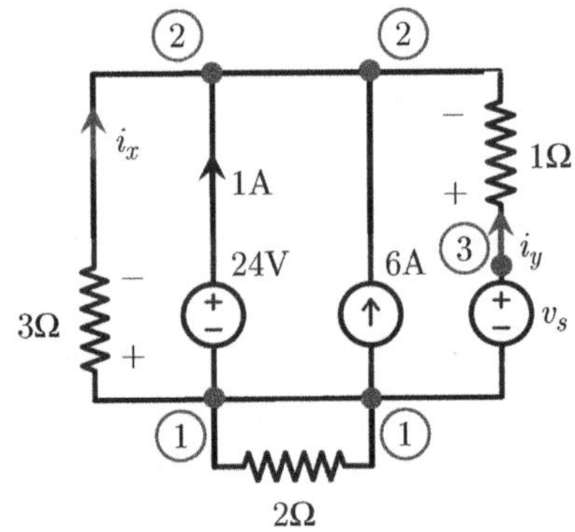

We note that the 2 Ω resistor is short-circuited and can be omitted in the analysis. Using KVL and KCL, we obtain

- KVL($1 \to 2 \to 1$): $3i_x + 24 = 0 \longrightarrow i_x = -8$ A,
- KCL(2): $i_x + 1 + 6 + i_y = 0 \longrightarrow i_y = 1$ A,
- KVL($1 \to 2 \to 3 \to 1$): $-24 - 1i_y + v_s = 0 \longrightarrow v_s = 25$ V.

As briefly discussed before, it is generally suggested to avoid using KCL at a node with a connection to a voltage source. However, in this case, the current across the voltage source v_s must be found in order to find the voltage value. Therefore, we apply KCL at node 2.

Example 16: Consider the following circuit involving six components, in addition to a given current along a path.

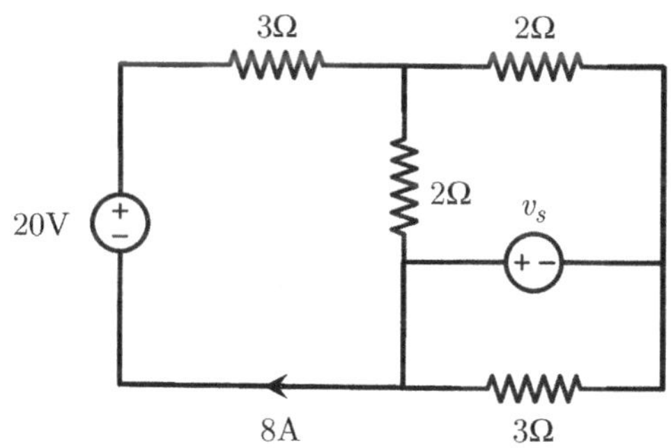

Find the value of v_s.

Solution: Once again, we process the circuit as follows.

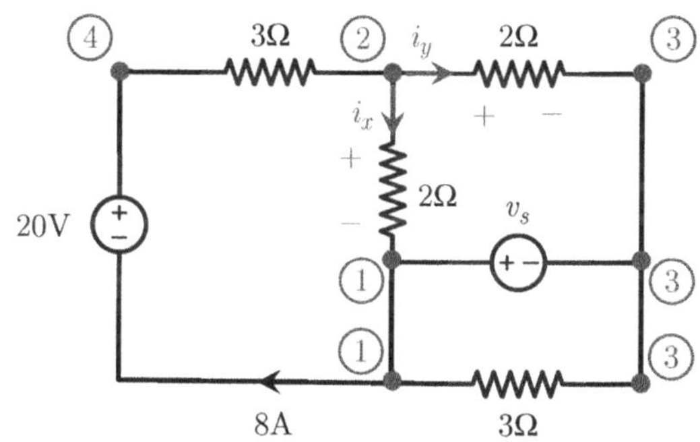

Using KVL and KCL, we derive

- KVL(1 → 4 → 2 → 1): $-20 + 3 \times 8 + 2i_x = 0 \longrightarrow i_x = -2$ A,
- KCL(2): $8 - i_x - i_y = 0 \longrightarrow i_y = 10$ A,
- KVL(1 → 2 → 3 → 1): $-2i_x + 2i_y - v_s = 0 \longrightarrow v_s = 24$ V.

Interestingly, the solution above does not depend on the 3 Ω resistor, and v_s is always 24 V if this resistor is not zero (not short-circuited). If this resistor was short-circuited, then the question would be inconsistent.

Exercise 2: In the following circuit, the device M works for currents in the range from 4 A to 6 A. Find the range of values for R.

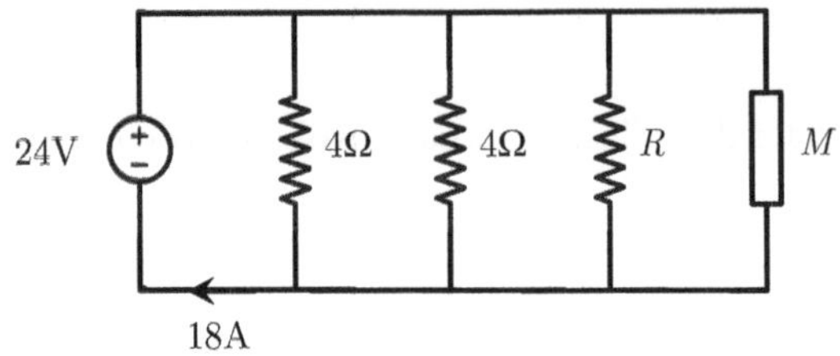

Exercise 3: In the following circuit, find the value of R.

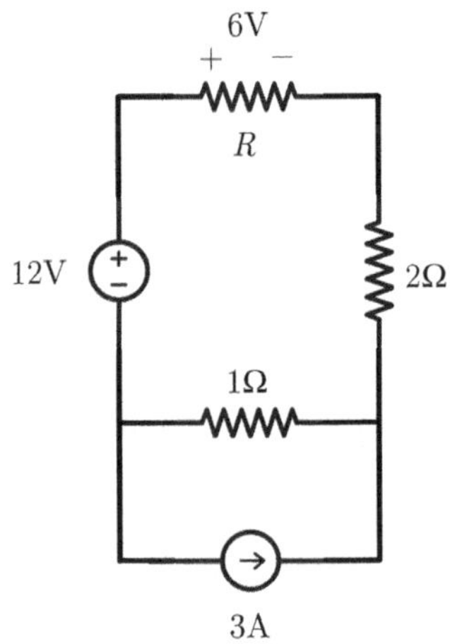

Exercise 4: In the following circuit, find the value of i_x.

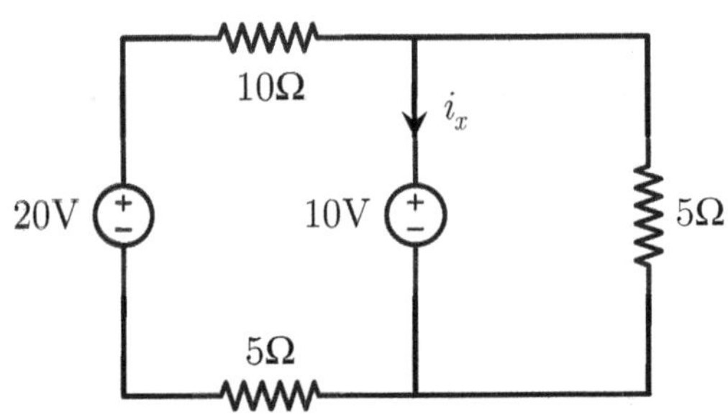

Exercise 5: In the following circuit, find the value of i_x.

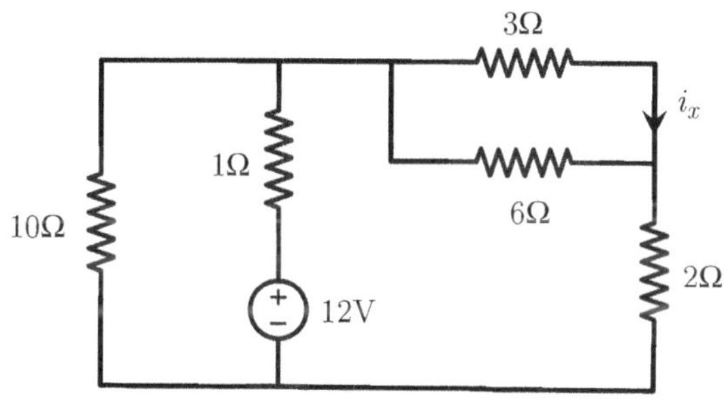

Exercise 6: In the following circuit, find the value of i_x.

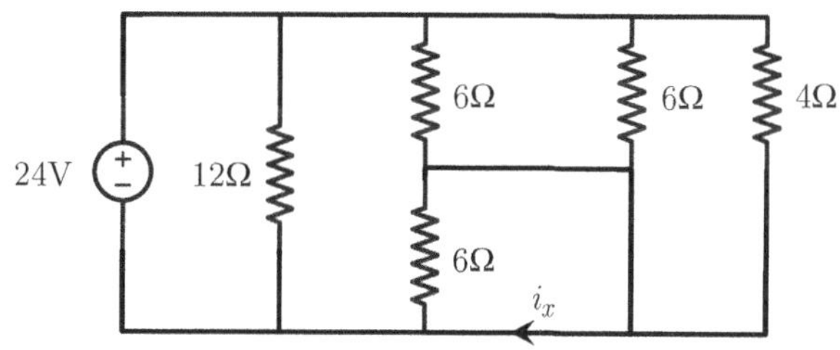

Exercise 7: In the following circuit, find the power of the device *D*.

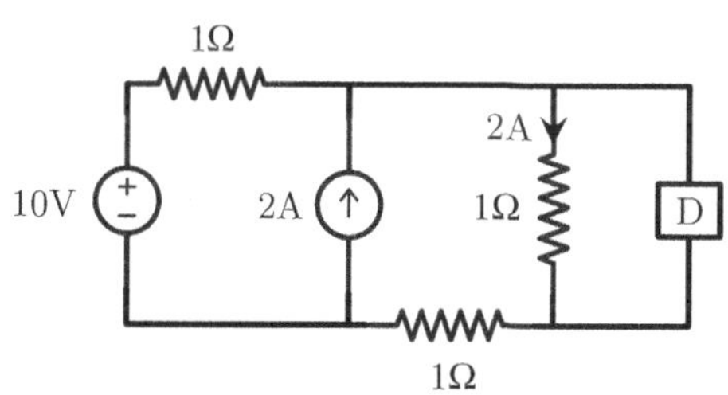

2.3 When Things Go Wrong with KCL and KVL

While KCL and KVL, in general, are easy to understand conceptually, they can be tricky to use, especially when circuits involve many components.

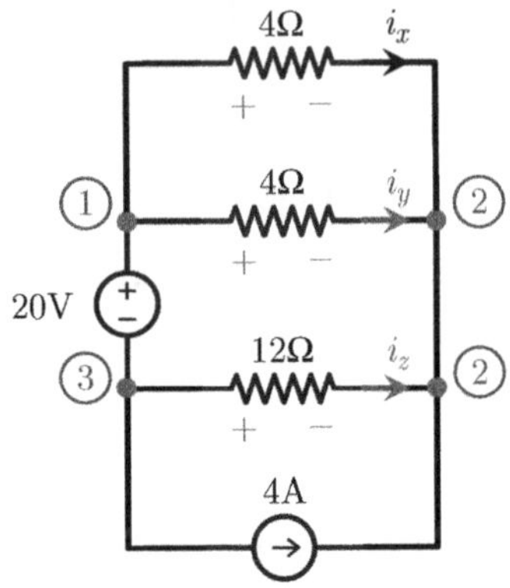

In the circuit above, we again would like to find i_x. As indicated before, when analyzing this circuit, applying KCL at node 1 or 3 would not be useful. For example, consider KCL at node 3. We must define a current between nodes 1 and 3 as follows.

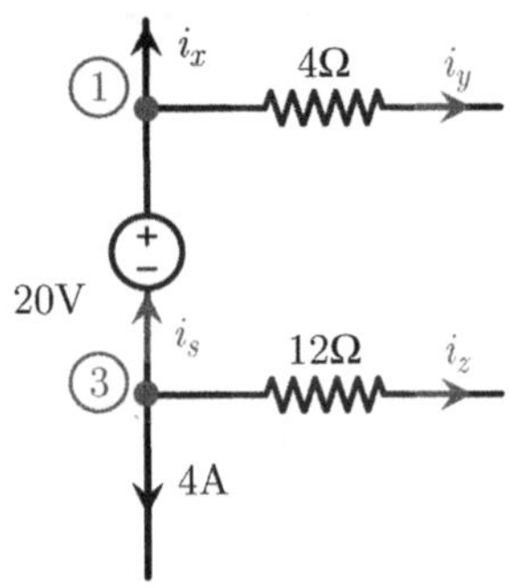

This way, we can write KCL as

- KCL(3): $-4 - i_z - i_s = 0$,

which does not provide new information as i_s must be defined in order to write this equation. The scenario becomes more interesting when we consider KCL on the other side, at node 1. We have

- KCL(1): $i_s - i_x - i_y = 0$.

Now, combining (adding) the two equations above, we further derive

$$-i_x - i_y - i_z = 4.$$

In this useful equation, we observe that the new variable i_s is eliminated. In other words, while KCL at node 1 or 3 would not be useful alone, they could be used together to derive a single useful equation. In the original solution this is not considered, as the necessary equations can already be obtained by means of KVL equations. However, in nodal analysis (see Chapter 3), where only KCL applications are allowed, we develop the supernode concept that effectively combines KCL equations at nodes, as practiced above.

On the other hand, in some cases, current along a voltage source may be required, needing KCL at one of its terminals. For example, consider the power of the voltage source above. Using KCL at node 3 and borrowing $i_z = -2$ A from the original solution, we obtain

$$i_s = -4 - i_z = -4 + 2 = -2 \text{ A}.$$

Hence, considering the sign convention, the power of the voltage source is found to be $p_s = v_s(-i_s) = 40$ W.

Problems also arise when applying KVL in a mesh involving a current source. We again consider the circuit above, while applying KVL in loops $1 \rightarrow 2 \rightarrow 3 \rightarrow 1$ containing the current source. Using one of the 4 Ω resistors, we have the following scenario.

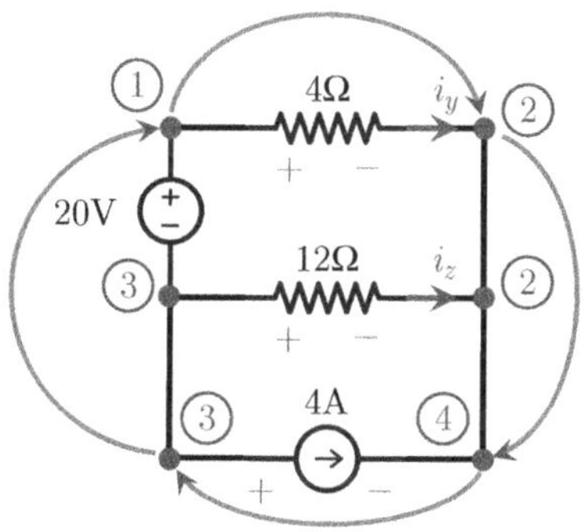

Then KVL leads to

- KVL($1 \rightarrow 2 \rightarrow 3 \rightarrow 1$): $4i_y - v_{4\text{A}} - 20 = 0$,

where $v_{4\text{ A}}$ is the voltage across the 4 A source, using the sign convention. Obviously, this equation is not useful alone, since it involves a new unknown, $v_{4\text{A}}$. Constructing KVL through the 12 Ω resistor, we can also obtain

- KVL($1 \rightarrow 2 \rightarrow 3 \rightarrow 1$): $12i_z - v_{4\text{A}} = 0$.

The two equations above can be combined to eliminate $v_{4\text{A}}$ as

$$4i_y - 12i_z = 20.$$

Nevertheless, the same equation could be found by applying KVL through the 4 Ω resistor, the 12 Ω resistor, and the voltage source (see the original solution, where i_x is used instead of i_y).

Confusion often occurs when dealing with short circuits. We reconsider the following circuit, where a 2 Ω resistor is short-circuited.

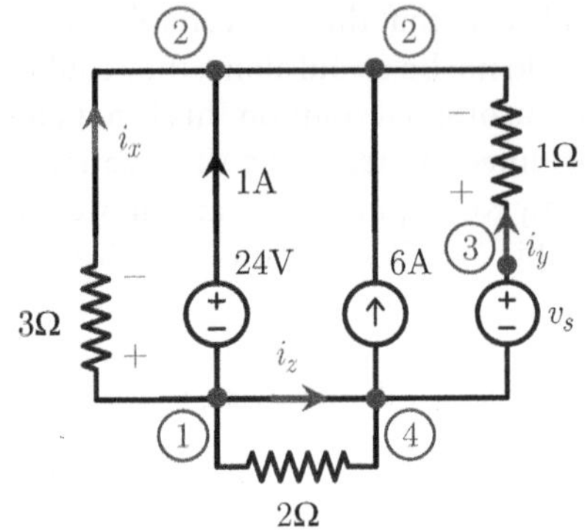

Our aim is to find the current through the short circuit, i_z. We label one of the related intersections node 4, whereas it is labeled node 1 in the original solution. Obviously, nodes 1 and 4 have the same voltage. However, this does not mean that there is no current between them. In fact, applying KCL at node 1, we derive

- KCL(1): $-i_x - i_z - 1 = 0$.

We note that there is no current flow through the 2 Ω resistor. Using $i_x = -8$ A, we obtain $i_z = 8 - 1 = 7$ A. One can check this value by applying KCL at the new node 4,

- KCL(4): $i_z - 6 - i_y = 0$,

where $i_y = 1$ A, as found before.

Dealing with too many current or voltage values may also lead to confusion in circuit analysis. We consider the following circuit, where the current values need to be found.

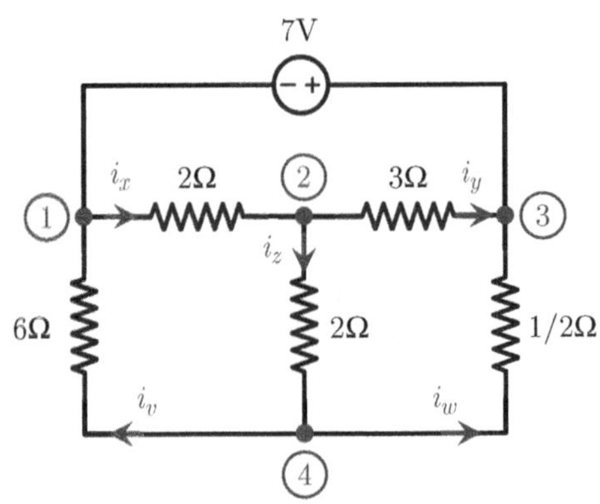

Applying KVL, we obtain the two equations

- KVL(1 → 2 → 4 → 1): $2i_x + 2i_z + 6i_v = 0 \longrightarrow 3i_v + i_x + i_z = 0$,
- KVL(2 → 3 → 4 → 2): $3i_y - i_w/2 - 2i_z = 0 \longrightarrow -i_w + 6i_y - 4i_z = 0$.

In addition, we obtain

- KVL(1 → 2 → 3 → 4 → 1): $2i_x + 3i_y - i_w/2 + 6i_v = 0$

or

$$12i_v - i_w + 4i_x + 6i_y = 0.$$

This final KVL equation is correct; unfortunately, it does not provide new information compared to the previous two equations. Indeed, combination of the first two KVL equations to eliminate i_z leads to

$$4(3i_v + i_x) - i_w + 6i_y = 0,$$

which is exactly the same as the whole KVL through $1 \rightarrow 2 \rightarrow 3 \rightarrow 4 \rightarrow 1$. To sum up, only two of the three KVL equations above provide useful information, while the third one is redundant.

In order to solve the problem above, we note that application of KCL at nodes 2 and 4 gives

- KCL(2): $i_x - i_y - i_z = 0$,
- KCL(4): $-i_v - i_w + i_z = 0$.

With these two equations, the total number of equations reaches four, while there are five unknowns (i_v, i_w, i_x, i_y, and i_z). The missing equation can be obtained by applying KVL through the voltage source. For example, one can derive

- KVL($1 \rightarrow 3 \rightarrow 2 \rightarrow 1$): $-7 - 3i_y - 2i_x = 0 \longrightarrow 2i_x + 3i_y = -7$.

At this stage, we can list all the useful equations once more as

$$\begin{aligned} 3i_v + i_x + i_z &= 0 \\ -i_w + 6i_y - 4i_z &= 0 \\ i_x - i_y - i_z &= 0 \\ -i_v - i_w + i_z &= 0 \\ 2i_x + 3i_y &= -7, \end{aligned}$$

or in matrix form,

$$\begin{bmatrix} 3 & 0 & 1 & 0 & 1 \\ 0 & -1 & 0 & 6 & -4 \\ 0 & 0 & 1 & -1 & -1 \\ -1 & -1 & 0 & 0 & 1 \\ 0 & 0 & 2 & 3 & 0 \end{bmatrix} \begin{bmatrix} i_v \\ i_w \\ i_x \\ i_y \\ i_z \end{bmatrix} = \begin{bmatrix} 0 \\ 0 \\ 0 \\ 0 \\ -7 \end{bmatrix}.$$

Solving the matrix equation leads to $i_v = 1\,\text{A}$, $i_w = -2\,\text{A}$, $i_x = -2\,\text{A}$, $i_y = -1\,\text{A}$, and $i_z = -1\,\text{A}$.

Obviously, the question above seems difficult to solve when considering five unknowns. In fact, a systematic application of KCL (see Chapter 3) or KVL (see Chapter 4) would lead to only three unknowns, as well as three equations to find them. As set out above, a solution with heuristic applications of KCL and KVL may lead to

- linearly dependent equations that often lead to an obvious equality $0 = 0$ after substitutions,
- insufficient numbers of equations, if the same variable is eliminated by substitution from all equations,

- looping through the same equalities if a node or mesh is excessively used in deriving alternative equations.

All these pitfalls can be avoided by choosing suitable strategies, such as nodal and mesh analysis, as described in the next chapters.

2.4 Series and Parallel Connections of Resistors

KCL and KVL provide all the information required to solve a given basic circuit. On the other hand, their application to alternative connections of resistors leads to shortcuts that can be employed to simplify the analysis.

2.4.1 Series Connection

Consider a series connection of two resistors with resistances R_1 and R_2.

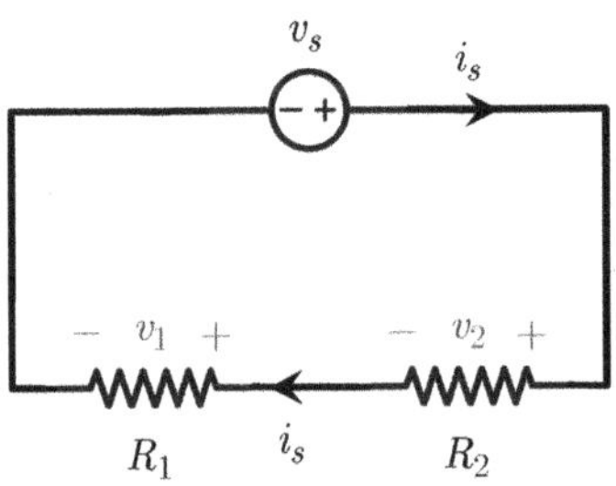

Using KVL, we have

$$-v_s + v_1 + v_2 = 0,$$

where $v_1 = R_1 i_s$ and $v_2 = R_2 i_s$. Then

$$v_s = R_{\text{eq}} i_s,$$

where

$$R_{\text{eq}} = R_1 + R_2$$

is the equivalent resistance of the combination of two resistors. We further note that the powers of the resistors are given by

$$p_1 = R_1 i_s^2,$$
$$p_2 = R_2 i_s^2.$$

Hence, the consumed power is divided into two parts that are proportional to the resistance values. Obviously, when there are n resistors all connected in series, we derive the equivalent resistance as

$$R_{\text{eq}} = \sum_{k=1}^{n} R_k.$$

2.4.2 Parallel Connection

Consider a parallel connection of two resistors with resistances R_1 and R_2.

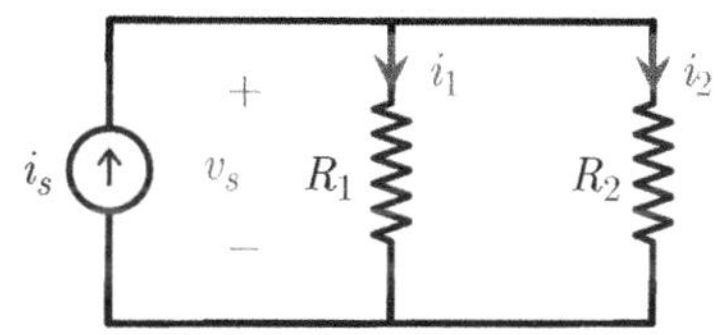

Using KCL, we derive

$$i_s - i_1 - i_2 = 0,$$

where $i_1 = v_s/R_1$ and $i_2 = v_s/R_2$. Hence, we obtain

$$i_s = \frac{v_s}{R_{eq}},$$

where

$$\frac{1}{R_{eq}} = \frac{1}{R_1} + \frac{1}{R_2}$$

represents the equivalent resistance of the combination of two resistors. One can also derive

$$R_{eq} = R_1 \parallel R_2 = \frac{R_1 R_2}{R_1 + R_2}$$

as the equivalent resistance. We note that, if $R_1 \neq 0$ and $R_2 \neq 0$, we have

$$\frac{R_1 R_2}{R_1 + R_2} < \frac{R_1 R_2 + R_2^2}{R_1 + R_2} = R_1$$

and

$$\frac{R_1 R_2}{R_1 + R_2} < \frac{R_1 R_2 + R_1^2}{R_1 + R_2} = R_2.$$

Therefore, the total resistance of a parallel connection is always smaller than the individual resistance of each resistor. It is remarkable that, when $R_2 = 0$, we obtain $R_{eq} = 0$, showing that R_1 is short-circuited. On the other hand, when $R_2 = \infty$, one can derive $R_{eq} \to R_1$ as the overall resistance.

When two resistors are connected in parallel, the total current i_s is divided into two parts as

$$i_1 = \frac{v_s}{R_1} = i_s \frac{R_2}{R_1 + R_2},$$

$$i_2 = \frac{v_s}{R_2} = i_s \frac{R_1}{R_1 + R_2}.$$

Obviously, the current tends to flow though the smaller resistance. Then the consumed power is distributed as inversely proportional to the resistance values as

$$p_1 = \frac{v_s^2}{R_1},$$
$$p_2 = \frac{v_s^2}{R_2}.$$

If there are n resistors all connected in parallel, the equivalent resistance can be written as

$$\frac{1}{R_{\text{eq}}} = \sum_{k=1}^{n} \frac{1}{R_k}.$$

Example 17: Consider the following circuit involving seven resistors connected to a 35 V voltage source.

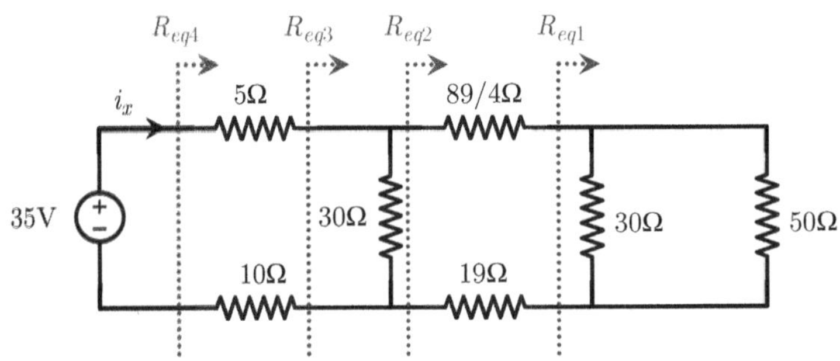

Find i_x.

Solution: By using the formulas for series and parallel connections, we systematically find equivalent resistances to reduce the circuit. We have

$$R_{\text{eq1}} = \frac{30 \times 50}{30 + 50} = \frac{75}{4}\ \Omega,$$
$$R_{\text{eq2}} = \frac{89}{4} + R_{\text{eq1}} + 19 = \frac{89}{4} + \frac{75}{4} + 19 = 60\ \Omega,$$
$$R_{\text{eq3}} = \frac{30 \times R_{\text{eq2}}}{30 + R_{\text{eq2}}} = \frac{30 \times 60}{30 + 60} = 20\ \Omega,$$
$$R_{\text{eq4}} = 5 + R_{\text{eq3}} + 10 = 35\ \Omega.$$

Then the current is simply

$$i_x = \frac{35}{35} = 1\ \text{A}.$$

Example 18: Consider the following circuit involving a one-dimensional infinite array of 1 Ω resistors connected to a 2 V voltage source.

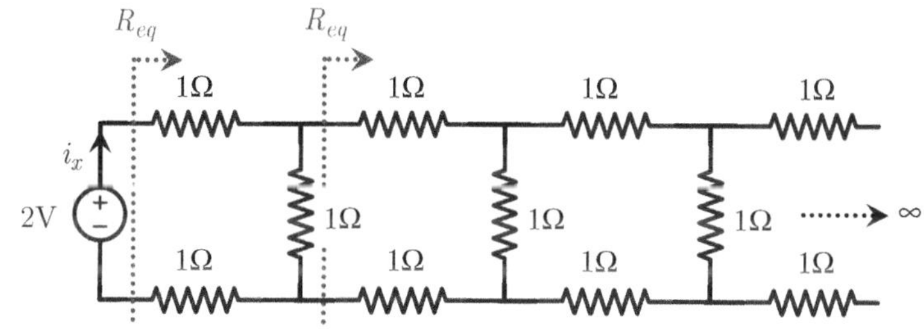

Find i_x.

Solution: First we assume that the equivalent resistance of the array is R_{eq}. Then, considering just the first loop, we have

$$(1 \parallel R_{\text{eq}}) + 2 = R_{\text{eq}},$$

which can be solved to obtain $R_{\text{eq}} = (1 + \sqrt{3})\ \Omega$. Therefore,

$$i_x = \frac{2}{1 + \sqrt{3}} = (\sqrt{3} - 1)\ \text{A}.$$

Example 19: Consider the following circuit involving a 60 V voltage source connected to a network of resistors.

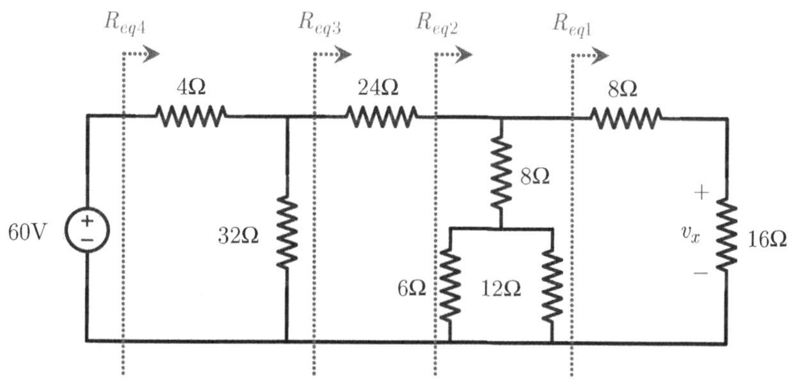

Find v_x.

Solution: Instead of applying KVL and KCL directly, we first reduce the circuit by considering series and parallel connections of resistors. We have

$$R_{\text{eq1}} = 8 + 16 = 24\ \Omega,$$

$$R_{\text{eq2}} = [8 + (6 \parallel 12)] \parallel R_{\text{eq1}} = 12 \parallel 24 = 8\ \Omega,$$

$$R_{\text{eq3}} = 24 + R_{\text{eq2}} = 24 + 8 = 32\ \Omega,$$

$$R_{eq4} = 4 + 32 \parallel R_{eq3} = 20\ \Omega.$$

Then, after finding the main current to be $i_s = 3$ A, we trace back the circuit as follows.

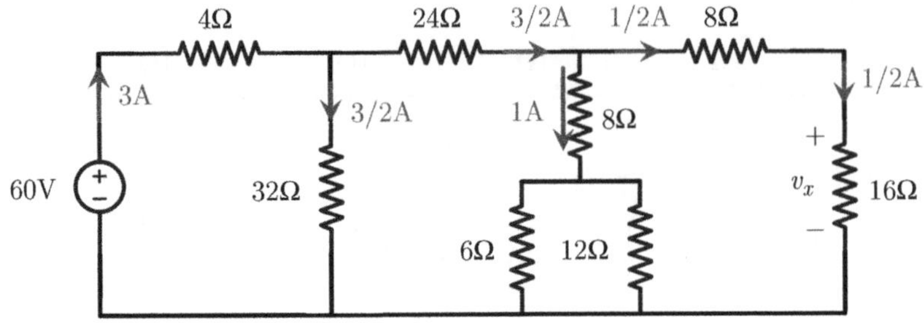

Hence, we obtain

$$i_x = \frac{1}{2}\ \text{A},$$
$$v_x = 16 i_x = 8\ \text{V}.$$

Exercise 8: In the following circuit, find the value of i_x.

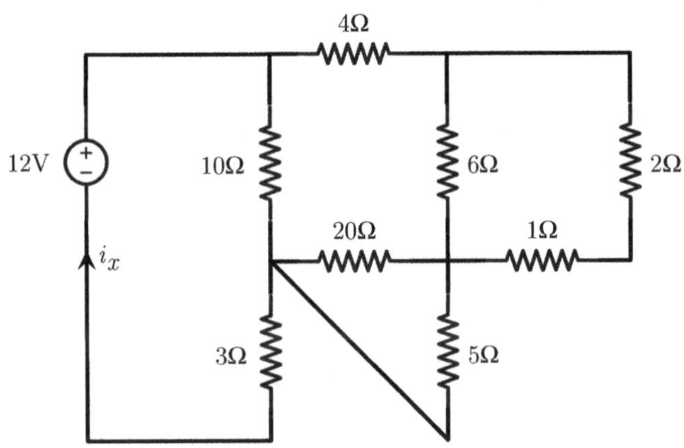

Exercise 9: In the following circuit, find the value of i_x.

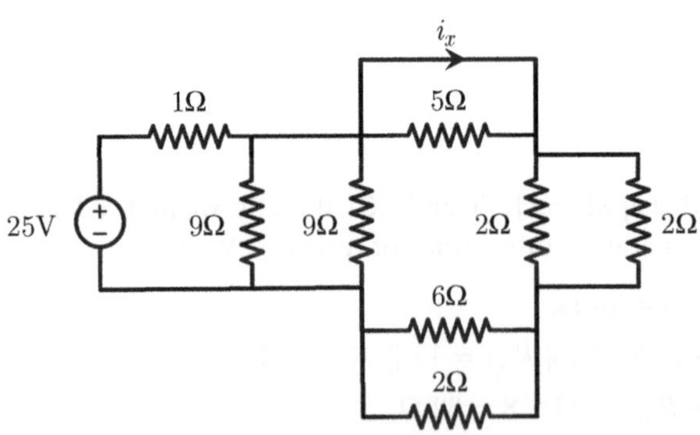

Exercise 10: In the following circuit, find the value of i_x.

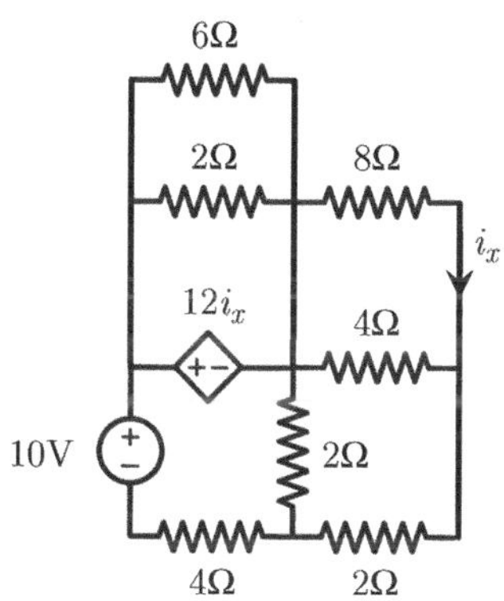

2.5 When Things Go Wrong with Series/Parallel Resistors

Major issues arise when considering resistors that appear to be connected in series or parallel, but in fact are not. We consider the following three different circuits.

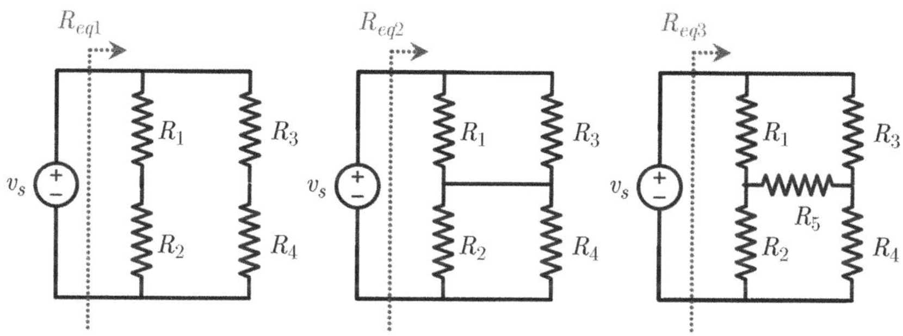

In the first case (the circuit on the left), R_1 and R_2 are connected in series, leading to $R_1 + R_2$ total resistance. Similarly, R_3 and R_4 are connected in series, leading to $R_3 + R_4$. Then the overall resistance is given by

$$R_{\text{eq1}} = (R_1 + R_2) \parallel (R_3 + R_4) = \frac{(R_1 + R_2)(R_3 + R_4)}{R_1 + R_2 + R_3 + R_4}.$$

In the second case (the circuit in the middle), R_1 and R_3 are connected in parallel, while R_2 and R_4 are similarly parallel. Therefore, the overall resistance is

$$R_{\text{eq2}} = (R_1 \parallel R_3) + (R_2 \parallel R_4) = \frac{R_1R_3}{R_1 + R_3} + \frac{R_2R_4}{R_2 + R_4}$$
$$= \frac{R_1R_3(R_2 + R_4) + R_2R_4(R_1 + R_3)}{(R_1 + R_3)(R_2 + R_4)},$$

which is obviously different from R_{eq1}. Only in some cases, such as when $R_1 = R_2 = R_3 = R_4 = R$, do we have $R_{\text{eq1}} = R_{\text{eq2}}$.

In the third case (the circuit on the right), where $R_5 \neq 0$ or $R_5 \neq \infty$, none of the resistor pairs are connected in series or parallel. This is because, none of the pairs share a common current or common voltage. For general values of the resistors, this circuit must be analyzed via KVL and KCL or via nodal/mesh methods. Interestingly, when $R_1 = R_2 = R_3 = R_4 = R$, we have $R_{eq3} = R$, independent of the value of R_5.

2.6 What You Need to Know before You Continue

Here are a few key points before proceeding to the next chapter.

- KCL: The sum of currents entering (+) and leaving (−) a node should be zero. When writing a KCL equation, we use plus and minus signs for entering and leaving currents, respectively.
- KVL: The sum of voltages in a closed loop is zero. When writing a KVL equation, we always choose the clockwise direction. Then the voltages of the components are added by considering their signs according to their first terminals.
- Current and voltage directions: When analyzing circuits, we define the current and voltage directions arbitrarily, while enforcing the sign convention for all components. When a current/voltage value is found to be negative, we understand that the initial assumption is not correct, although this is not a problem.
- Useless equations: We usually avoid applying KCL at a node to which a voltage source is connected. Similarly, a KVL in a mesh containing a current source is usually not useful.
- Series/parallel resistors: Combinations of resistors can often be simplified by considering series and parallel connections before the circuit is analyzed via KCL and KVL.

In the next chapter, we focus on nodal analysis, which is a systematic way of applying KCL at nodes. Such methods are essential for deriving required numbers of equations, so as to avoid redundant equations and variables, when analyzing circuits.

3

Analysis of Resistive Networks: Nodal Analysis

KCL and KVL are fundamental laws for analyzing circuits. However, as discussed in the previous chapter, their application may not be always clear. Specifically, for large circuits involving many components and connections, it may be difficult to derive necessary relationships between voltages and currents, while avoiding duplications and linearly dependent equations. In this chapter, we focus on nodal analysis, which is a higher-level tool based on a systematic application of KCL. Given any circuit, a proper application of nodal analysis guarantees the derivation of necessary equations for the analysis. We also discuss the generalization of nodes to further use the benefits of nodal analysis for complex circuits.

3.1 Application of Nodal Analysis

We start by briefly listing the major steps of nodal analysis as follows.

- Ground selection: We select a reference node with zero voltage. Any node can be selected, but it usually better to choose one with more connections than the others. The node selected is called the ground of the circuit. All voltages at other nodes are defined with respect to the ground.
- Constructing equations: We use only KCL at nodes, except the ground, to derive all equations. KVL is not preferred in nodal analysis unless necessary. We write all equations in terms of node voltages.
- Solution: Next, we solve the equations to find the node voltages.
- Analysis: Finally, by using the node voltages, we find the desired voltage, current, and/or power values.

Once again, we emphasize that a formal voltage definition requires two points. On the other hand, if there is a node where the voltage is defined as zero, it becomes practical to define node voltages as if they are independent values.

Introduction to Electrical Circuit Analysis, First Edition. Özgür Ergül.

Companion Website: www.wiley.com/go/ergul4412

Example 20: Consider the following circuit involving a current-dependent current source, a voltage source, and two resistors.

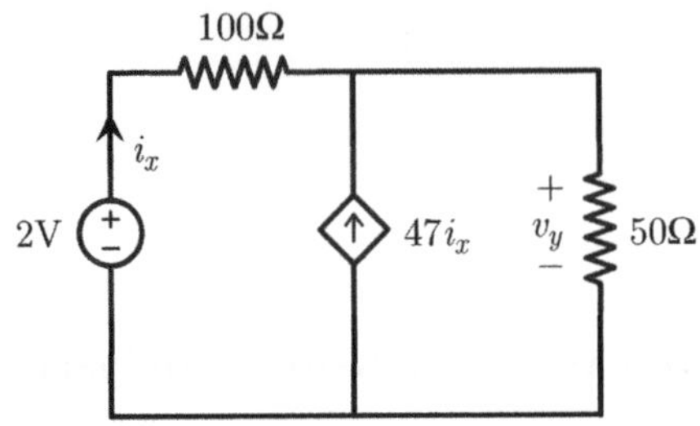

Find i_x, which flows through the voltage source.

Solution: This circuit can be analyzed using KCL and KVL, as usual. After labeling nodes and defining directions, we have the following circuit.

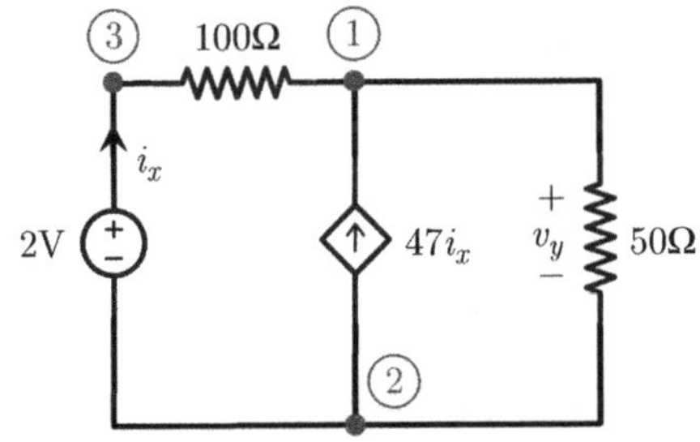

Then, applying KCL and KVL, we derive

- KCL(1): $i_x + 47i_x - i_y = 0 \longrightarrow i_y = 48i_x$,
- KVL($1 \to 2 \to 3 \to 1$): $v_y - 2 + v_x = 0 \longrightarrow v_x + v_y = 2$.

Finally, using Ohm's law,

$$100i_x + 50i_y = 2 \longrightarrow 2500i_x = 2 \longrightarrow i_x = 2/2500 \text{ A}.$$

As an alternative solution, we now apply nodal analysis, selecting node 2 as the reference node.

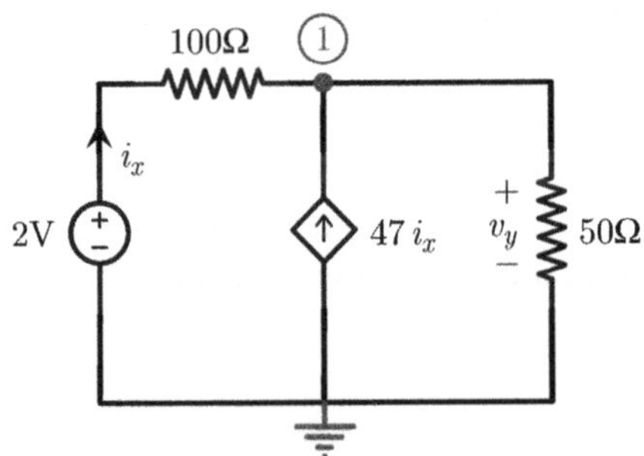

Using Ohm's law, we have $i_x = (2 - v_1)/100$. Then, using KCL,

- KCL(1): $48(2 - v_1)/100 - v_1/50 = 0 \longrightarrow 96 - 48v_1 - 2v_1 = 0$,

leading to $v_1 = 48/25$ V. Therefore, $i_x = (2 - 48/25)/100 = 2/2500$ A. We note that only one KCL equation has been sufficient to analyze the circuit. Specifically, two of the nodes in the earlier analysis are not used in nodal analysis. Some important points are as follows.

- Node 2 is made the reference node (ground) with zero voltage. In general, KCL need not be applied at a ground. In fact, needing to apply KCL at a ground is usually an indicator that another node has been skipped by mistake in the analysis.
- Node 3 is also not used directly in nodal analysis, because its voltage is already known due to the voltage source. In general, if the voltage at a node is easily defined, one does not need to write a KCL equation at that node. In fact, applying KCL at a node with a directly connected voltage source should be avoided, unless it is mandatory (e.g., if one must find the current through the voltage source).
- Application of KVL should be avoided in nodal analysis since a proper set of KCL equations should be sufficient to solve the circuit. KVL can be used to find other quantities after all node voltages are obtained.

Two facts provide a deeper understanding of nodal analysis.

- Setting zero voltage at the ground is merely a choice. Indeed, one could assign any voltage (e.g., 10 V), which would shift voltage values at all other nodes by 10. On the other hand, real quantities, such as component voltages, currents, and powers, do not depend on this selection.
- Selecting a node as ground is also completely arbitrary. One can choose any node as a reference, provided that the voltages are defined accordingly. As mentioned above, certain selections (e.g., choosing nodes with more connections) can simplify nodal analysis.

Example 21: Consider the following circuit.

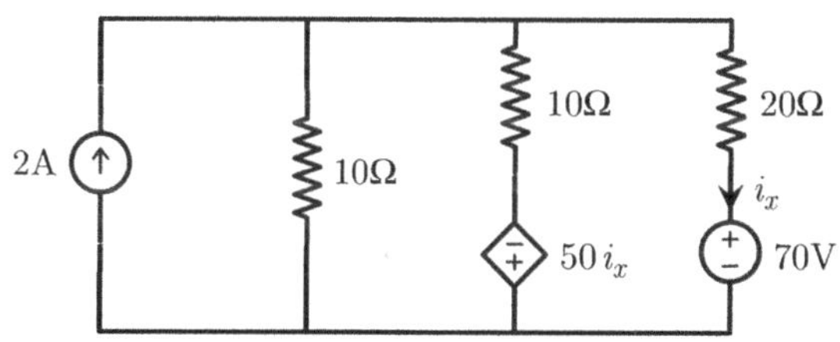

Find i_x.

Solution: Applying nodal analysis, we select a ground and define other node voltages accordingly.

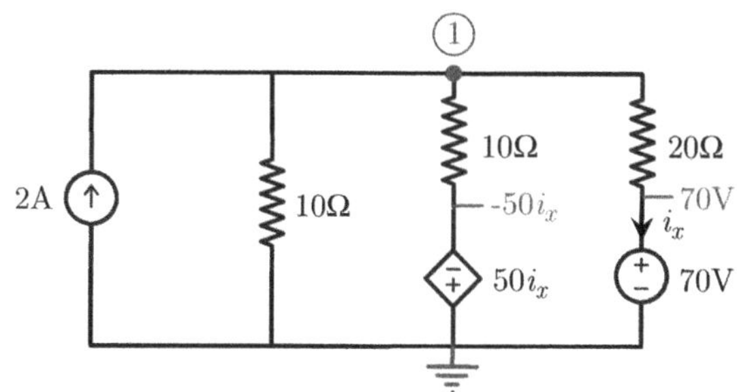

We note that only one node needs to be defined and used in the analysis, since two other nodes already have well-defined voltage values (i.e., $-50i_x$ and 70 V). We also have $i_x = (v_1 - 70)/20$ using Ohm's law. Applying KCL, we have

- KCL(1): $2 - v_1/10 - (v_1 + 50i_x)/10 - (v_1 - 70)/20 = 0$,

leading to $40 - 2v_1 - 2v_1 - 100i_x - v_1 + 70 = 0$ or

$$5v_1 = -100i_x + 110 \longrightarrow v_1 = -20i_x + 22.$$

Then, using the relation between v_1 and i_x, we obtain

$$v_1 = 70 - v_1 + 22 \longrightarrow 2v_1 = 92 \longrightarrow v_1 = 46 \text{ V}.$$

Finally, we have

$$i_x = (v_1 - 70)/20 = -24/20 = -6/5 \text{ A}.$$

Example 22: Consider the following circuit.

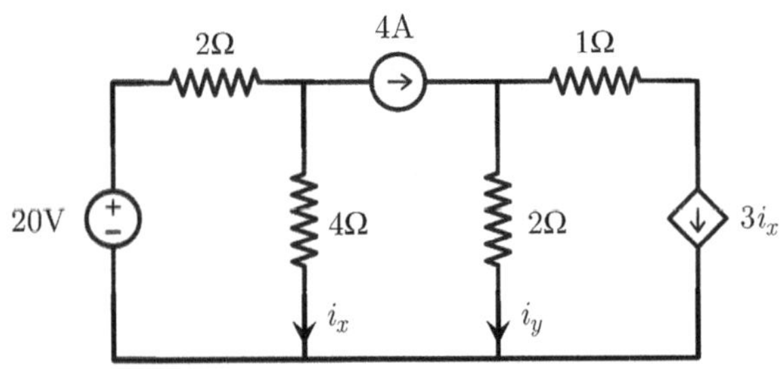

Find i_y.

Solution: As in the previous examples, we select a ground and define other node voltages accordingly.

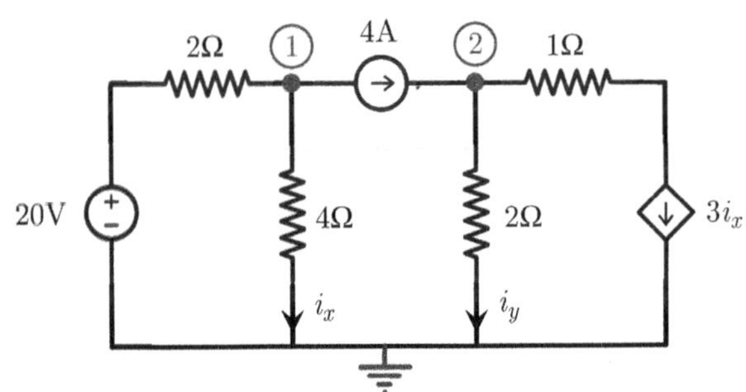

Using Ohm's law, we have $i_x = v_1/4$. Then, applying KCL at node 1, we derive

- KCL(1): $(20 - v_1)/2 - v_1/4 - 4 = 0 \longrightarrow v_1 = 8\text{ V} \longrightarrow i_x = 2\text{ A}$.

In the above, i_x and 4 A currents are leaving node 1 so that they are written as negative on the left-hand side of the equation. For the current flowing in the first branch, we have two options, leading to the same result.

- Defining the current as entering node 1, its value should be $(20 - v_1)/2$. This value should be treated as positive in the KCL equation.
- Defining the current as leaving node 1, its value should be $(v_1 - 20)/2$. This value should be treated as negative in the KCL equation, making it $(20 - v_1)/2$ overall.

Hence, the choice of current directions will change neither the KCL equation nor the result. In order to complete the solution and find i_y, we appply KCL at node 2, yielding

- KCL(2): $4 - v_2/2 - 3i_x = 0 \longrightarrow v_2 = 8 - 6i_x = -4\,\text{V}$,

leading to

$$i_y = v_2/2 = -2\ \text{A}.$$

Example 23: Consider the following circuit.

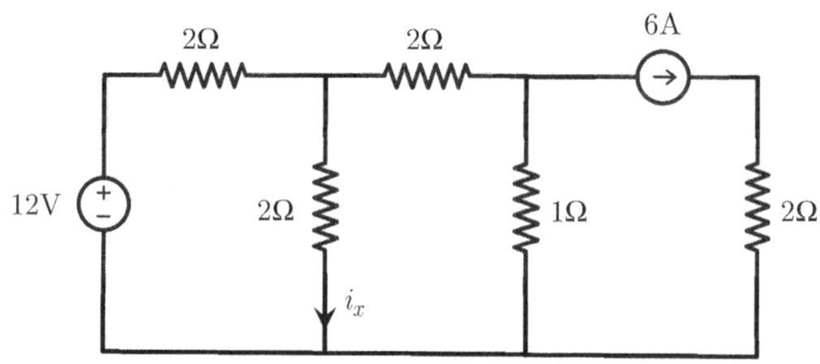

Find i_x.

Solution: Using nodal analysis, we again select a ground and define other node voltages accordingly.

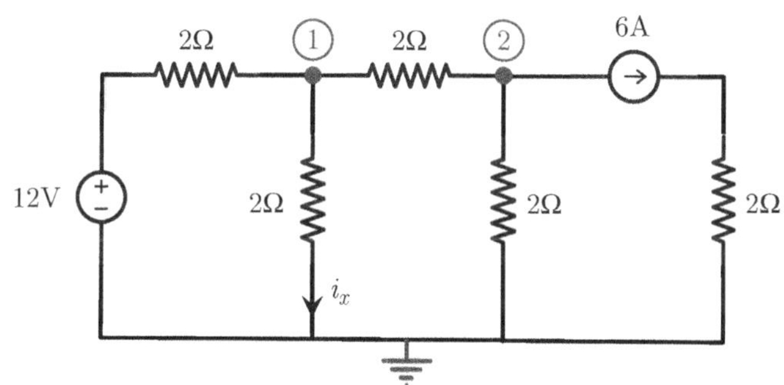

Using KCL at nodes 1 and 2, we obtain two equations with two unknowns as

- KCL(1): $(12 - v_1)/2 - v_1/2 - (v_1 - v_2)/2 = 0 \longrightarrow 3v_1 - v_2 = 12$,
- KCL(2): $(v_1 - v_2)/2 - 6 - v_2 = 0 \longrightarrow v_1 - 3v_2 = 12$.

Finally, solving the equations, we get $v_1 = 3\,\text{V}$, $v_2 = -3\,\text{V}$, and $i_x = v_1/2 = 3/2\,\text{A}$.

Example 24: Consider the following circuit.

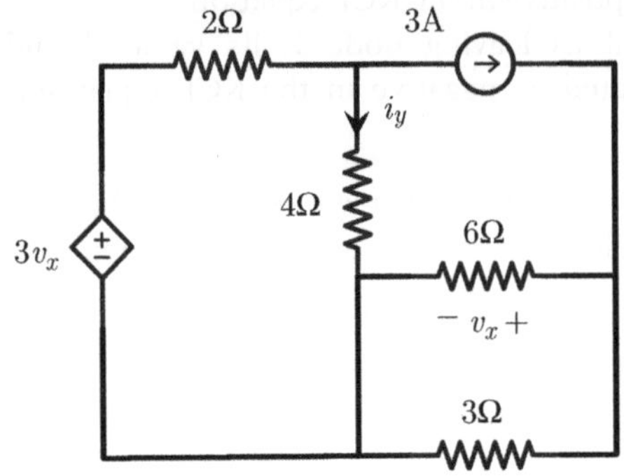

Find i_y.

Solution: We again start by selecting a ground and defining the other node voltages accordingly.

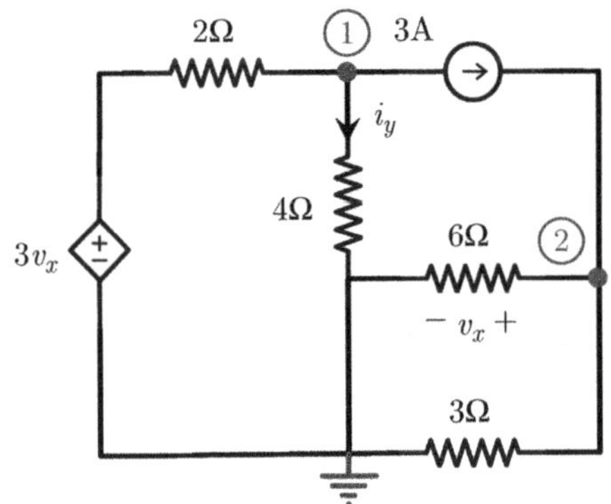

Following the selections above, we note that $v_x = v_2$. Applying KCL at nodes 1 and 2, we derive

- KCL(1): $(3v_x - v_1)/2 - v_1/4 - 3 = 0 \longrightarrow 2v_2 - v_1 = 4$,
- KCL(2): $3 - v_2/6 - v_2/3 = 0 \longrightarrow v_2 = 6\,\text{V}$.

Then we have $v_1 = 8\,\text{V}$ and $i_y = 8/4 = 2\,\text{A}$.

Example 25: Consider the following circuit with three sources and four resistors.

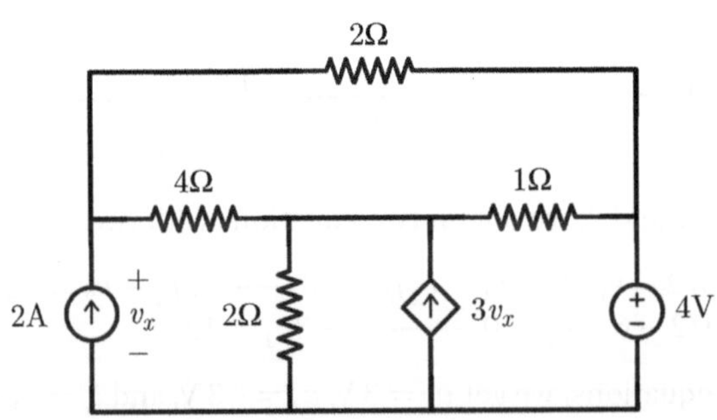

Find the power of the 4 Ω resistor.

Solution: Using nodal analysis, we define only two nodes to construct equations.

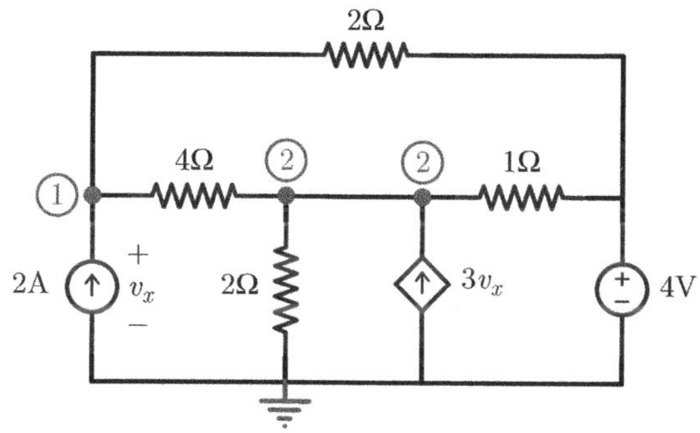

In the above, $v_x = v_1$. Applying KCL, we have

- KCL(1): $2 - (v_1 - v_2)/4 - (v_1 - 4)/2 = 0 \longrightarrow 3v_1 - v_2 = 16,$
- KCL(2): $(v_1 - v_2)/4 - v_2/2 + 3v_1 - (v_2 - 4)/1 = 0 \longrightarrow 13v_1 - 7v_2 = -16.$

Then we obtain $v_1 = 16$ V, $v_2 = 32$ V, and $p_{4\Omega} = (32 - 16)^2/4 = 64$ W.

Example 26: Consider the following circuit.

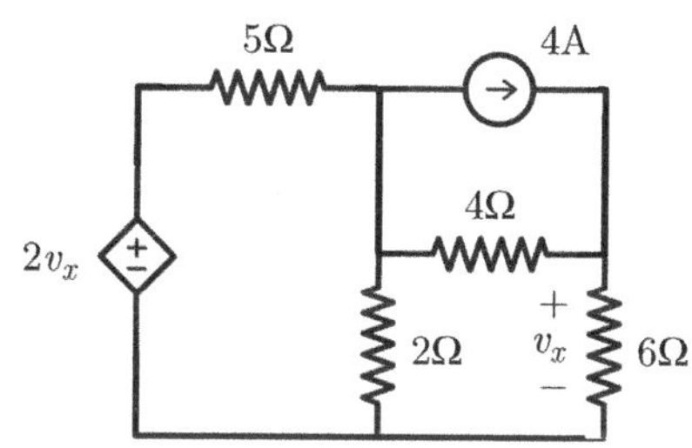

Find v_x.

Solution: Using nodal analysis, we have just two nodes where we apply KCL.

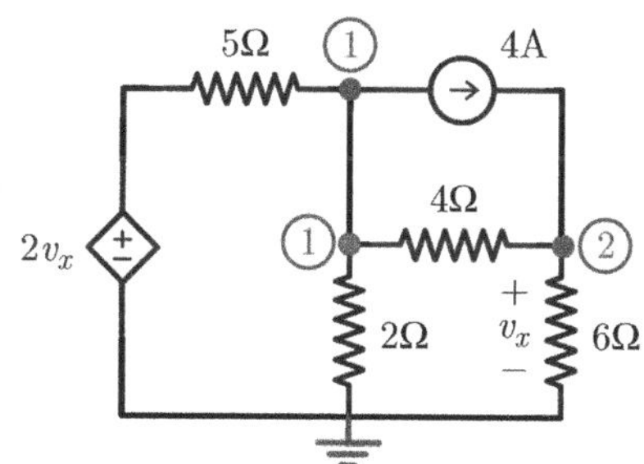

Applying KCL at node 1, we have

- KCL(1): $(2v_x - v_1)/5 - 4 - v_1/2 - (v_1 - v_2)/4 = 0 \longrightarrow -19v_1 + 13v_2 = 80,$

where $v_x = v_2$ is used. Similarly, at node 2, we have

- KCL(2): $4 + (v_1 - v_2)/4 - v_2/6 = 0 \longrightarrow 3v_1 - 5v_2 = -48$.

Solving the equations, we obtain $v_1 = 4$ V and

$$v_x = v_2 = (3v_1 + 48)/5 = 12 \text{ V}.$$

Example 27: Consider the following circuit.

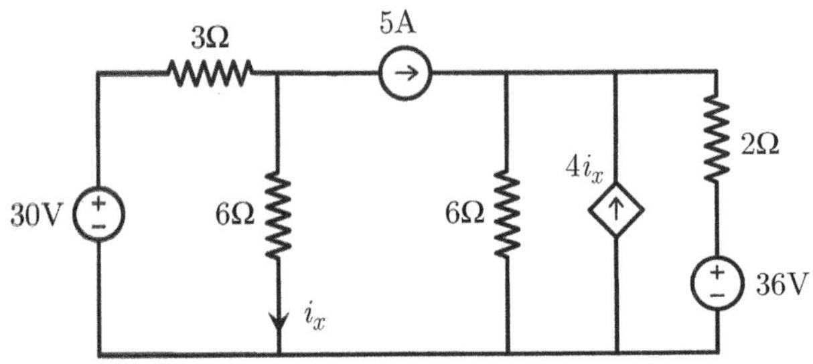

Find the power of the current-dependent current source.

Solution: Using nodal analysis, we again have two nodes where we apply KCL.

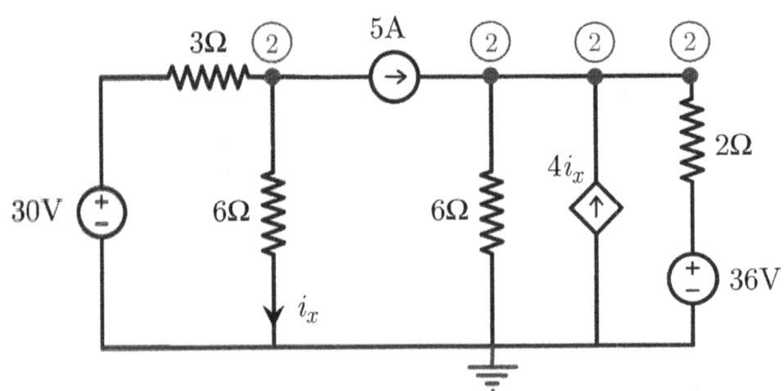

First, we note that $i_x = v_1/6$, according to the selection of the reference node. Using KCL at nodes 1 and 2, we derive

- KCL(1): $(30 - v_1)/3 - v_1/6 - 5 = 0 \longrightarrow v_1 = 10$ V and $i_x = 5/3$ A,
- KCL(2): $5 - v_2/6 + 4i_x - (v_2 - 36)/2 = 0 \longrightarrow v_2 = 89/2$ V.

The power of the current-dependent current source can be found by multiplying its voltage and current (keeping consistent with the sign convention), yielding

$$p_s = (0 - v_2)4i_x = (-89/2) \times (20/3) = -890/3 \text{ W}.$$

Example 28: Consider the following circuit involving four different kinds of sources.

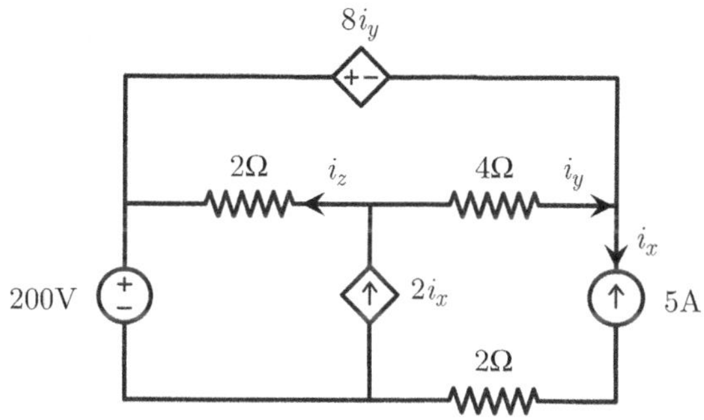

Find i_z.

Solution: First, we note that $i_x = -5$ A. Using nodal analysis, we can define the ground and label the nodes as follows.

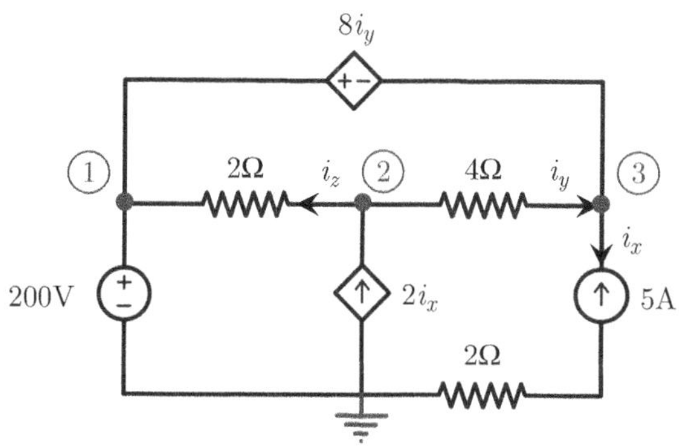

We further note that KCL is not required (and not useful) at node 1, since $v_1 = 200$ V. In addition, one should avoid applying KCL at node 3 since a voltage source (dependent source in this case) is connected to this node. Instead, using the definition of the voltage source, we have

$$v_3 = v_1 - 8i_y = 200 - 8i_y,$$

where $i_y = (v_2 - v_3)/4$. Therefore,

$$v_3 = 200 - 2v_2 + 2v_3$$

or

$$2v_2 - v_3 = 200$$

as the first equation. For the solution, we apply KCL at the only remaining node (node 2), yielding

- KCL(2): $(200 - v_2)/2 - 10 + (v_3 - v_2)/4 = 0 \longrightarrow 3v_2 - v_3 = 360\,\text{V}$.

Finally, we obtain $v_2 = 160\,\text{V}$ and $i_z = -20\,\text{A}$.

Example 29: Consider the following circuit.

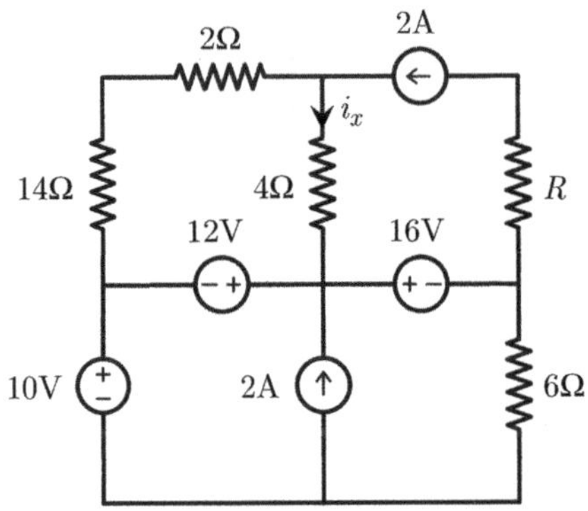

Find i_x.

Solution: The circuit looks complex at first glance, but can easily be solved using nodal analysis.

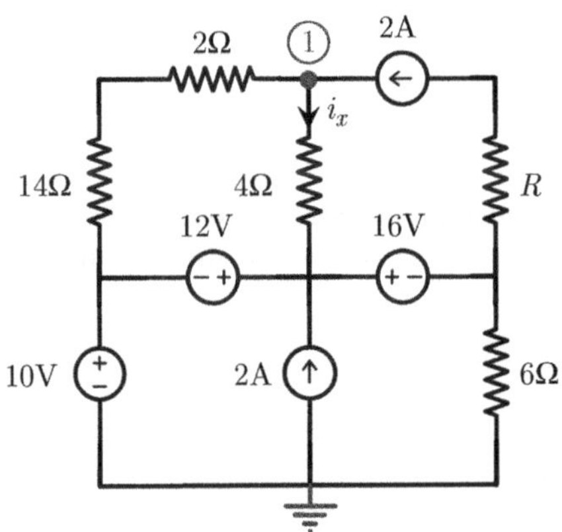

As shown above, using a suitable ground location, we have only one node where we need to apply KCL, and voltages at all other nodes are already known. We have

- KCL(1): $(10 - v_1)/16 + 2 - (v_1 - 22)/4 = 0 \longrightarrow v_1 = 26\,\text{V}$, leading to

$$i_x = (v_1 - 22)/4 = 1 \text{ A}.$$

Exercise 11: In the following circuit, find the value of i_x.

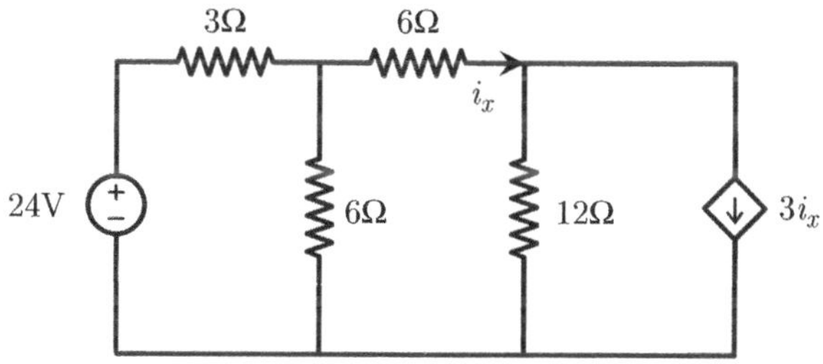

Exercise 12: In the following circuit, find the power of the 24 Ω resistor.

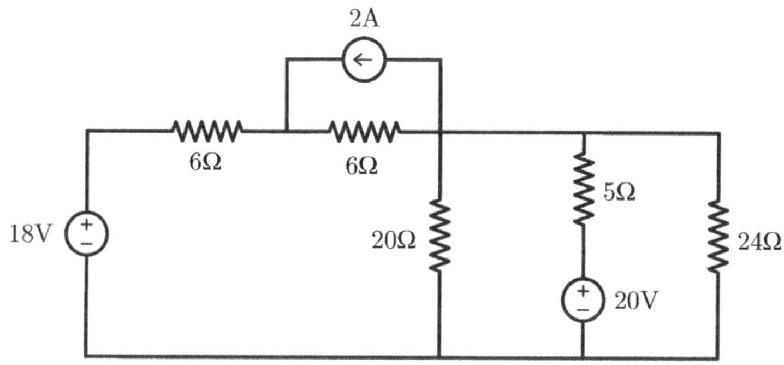

Exercise 13: In the following circuit, find the power of the 5 Ω resistor.

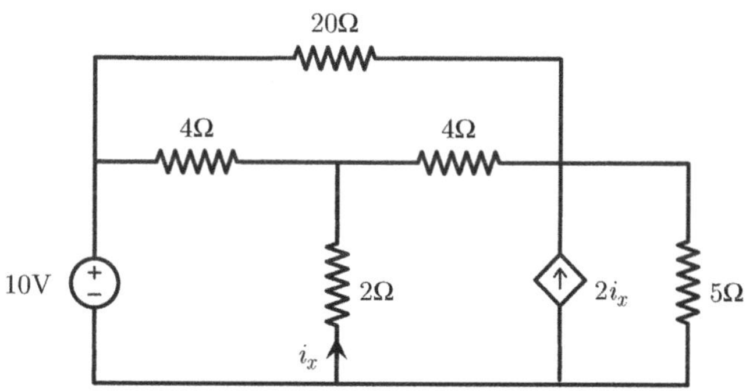

Exercise 14: In the following circuit, find v_s given that $i_x = 2\,\text{A}$.

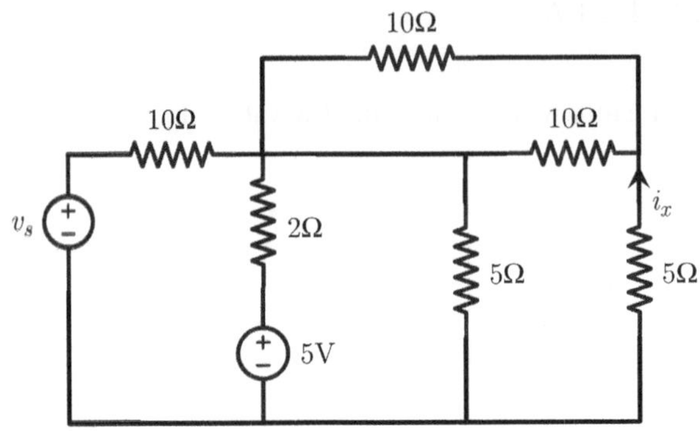

Exercise 15: In the following circuit, find the value of i_x.

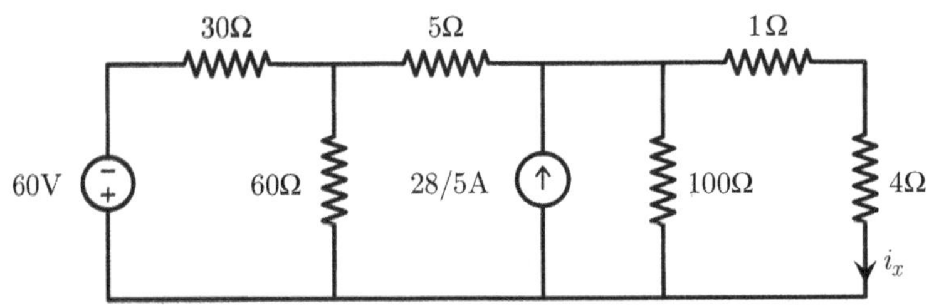

Exercise 16: In the following circuit, find the power of the 8 A source.

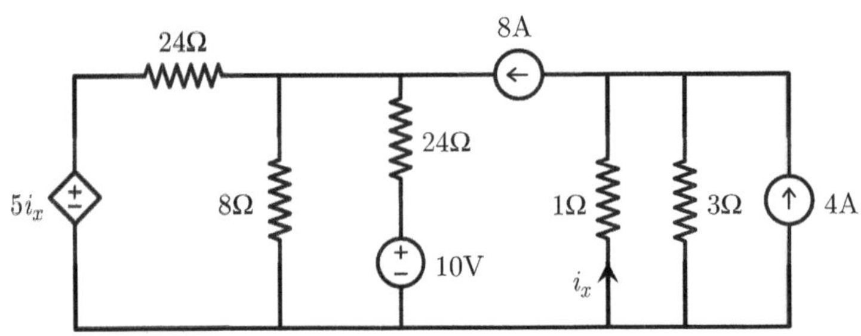

Exercise 17: In the following circuit, find the value of v_y.

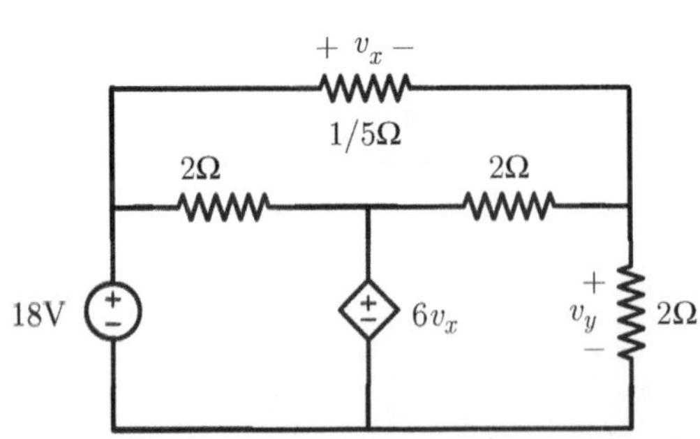

Exercise 18: In the following circuit, find the value of i_x.

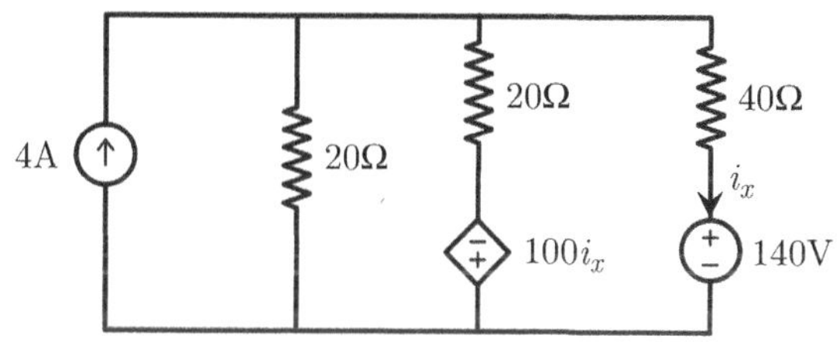

Exercise 19: In the following circuit, find the value of i_x.

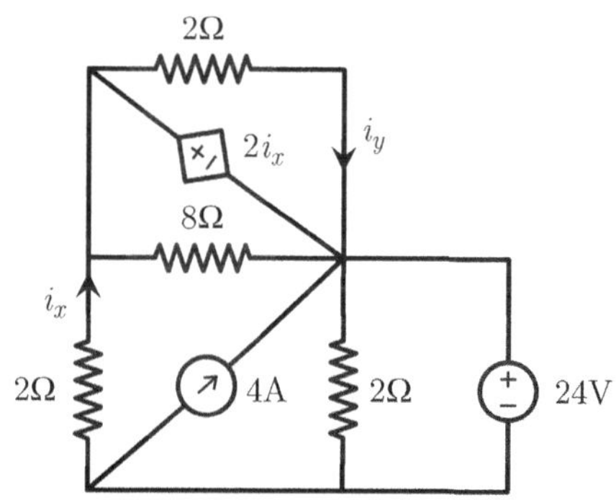

Exercise 20: In the following circuit, find the value of i_x.

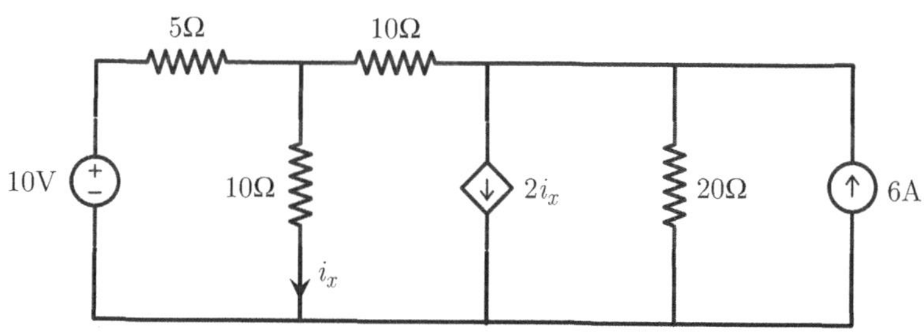

3.2 Concept of Supernode

KCL can be generalized to any arbitrary surface: the sum of currents entering and leaving a closed surface must be zero. An application of this generalized form is to define supernodes that contain several nodes, as well as components, in nodal analysis. A supernode enclosing a voltage source is particularly useful, especially when there are nodes (attached to this voltage source) at which voltages cannot be defined easily.

Example 30: Consider the following circuit.

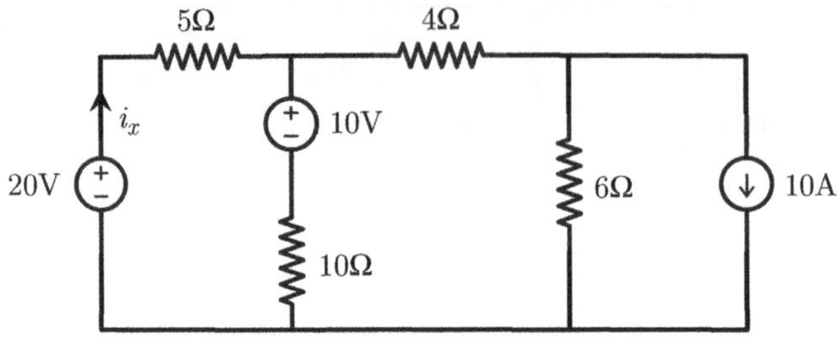

Find i_x.

Solution: Using nodal analysis, we define the ground and label the nodes as follows.

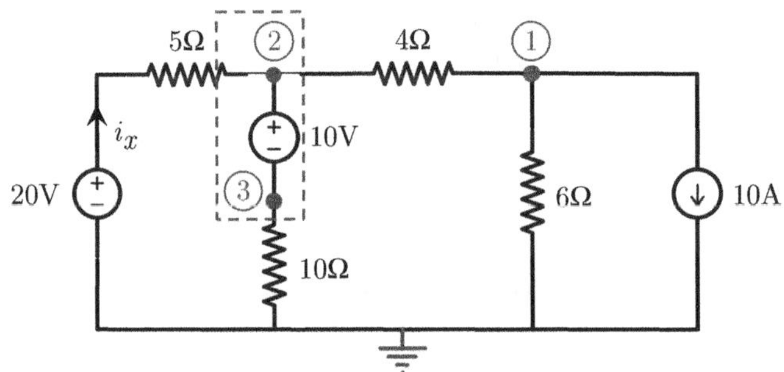

In the above, it is straightforward to write a KCL equation at node 1. On the other hand, a KCL equation at node 2 requires the current along the 10 V source, which cannot be defined in terms of node voltages. Interestingly, for this circuit, there is a shortcut without resorting to a supernode. Considering node 3, one can claim that the current along the voltage source is actually $v_3/10$, since the source is serially connected to a 10 Ω resistor. However, this is a very special case, and we often find nodes where KCL is not trivial to write.

In order to facilitate nodal analysis for this circuit, we combine nodes 2 and 3, leading to a supernode that contains the voltage source. This combination is shown as a dashed box in the figure above. We start by writing KCL at node 1 as

- KCL(1): $(v_2 - v_1)/4 - v_1/6 - 10 = 0 \longrightarrow -5v_1 + 3v_2 = 120.$

Then, using KCL at the supernode, we also have

- KCL(2&3): $(20 - v_2)/5 - (v_2 - v_1)/4 - v_3/10 = 0 \longrightarrow 5v_1 - 9v_2 - 2v_3 = -80.$

We note that KCL at the supernode involving nodes 2 and 3 does not involve the current between these nodes. It only consists of the currents entering and leaving the box that defines the supernode.

When using a supernode in nodal analysis, the number of equations derived from the application of KCL at nodes is usually one less than the number of unknowns, that is, node voltages. The extra equation needed is always derived from the supernode itself. Specifically, looking inside the node for the circuit above, we have

$$v_3 = v_2 - 10,$$

leading to

$$5v_1 - 11v_2 = -100.$$

Then, solving the equations, we obtain $v_1 = -5/2\,\text{V}$, $v_2 = -5/2\,\text{V}$, $v_3 = -25/2\,\text{V}$, and $i_x = 9/2\,\text{A}$.

Example 31: Consider the following circuit.

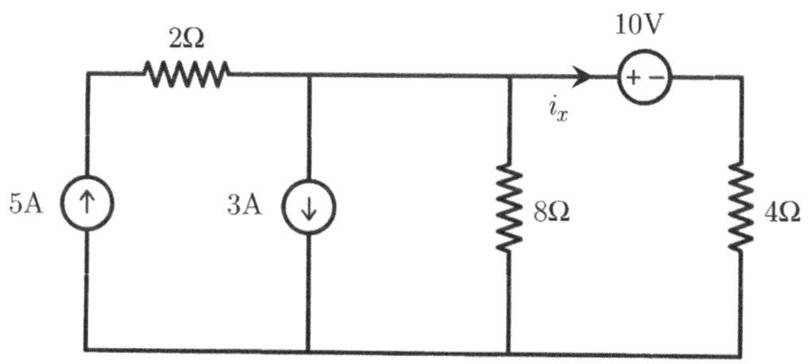

Find i_x.

Solution: Using nodal analysis, we define the ground, and then label the nodes.

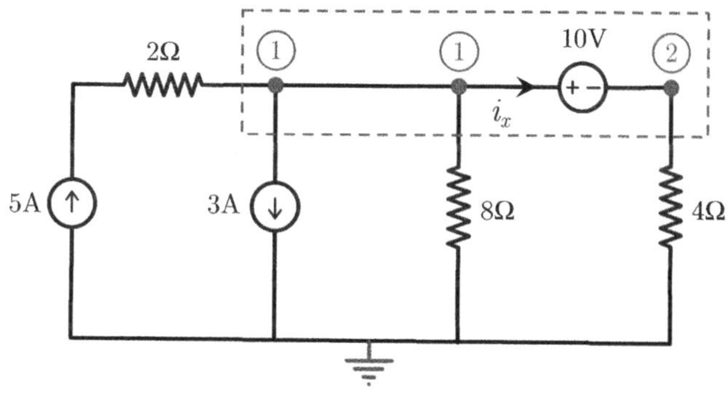

In addition, we combine the nodes to obtain a supernode, leading to a single application of KCL,

- KCL(1&2): $5 - 3 - v_1/8 - v_2/4 = 0 \longrightarrow v_1 + 2v_2 = 16$.

Using the supernode, the node voltages are related by $v_1 = v_2 + 10$. Hence, we obtain $v_1 = 12\,\text{V}$, $v_2 = 2\,\text{V}$, and $i_x = 2/4 = 1/2\,\text{A}$.

Example 32: Consider the following circuit involving a voltage-dependent voltage source.

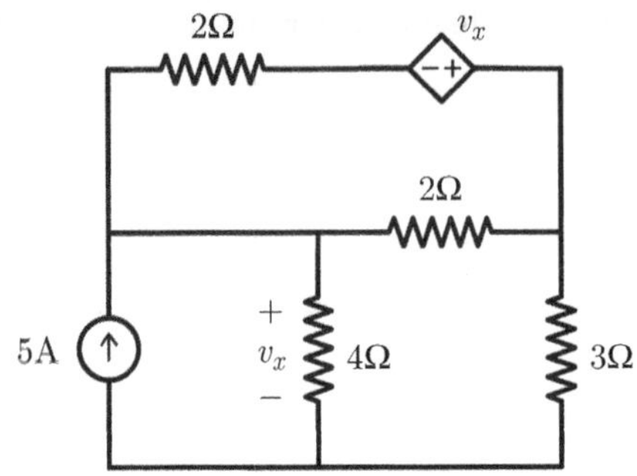

Find the power of the 3 Ω resistor.

Solution: Using nodal analysis, we define the ground and label the nodes as follows.

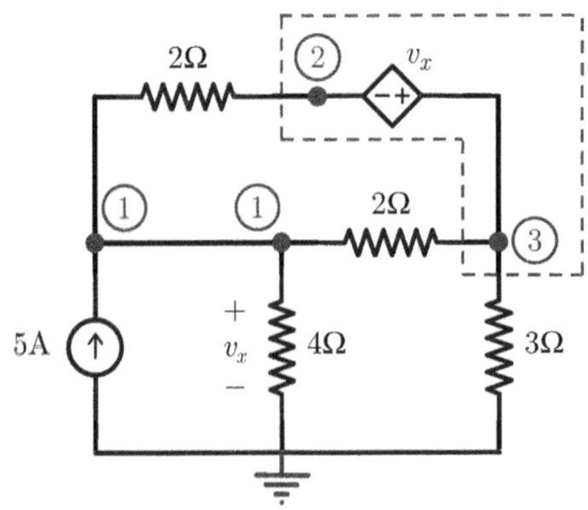

In the above, nodes 2 and 3 are combined as a supernode. We have

$$v_x = v_1$$

and

$$v_3 - v_2 = v_x = v_1$$

using the supernode. Applying KCL at node 1, we derive

- KCL(1): $5 + (v_2 - v_1)/2 - v_1/4 - (v_1 - v_3)/2 = 0$,

leading to

$$5v_1 - 2v_2 - 2v_3 = 20$$
$$3v_3 - 7v_2 = 20.$$

Then, applying KCL at the supernode, we have

- KCL(2&3): $(v_1 - v_2)/2 + (v_1 - v_3)/2 - v_3/3 = 0$,

leading to

$$6v_1 - 3v_2 - 5v_3 = 0$$
$$-9v_2 + v_3 = 0.$$

Solving the equations, we obtain $20v_2 = 20$ or $v_2 = 1\,\text{V}$. Finally, we get $v_3 = 9\,\text{V}$ and $p_{3\Omega} = 9^2/3 = 27\,\text{W}$.

Example 33: Consider the following circuit.

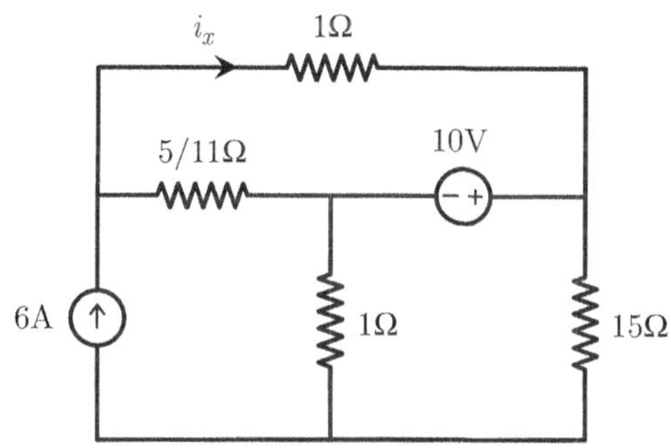

Find i_x.

Solution: Using nodal analysis, we define the ground and label the nodes.

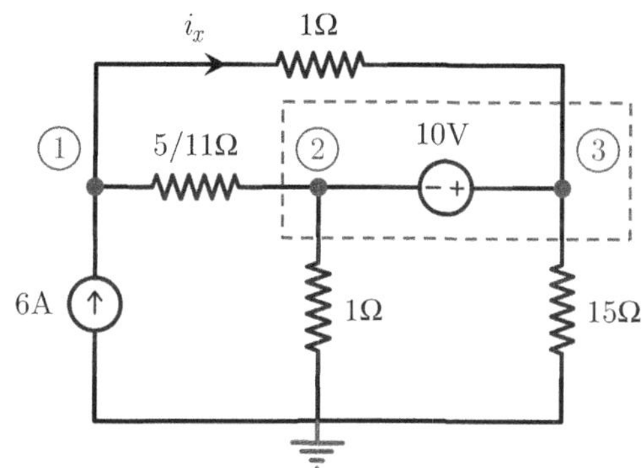

In the above, nodes 2 and 3 are combined as a supernode. We apply KCL at node 1 and at the supernode, yielding

- KCL(1): $6 - (v_1 - v_2)/(5/11) - (v_1 - v_3) = 0 \longrightarrow 16v_1 - 11v_2 - 5v_3 = 30$
- KCL(2&3): $(v_1 - v_2)/(5/11) + (v_1 - v_3) - v_2 - v_3/15 = 0$
  $\longrightarrow 3v_1 - 3v_2 - v_3 = 0.$

Furthermore, the supernode gives $v_3 = v_2 + 10$, leading to

$$16v_1 - 16v_2 = 80 \longrightarrow v_1 - v_2 = 5$$
$$3v_1 - 4v_2 = 10.$$

Solving the equations, we get $v_1 = 10$ V, $v_2 = 5$ V, $v_3 = 15$ V, and

$$i_x = (v_1 - v_3)/1 = -5 \text{ A}.$$

Example 34: Consider the following circuit.

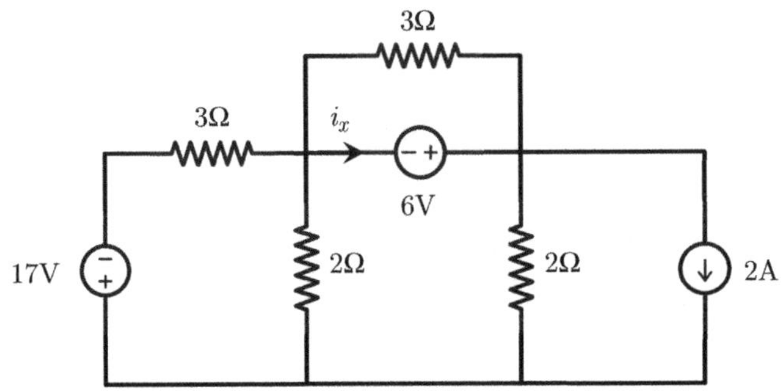

Find i_x.

Solution: Using nodal analysis, we define the ground and label the nodes. In this case, we form a supernode that contains a voltage source and a resistor.

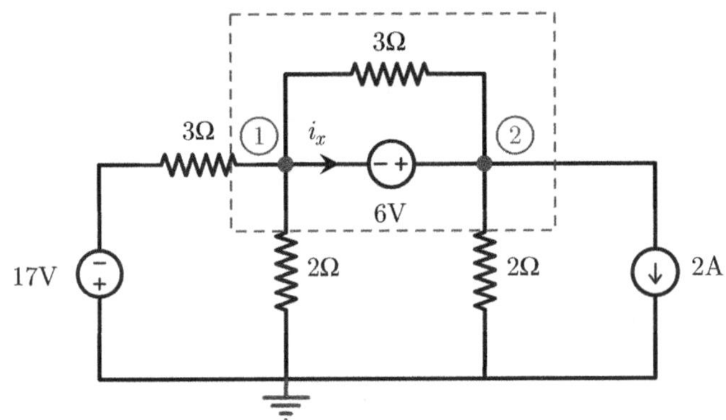

Applying KCL at the supernode, we derive

- KCL(1&2): $(-17 - v_1)/3 - v_1/2 - v_2/2 - 2 = 0 \longrightarrow 5v_1 + 3v_2 = -46.$

In addition, we have $v_2 - v_1 = 6$ V (using the supernode), leading to $v_1 = -8$ V and $v_2 = -2$ V. On the other hand, in order to find the value of i_x, we need to go further and apply KCL again at node 1 to obtain

- KCL(1): $(-17 - v_1)/3 - v_1/2 - i_x - (v_1 - v_2)/3 = 0,$

leading to $i_x = 3$ A. We note that the final KCL at node 1 is required to find the value of i_x, while we already have all node voltages without it.

Example 35: Consider the following circuit.

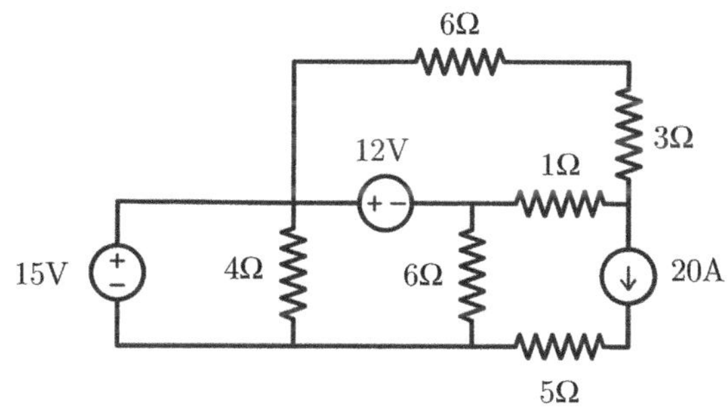

Find the power of the 15 V voltage source.

Solution: Using nodal analysis, we define the ground and label the nodes, as usual.

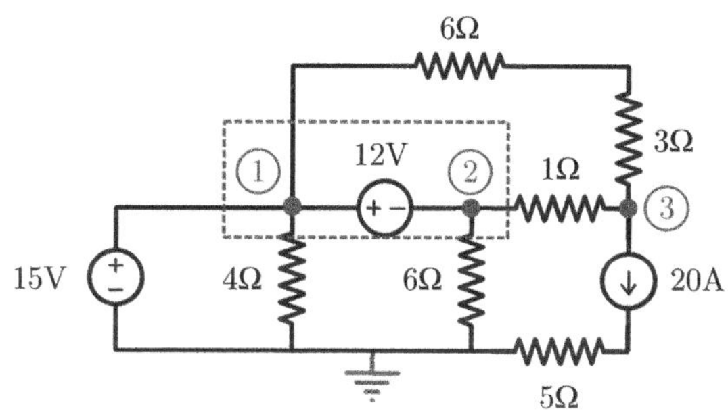

We again combine two nodes, which are connected via a voltage source, as a supernode. At the same time, with a correct placement of the ground, we already have $v_1 = 15$ V and $v_2 = v_1 - 12 = 3$ V. Applying KCL at node 3, we find

- KCL(3): $(v_2 - v_3)/1 + (v_1 - v_3)/9 - 20 = 0 \longrightarrow v_3 = -138/10$ V.

At this stage, we have all node voltages without using KCL at the supernode. On the other hand, the current along the 15 V source is needed to find its power; hence, another application of KCL is required either at node 1 or at the supernode. For example, using the supernode, we have

- KCL(1&2): $-i_{15V} - v_1/4 - v_2/6 - (v_2 - v_3)/1 - (v_1 - v_3)/9 = 0$,

which leads to $i_{15V} = -97/4$ A and $p_{15V} = -1455/4$ W.

Example 36: Consider the following circuit.

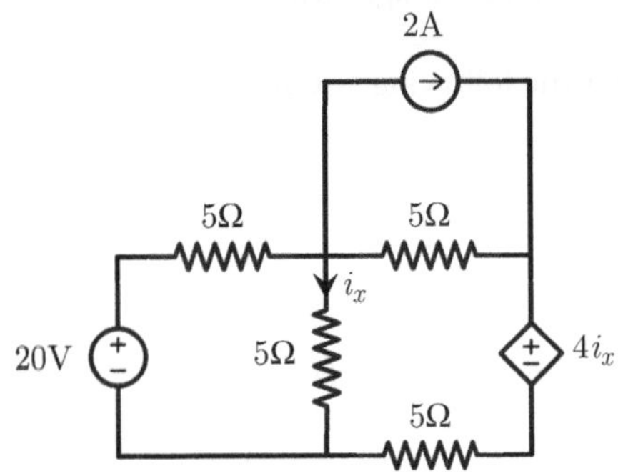

Find the power of the current-dependent voltage source.

Solution: This circuit can be solved via nodal analysis as follows.

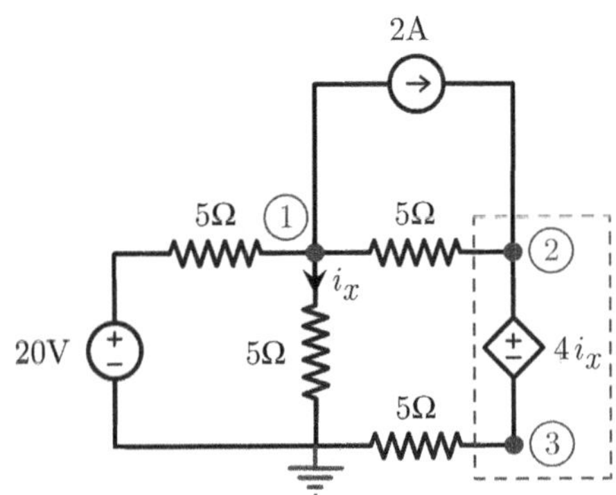

Considering the selection of the ground, we note that $i_x = v_1/5$. Then, applying KCL at node 1, we obtain

- KCL(1): $(20 - v_1)/5 - v_1/5 - 2 - (v_1 - v_2)/5 = 0 \longrightarrow 3v_1 - v_2 = 10.$

In addition, applying KCL at the supernode, we have

- KCL(2&3): $(v_1 - v_2)/5 + 2 - v_3/5 = 0 \longrightarrow v_1 - v_2 - v_3 = -10.$

Finally, considering the inside of the supernode, an additional equation can be derived as $v_2 - v_3 = 4i_x$or

$$5v_2 - 5v_3 - 4v_1 = 0.$$

Using all three equations, we find $v_1 = 50/7$ V, $v_2 = 80/7$ V, and $v_3 = 40/7$ V. Then the power of the dependent source can be found by considering the current through it, $i_s = v_3/5 = 8/7$ A, leading to

$$p_s = 4i_x \times \frac{8}{7} = \frac{40}{7} \times \frac{8}{7} = \frac{320}{49} \text{ W}.$$

Example 37: Consider the following circuit.

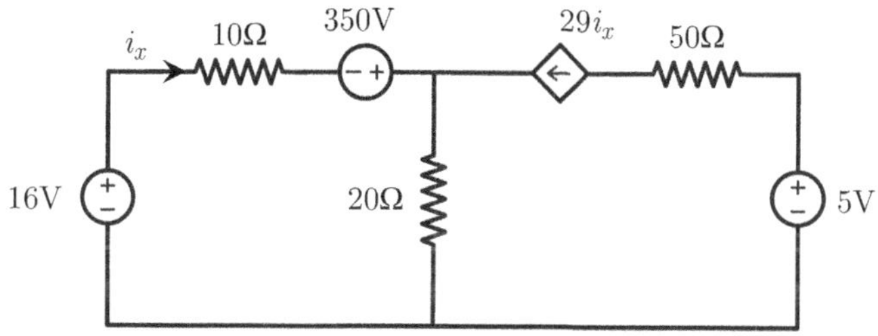

Find the power of the 5 V voltage source.

Solution: This circuit can be solved via nodal analysis as follows.

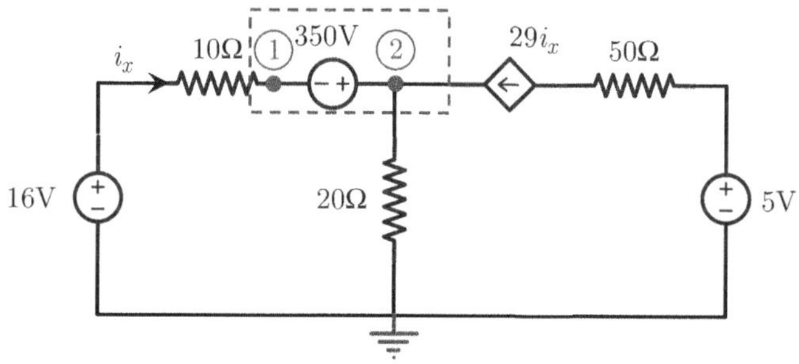

First, we note that $i_x = (16 - v_1)/10$. Then, applying KCL at the supernode, we obtain

- KCL(1&2): $(16 - v_1)/10 - v_2/20 + 29i_x = 0 \longrightarrow 60v_1 + v_2 = 960.$

Also considering that $v_2 = v_1 + 350$, we get $v_1 = 10\,\text{V}$, $v_2 = 360\,\text{V}$, and $i_x = 3/5\,\text{A}$. Finally, the power of the 5 V voltage source can be found as

$$p_{5\text{V}} = 5 \times (-29i_x) = -87\ \text{W}.$$

Example 38: Consider the following circuit.

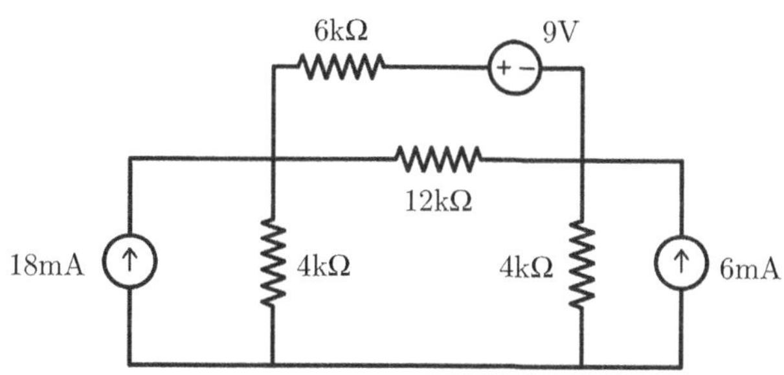

Find the power of the 12 kΩ resistor.

Solution: Using nodal analysis, we define the nodes as follows.

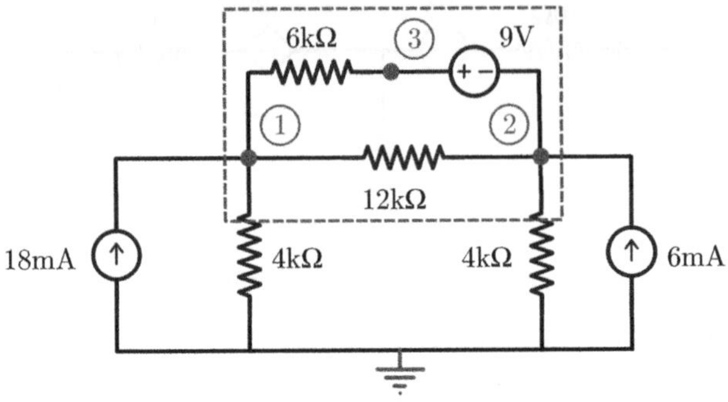

Then, applying KCL at the supernode formed of nodes 1, 2, and 3, we derive

- KCL(1&2&3): $18 - v_1/4 - v_2/4 + 6 = 0 \longrightarrow v_1 + v_2 = 96$,

where the voltage and current values are written in terms of volts and milliamperes, respectively. Furthermore, applying KCL at node 1, we get

- KCL(1): $18 - v_1/4 + (v_2 - v_1)/12 + (v_3 - v_1)/6 = 0 \longrightarrow 6v_1 - v_2 - 2v_3 = 216$.

Using $v_3 = v_2 + 9$, the final equation can be rewritten as

$$2v_1 - v_2 = 78.$$

Then, solving the equations, we obtain $v_1 = 58$ V and $v_2 = 38$ V. Hence, the power of the 12 kΩ resistor can be found as

$$p_{12\text{k}\Omega} = \frac{(58-38)^2}{12} = \frac{100}{3} \text{ mW} = \frac{1}{30} \text{ W}.$$

Example 39: Consider the following circuit.

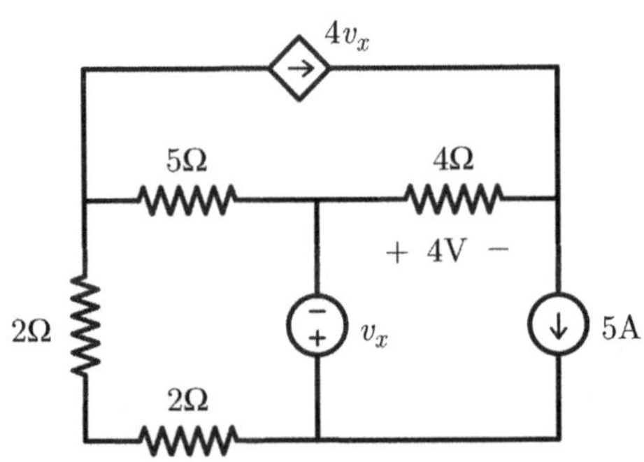

Find the power of the voltage source.

Solution: This circuit can be solved via nodal analysis as follows.

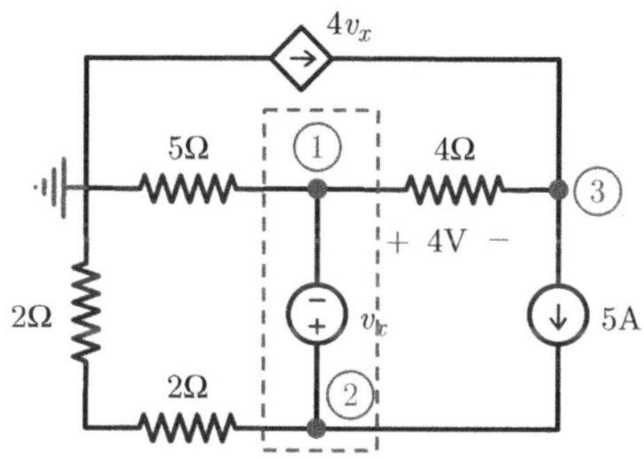

First, we note that $v_1 - v_3 = 4$, leading to 1 A current from node 1 to node 3. Applying KCL at the supernode, we have

- KCL(1&2): $-v_1/5 - v_2/4 + 5 - 1 = 0 \longrightarrow 4v_1 + 5v_2 = 80.$

Moreover, applying KCL at node 3, we derive

- KCL(3): $4v_x + 1 - 5 = 0 \longrightarrow v_x = 1\,\text{V}.$

Inside the supernode, we further have $v_2 - v_1 = v_x$; hence,

$$v_2 - v_1 = 1.$$

Solving the equations, one obtains $v_1 = 25/3\,\text{V}$ and $v_2 = 84/9\,\text{V}$. In order to find the power of the voltage source, one needs to apply KCL either at node 1 or node 2. For example, we have

- KCL(2): $5 - i_s - v_2/4 = 0 \longrightarrow i_s = 8/3\,\text{A},$

leading to

$$p_s = v_x \times i_s = 8/3\,\text{W}.$$

Exercise 21: In the following circuit, find the value of i_y.

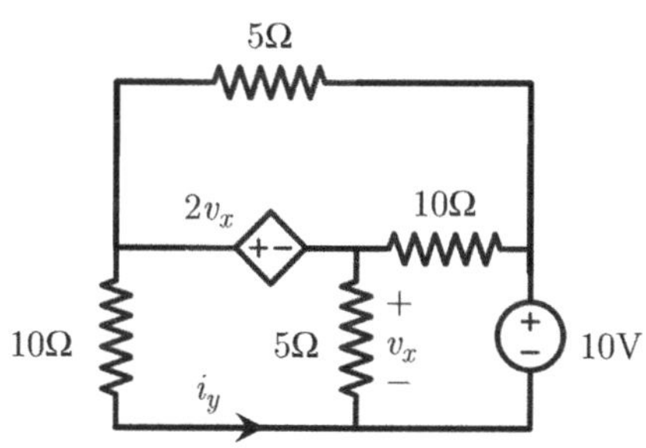

Exercise 22: In the following circuit, find the power of the 2 Ω resistor.

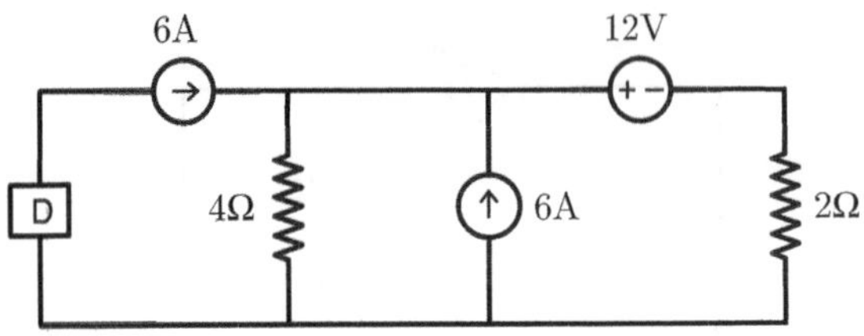

Exercise 23: In the following circuit, find the power of the top 10 Ω resistor.

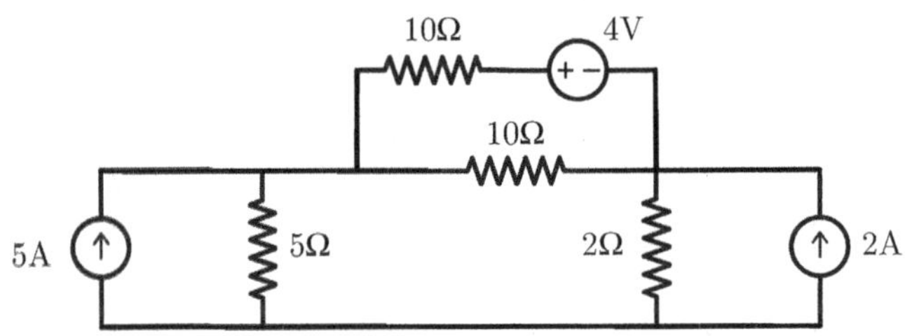

Exercise 24: In the following circuit, find i_y.

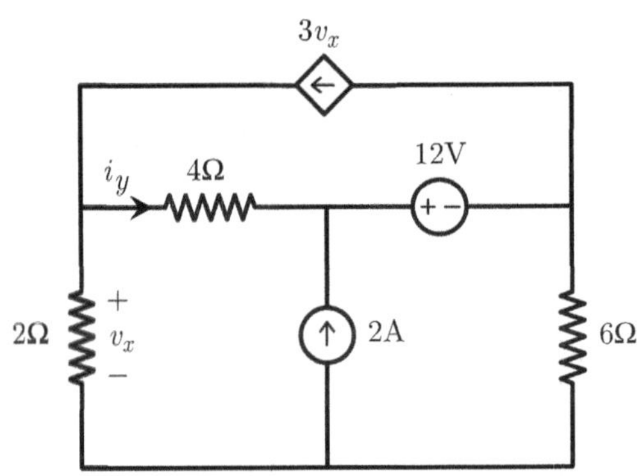

Exercise 25: In the following circuit, find v_s.

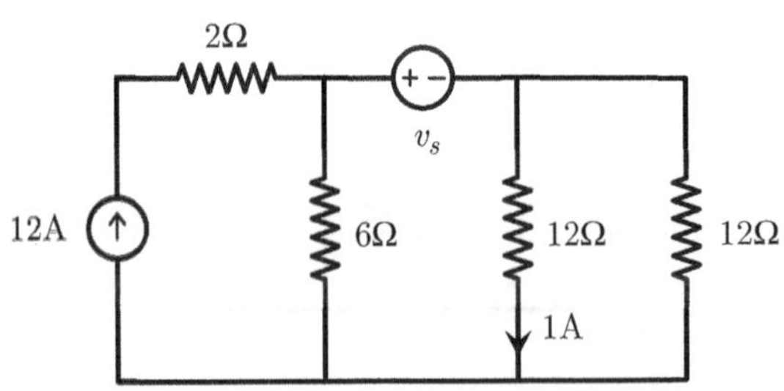

Exercise 26: In the following circuit, find the power of the current-dependent voltage source.

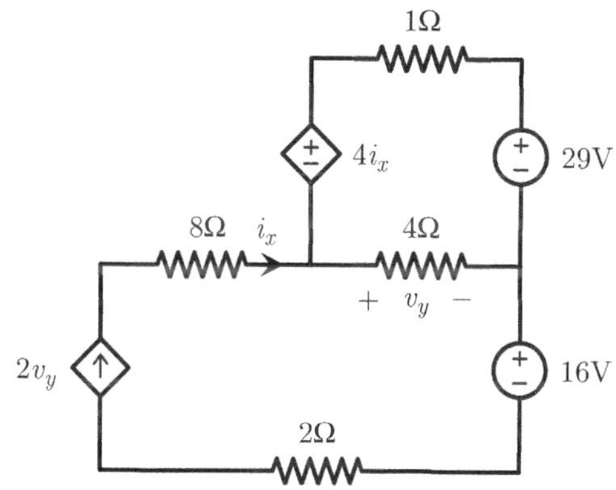

Exercise 27: In the following circuit, find v_y.

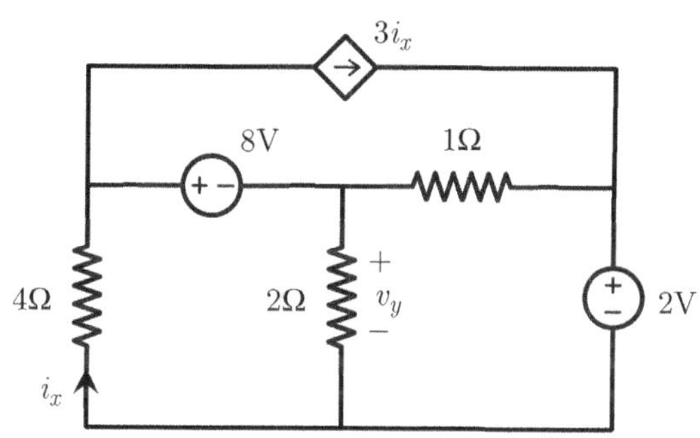

Exercise 28: In the following circuit, find the power of the 5 A source.

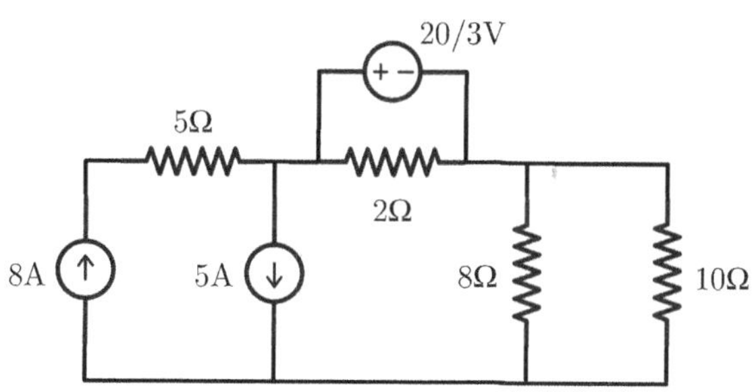

Exercise 29: In the following circuit, find i_x.

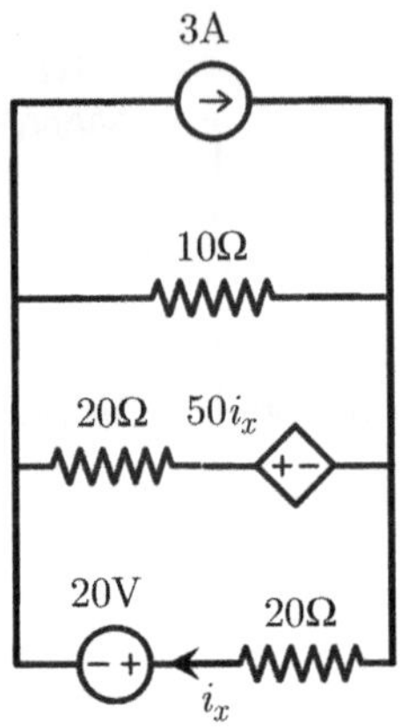

Exercise 30: In the following circuit, find v_x.

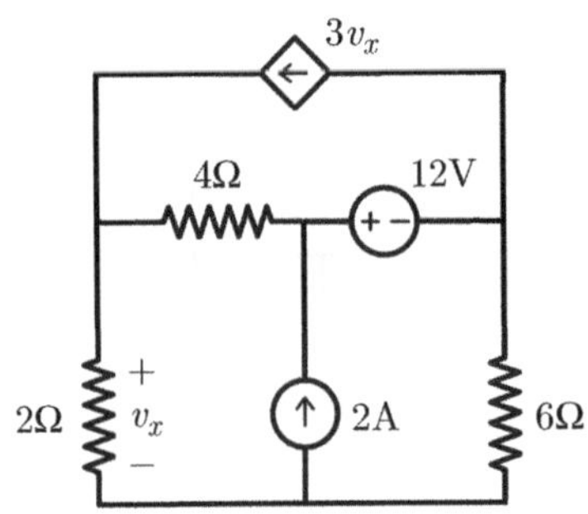

3.3 Circuits with Multiple Independent Voltage Sources

In the next chapter, we present another systematic technique, namely mesh analysis, for the analysis of circuits. Then an important question arises: for a given circuit, which method is better than the others—nodal analysis, mesh analysis, or simply using KCL and KVL together. While there is not a simple answer to this question, we can claim that nodal analysis is particularly useful when there are multiple voltage sources that are connected to each other, leading to trivial determination of node voltages. By way of demonstration, consider the following circuit.

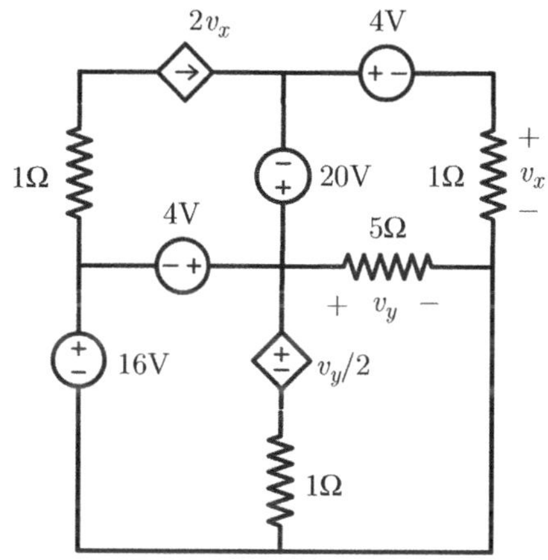

We would like to find the power of the dependent sources. After selecting a suitable ground, one can actually define most of the node voltages as follows.

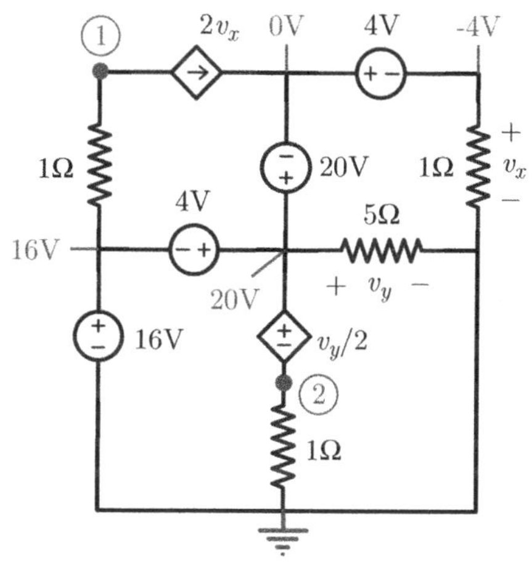

Now, considering the node voltages, we also have $v_x = -4$ V and $v_y = 20$ V. In addition, considering KCL at node 1, we derive

$$16 - v_1 = 2v_x = -8,$$

leading to $v_1 = 24$ V. Consequently, the power of the voltage-dependent current source can be found as

$$p_{s1} = v_{s1}i_{s1} = v_1 \times 2v_x = 24 \times (-8) = -192 \text{ W}.$$

The voltage at node 2 can also be found easily as

$$v_2 = 20 - v_y/2 = 10 \text{ V}.$$

Then the current through the voltage-dependent voltage source can be obtained as

$$i_{s2} = i_{1\Omega} = v_2/1 = 10 \text{ A}.$$

Finally, the power of this source can be found as

$$p_{s2} = v_{s2} i_{s2} = (20 - v_2)10 = 100 \text{ W}.$$

Analysis of this problem using KCL/KVL or mesh analysis could be quite complicated, while nodal analysis leads to relatively simple solution.

3.4 Solving Challenging Problems Using Nodal Analysis

In this section, we solve more complex and challenging problems using nodal analysis. In some cases, the solution of the problem may require careful organization of multiple equations. In other cases, however, the solution of the problem is not long, provided that the reference node is selected wisely.

Example 40: Consider the following circuit.

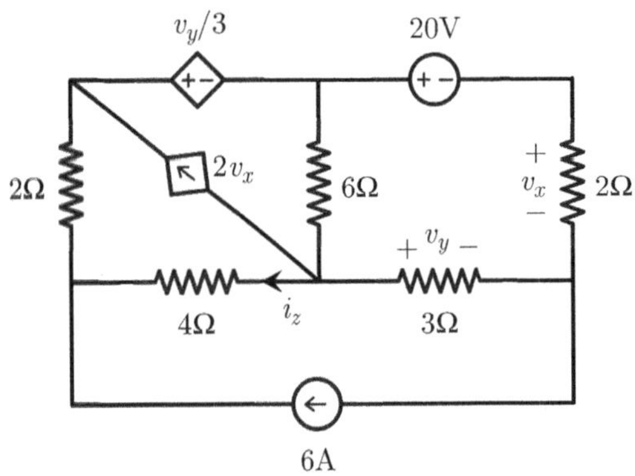

Find the value of i_z.

Solution: Using nodal analysis, we define the ground and label the nodes as follows.

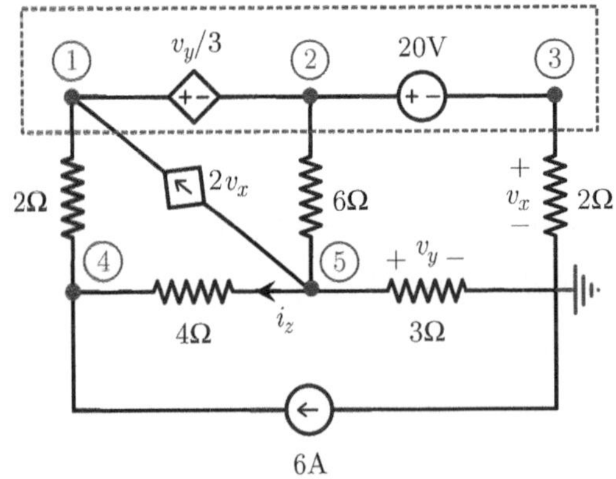

In this circuit, we combine three of the nodes as a supernode. We also note that, based on the selection of the reference node, $v_x = v_3$ and $v_y = v_5$. Inside the supernode, we further have

$$v_2 - v_3 = 20,$$
$$v_1 - v_2 = v_y/3 \longrightarrow 3v_1 - 3v_2 - v_5 = 0.$$

Next, we apply KCL at node 4, node 5, and the supernode as

- KCL(4): $6 + (v_5 - v_4)/4 - (v_4 - v_1)/2 = 0 \longrightarrow 2v_1 - 3v_4 + v_5 = -24,$
- KCL(5): $-(v_5 - v_4)/4 - 2v_x + (v_2 - v_5)/6 - v_5/3 = 0$
  $\longrightarrow 2v_2 - 24v_3 + 3v_4 - 9v_5 = 0,$
- KCL(1&2&3): $(v_4 - v_1)/2 + 2v_x - (v_2 - v_5)/6 - v_3/2 = 0$
  $\longrightarrow -3v_1 - v_2 + 9v_3 + 3v_4 + v_5 = 0.$

At this stage, we have five equations and five unknowns. These can be written in matrix form as

$$\begin{bmatrix} 0 & 1 & -1 & 0 & 0 \\ 3 & -3 & 0 & 0 & -1 \\ 2 & 0 & 0 & -3 & 1 \\ 0 & 2 & -24 & 3 & -9 \\ -3 & -1 & 9 & 3 & 1 \end{bmatrix} \begin{bmatrix} v_1 \\ v_2 \\ v_3 \\ v_4 \\ v_5 \end{bmatrix} = \begin{bmatrix} 20 \\ 0 \\ -24 \\ 0 \\ 0 \end{bmatrix}.$$

Solving the equations, we obtain $v_1 = 660/27$ V, $v_2 = 456/27$ V, $v_3 = -84/27$ V, $v_4 = 860/27$ V, and $v_5 = 612/27$ V. Therefore, we obtain

$$i_z = \frac{v_5 - v_4}{4} = \frac{612 - 860}{4 \times 27} = -62/27 \text{ A}.$$

Example 41: Consider the following circuit.

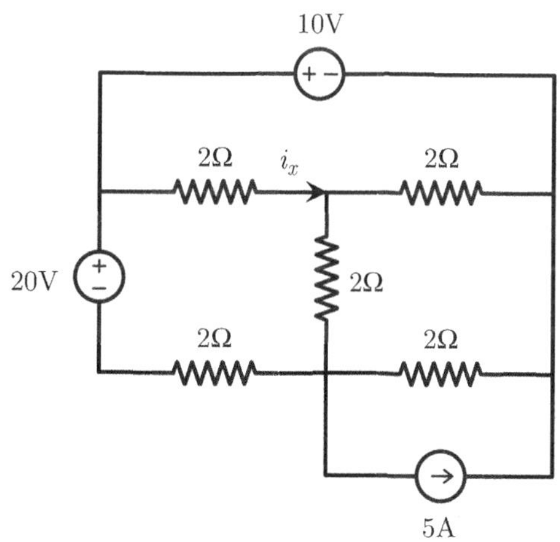

Find the value of i_x.

Solution: Using nodal analysis, we define the ground and label the nodes as follows.

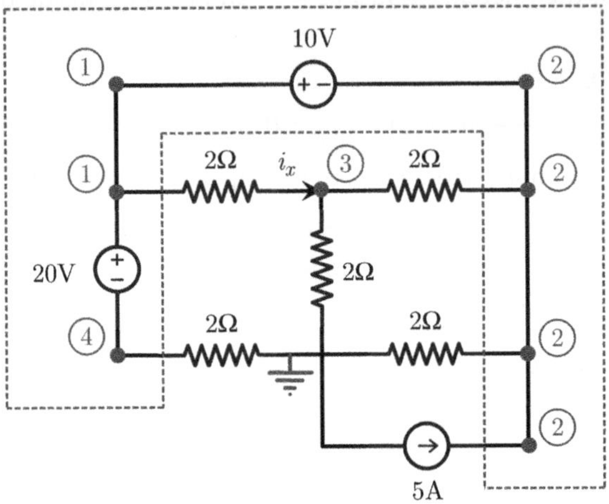

In the setup above, we combine three of the nodes as a supernode. Considering the inside, we also have

$$v_4 = v_1 - 20,$$
$$v_2 = v_1 - 10.$$

Then, applying KCL for the supernode, we derive

- KCL(1&2&4): $-v_4/2 - (v_1 - v_3)/2 - (v_2 - v_3)/2 - v_2/2 + 5 = 0$
  $\longrightarrow 2v_1 - v_3 = 25.$

KCL can again be applied at node 3 to obtain

- KCL(3): $(v_1 - v_3)/2 + (v_2 - v_3)/2 - v_3/2 = 0 \longrightarrow 2v_1 - 3v_3 = 10.$

Solving the equations, we obtain $v_1 = 65/4\,\text{V}$, $v_2 = 25/4\,\text{V}$, and $v_3 = 30/4\,\text{V}$, leading to

$$i_x = \frac{v_1 - v_3}{2} = 35/8 \text{ A}.$$

Example 42: Consider the following circuit.

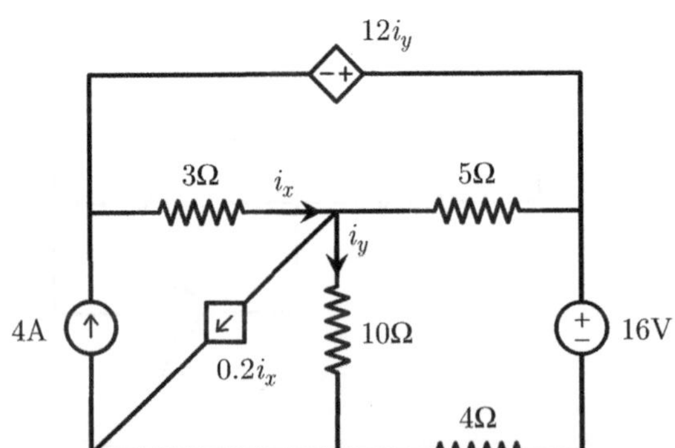

Find i_x.

Solution: Using nodal analysis, we once again define the ground and label the nodes.

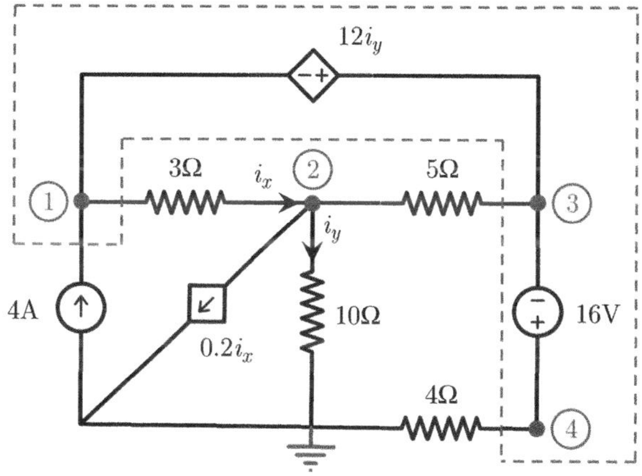

In the above, three of four nodes are combined as a supernode. We have

$$v_3 = v_1 + 12i_y = v_1 + 6/5v_2$$

$$v_4 = v_3 + 16 = v_1 + 6/5v_2 + 16,$$

considering that $i_y = v_2/10$. Applying KCL at the supernode and at node 2, we derive

- KCL(1&3&4): $4 - (v_1 - v_2)/3 - (v_3 - v_2)/5 - v_4/4 = 0 \longrightarrow v_2 = -235v_1/2,$
- KCL(2): $(v_1 - v_2)/3 - 0.2i_x - (v_2 - v_3)/5 - v_2/10 = 0 \longrightarrow v_2 = 10v_1/7.$

Then we have $v_1 = 0$ V, $v_2 = 0$ V, $v_3 = 0$ V, and $v_4 = 16$ V. Finally,

$$i_x = (v_1 - v_2)/3 = 0 \text{ A}.$$

Example 43: Consider the following circuit.

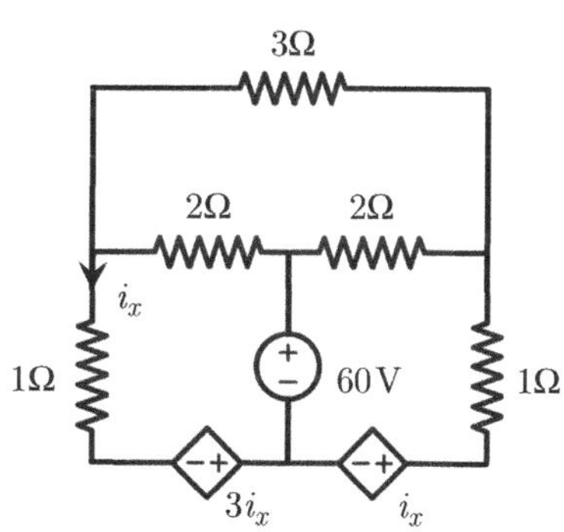

Find the value of i_x.

Solution: We again select a suitable ground and define some of the node voltages as follows.

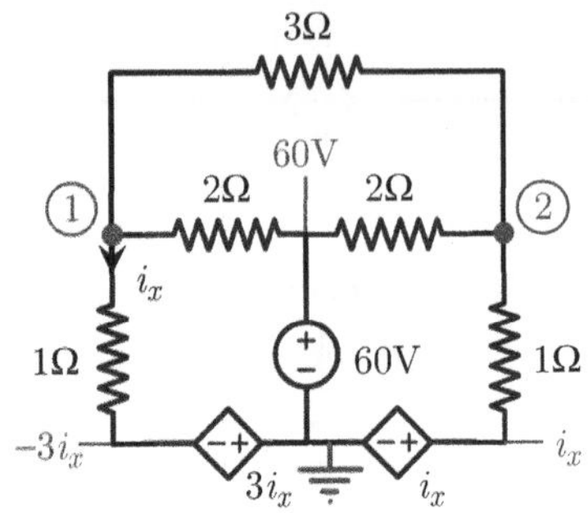

In addition, considering the definition of i_x, we have $v_1 = -3i_x + i_x = -2i_x$. Next, applying KCL at node 2, we derive

- KCL(2): $(-2i_x - v_2)/3 + (60 - v_2)/2 - (v_2 - i_x)/1 = 0$
  $\longrightarrow -11v_2 + 2i_x = -180$.

At this stage, we are forced to apply KCL at node 1; this may be confusing as v_1 is already known in terms of i_x. On the other hand, i_x is not a standard unknown (node voltage) of nodal analysis. We obtain

- KCL(1): $-i_x - (v_1 - 60)/2 - (v_1 - v_2)/3 = 0 \longrightarrow v_2 + 2i_x = -90$,

leading to $v_2 = 15/2$ V and $i_x = -195/4$ A.

Example 44: Consider the following circuit.

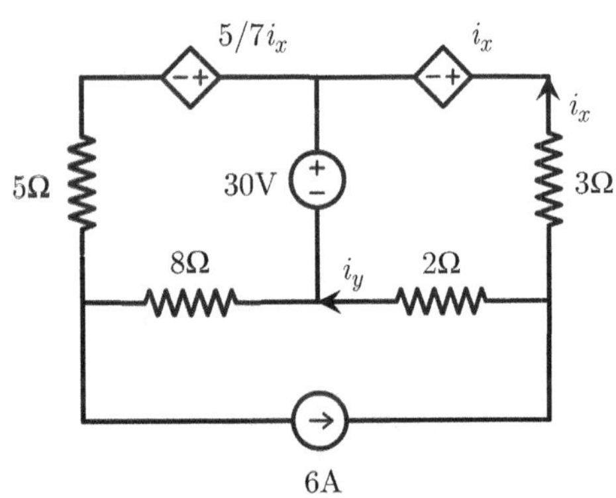

Find the value of i_y.

Solution: This circuit can be solved via nodal analysis as follows.

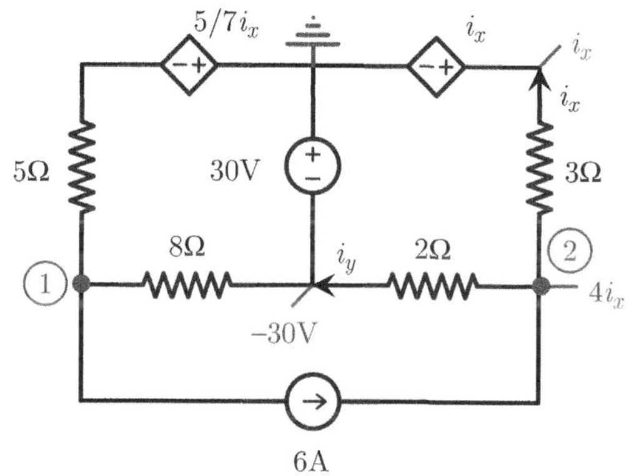

By a proper choice of the ground, most of the node voltages are defined easily. We apply KCL at node 2 to derive

- KCL(2): $-i_x - (4i_x + 30)/2 + 6 = 0 \longrightarrow i_x = -3$ A.

Hence, i_y is found to be

$$i_y = \frac{4i_x + 30}{2} = 9 \text{ A}.$$

Example 45: Consider the following circuit.

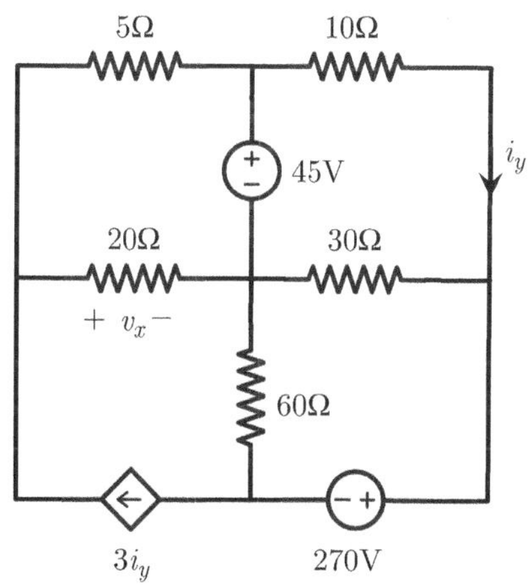

Find v_x.

Solution: Using nodal analysis, we can define the ground and label the nodes as follows.

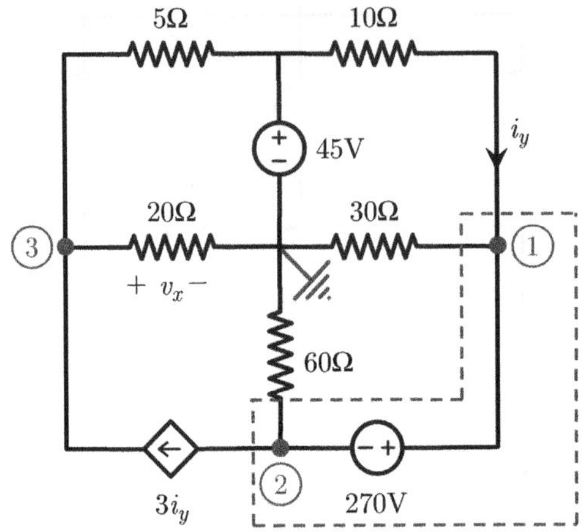

We note that $i_y = (45 - v_1)/10$. Then we apply KCL at the supernode to derive

- KCL(1&2): $(45 - v_1)/10 - v_1/30 - v_2/60 - 3i_y = 0 \longrightarrow 10v_1 - v_2 = 540$.

In addition, using $v_1 - v_2 = 270$, we obtain $v_1 = 30$ V, $v_2 = -240$ V, and $i_y = 3/2$ A. In order find v_x, we further apply KCL at node 3, obtaining

- KCL(3): $3i_y - v_3/20 - (v_3 - 45)/5 = 0$,

leading to $v_x = v_3 = 54$ V.

Example 46: Consider the following circuit.

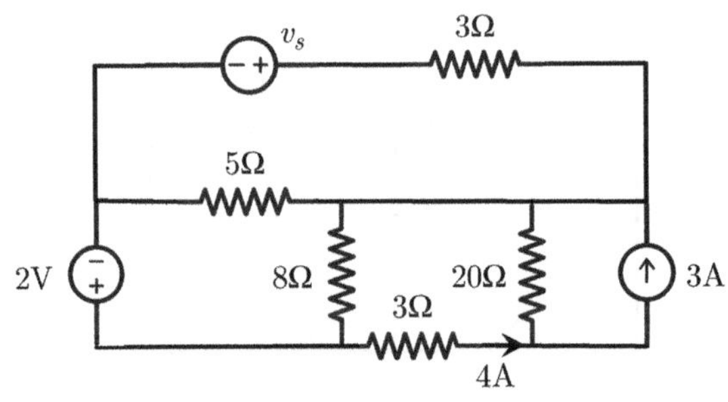

Find v_s.

Solution: Using nodal analysis, we solve the circuit as follows.

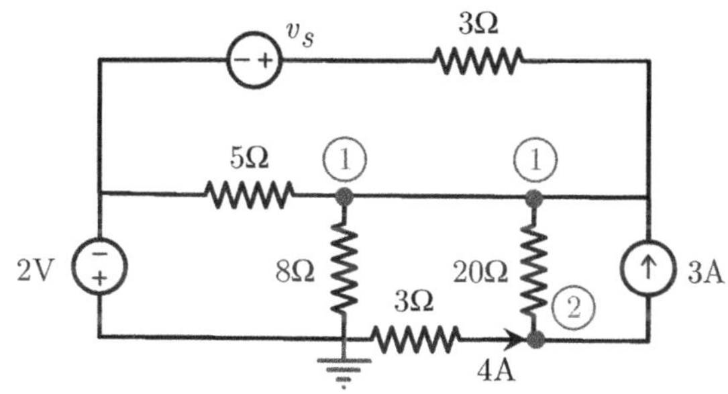

First, we note that, by using the selection of the ground, $v_2 = -12$ V. Then we apply KCL at node 2 to derive

- KCL(2): $4 - 3 - (v_2 - v_1)/20 = 0 \longrightarrow v_1 = -32$ V.

Finally, we can apply KCL at node 1, by carefully considering all connections, to obtain

- KCL(1): $(-2 - v_1)/5 - v_1/8 - (v_1 - v_2)/20 + 3 - (v_1 - v_s + 2)/3 = 0$,

leading to

$$(-v_s - 30)/3 = 14$$

or $v_s = -72$ V.

Example 47: Consider the following circuit.

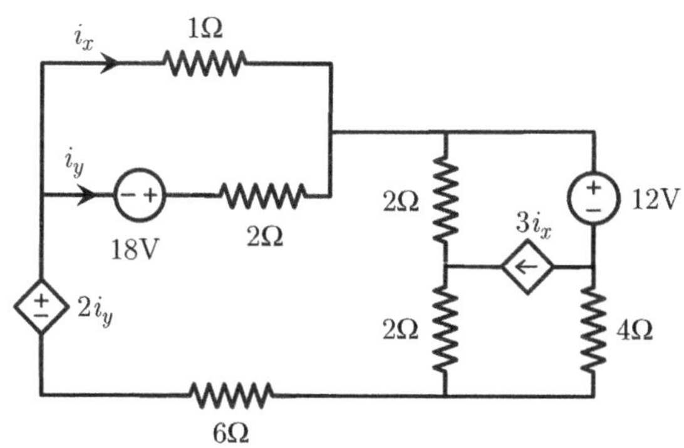

Find the power of the current-dependent current source.

Solution: Using nodal analysis, this circuit can be solved as follows.

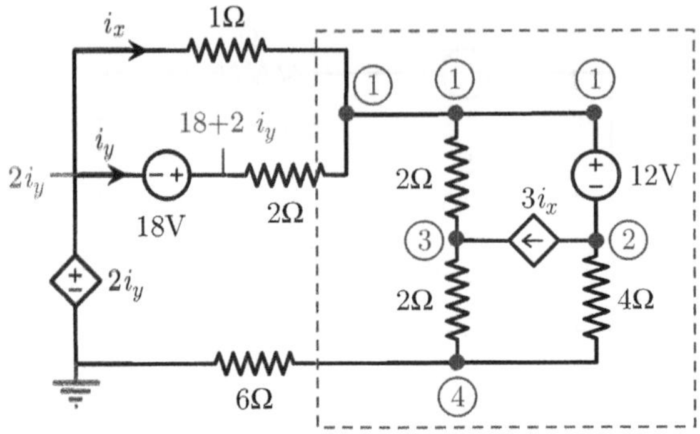

First, we note that $18 + 2i_y - v_1 = 2i_y$, leading to $v_1 = 18$ V. This leads in turn to $i_x = 2i_y - 18$. In addition, we have $v_2 = v_1 - 12 = 6$ V. Hence, we already know two node voltages without applying KCL. Next, we can apply KCL at nodes 3 and 4 to derive

- KCL(3): $(v_1 - v_3)/2 + 3i_x - (v_3 - v_4)/2 = 0 \longrightarrow 2v_3 - 6i_x - v_4 = 18$ $\longrightarrow 2v_3 - 12i_y - v_4 = -90$,
- KCL(4): $(v_3 - v_4)/2 + (v_2 - v_4)/4 - v_4/6 = 0 \longrightarrow 11v_4 - 6v_3 = 18$.

Obviously, we need another equation to solve this problem. While there are alternative choices, we apply KCL in the supernode formed of nodes 1, 2, 3, and 4 as

- KCL(1&2&3&4): $i_x + i_y - v_4/6 = 0 \longrightarrow 18i_y - v_4 = 108$.

Combining two of the three equations, we get

$$6v_3 - 5v_4 = -54,$$

leading to $v_3 = -14$ V, $v_4 = -6$ V, $i_y = 17/3$ A, and $i_x = -20/3$ A. Therefore, the power of the dependent source is found to be

$$p_s = (v_2 - v_3) \times 3i_x = 20 \times (-20) = -400 \text{ W}.$$

Example 48: As a final example, consider the following circuit.

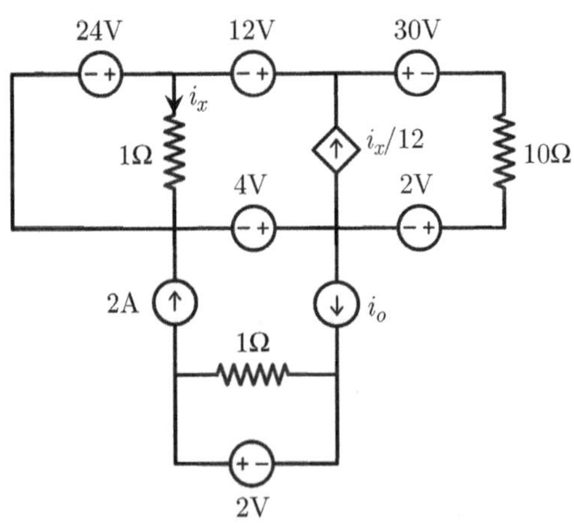

Find the power of the 12 V voltage source.

Solution: This circuit can be solved via nodal analysis as follows.

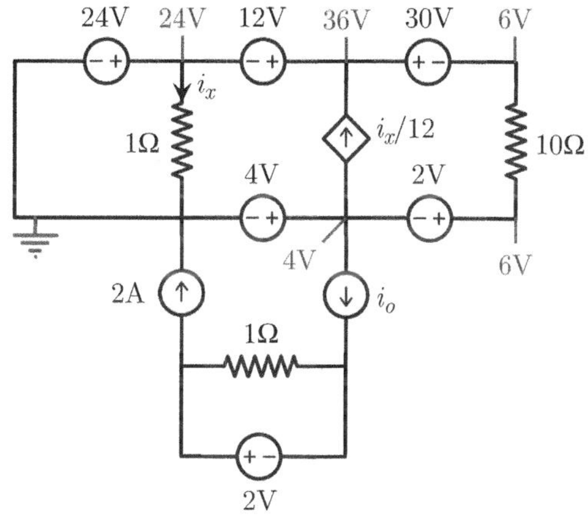

We note that most of the node voltages are known. In addition, $i_x = 24/1 = 24$ A, $i_{30V} = 0$ A, and

$$i_{12V} = i_x/12 = 2 \text{ A},$$

leading to

$$p_{12V} = 12 \times 2 = 24 \text{ W}.$$

Exercise 31: In the following circuit, find the power of the current-dependent voltage source.

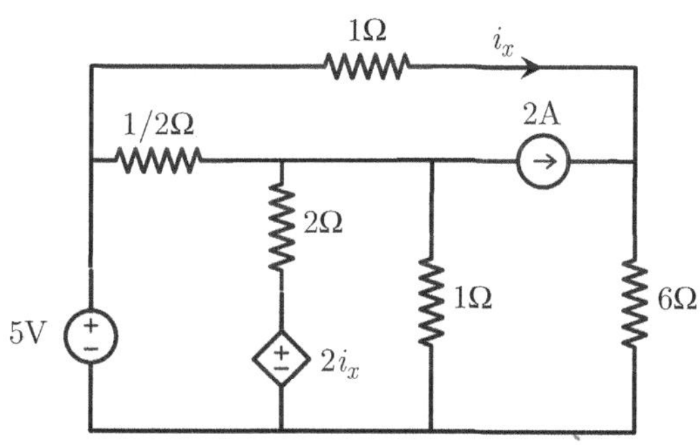

Exercise 32: In the following circuit, find i_z.

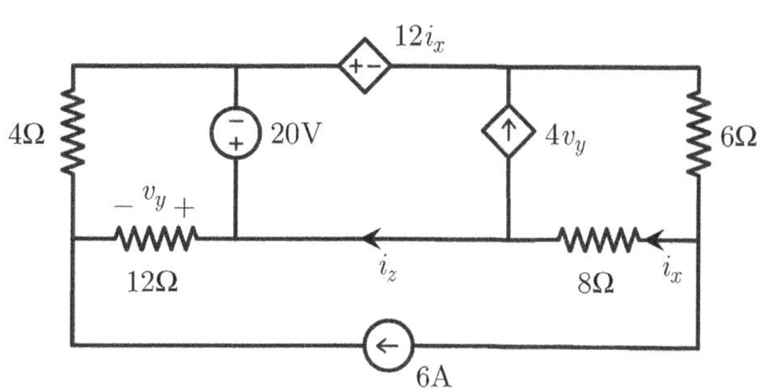

Exercise 33: In the following circuit, find i_x.

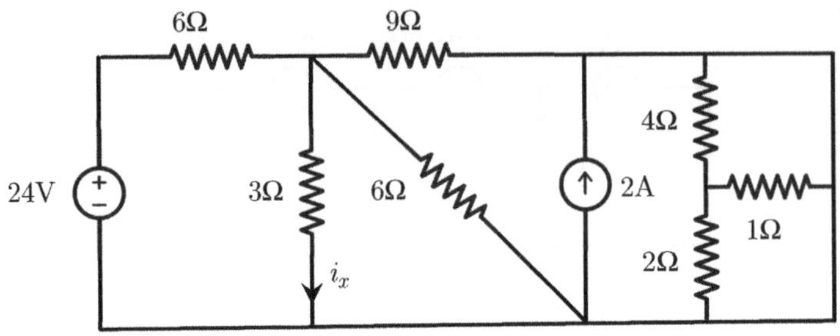

Exercise 34: In the following circuit, find the power of the voltage-dependent voltage source.

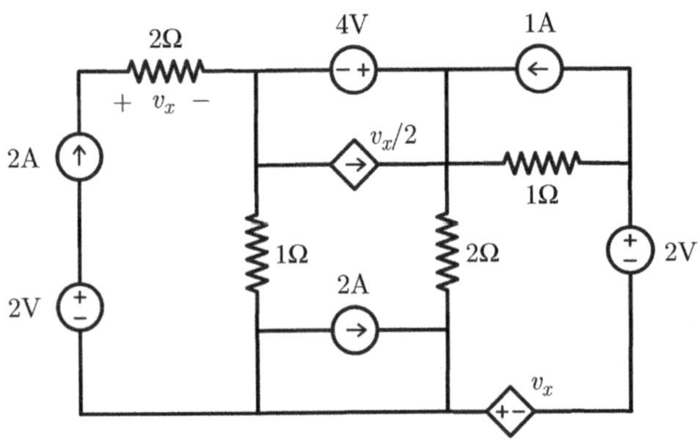

Exercise 35: In the following circuit, find the power of the current-dependent voltage source.

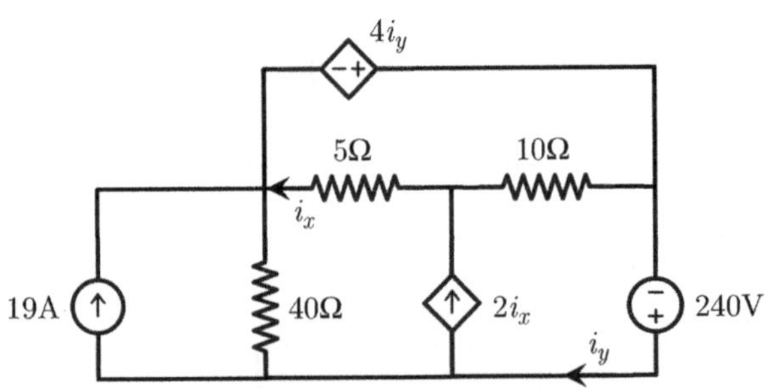

Exercise 36: In the following circuit, find the power of the 5 A current source.

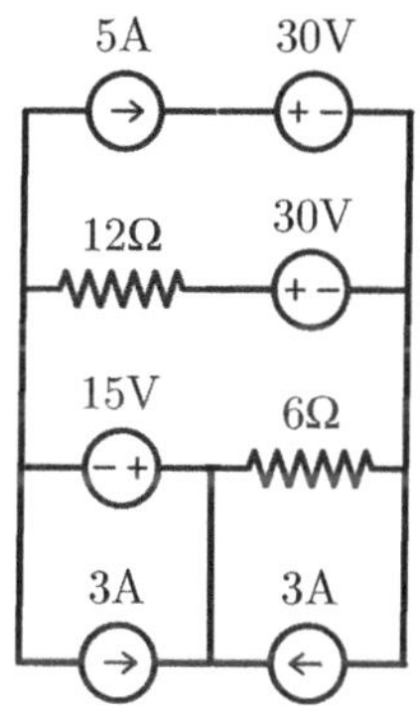

Exercise 37: In the following circuit, find i_x.

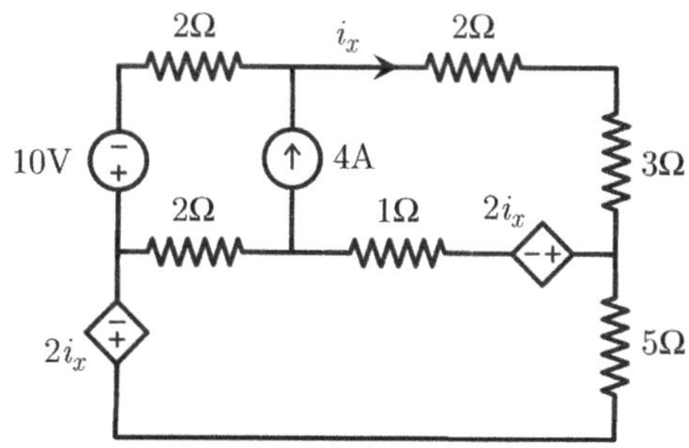

Exercise 38: In the following circuit, find the power of the voltage source.

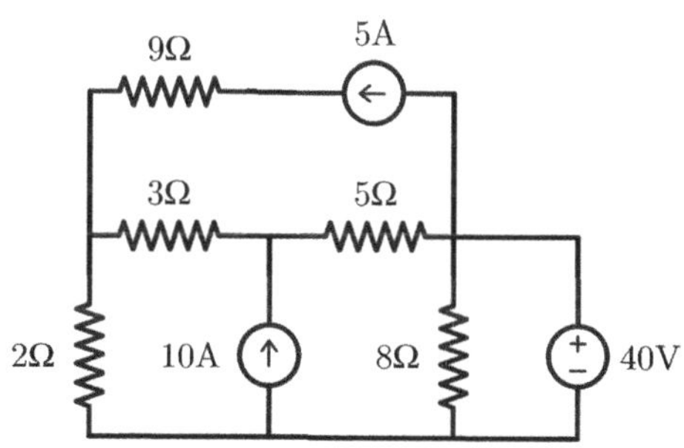

Exercise 39: In the following circuit, find i_x.

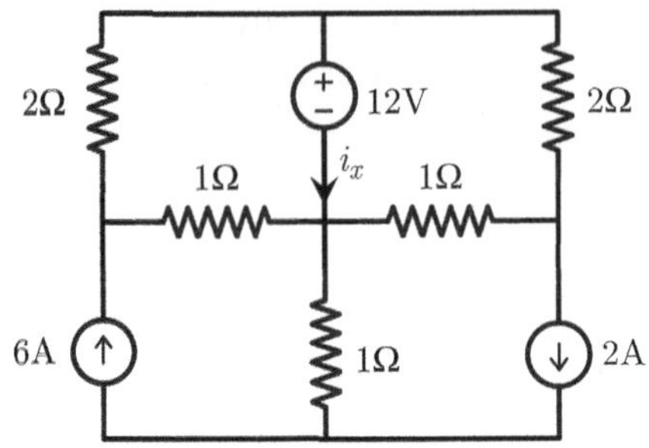

Exercise 40: In the following circuit, find i_x.

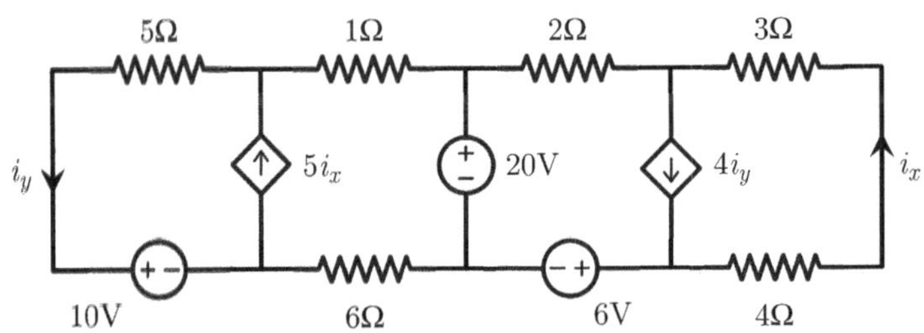

3.5 When Things Go Wrong with Nodal Analysis

In nodal analysis, the selection of the ground is an important issue. In general, if it is not given, one can select any node as a ground. But one must be consistent with this selection, that is,

- all other node voltages must be defined with respect to the selected ground,
- application of KCL should be avoided at the ground node as the application of KCL at the other nodes should be sufficient to solve the problem.

In addition, mixing KVL in nodal analysis is unnecessary and often confusing unless it is mandatory to find the voltage of a component.

In order to discuss some common sources of confusion, as well as to gain more insight on nodal analysis, we reconsider the following circuit, where we wish to find the value of i_x.

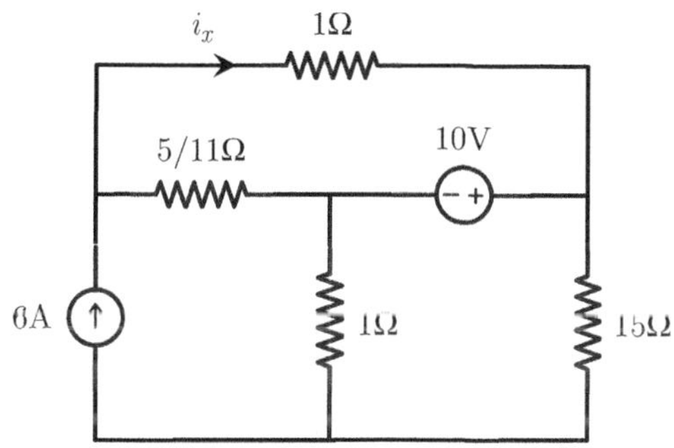

In comparison to the original analysis of this circuit, we select a different ground as follows.

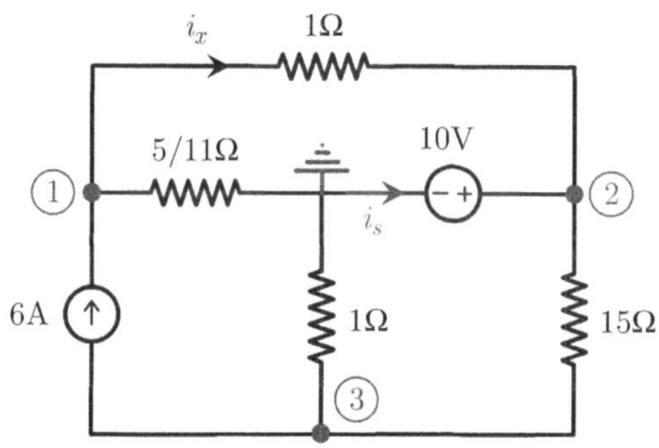

Now one can start with KCL at node 1 which yields

- KCL(1): $6 - v_1/(5/11) - (v_1 - v_2)/1 = 0 \longrightarrow 16v_1 - 5v_2 = 30.$

Similarly, at node 3, we have

- KCL(3): $-6 - v_3/1 - (v_3 - v_2)/15 = 0 \longrightarrow v_2 - 16v_3 = 90.$

But what about applying KCL at node 2 in order to obtain a third equation? If we are forced to do this, we have

- KCL(2): $i_s - (v_2 - v_3)/15 - (v_2 - v_1)/1 = 0,$

which is not useful since we need i_s as another variable. In fact, we should never apply KCL at node 2, because the voltage of this node is already known, $v_2 = 10$ V, considering the selection of the ground. We also note that the voltage of this node is 15 V in the original solution, but now it is different due to the different selection of the ground. Using $v_2 = 10$ V, we have $v_1 = 5$ V and $v_3 = -5$ V. Furthermore, the value of i_x can be found to be

$$i_x = (v_1 - v_2)/1 = -5 \text{ A}.$$

We emphasize that the value of i_x should be independent of the selection of the ground. Similarly, the voltage across the related resistor, $v_{1\Omega} = v_1 - v_2 = -5\,\text{V}$, should not depend on the selection of the ground while v_1 and v_2 may change.

In the circuit above, KCL must be applied at node 2 if i_s must be found. Using KCL(2), we derive

$$i_s = (v_2 - v_3)/15 + (v_2 - v_1)/1 = 1 + 5 = 6\ \text{A},$$

which must again be independent of the selection of the ground.

We now consider the following circuit, where the value of i_x needs to be found.

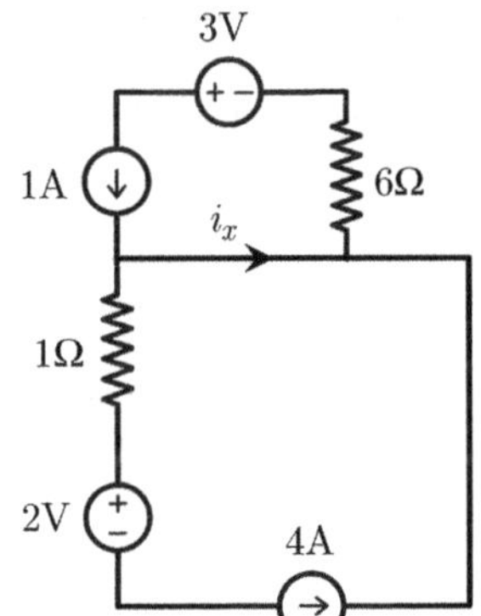

We particularly consider the case below, where the ground is selected at the middle.

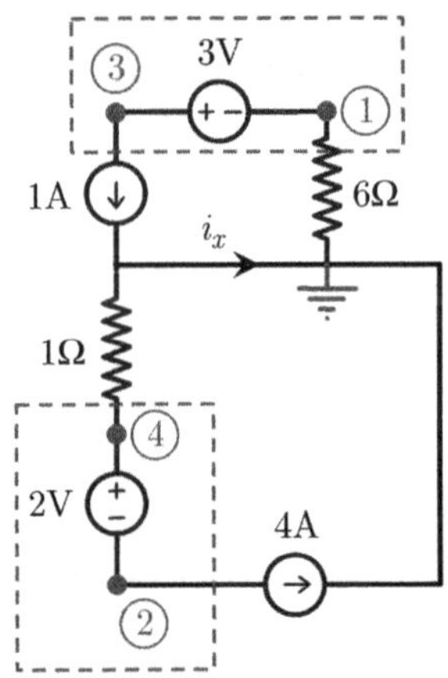

Interestingly, this selection of the ground leads to a large number of nodes that need to be defined in nodal analysis. In addition, following a formal procedure, we must combine two pairs as supernodes. Using the definitions of the supernodes, we have two equations

$$v_4 = v_2 + 2,$$
$$v_3 = v_1 + 3.$$

Applying KCL at the top supernode, we further derive

- KCL(1&3): $-1 - v_1/6 = 0$,

leading to $v_1 = -6$ V. Interestingly, and perhaps confusingly, the value of the node voltage v_1 does not depend on the 3 V voltage source. Specifically, v_1 is fixed to -6 V due to the 1 A current source, whatever the value of the voltage source. However, the voltage across the current source, v_3 in this nodal analysis, depends on the voltage source, as indicated above. We have

$$v_3 = v_1 + 3 = -3 \text{ V}.$$

The supernode at the bottom leads to another equation for completing the analysis. We derive

- KCL(2&4): $-v_4/1 - 4 = 0 \longrightarrow v_4 = -4$ V,

and

$$v_2 = v_4 - 2 = -6 \text{ V}.$$

At this stage, we have all the node voltages. On the other hand, they do not give complete information on the value of i_x. In order to find i_x, we are forced to apply KCL at a ground node. More specifically, we have an exception to the general rule that applying KCL at a ground is unnecessary and useless. We note that this extraordinary case occurs due to a specific selection of the ground.

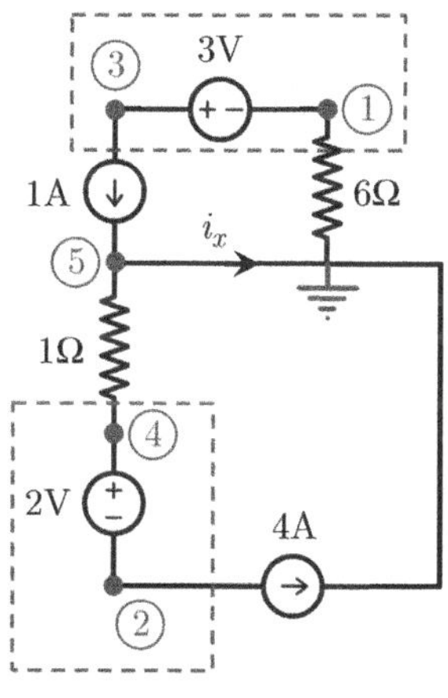

Considering the updated circuit, we have

- KCL(5): $(v_4 - v_5)/1 + 1 - i_x = 0$,

where $v_5 = 0$, leading to

$$i_x = -3 \text{ A}.$$

What about selecting another ground for the circuit above? We now consider the following case, where the node labels are also changed.

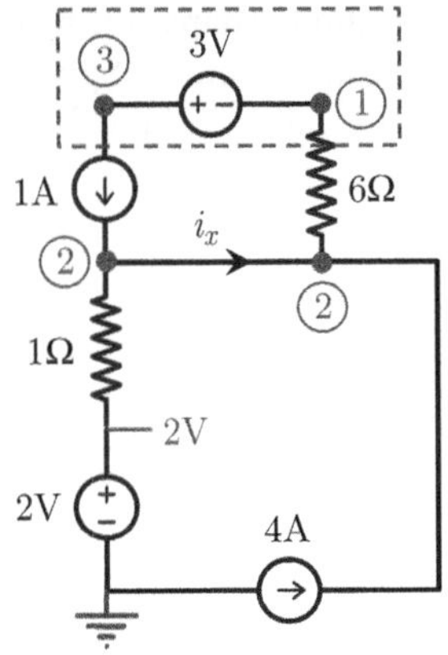

First, we note that the voltage at a node (not labeled) is immediately known to be 2 V. Considering all entering and leaving currents at node 2, we have

- KCL(2): $1 - (v_2 - 2)/1 + (v_1 - v_2)/6 + 4 = 0 \longrightarrow v_1 - 7v_2 = -42$.

In addition, KCL at the supernode leads to

- KCL(1&3): $-1 - (v_1 - v_2)/6 = 0 \longrightarrow v_1 - v_2 = -6$.

Solving the equations, we obtain $v_1 = 0$ V and $v_2 = 6$ V. Once again, we have all node voltages, but obviously we still need to apply KCL again to find i_x. We have

- KCL(2-left): $1 + (2 - v_2)/1 - i_x = 0$,

leading to $i_x = 1 - 4 = -3$ A.

3.6 What You Need to Know before You Continue

As discussed in this chapter, nodal analysis is a systematic way of analyzing circuits. We emphasize the following points on this technique.

- Equations in nodal analysis: Only KCL is used in nodal analysis, while KVL is not required. All equations are written by using node voltages as unknowns. Once the analysis is completed, KVL may be applied to find the voltage of a component, if it is not already known.
- Effect of the ground: Voltages and currents through components do not depend on the selection of the ground. The node voltages, however, may depend on the selection of the ground.
- It is not necessary to use KCL at a ground for nodal analysis, except for special cases. Therefore, the number of equations derived in nodal analysis is usually the same as the number of nodes other than the ground.

- A supernode, which involves a combination of multiple nodes and components, is often required to simplify nodal analysis. KCL is applied by considering the supernode as a closed surface with entering and leaving currents summing to zero. The extra equation required is obtained from the supernode itself.

In the next chapter, we focus on another systematic method, namely, mesh analysis based on KVL instead of KCL.

4

Analysis of Resistive Networks: Mesh Analysis

We now turn our attention to mesh analysis, which is based on a systematic application of KVL. Similar to nodal analysis, mesh analysis guarantees the development of sufficient numbers of equations, without any redundant equations or unknowns. However, in contrast to nodal analysis, where node voltages are major unknowns, loop currents are the unknowns of mesh analysis. These loop currents are not real currents but defined to facilitate the separation of meshes from each other. In addition to simple loops, we consider the generalization of the loops and the advantages of mesh analysis in detail.

4.1 Application of Mesh Analysis

The steps involved in mesh analysis can be listed as follows.

- Mesh current definitions: We define mesh currents inside closed loops, namely meshes, which contain only a sequence of components (but not an inner component). A mesh current is not a real current; it is defined as if it flows inside the loop, when the loop is not affected by the other parts of the circuit. A real current through a branch is the combination of the mesh currents sharing the branch.
- Constructing equations: We use only KVL inside meshes to derive all equations. KCL is not used in mesh analysis unless necessary. We write equations in terms of mesh currents.
- Solution: We solve the equations to find the mesh currents.
- Analysis: Finally, by using the mesh currents, we find the desired voltage, current, and/or power values.

As usual, we start with simple techniques, and then consider more complex circuits that are solved via mesh analysis.

Introduction to Electrical Circuit Analysis, First Edition. Özgür Ergül.

Companion Website: www.wiley.com/go/ergul4412

Example 49: Consider the following circuit.

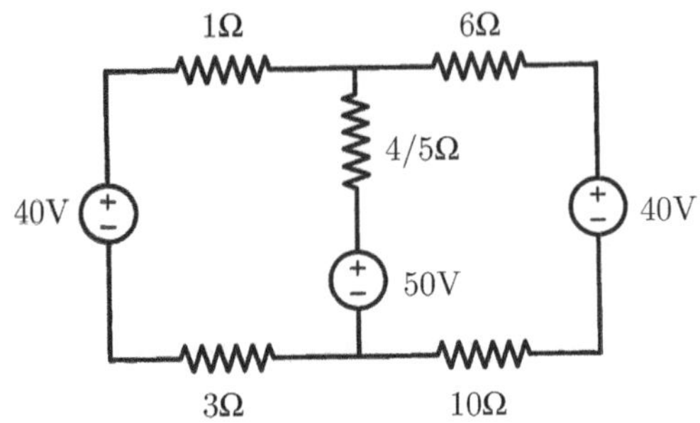

Find the power of the 50 V source.

Solution: Using mesh analysis, we define two mesh currents, namely, i_a and i_b. While the mesh currents can flow in an arbitrary direction, we only use the clockwise direction, to be consistent with KVL operations. In the following figure, we also define real currents i_1, i_2, and i_3, in order to demonstrate the details of the analysis.

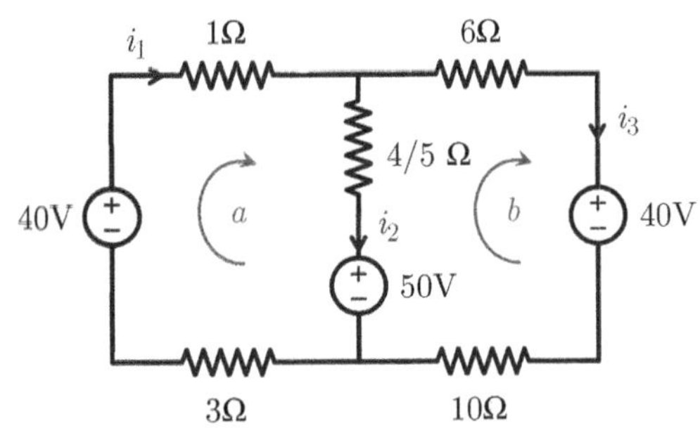

First, we note that the mesh and real currents are related to each other by

$$i_1 = i_a,$$
$$i_2 = i_a - i_b,$$
$$i_3 = i_b.$$

In order to understand these relationships, consider the 1 Ω resistor. The real current through this resistor is defined as i_1. On the other hand, this real current is only formed by the mesh current i_a, which is flowing in the same direction. Therefore, i_1 must be the same as i_a. Consider now the real current i_2 that is flowing through the 4/5 Ω resistor and the 50 V voltage source. In this case, two mesh currents are related to the branch: i_a flowing in the same direction as i_2 and i_b flowing in the opposite direction. In other words, i_2 must be formed by the combination of two mesh currents, i_a and i_b, while the contribution of i_b is negative, leading to $i_2 = i_a - i_b$. This important relationship is shown in detail as follows.

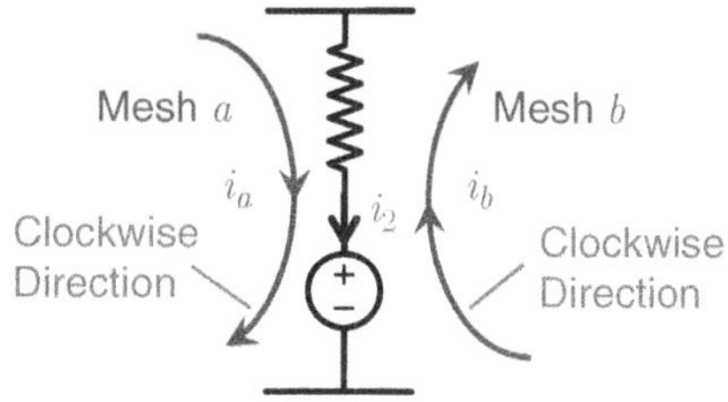

At this stage, after defining the mesh currents, we apply KVL in mesh a and mesh b. For mesh a, one can derive

- KVL(a): $-40 + 1i_1 + (4/5)i_2 + 50 + 3i_1 = 0$.

However, in mesh analysis, we write equations in terms of mesh currents; hence,

- KVL(a): $-40 + 1i_a + (4/5)(i_a - i_b) + 50 + 3i_a = 0 \longrightarrow 12i_a - 2i_b = -25$.

Applying KVL in mesh b, we also have

- KVL(b): $-50 - (4/5)i_2 + 6i_3 + 40 + 10i_3 = 0$

or

- KVL(b): $-50 + (4/5)(i_b - i_a) + 6i_b + 40 + 10i_b = 0 \longrightarrow -2i_a + 42i_b = 25$.

Solving the equations, we obtain $i_a = -2$ A and $i_b = 1/2$ A. Finally, $p_{50\text{V}} = 50 \times (-2 - 1/2) = -125$ W, indicating that the source delivers power.

Writing equations in terms of mesh currents guarantees that linearly independent variables are used. Specifically, the mesh currents i_a and i_b in the above are linearly independent, while the real currents i_1, i_2, i_3 are not. This dependency can be seen easily by considering KCL at one of the nodes, $i_1 - i_2 - i_3 = 0$. While it is trivial for this circuit, such dependencies may not be obvious in many cases. Therefore, using mesh currents as variables is always safe, making it possible to solve the circuit by using only KVL.

Example 50: Consider the following circuit.

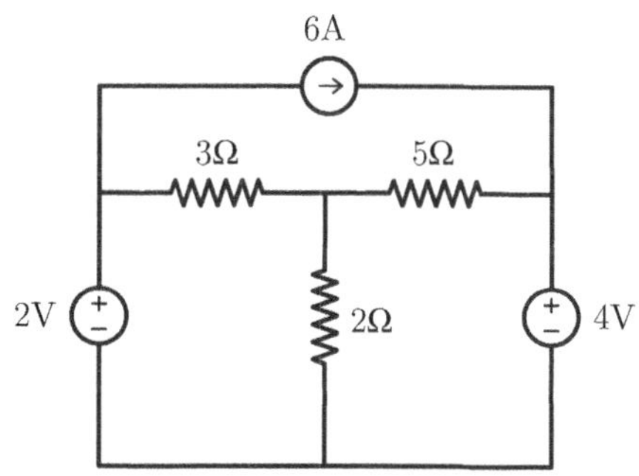

Find the power of the 2 Ω resistor.

Solution: Using mesh analysis, we define mesh currents, i_a, i_b, and i_c.

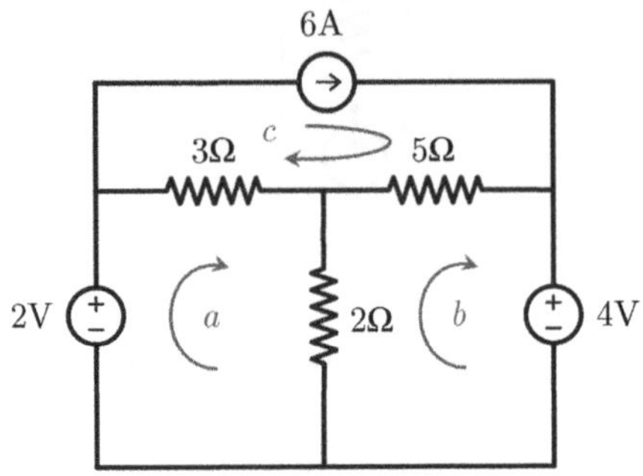

As we avoid using KCL at a node with a voltage source connected, we avoid using KVL in a mesh involving a current source. This is because the voltage across the current source cannot be written in terms of mesh currents. On the other hand, in most cases, the related mesh currents are easily found by considering the current source that directly corresponds to the branch current. In the above, we have

$$i_c = 6 \text{ A}$$

since there is no any other mesh sharing the branch of the 6 A source. Then, applying KVL in mesh a and mesh b, we obtain

- KVL(a): $-2 + 3(i_a - i_c) + 2(i_a - i_b) = 0 \longrightarrow 5i_a - 2i_b = 20$,
- KVL(b): $2(i_b - i_a) + 5(i_b - i_c) + 4 = 0 \longrightarrow -2i_a + 7i_b = 26$.

Solving the equations, we obtain $i_a = 192/31$ A, $i_b = 170/31$ A, and $p_{2\Omega} = 2(192/31 - 170/31)^2 = 968/961$ W.

Example 51: Consider the following circuit.

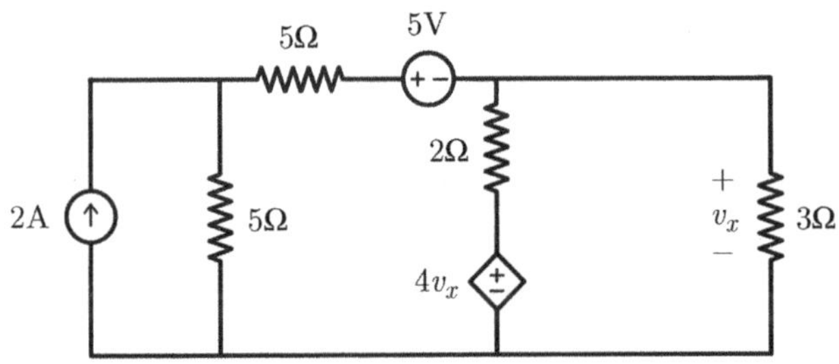

Find v_x.

Solution: To compare different kinds of solutions of the same circuit, we first consider nodal analysis.

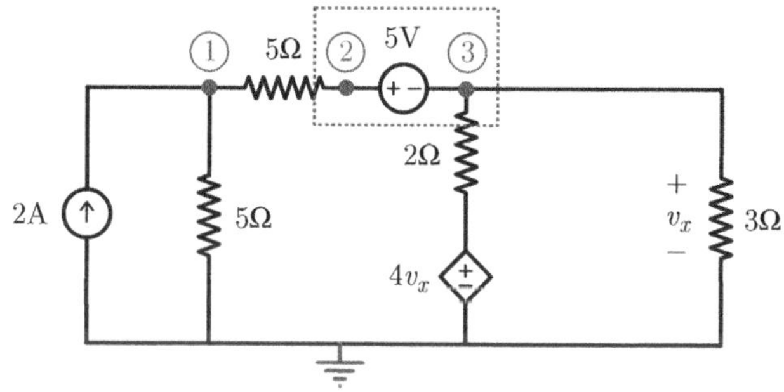

Note that we combine nodes 2 and 3 as a supernode. Applying KCL at node 1 and at the supernode, we obtain

- KCL(1): $2 - v_1/5 - (v_1 - v_2)/5 = 0 \longrightarrow 2v_1 - v_2 = 10$,
- KCL(2&3): $(v_1 - v_2)/5 - (v_3 - 4v_x)/2 - v_3/3 = 0 \longrightarrow 6v_1 - 6v_2 + 35v_3 = 0$.

In addition, using the supernode, we have $v_3 = v_2 - 5$. Then the KCL equation at the supernode becomes

$$6v_1 + 29v_2 = 175,$$

leading to $v_1 = 465/64$ V, $v_2 = 145/32$ V, and $v_3 = -15/32$ V. Finally, we obtain

$$v_x = v_3 = -15/32\text{ V}.$$

Now, applying mesh analysis to the same problem, we define three mesh currents.

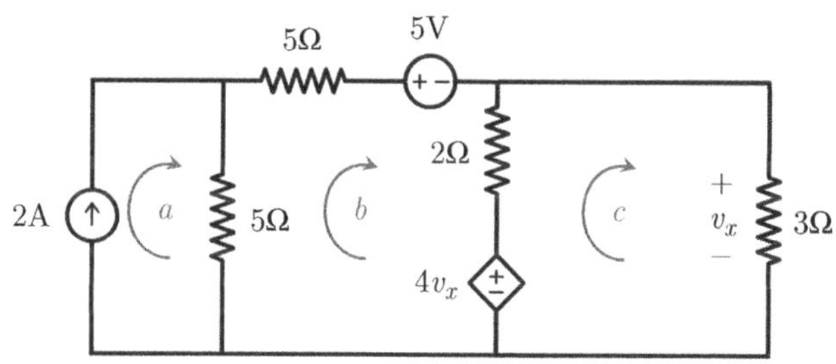

First, we have $i_a = 2$ A and $v_x = 3i_c$. Applying KVL in meshes *b* and *c*, we further derive

- KVL(*b*): $5(i_b - i_a) + 5i_b + 5 + 2(i_b - i_c) + 4v_x = 0 \longrightarrow 12i_b + 10i_c = 5$,
- KVL(*c*): $-4v_x + 2(i_c - i_b) + 3i_c = 0 \longrightarrow 2i_b + 7i_c = 0$.

Solving the equations, we obtain $i_b = 35/64$ A, $i_c = -10/64$ A, and $v_x = -15/32$ V.

For the circuit above, nodal analysis requires a supernode definition, while mesh analysis can handle the problem with two standard applications of KVL. The equations that need to be solved have more or less the same complexity. In some circuits, however, the selection of the analysis method can be more critical.

Example 52: Consider the following circuit, previously studied using nodal analysis (see Example 45).

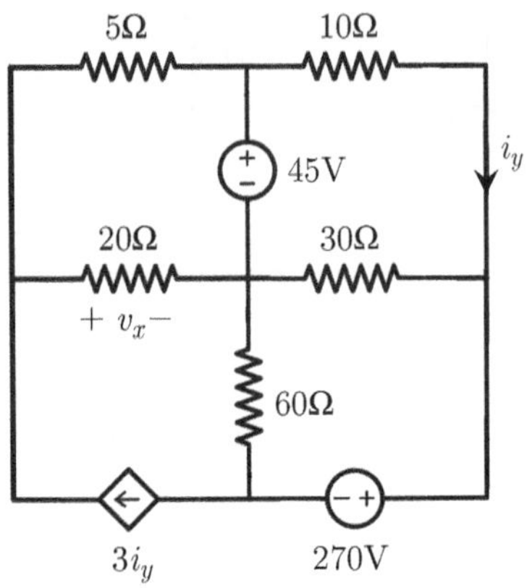

Find v_x.

Solution: Using mesh analysis, we define four different mesh currents.

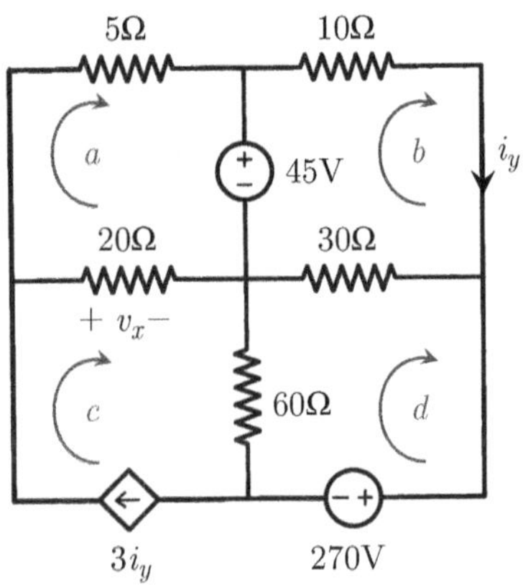

First, we note that $i_b = i_y$, as i_y is defined on a branch that is used only by mesh b. Similarly, considering mesh c, we have $i_c = 3i_y$, leading to

$$i_c = 3i_b.$$

Then, applying KVL in meshes a, b, and d, we get

- KVL(a): $5i_a + 45 + 20(i_a - i_c) = 0 \longrightarrow 5i_a - 12i_b = -9$,
- KVL(b): $-45 + 10i_b + 30(i_b - i_d) = 0 \longrightarrow 8i_b - 6i_d = 9$,
- KVL(d): $30(i_d - i_b) + 270 + 60(i_d - i_c) = 0 \longrightarrow 7i_b - 3i_d = 9$.

Solving the three equations above, we obtain mesh currents $i_a = 9/5\,\text{A}$, $i_b = 3/2\,\text{A}$, $i_c = 9/2\,\text{A}$, and $i_d = 1/2\,\text{A}$. Finally, we find v_x to be

$$v_x = 20 \times (i_c - i_a) = 20 \times (9/2 - 9/5) = 54\,\text{V}.$$

Example 53: Consider the following circuit, also previously studied using nodal analysis (see Example 42).

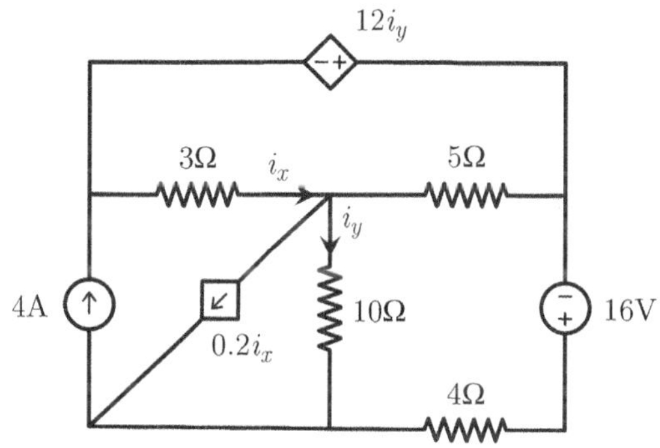

Find i_y.

Solution: As previously shown, the solution of this problem via nodal analysis is quite challenging and involves a delicate use of the supernode concept. On the other hand, using mesh analysis, we define four standard mesh currents. Once again, we remember that the mesh currents are linearly independent; hence, we need four equations to solve them.

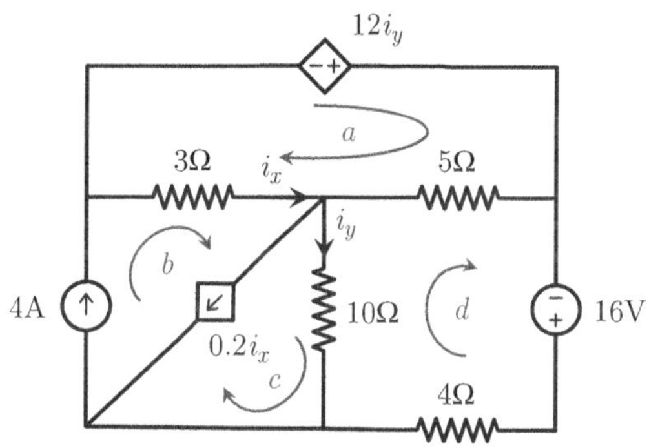

In the above, considering mesh and real currents, we have

$$i_b = 4\text{ A},$$
$$i_x = i_b - i_a,$$
$$i_y = i_c - i_d,$$
$$i_b - i_c = 0.2i_x \longrightarrow 5i_b - 5i_c = i_b - i_a \longrightarrow i_a + 4i_b - 5i_c = 0.$$

Now, we need to apply KVL to only two meshes,

- KVL(a): $-12i_y + 5(i_a - i_d) + 3(i_a - i_b) = 0 \longrightarrow 8i_a - 12i_c + 7i_d = 12,$
- KVL(d): $10(i_d - i_c) + 5(i_d - i_a) - 16 + 4i_d = 0 \longrightarrow -5i_a - 10i_c + 19i_d = 16.$

Solving three equations in three unknowns, we find $i_a = 4\,\text{A}$, $i_c = 4\,\text{A}$, $i_d = 4\,\text{A}$. Finally, we obtain

$$i_y = i_c - i_d = 4 - 4 = 0\,\text{A}.$$

Example 54: Consider the following circuit involving four different kinds of sources, which was previously studied in Example 28.

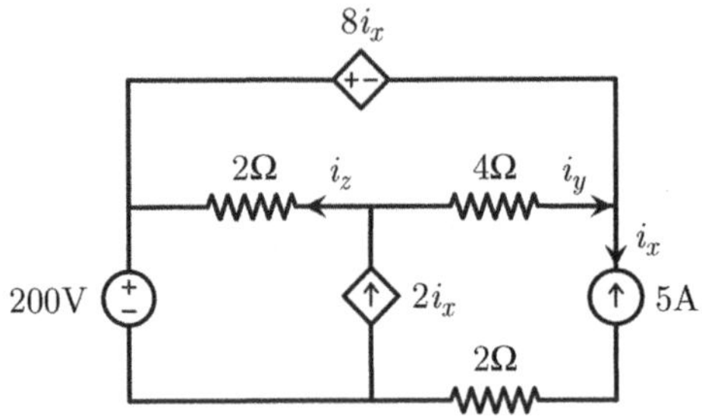

Find i_z.

Solution: This circuit was also previously solved using nodal analysis. In this case, using mesh analysis, we define three mesh currents.

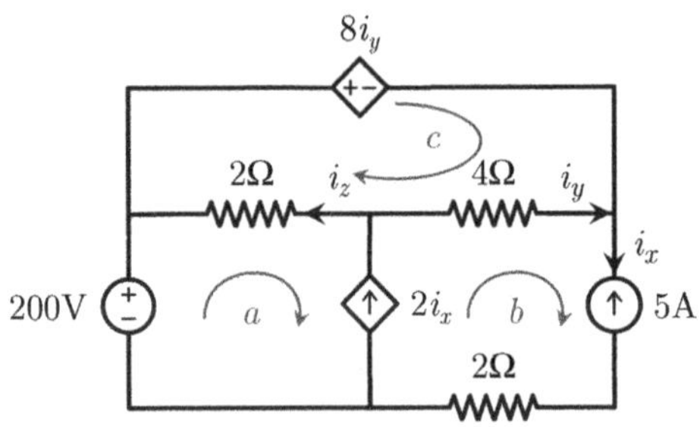

Then, considering real currents through branches, we realize that

$$i_b = -5\,\text{A},$$
$$i_x = -5\,\text{A},$$
$$2i_x = i_b - i_a \longrightarrow i_a = 5\,\text{A}.$$

In addition, $i_y = i_b - i_c = -5 - i_c$. KVL is only required for mesh c as

- KVL(c): $8i_y + 4(i_c - i_b) + 2(i_c - i_a) = 0$,

leading to $i_c = -15\,\text{A}$. Finally, we obtain

$$i_z = i_c - i_a = -20\,\text{A}.$$

Example 55: Consider the following circuit.

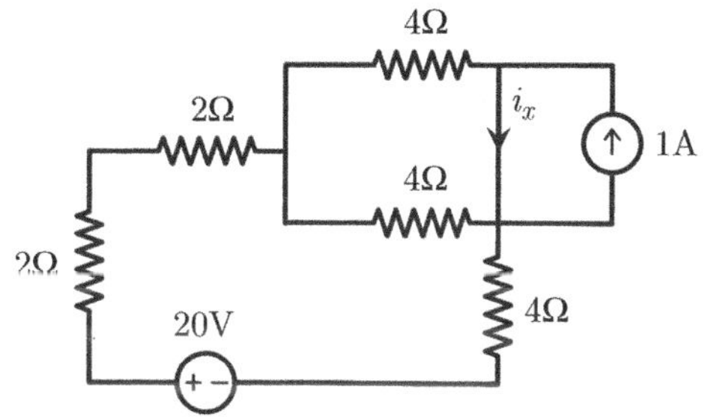

Find i_x.

Solution: Using mesh analysis, we define two different mesh currents, while noting that $i_x = i_a + 1$.

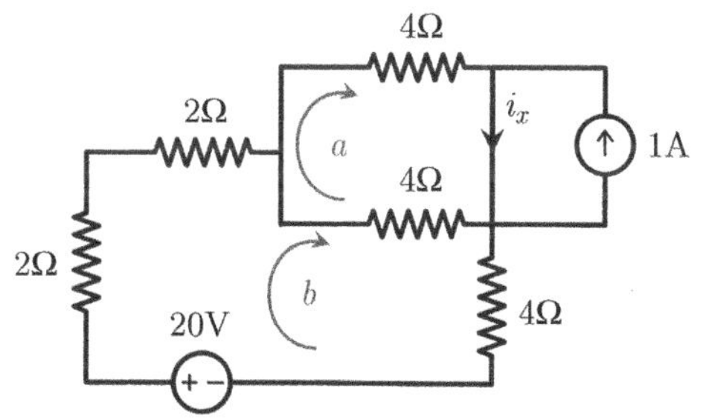

Applying KVL in meshes *a* and *b*, we derive

- KVL(*a*): $4i_a + 4(i_a - i_b) = 0 \longrightarrow i_b = 2i_a$,
- KVL(*b*): $-20 + 4i_b + 4(i_b - i_a) + 4i_b = 0 \longrightarrow i_b = 2\,\text{A}$.

Then we obtain $i_a = 1\,\text{A}$ and $i_x = i_a + 1 = 2\,\text{A}$.

Example 56: Consider the following circuit, previously solved using nodal analysis (see Exercise 26).

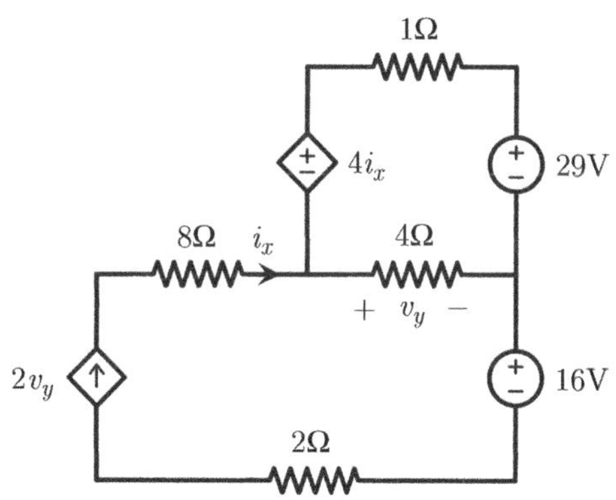

Find the power of the current-dependent voltage source.

Solution: Using mesh analysis, we define only two different mesh currents.

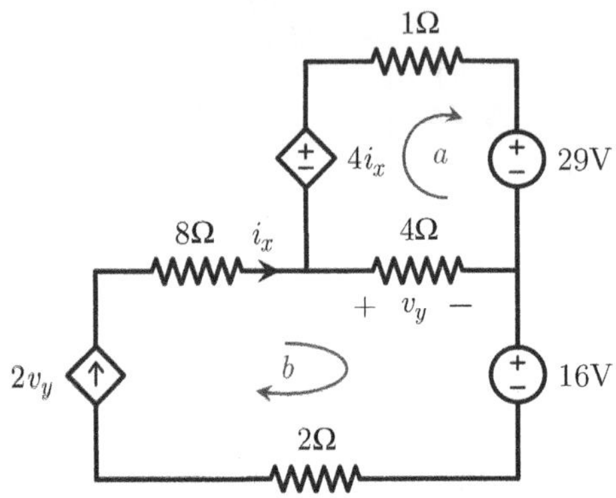

First, we note that $i_b = i_x$. At the same time, $i_b = 2v_y$ and $v_y = 4(i_b - i_a)$. Then

$$4i_a = 4i_b - v_y = 7i_b/8 = 7i_x/8.$$

Applying KVL in mesh a, we derive

- KVL(a): $-4i_x + i_a + 29 + 4(i_a - i_b) = 0$,

which leads to the solution of i_x as

$$-4i_x + 7i_x/8 + 29 + 4(7i_x/8 - i_x) = 0 \longrightarrow i_x = 8\text{ A}.$$

Therefore, $i_a = 7$ A and $i_b = 8$ A. Finally, the power of the current-dependent voltage source can be found to be

$$p_{4i_x} = 4i_x \times (-i_a) = 32 \times (-7) = -224\text{ W}.$$

Example 57: Consider the following circuit.

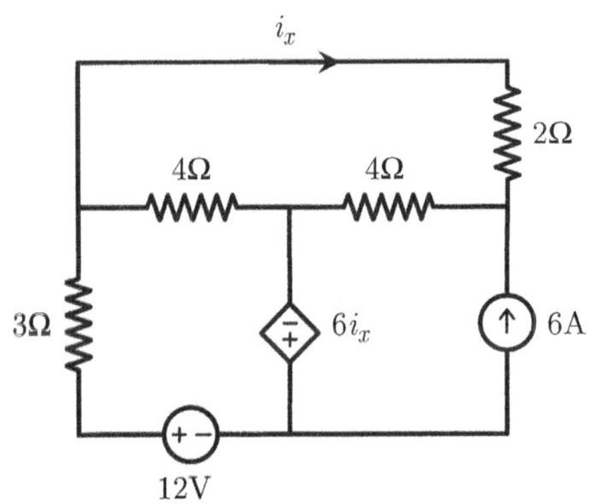

Find the power of the current-dependent voltage source.

Solution: Using mesh analysis, we define three different mesh currents.

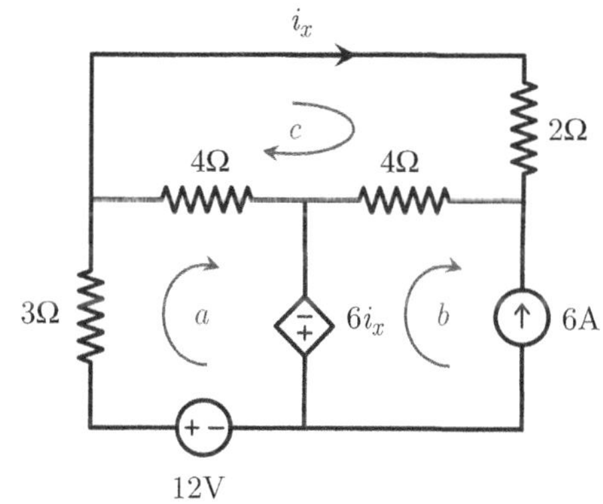

Considering meshes b and c, we immediately obtain $i_b = -6$ A and $i_x = i_c$. Then, applying KVL in meshes a and c, we derive two equations

- KVL(a): $-12 + 3i_a + 4(i_a - i_c) - 6i_c = 0 \longrightarrow 7i_a - 10i_c = 12$,
- KVL(c): $2i_c + 4(i_c - i_b) + 4(i_c - i_a) = 0 \longrightarrow 2i_a - 5i_c = 12$.

Solving the equations, we obtain $i_a = -4$ A, $i_c = -4$ A, and

$$p_s = 6i_x(i_b - i_a) = 48 \text{ W}.$$

The positive value indicates that this source consumes power.

Example 58: Consider the following circuit.

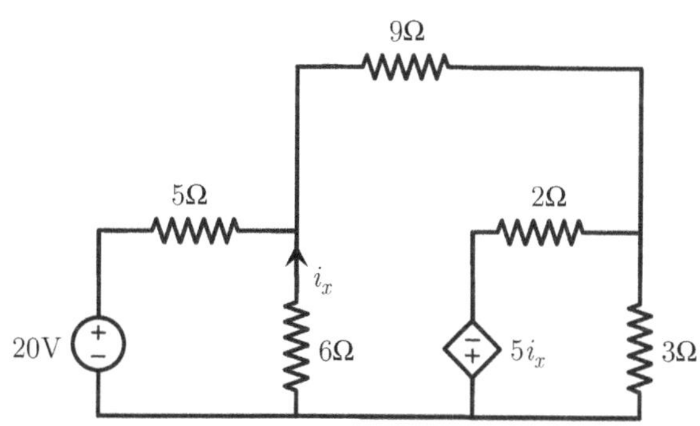

Find i_x.

Solution: Using mesh analysis, we define three different mesh currents.

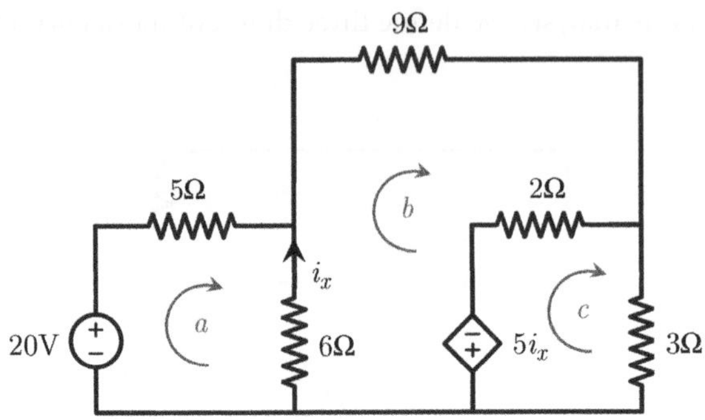

Then we have $i_x = i_b - i_a$ and we apply KVL to derive three equations

- KVL(a): $-20 + 5i_a + 6(i_a - i_b) = 0 \longrightarrow 11i_a - 6i_b = 20$,
- KVL(b): $6(i_b - i_a) + 9i_b + 2(i_b - i_c) - 5i_x = 0 \longrightarrow -i_a + 12i_b - 2i_c = 0$,
- KVL(c): $5i_x + 2(i_c - i_b) + 3i_c = 0 \longrightarrow -5i_a + 3i_b + 5i_c = 0$.

Solving the equations, we find $i_a = 22i_b/5$, $i_b = 25/53$ A, and $i_x = (-17/5)(25/53) = -85/53$ A.

Exercise 41: In the following circuit, previously analyzed using the nodal approach (see Exercise 39), find i_x.

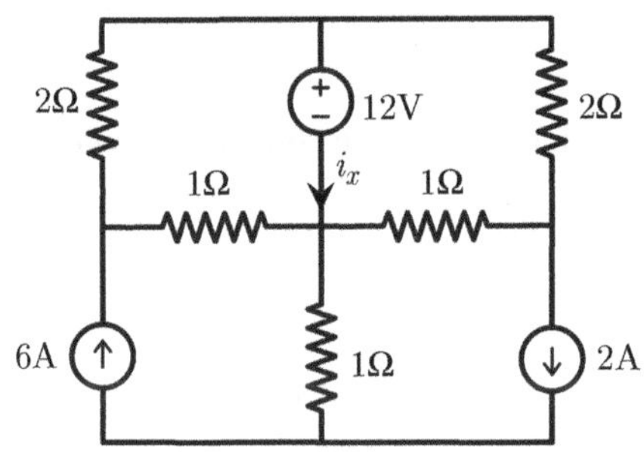

Exercise 42: In the following circuit, find the power of the 10 V voltage source.

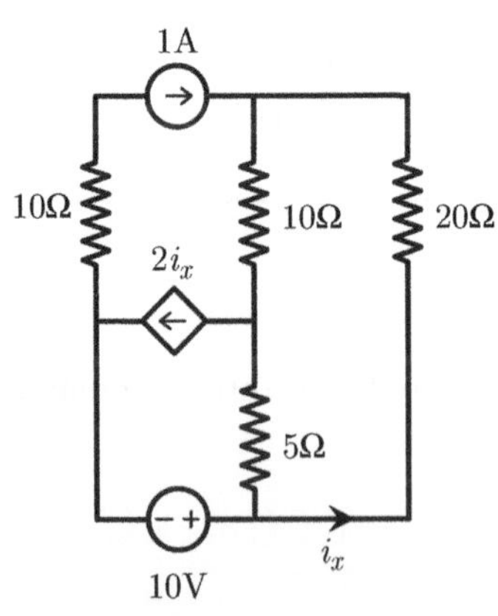

Exercise 43: In the following circuit, previously solved via nodal analysis (see Example 44), find i_y.

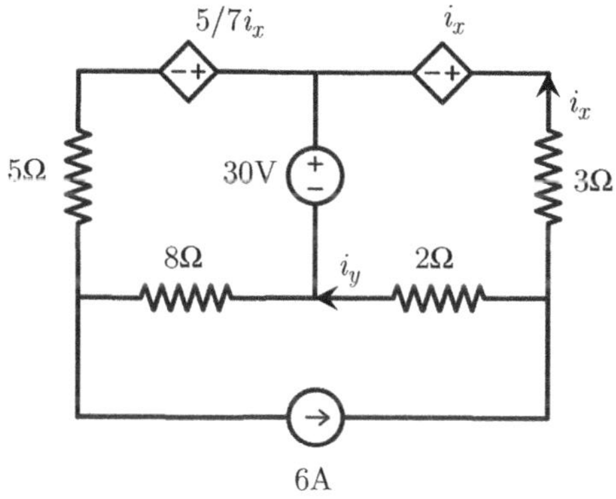

Exercise 44: In the following circuit, previously considered using nodal analysis (see Exercise 29), find i_x.

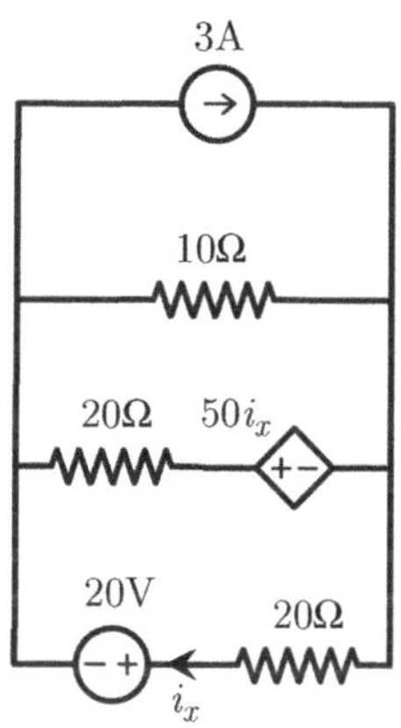

Exercise 45: In the following circuit, previously analyzed using nodal analysis (see Exercise 24), find i_y.

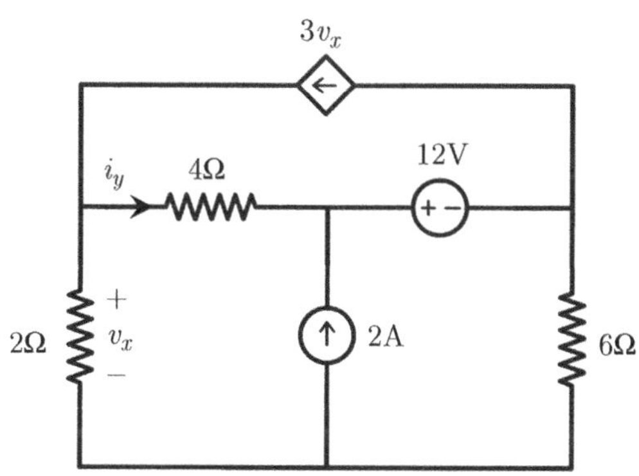

Exercise 46: In the following circuit, find v_s.

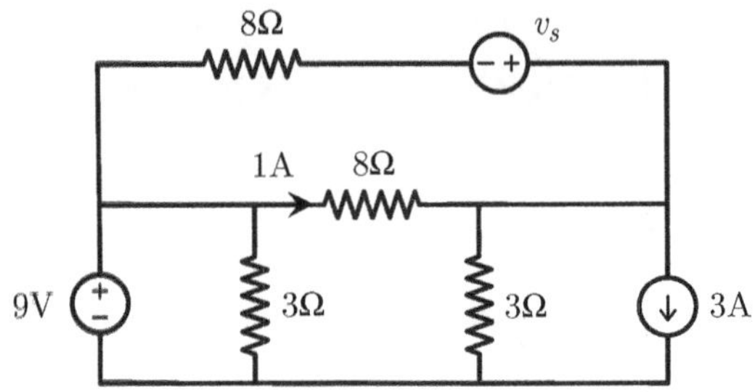

Exercise 47: In the following circuit, previously solved using nodal analysis (see Exercise 27), find v_y.

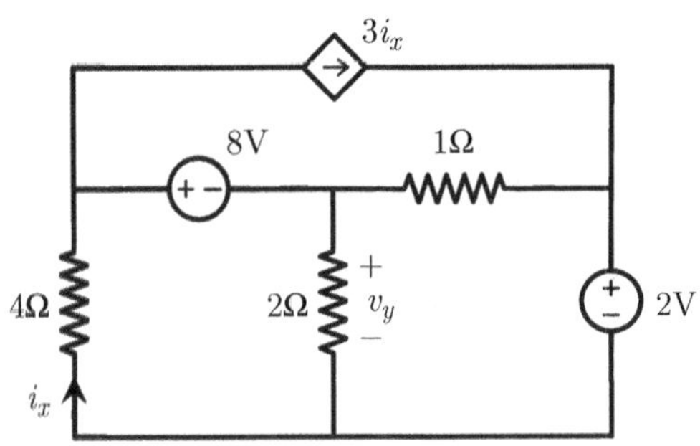

Exercise 48: In the following circuit, find the power of the 5 Ω resistor.

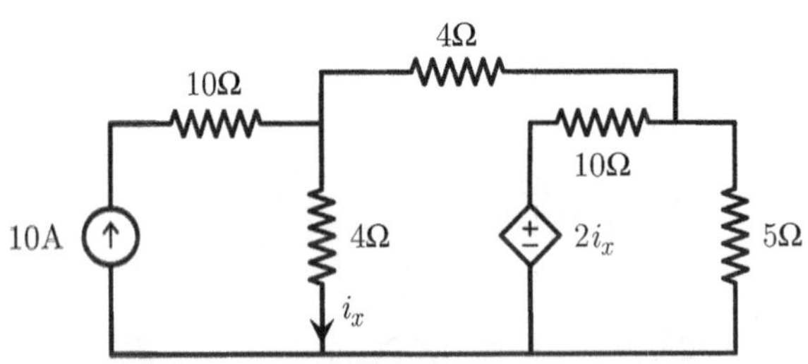

Exercise 49: In the following circuit, find i_s.

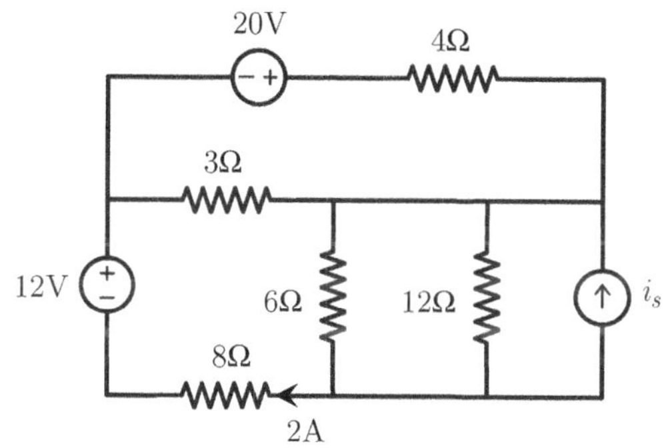

Exercise 50: In the following circuit, previously solved using nodal analysis (see Example 46), find v_s.

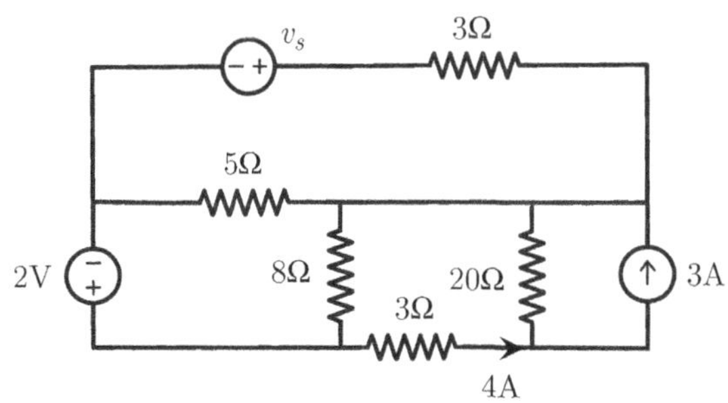

4.2 Concept of Supermesh

KVL can be generalized to any arbitrary loop, that is, the sum of voltages through a closed loop must be zero, being independent of the shape of the loop itself. An application of such a generalized form is to define supermeshes that contain several meshes, as well as components, between them. A supermesh enclosing a current source is particularly useful, since such a source just indicates the difference between the mesh currents but does not give full information on both of them.

It is important to note that a supermesh does not indicate a higher-level mesh current that can flow around multiple meshes. Mesh currents are always defined in meshes, that is, sequences of components without enclosing any inner component. Hence, a mesh current cannot enclose a part or the whole of another mesh. This is also the reason why mesh analysis is restricted to planar circuits (nonplanar circuits are not considered in this book). But a supermesh indicates the application of KVL (a universal law based on the conservation of energy) around multiple meshes using their mesh currents. In the following, we discuss the application of supermeshes by solving various complex circuits.

Example 59: Consider the following circuit.

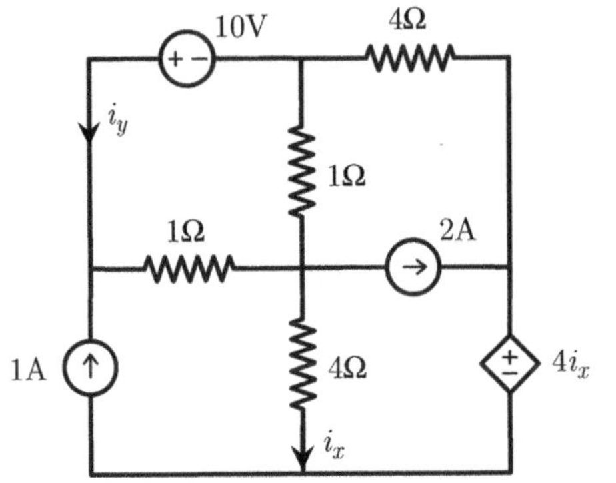

Find i_x, i_y, and the power of the 1 A source.

Solution: Using mesh analysis, we define four mesh currents.

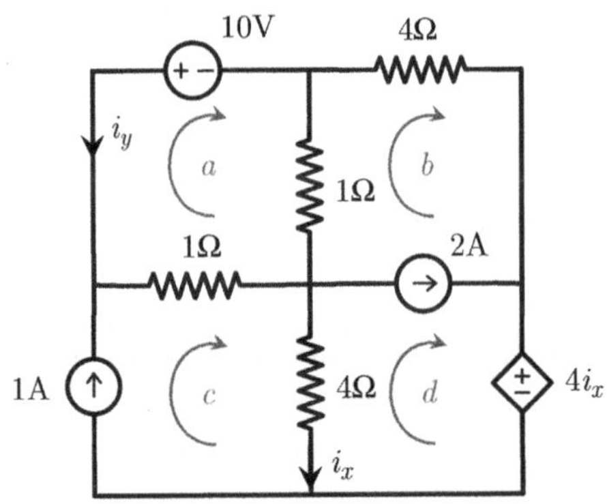

In the above, it is straightforward to write a KVL in mesh *a*,

- KVL(*a*): $10 + 1(i_a - i_b) + 1(i_a - i_c) = 0 \longrightarrow 2i_a - i_b - i_c = -10.$

Also, inspecting mesh *c*, we have

$$i_c = 1\text{ A},$$

which simplifies the equation above to

$$2i_a - i_b = -9.$$

In addition,

$$i_d - i_b = 2,$$
$$i_c - i_d = i_x \longrightarrow i_d = 1 - i_x.$$

The last equation above does not give new useful information; it is merely a connection between a mesh current and a real current. Considering only mesh currents, we have three unknowns (i_a, i_b, and i_d) but only two equations. Now, we inspect which meshes have been used for information.

- Mesh a has already been used to derive an equation.
- Mesh b has not been used directly, because it involves a current source whose voltage is not known in terms of mesh currents.
- Mesh c has been used to write i_c.
- Mesh d has also not been used due to the 2 A current source.
- The current source between meshes b and d is used to write the relation between i_b and i_d.

Overall, it can be understood that meshes b and d generate only one equation overall, while we need two from them. The missing information should come from the combination of these meshes, that is, a supermesh enclosing b and d.

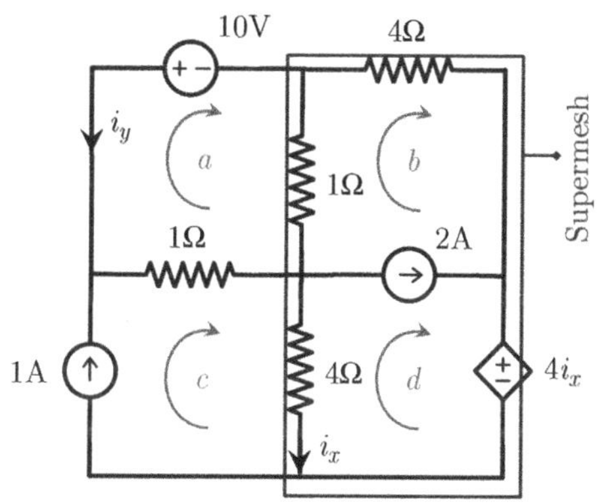

Considering the supermesh shown above, we obtain

- KVL(b&d): $4i_b + 4i_x + 4(i_d - i_c) + 1(i_b - i_a) = 0 \longrightarrow i_a = 5i_b$,

leading to $i_a = -5$ A, $i_b = -1$ A, and $i_d = 1$ A. Then we get

$$i_x = 1 - i_d = 0 \text{ A},$$
$$i_y = -i_a = 5 \text{ A}.$$

Finally, in order to find the power of the 1 A source, we need to know the voltage across it. This can be done by applying KVL in mesh c,

- KVL(c): $v_{1\text{ A}} + 1(i_c - i_a) + 4(i_c - i_d) = 0 \longrightarrow v_{1\text{ A}} + 6 = 0$,

leading to

$$v_{1\text{ A}} = -6 \text{ V}$$
$$p_{1\text{ A}} = -6 \times 1 = -6 \text{ W}.$$

Example 60: Consider the following circuit.

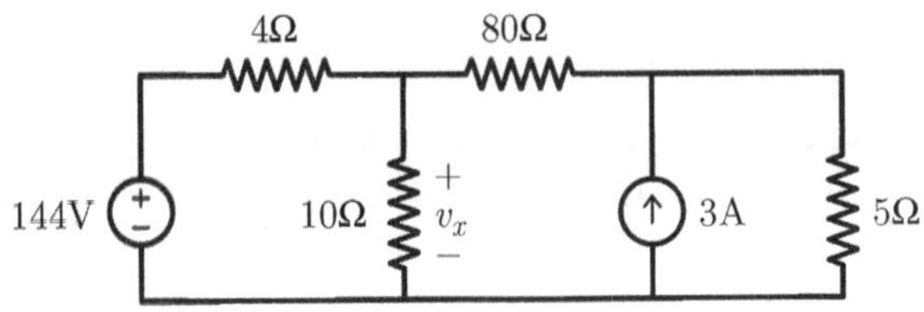

Find v_x.

Solution: Using mesh analysis, we define three mesh currents as follows.

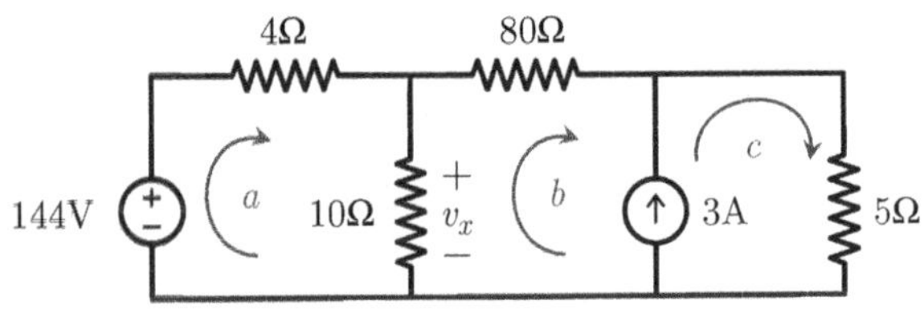

Then, using KVL in mesh *a*, we have

- KVL(*a*): $-144 + 4i_a + 10(i_a - i_b) = 0 \longrightarrow 7i_a - 5i_b = 72.$

Furthermore, we apply KVL in a supermesh containing meshes *b* and *c* to obtain

- KVL(*b*&*c*): $10(i_b - i_a) + 80i_b + 5i_c = 0 \longrightarrow 18i_b - 2i_a + i_c = 0.$

The extra information comes from the supermesh itself (i.e., $i_c - i_b = 3$), considering the current source. Then we have

$$19i_b - 2i_a = -3.$$

Solving the equations, we get $i_a = 11$ A, $i_b = 1$ A, and $i_c = 4$ A. Finally, we obtain

$$v_x = 10(i_a - i_b) = 100 \text{ V}.$$

Example 61: Consider the following circuit.

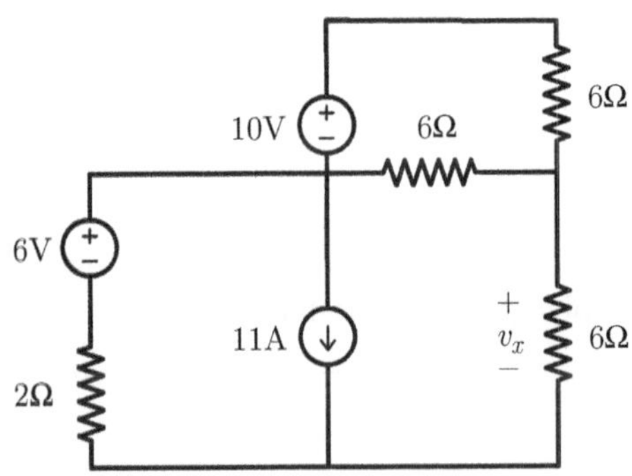

Find v_x.

Solution: Using mesh analysis, we define three mesh currents.

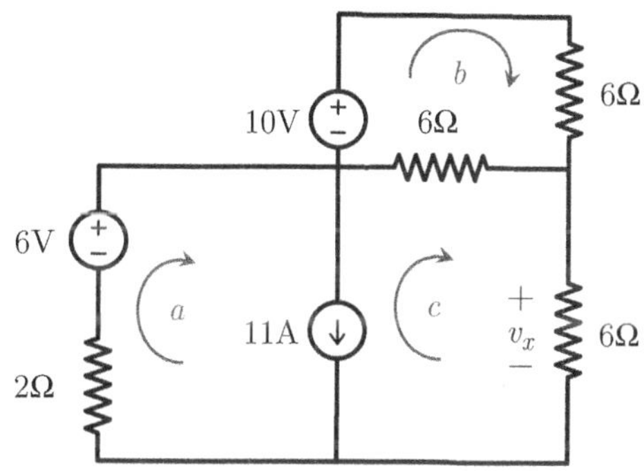

First, using KVL in mesh b, we get

- KVL(b): $-10 + 6i_b + 6(i_b - i_c) = 0 \longrightarrow 6i_b - 3i_c = 5.$

Then, using KVL in a supermesh involving meshes a and c together, we have

- KVL(a&c): $2i_a - 6 + 6(i_c - i_b) + 6i_c = 0 \longrightarrow i_a - 3i_b + 6i_c = 3.$

The extra information needed is again provided by the current source (supermesh), $i_a - i_c = 11$ A. Using this information, the equation above can be simplified to

$$-3i_b + 7i_c = -8.$$

Solving the equations, we obtain $i_c = -1$ A, $i_b = 1/3$ A, and $i_a = 10$ A. Then

$$v_x = 6i_c = -6\text{ V}.$$

Example 62: Consider the following circuit.

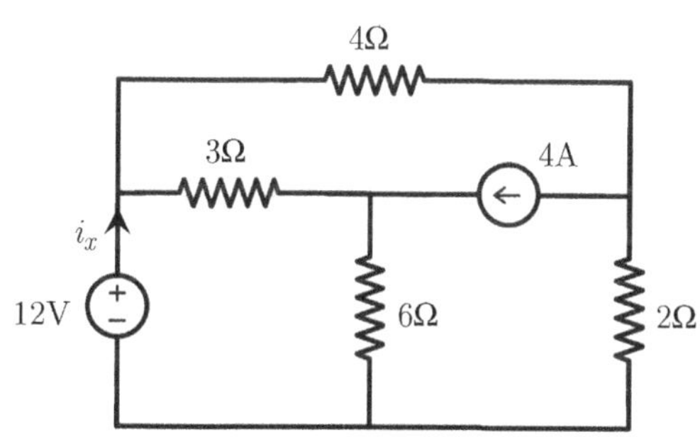

Find i_x.

Solution: Using mesh analysis, we define three mesh currents as follows.

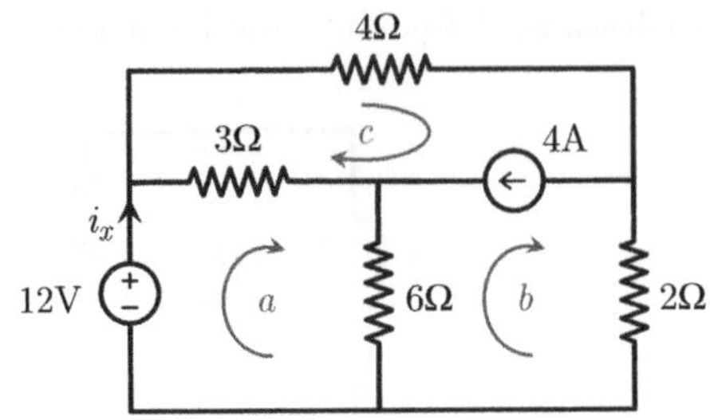

First, considering the current source, we have $i_c - i_b = 4$ A. Then applying KVL in mesh a gives one of the required equations as

- KVL(a): $-12 + 3(i_a - i_c) + 6(i_a - i_b) = 0 \longrightarrow 3i_a - 2i_b - i_c = 4$.

For the missing equation, we consider a supermesh involving meshes b and c,

- KVL(b&c): $6(i_b - i_a) + 3(i_c - i_a) + 4i_c + 2i_b = 0 \longrightarrow -9i_a + 8i_b + 7i_c = 0$.

The equations above can be simplified via substitution to

$$3i_a - 2i_b - i_c = 4 \longrightarrow 3i_a - 3i_b = 8,$$
$$-9i_a + 8i_b + 7i_c = 0 \longrightarrow -9i_a + 15i_b = -28.$$

Then we obtain $i_b = -2/3$ A, $i_a = 2$ A, and $i_x = i_a = 2$ A.

Example 63: Consider the following circuit.

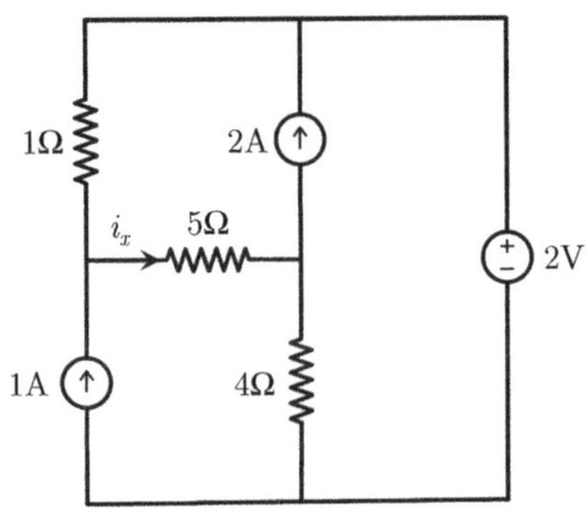

Find i_x.

Solution: Using mesh analysis, we define three mesh currents.

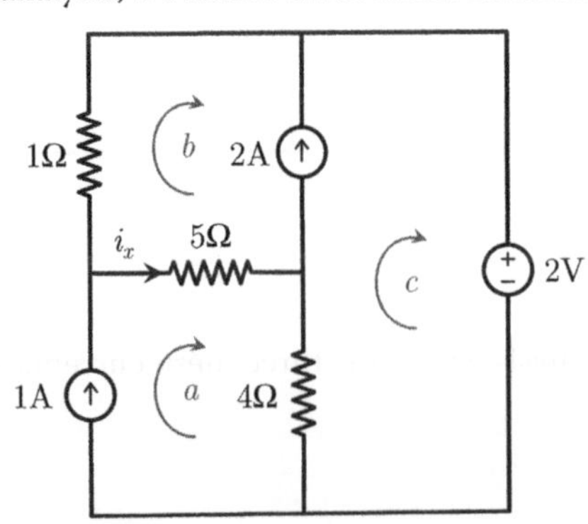

In mesh a, we immediately find $i_a = 1$ A. In addition, $i_c - i_b = 2$, considering the 2 A current source between meshes b and c. In order to solve the problem, we need a KVL equation simultaneously in meshes b and c, namely,

- KVL(b&c): $i_b + 2 + 4(i_c - i_a) + 5(i_b - i_a) = 0 \longrightarrow 6i_b + 4i_c = 7.$

Solving the equations, we get $i_b = -1/10$ A and $i_c = 19/10$ A. Finally, we obtain the value of i_x as

$$i_x = i_a - i_b = 11/10 \text{ A}.$$

Example 64: Consider the following circuit, previously analyzed using the nodal approach (see Exercise 38).

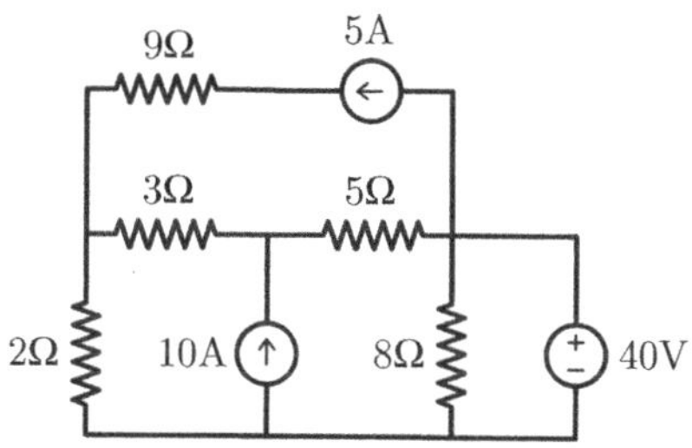

Find the power of the voltage source.

Solution: Using mesh analysis, we define four mesh currents.

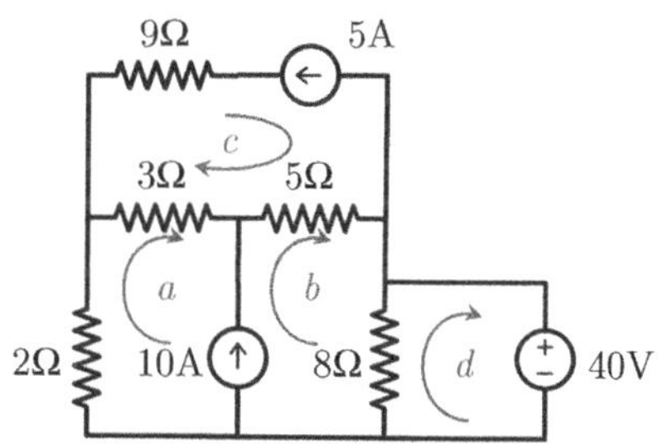

Then it can be seen that $i_c = -5$ A and $i_b - i_a = 10$. Applying KVL in a supermesh consisting of meshes a, b, and d, we derive

- KVL(a&b&d): $2i_a + 3(i_a - i_c) + 5(i_b - i_c) + 40 = 0 \longrightarrow 5i_a + 5i_b - 8i_c = -40$

or

$$i_a + i_b = -16.$$

We note that, by using a supermesh involving three meshes, the voltage across the 8 Ω resistor is directly written as 40 V, without resorting to an extra equation. We further obtain $i_b = -3$ A and $i_d = i_b - 5 = -8$ A since the real current across the 8 Ω resistor is 5 A. Finally, the power of the voltage source can be found to be

$$p_{40\text{ V}} = 40 \times (-8) = -320 \text{ W}.$$

Example 65: Consider the following circuit.

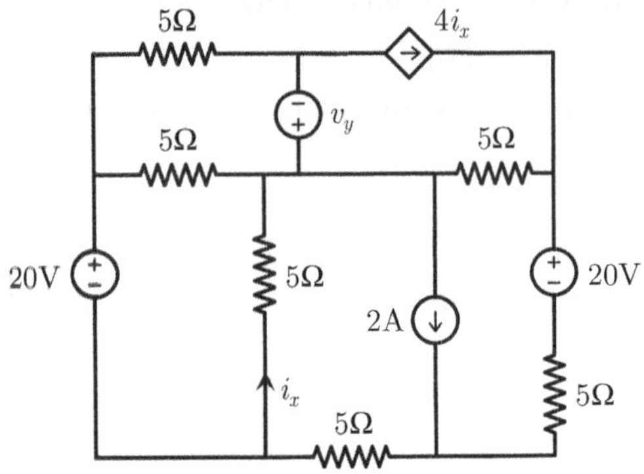

Find the power of the top 5 Ω resistor if $v_y = 20$ V.

Solution: Using mesh analysis, we define five mesh currents.

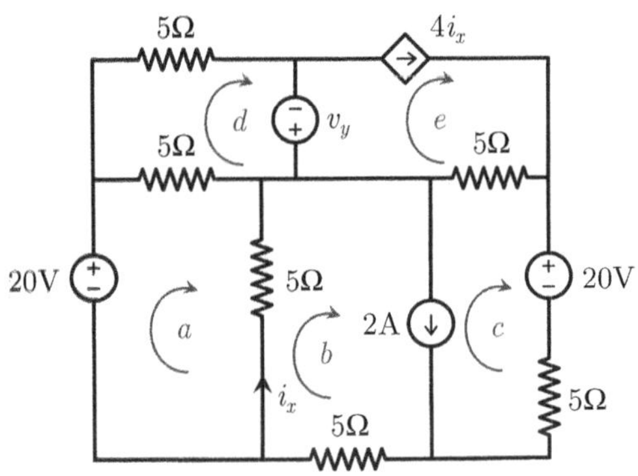

With large circuits of this type, it can be easy to get lost. In order to analyze the circuit systematically, we first consider the given real currents to find relationships between the mesh currents. We have

- $i_x = i_b - i_a$,
- $i_e = 4i_x = 4i_b - 4i_a$,
- $i_b - i_c = 2$.

Then, applying KVL in mesh a, we derive

- KVL(a): $-20 + 5(i_a - i_d) + 5(i_a - i_b) = 0 \longrightarrow 2i_a - i_b - i_d = 4.$

None of the pairs of equations above are solvable. As usual, the required information is available in a supermesh:

- KVL(b&c): $5(i_b - i_a) + 5(i_c - i_e) + 20 + 5i_c + 5i_b = 0 \longrightarrow i_a - 2i_b - 2i_c + i_e = 4.$

Substituting i_e in the above, we get

$$i_a - 2i_b - 2i_c + 4i_b - 4i_a = -3i_a + 2i_b - 2i_c = 4.$$

Furthermore, using $i_b - i_c = 2$, we obtain $i_a = 0$. In order to obtain i_d, we consider KVL in mesh d,

- KVL(d): $5i_d - v_y + 5(i_d - i_a) = 0 \longrightarrow i_d = 2\,\text{A}$.

Hence, $i_b = 2i_a - i_d - 4 = -6\,\text{A}$ and $i_c = i_b - 2 = -8\,\text{A}$. We note that finding i_b and i_c requires extra calculations, since the power of the top 5 Ω resistor can be found via i_d as

$$p_{5\,\Omega} = 5i_d^2 = 20\,\text{W}.$$

Example 66: Consider the following circuit, previously solved using nodal analysis (see Exercise 20).

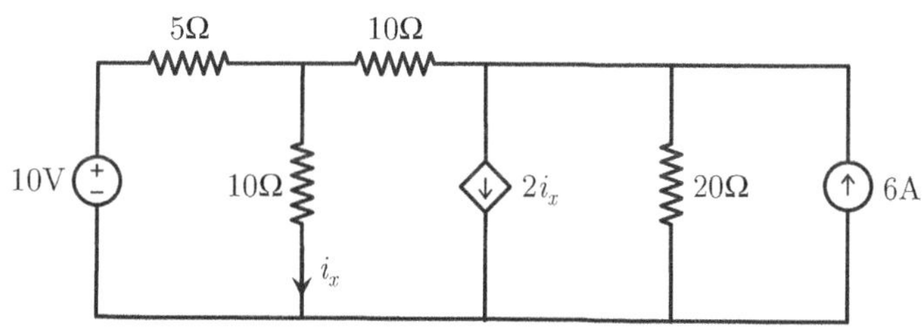

Find the value of i_x.

Solution: Using mesh analysis, we define four different mesh currents.

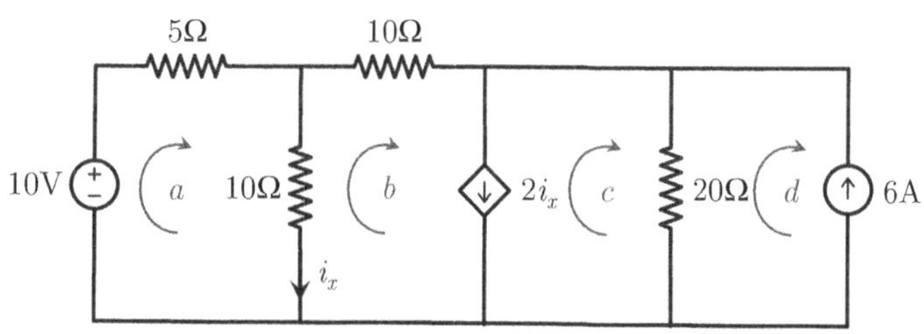

We immediately find that $i_d = -6\,\text{A}$. In addition, using KVL in mesh a, we get

- KVL(a): $-10 + 5i_a + 10(i_a - i_b) = 0 \longrightarrow 3i_a - 2i_b = 2$.

At this stage, KVL in a supermesh involving meshes b and c should be considered:

- KVL(b&c): $10(i_b - i_a) + 10i_b + 20(i_c - i_d) = 0 \longrightarrow i_a - 2i_b - 2i_c = 12$.

Furthermore, inside the supermesh, we have

$$i_b - i_c = 2i_x = 2i_a - 2i_b \longrightarrow 2i_a - 3i_b + i_c = 0.$$

For the solution, the last two equations can be combined to eliminate i_c,

$$5i_a - 8i_b = 12.$$

Solving the equations, we obtain $i_a = -4/7$ A, $i_b = -13/7$ A, $i_c = -31/7$ A, and finally,

$$i_x = i_a - i_b = 9/7 \text{ A}.$$

Example 67: Consider the following circuit.

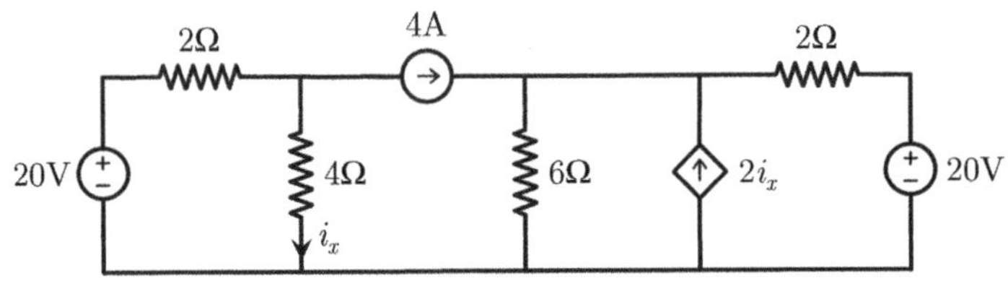

Find the power of the current-dependent current source.

Solution: For this circuit, we again define four different mesh currents.

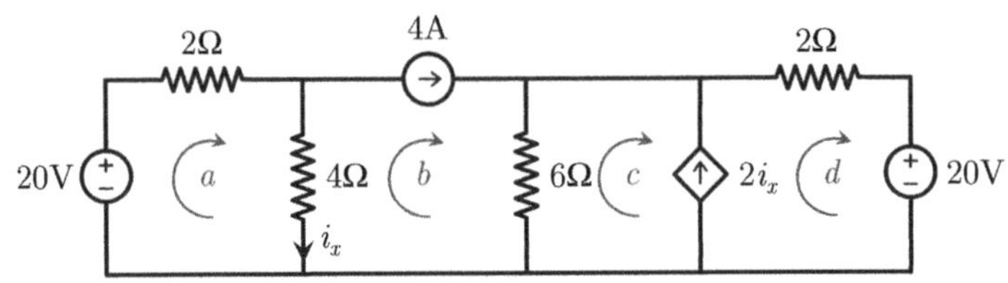

We have $i_b = 4$ A and

$$i_d - i_c = 2i_x = 2(i_a - i_b).$$

Applying KVL in mesh a, we derive

- KVL(a): $-20 + 2i_a + 4(i_a - i_b) = 0 \longrightarrow 3i_a - 2i_b = 10.$

Then, using $i_b = 4$ A, we find $i_a = 6$ A. Therefore, the relationship between i_c and i_d can be updated as

$$i_d - i_c = 4.$$

Furthermore, applying KVL in a supermesh involving meshes c and d, we obtain

- KVL(c&d): $6(i_c - i_b) + 2i_d + 20 = 0 \longrightarrow 3i_c - 3i_b + i_d = -10,$

leading to

$$3i_c + i_d = 2.$$

Solving the equations, we obtain $i_c = -1/2$ A and $i_d = 7/2$ A.

At this stage, we need to find the voltage across the dependent source. This can be done by considering KVL in mesh c:

$$6(i_c - i_b) - v_s = 0 \longrightarrow v_s = 6 \times (-9/2) = -27\text{ V},$$

where the direction of v_s is determined via the sign convention. Finally, the required power is found to be

$$p_s = 2i_x v_s = 4 \times (-27) = -108\text{ W}.$$

Example 68: Consider the following circuit, previously solved via nodal analysis (see Example 47).

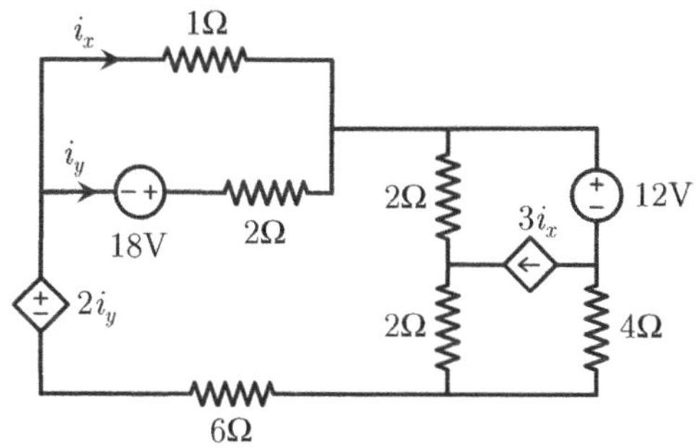

Find the power of the current-dependent current source.

Solution: This circuit can be analyzed by considering four mesh currents as follows.

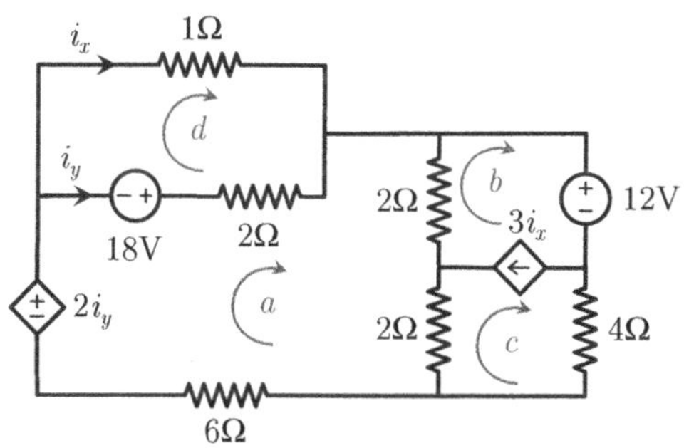

First, we have

$$i_b - i_c = 3i_x = 3i_d,$$
$$i_a - i_d = i_y.$$

Applying KVL in meshes a and d, we derive

- KVL(a): $-2i_y - 18 + 2(i_a - i_d) + 2(i_a - i_b) + 2(i_a - i_c) + 6i_a = 0$
  $\longrightarrow 5i_a - i_b - i_c = 9$

and

- KVL(d): $i_d + 2(i_d - i_a) + 18 = 0 \longrightarrow 2i_a - 3i_d = 18.$

Furthermore, applying KVL in the supermesh formed by meshes b and c, we obtain

- KVL(b&c): $2(i_b - i_a) + 12 + 4i_c + 2(i_c - i_a) = 0 \longrightarrow 2i_a - i_b - 3i_c = 6.$

Using the equations above yields $i_a = -1$ A, $i_b = -17$ A, $i_c = 3$ A, and $i_d = -20/3$ A. In addition, we have $i_x = -20/3$ A and $i_y = 17/3$ A. The voltage of the current-dependent current source can be obtained by considering KVL in mesh b:

$$12 + v_s + 2(i_b - i_a) = 0 \longrightarrow v_s = -12 - 2(-17 + 1) = 20 \text{ V}.$$

Therefore, its power is found to be

$$p_s = 3i_x v_s = -20 \times 20 = -400 \text{ W}.$$

Exercise 51: In the following circuit, find the power of the 5 A voltage source.

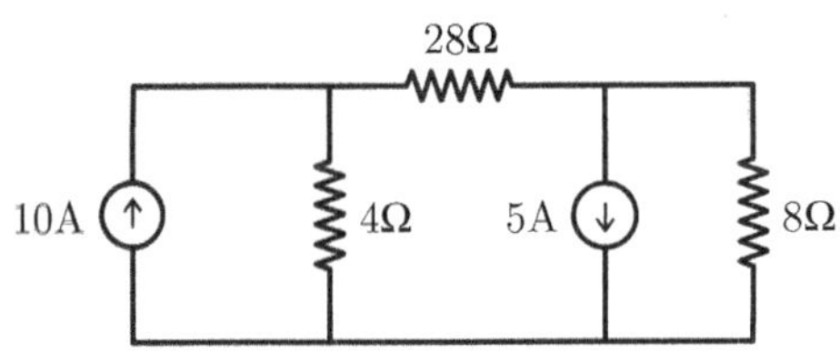

Exercise 52: In the following circuit, find the value of i_x.

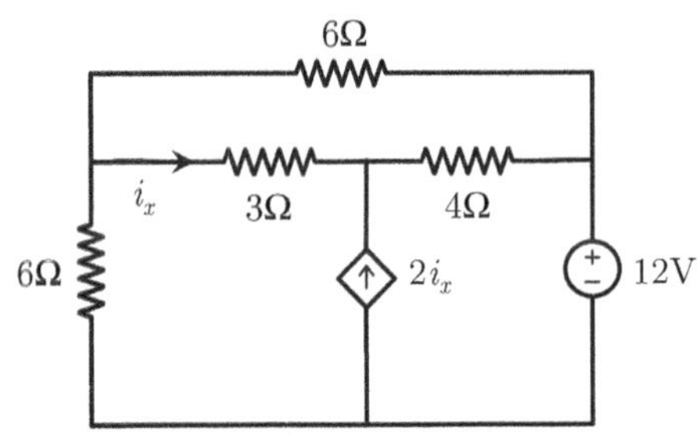

Exercise 53: In the following circuit, find the power of the 100 V voltage source.

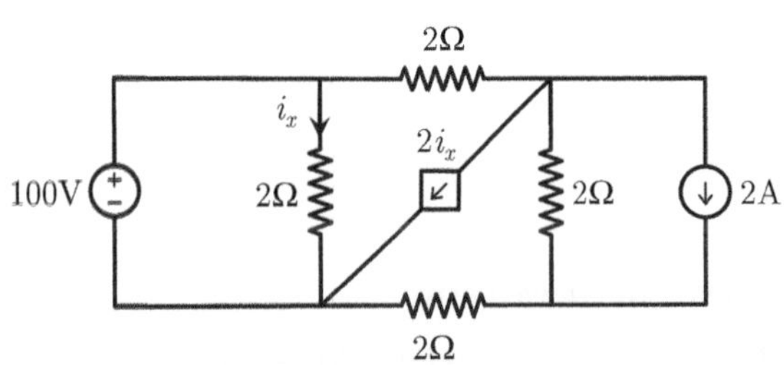

Exercise 54: In the following circuit, find i_x.

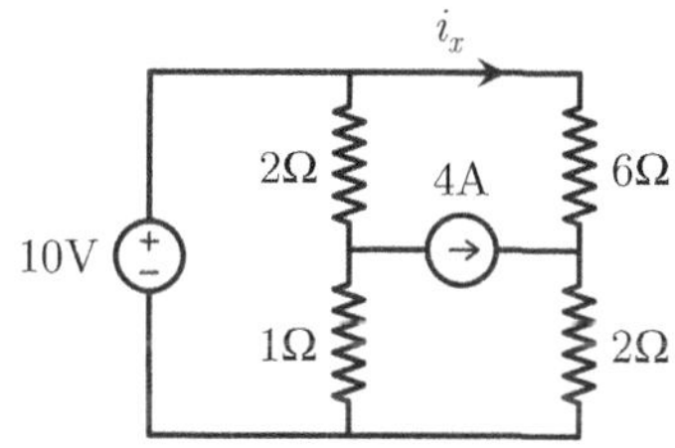

Exercise 55: In the following circuit, previously solved using nodal analysis (see Exercise 35), find the power of the current-dependent voltage source.

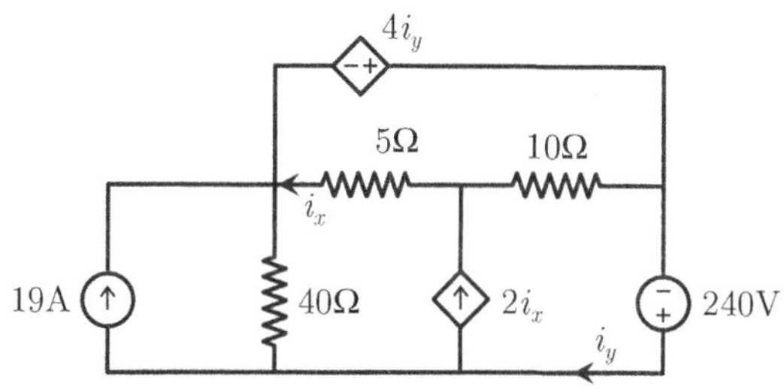

Exercise 56: In the following circuit, find the power of the independent current source.

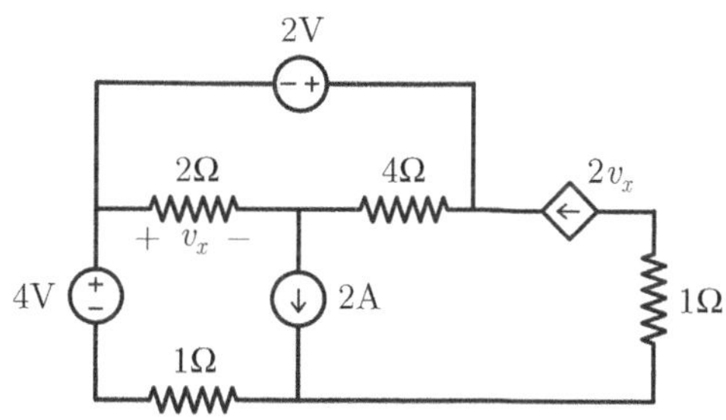

Exercise 57: In the following circuit, find the power of the 2 A current source.

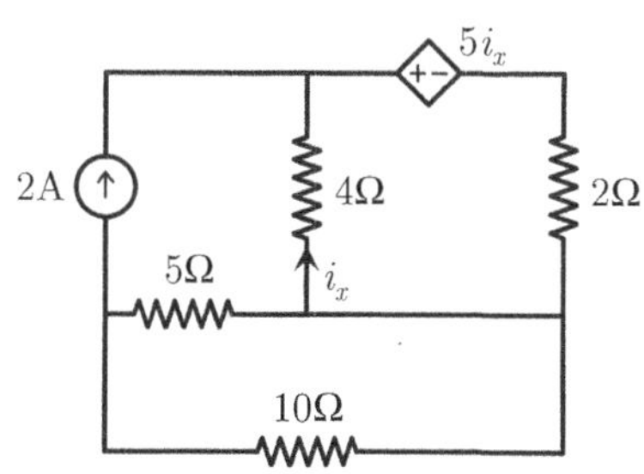

Exercise 58: In the following circuit, previously solved using nodal analysis (see Exercise 37), find i_x.

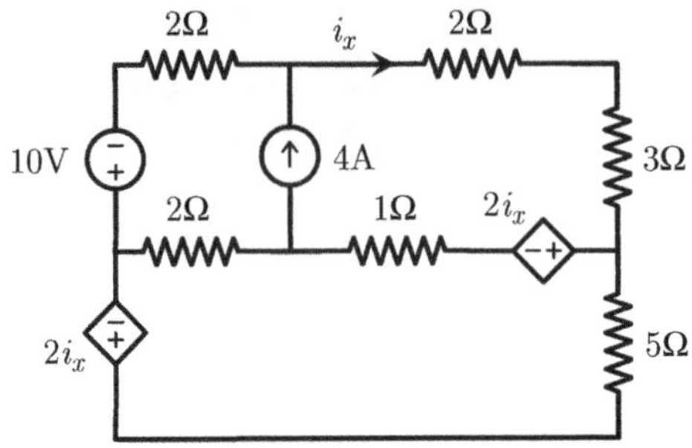

Exercise 59: In the following circuit, previously solved using nodal analysis (see Exercise 30), find v_x.

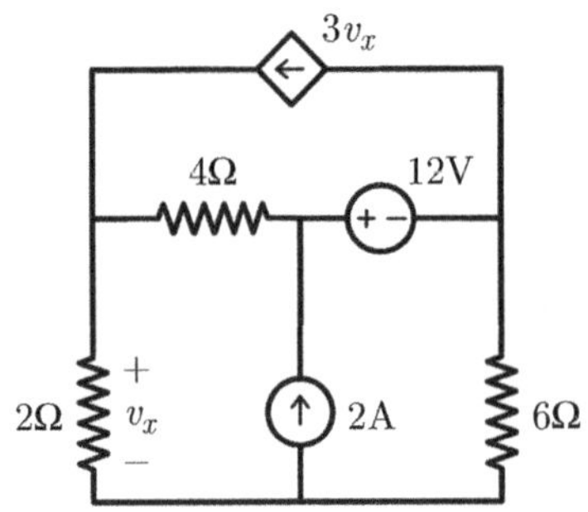

Exercise 60: In the following circuit, previously solved using nodal analysis (see Example 29), find the value of i_x.

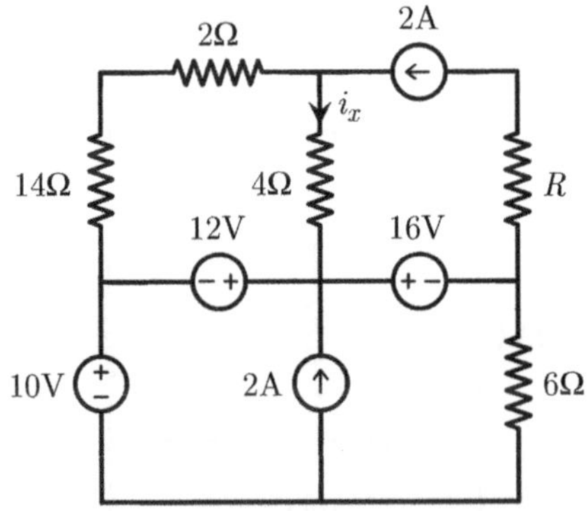

4.3 Circuits with Multiple Independent Current Sources

As discussed in Section 3.3, nodal analysis is particularly useful when multiple voltage sources are involved in the circuit. For such a circuit, a wise selection of the ground leads to trivial voltage values at some of the nodes. Similarly, mesh analysis can be very useful for circuits with multiple independent current sources.

By way of demonstration, we consider the following circuit, where the power of the 3 Ω resistor needs to be found.

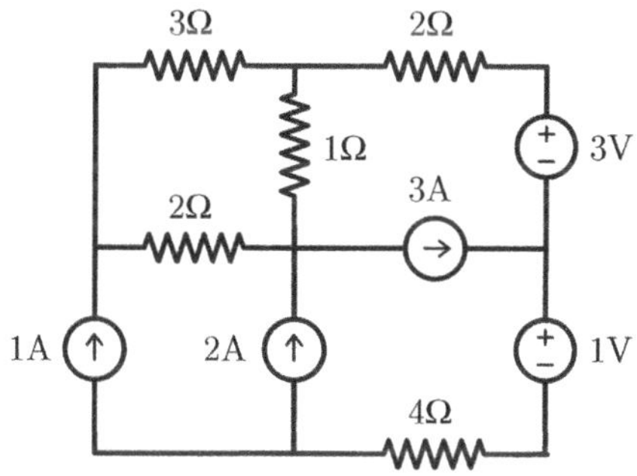

Using mesh analysis, we label the loops as follows.

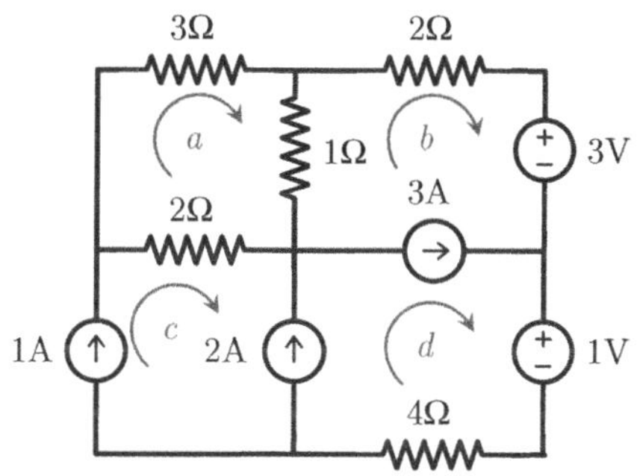

Now, considering mesh c, we immediately have

$$i_c = 1\text{ A}.$$

Similarly, using mesh d, we get

$$i_d - i_c = 2 \longrightarrow i_d = 3\text{ A}.$$

Interestingly, the 3 A current source yields the value of i_b,

$$i_d - i_b = 3 \longrightarrow i_b = 0\text{ A}.$$

At this stage, we know three mesh currents without applying KVL. We only need KVL in mesh a:

- KVL(a): $3i_a + 1(i_a - i_b) + 2(i_a - i_c) = 0 \longrightarrow 6i_a - i_b - 2i_c = 0,$

leading to $i_a = 1/3$ A. Therefore, the power of the 3 Ω resistor can be found as

$$p_{3\,\Omega} = 3 \times (1/3)^2 = 1/3 \text{ W}.$$

4.4 Solving Challenging Problems Using the Mesh Analysis

In theory, mesh analysis should provide a solution to any (planar) resistive network. On the other hand, difficulties may arise in some circuits, for example, when a mesh involves large number of components or when a loop shape is deformed such that it may be easy to lose orientation. This section is dedicated to some challenging problems and their solutions with mesh analysis.

Example 69: Consider the following circuit.

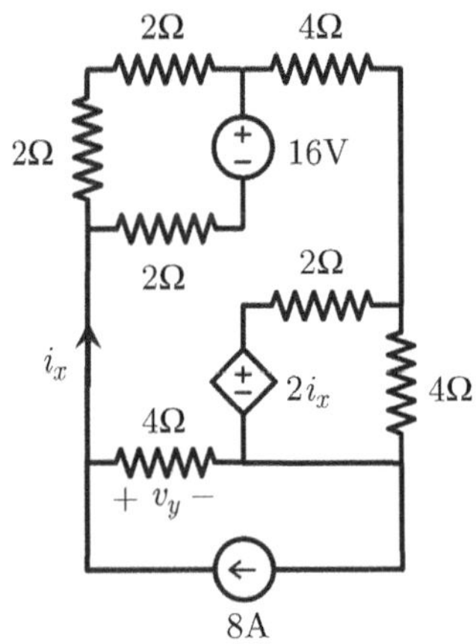

Find i_x.

Solution: Using mesh analysis, we define four mesh currents as follows.

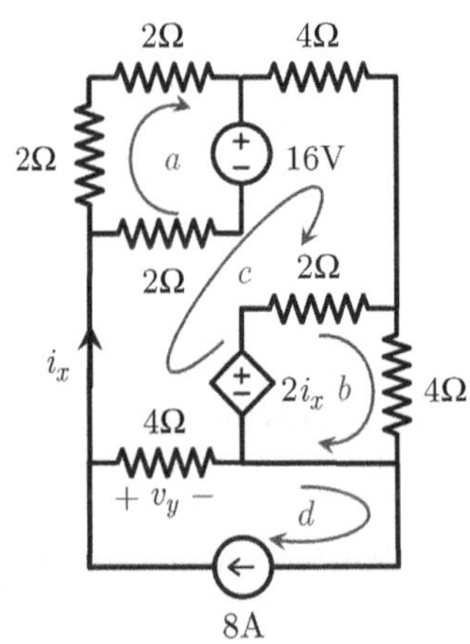

First, we note that $i_x = i_c$ and $i_d = 8$ A. Applying KCL in meshes a and b, we derive

- KVL(a): $4i_a + 16 + 2(i_a - i_c) = 0 \longrightarrow 3i_a - i_c = -8$,
- KVL(b): $-2i_x + 2(i_b - i_c) + 4i_b = 0 \longrightarrow 3i_b - 2i_c = 0$.

A relatively difficult application of KVL in mesh c is required to solve the problem. We obtain

- KVL(c): $2(i_c - i_a) - 16 + 4i_c + 2(i_c - i_b) + 2i_x + 4(i_c - 8) = 0$,

leading to

$$-i_a - i_b + 7i_c = 24.$$

Combining the equations, we have

$$-(i_c - 8)/3 - 2i_c/3 + 7i_c = 24 \longrightarrow i_c = 32/9 \text{ A}.$$

Therefore,

$$i_x = i_c = 32/9 \text{ A}.$$

Example 70: Consider the following circuit, previously solved using nodal analysis (see Example 40).

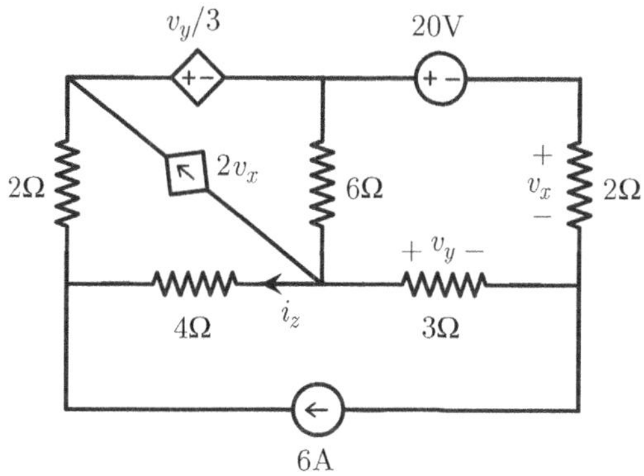

Find i_z.

Solution: Using mesh analysis, we define four different meshes.

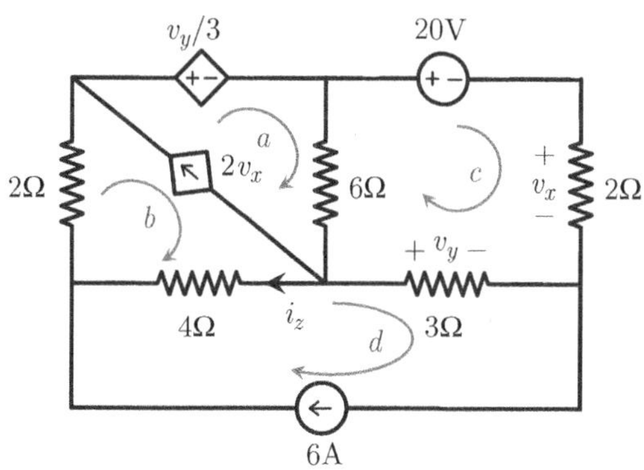

First, we have $i_d = 6$ A, $v_x = 2i_c$, and $v_y = 3(i_d - i_c)$. Then, using KVL in mesh c and in the supermesh (formed by combining meshes a and b), we obtain

- KVL(c): $20 + 2i_c + 3(i_c - i_d) + 6(i_c - i_a) = 0 \longrightarrow 6i_a - 11i_c = 2$,
- KVL(a&b): $v_y/3 + 6(i_a - i_c) + 4(i_b - i_d) + 2i_b = 0 \longrightarrow 6i_a + 6i_b - 7i_c = 18$.

Furthermore, considering the supermesh, we have

$$i_a - i_b = 2v_x = 4i_c.$$

Combining two of the equations, i_b can be eliminated to give

$$12i_a - 31i_c = 18.$$

Solving the equations, we get $i_a = -68/27$ A, $i_b = 100/27$ A, and $i_c = -42/27$ A. Hence,

$$i_z = i_b - i_d = -62/27 \text{ A}.$$

Example 71: Consider the following circuit, previously solved using nodal analysis (see Exercise 32).

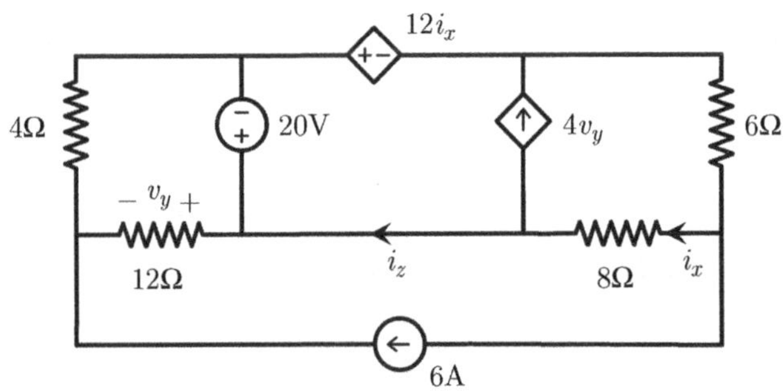

Find i_z.

Solution: Using mesh analysis, we define four different meshes as follows.

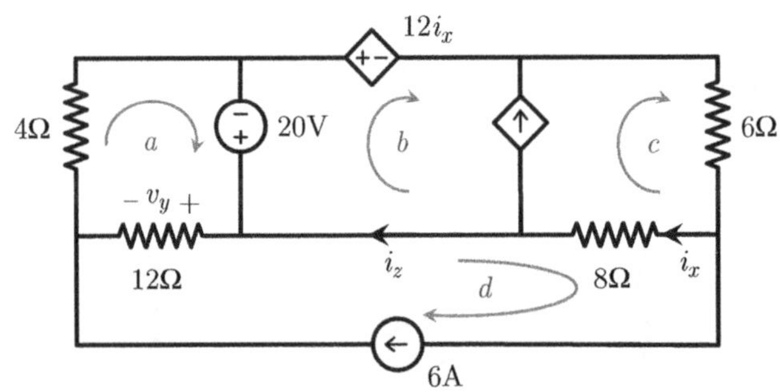

We have $i_d = 6$ A and $i_c - i_b = 4v_y$. Applying KVL in mesh a, we obtain

- KVL(a): $4i_a - 20 + 12(i_a - i_d) = 0 \longrightarrow i_a = 23/4$ A.

Hence, v_y is found to be

$$v_y = 12(i_a - i_d) = 12(23/4 - 6) = -3\text{ V},$$

leading to

$$i_c - i_b = -12.$$

Considering $i_x = i_c - i_d$, KVL in the supermesh (b and c) can be used to find i_c:

- KVL(b&c): $20 + 12i_x + 6i_c + 8(i_c - i_d) = 0 \longrightarrow 20 + 26i_c - 20i_d = 0$,

leading to

$$i_c = 50/13\text{ A}.$$

Then $i_b = i_c + 12 = 206/13$ A and

$$i_z = i_b - i_d = 206/13 - 6 = 128/13\text{ A}.$$

Example 72: A challenging problem need not involve a large circuit. Consider the following circuit.

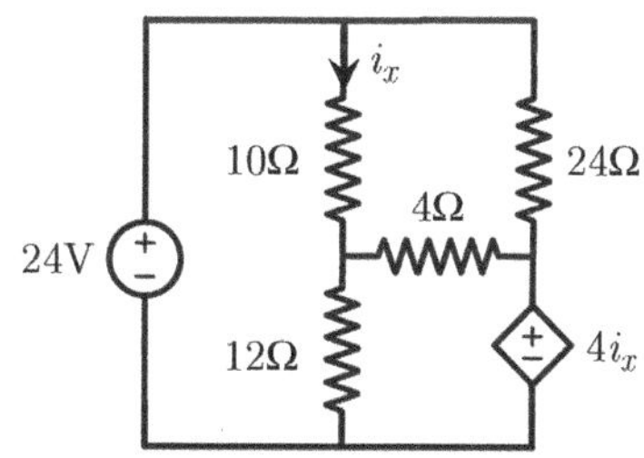

Find i_x.

Solution: This example contains only three loops but needs a careful arrangement of equations for the solution. Using mesh analysis, we define three different meshes.

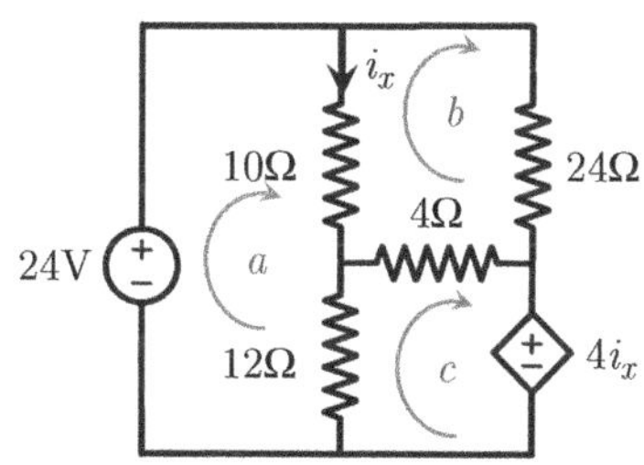

We apply KVL in all three meshes to derive three equations:

- KVL(a): $-24 + 10(i_a - i_b) + 12(i_a - i_c) = 0 \longrightarrow 11i_a - 5i_b - 6i_c = 12$,
- KVL(b): $24i_b + 4(i_b - i_c) + 10(i_b - i_a) = 0 \longrightarrow -5i_a + 19i_b - 2i_c = 0$,
- KVL(c): $12(i_c - i_a) + 4(i_c - i_b) + 4i_x = 0 \longrightarrow i_a + i_b = 2i_c$,

using $i_x = i_a - i_b$. At this stage, the equations must be solved using a method such as substitution. For example, using $i_a = 2i_c - i_b$, we have

$$22i_c - 11i_b - 5i_b - 6i_c = -16i_b + 16i_c = 12 \longrightarrow -4i_b + 4i_c = 3,$$
$$-10i_c + 5i_b + 19i_b - 2i_c = 24i_b - 12i_c = 0.$$

The final equation states that $i_c = 2i_b$; hence, we have $i_b = 3/4\,\text{A}$, $i_c = 3/2\,\text{A}$, and $i_a = 9/4\,\text{A}$. Finally, we obtain

$$i_x = i_a - i_b = 9/4 - 3/4 = 3/2\,\text{A}.$$

Example 73: Consider the following circuit.

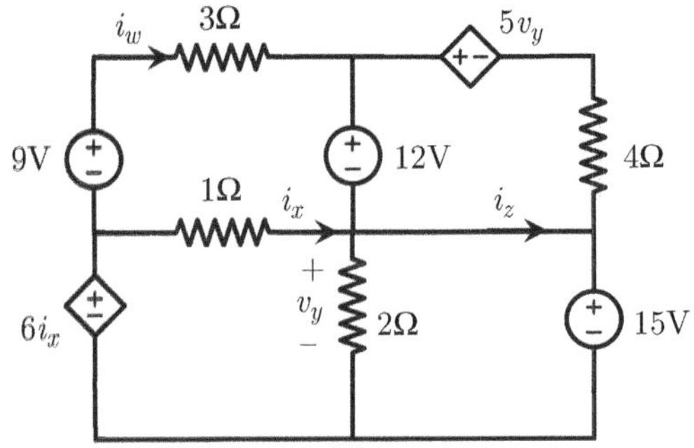

Find i_z.

Solution: Using mesh analysis, we define four different meshes.

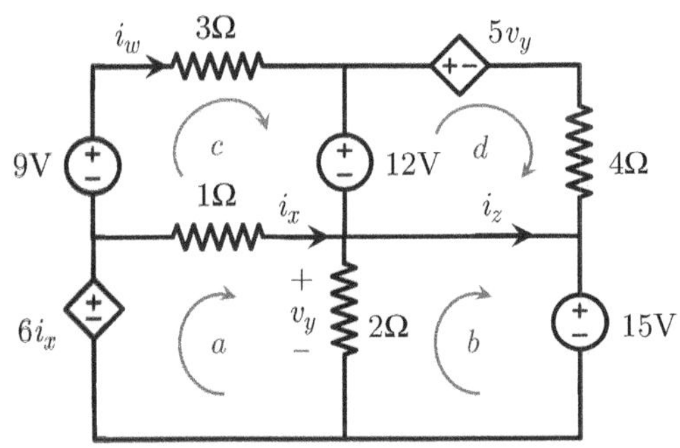

First, we note that $i_x = i_a - i_c$ and $v_y = 2(i_a - i_b) = 15\,\text{V}$, which can be found by considering KVL in mesh *b*. Applying KVL in meshes *a* and *c*, we obtain

- KVL(a): $-6i_x + (i_a - i_c) + 2(i_a - i_b) = 0 \longrightarrow i_a - i_c = 3,$
- KVL(c): $-9 + 3i_c + 12 + (i_c - i_a) = 0 \longrightarrow 4i_c - i_a = -3.$

Solving the equations, we obtain $i_c = 0$ A, $i_a = 3$ A, and $i_b = -9/2$ A. In order to find i_z, we further need KVL in mesh d,

- KVL(d): $-12 + 5v_y + 4i_d = 0 \longrightarrow i_d = -63/4$ A.

Finally, i_z is found to be

$$i_z = i_b - i_d = 45/4 \text{ A}.$$

Example 74: In general, given a basic circuit, it may not be obvious which analysis method (nodal or mesh) leads to an easier solution. In some circuits (see Section 4.3), however, there are consecutive meshes, each involving one or more independent current sources, leading to trivial determination of mesh currents. Consider the following circuit.

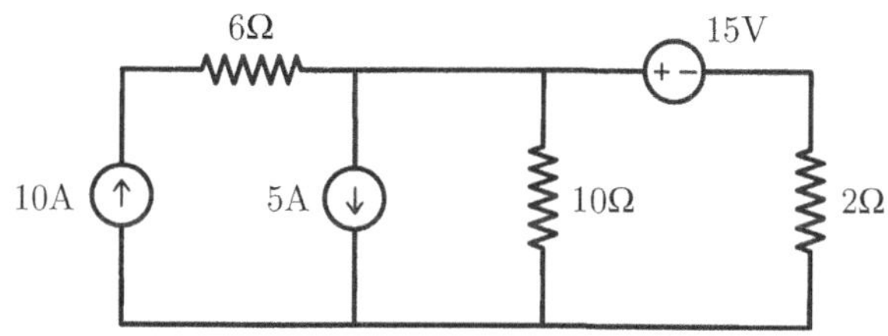

Find the power of the 5 A source.

Solution: Using mesh analysis, we define three mesh currents.

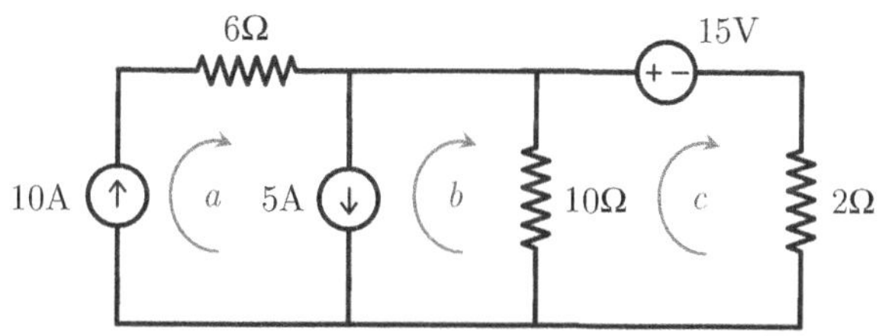

Considering meshes a and b, we have $i_a = 10$ A and $i_b = i_a - 5 = 5$ A, respectively. KVL is required for only one mesh (mesh c):

- KVL(c): $10(i_c - i_b) + 15 + 2i_c = 0 \longrightarrow i_c = 35/12$ A.

Then $v_{5\text{ A}} = 10(i_b - i_c) = 125/6$ V and

$$p_{5\text{ A}} = v_{5\text{ A}} i_{5\text{ A}} = (125/6) \times 5 = 625/6 \text{ W}.$$

Example 75: Consider the following circuit.

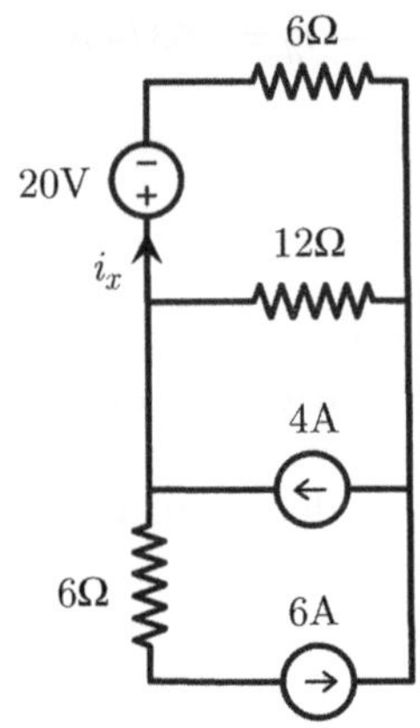

Find i_x.

Solution: Using mesh analysis, we again define three mesh currents as follows.

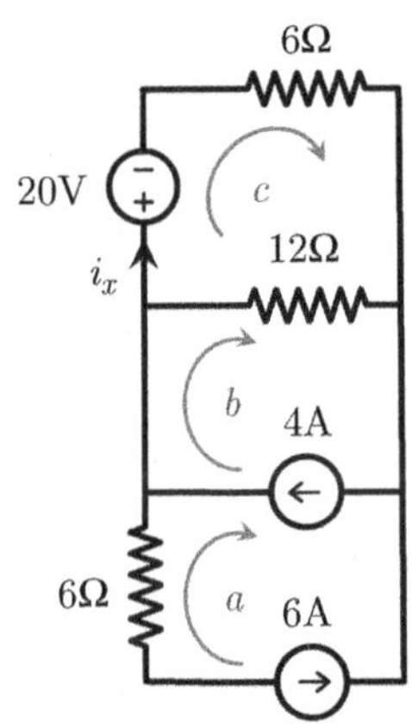

We immediately find $i_a = -6$ A and $i_b = 4 + i_a = -2$ A. Then, applying KVL in mesh c, we get

- KVL(c): $20 + 6i_c + 12(i_c - i_b) = 0 \longrightarrow i_c = -22/9$ A.

Therefore, we have $i_x = i_c = -22/9$ A.

Example 76: Consider the following circuit, previously solved using nodal analysis.

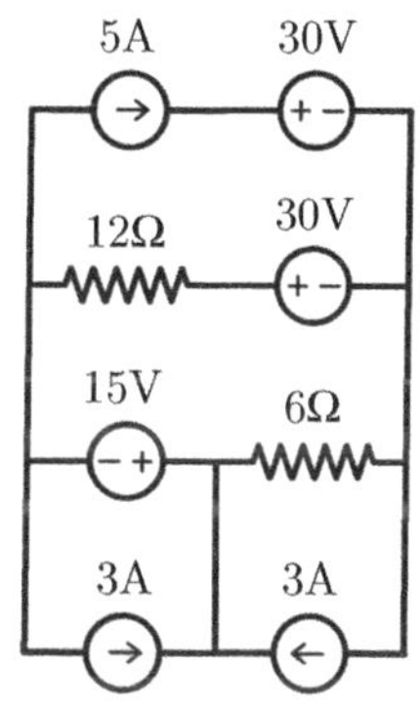

Find the power of the 5 A current source.

Solution: Using mesh analysis, we define three mesh currents.

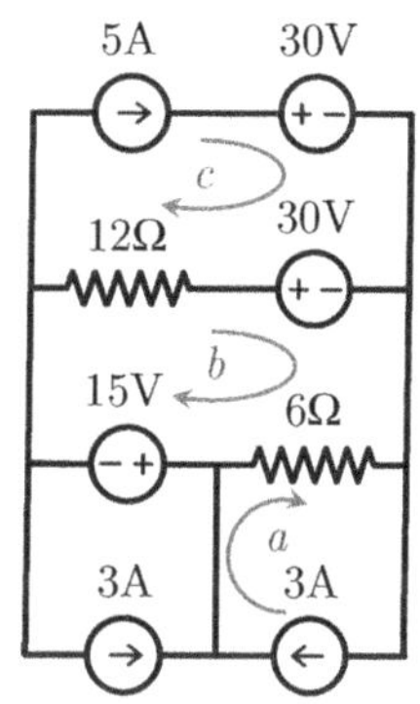

First, we note that $i_a = 3$ A and $i_c = 5$ A. Second, applying KVL in mesh b, we obtain

- KVL(b): $6(i_b - i_a) + 15 + 12(i_b - i_c) + 30 = 0 \longrightarrow i_b = 11/6$ A.

In order to find the power of the 5 A source, we need its voltage, which can be obtained via KVL as

- KVL(c): $v_{5A} + 30 - 30 + 12(i_c - i_b) = 0 \longrightarrow v_{5A} = -38$ V.

Then we have

$$p_{5A} = -38 \times 5 = -190 \text{ W}.$$

We also note that no mesh current is defined (and needed) for the mesh involving the 15 V source and one of the 3 A sources.

Example 77: Consider the following circuit, previously analyzed using the nodal approach (see Exercise 34).

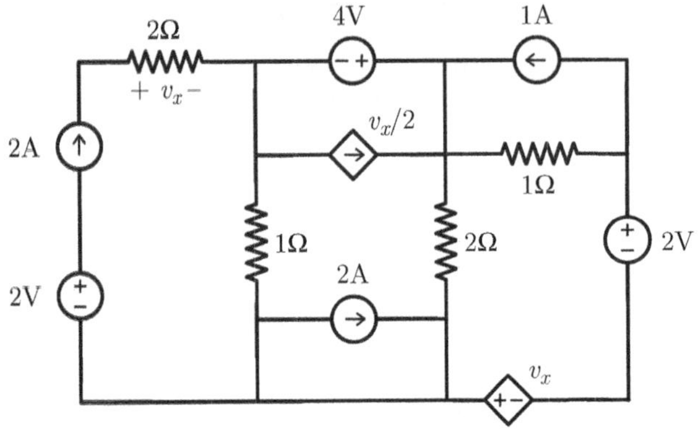

Find the power of the voltage-dependent voltage source.

Solution: For this circuit, we define five mesh currents. We also note that mesh *c* is defined through the short-circuit line rather than the 2 A current source.

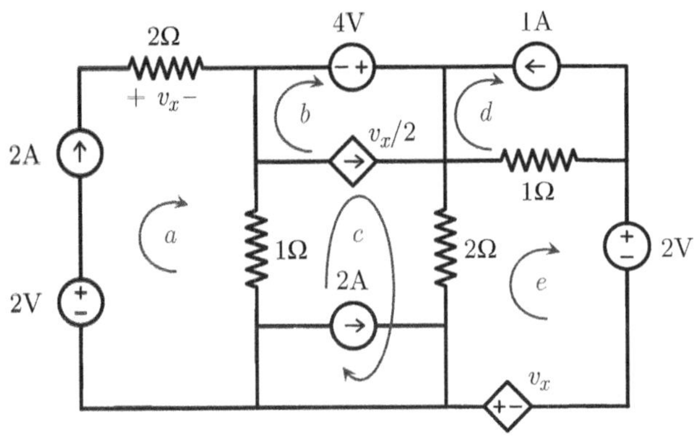

We have $i_a = 2$ A, $i_d = -1$ A and $v_x = 2 \times i_a = 4$ V. Applying KVL in the supermesh involving meshes *b* and *c*, we obtain

- KVL(*b*&*c*): $-4 + 2(i_c - i_e) + (i_c - i_a) = 0 \longrightarrow 3i_c - 2i_e = 6.$

In addition, applying KVL in mesh *e*, we get

- KVL(*e*): $-v_x + 2(i_e - i_c) + (i_e - i_d) + 2 = 0 \longrightarrow 2i_c - 3i_e = -1.$

Then, using the derived equations, we find $i_c = 4$ A and $i_e = 3$ A. Moreover, i_b can be found from

$$i_c - i_b = v_x/2 \longrightarrow i_b = 4 - 2 = 2\text{ A}.$$

Finally, the power of the voltage-dependent voltage source can be obtained as

$$p_s = v_x \times (-i_e) = 4 \times (-3) = -12\text{ W}.$$

Example 78: Consider the following circuit, previously studied using nodal analysis (see Exercise 15).

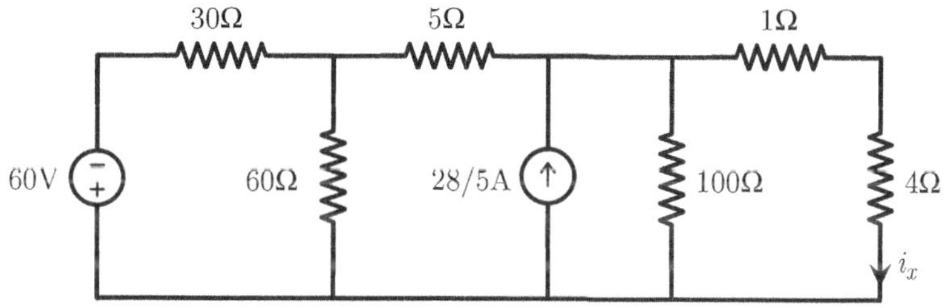

Find the value of i_x.

Solution: Using mesh analysis, we define four mesh currents as follows.

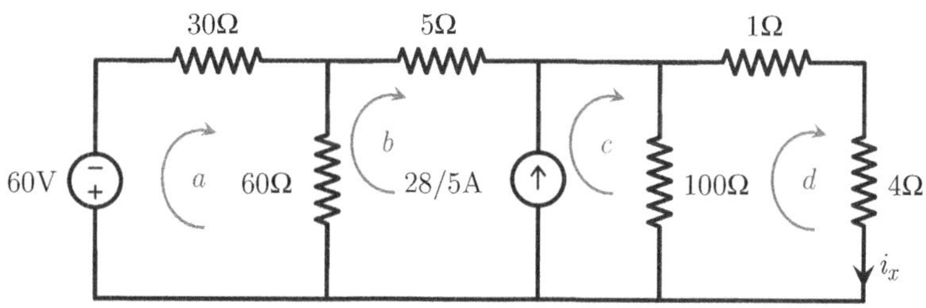

First, we note that $i_c - i_b = 28/5$. Applying KVL in meshes a and d, we obtain

- KVL(a): $60 + 30i_a + 60(i_a - i_b) = 0 \longrightarrow 3i_a - 2i_b = -2$,
- KVL(d): $100(i_d - i_c) + 5i_d = 0 \longrightarrow 20i_c = 21i_d$.

In addition, applying KVL in the supermesh involving b, c, and d, we have

- KVL(b&c&d): $60(i_b - i_a) + 5i_b + 5i_d = 0 \longrightarrow -12i_a + 13i_b + i_d = 0$.

Solving the equations yields $i_x = i_d = 16/5$ A.

Exercise 61: In the following circuit, previously solved using nodal analysis (see Example 43), find the value of i_x.

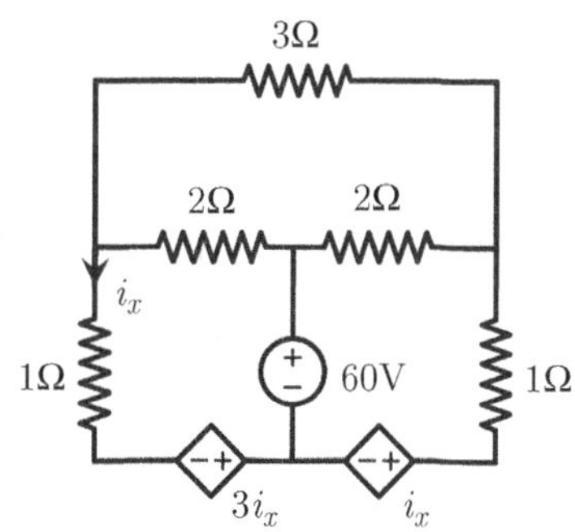

Exercise 62: In the following circuit, previously solved using nodal analysis (see Exercise 40), find i_x.

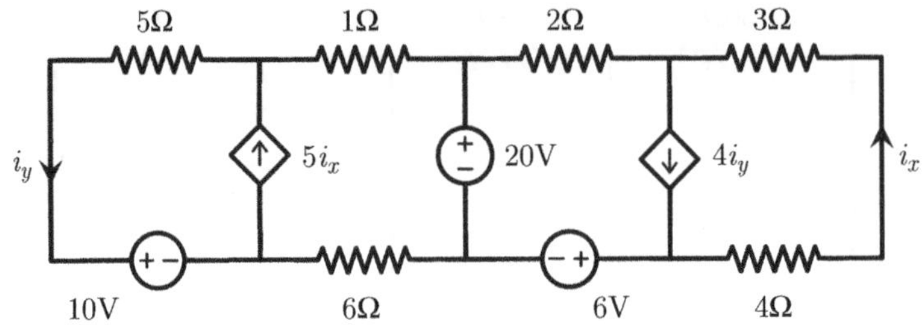

Exercise 63: In the following circuit, find the total power consumed by all resistors.

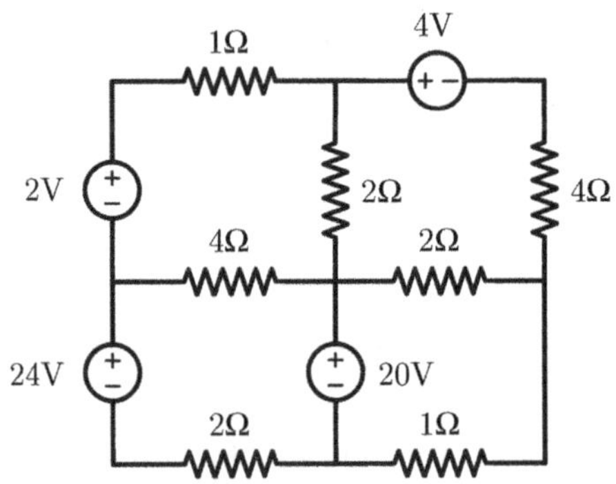

Exercise 64: In the following circuit, find the power of the 10 V source.

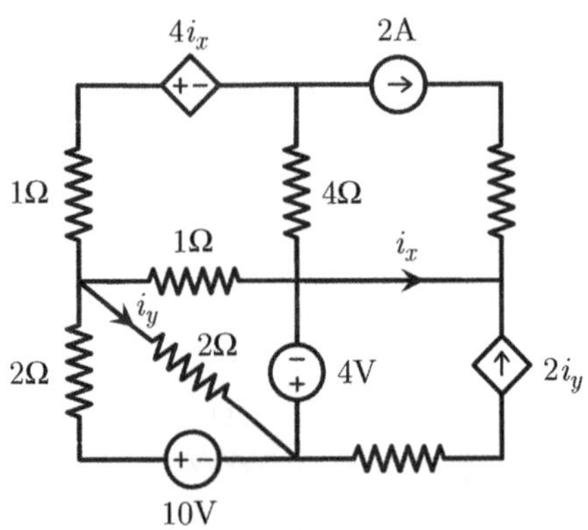

Exercise 65: In the following circuit, previously solved using nodal analysis (see Example 41), find i_x.

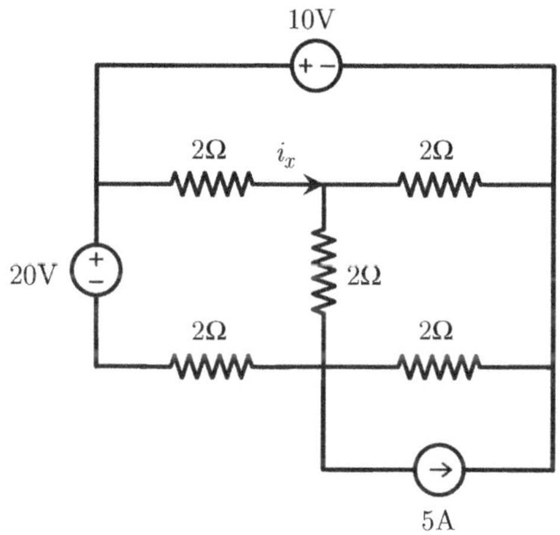

Exercise 66: In the following circuit, find v_x.

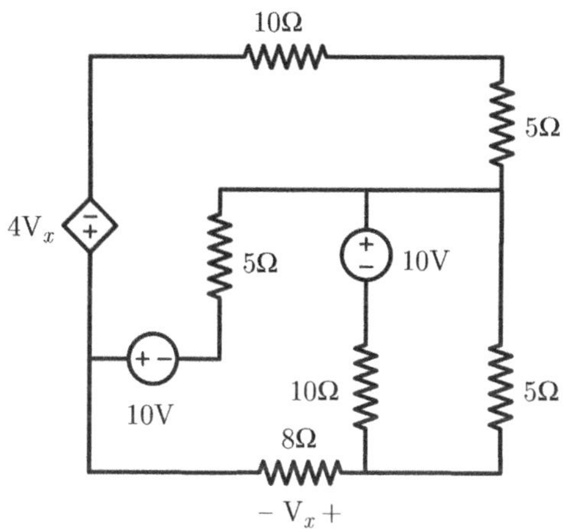

Exercise 67: In the following circuit, find i_x.

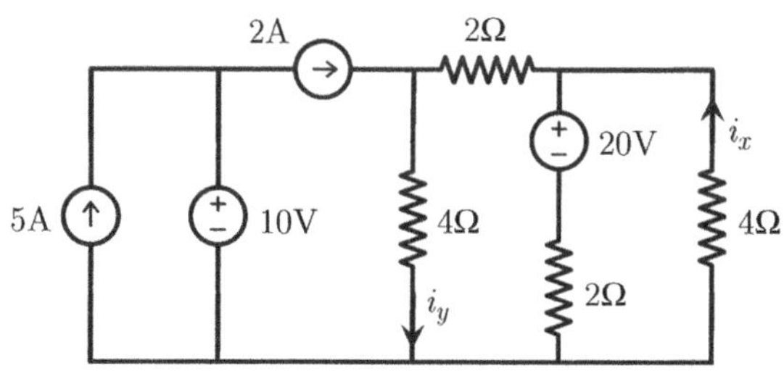

Exercise 68: In the following circuit, previously solved using nodal analysis (see Exercise 28), find the power of the 5 A source.

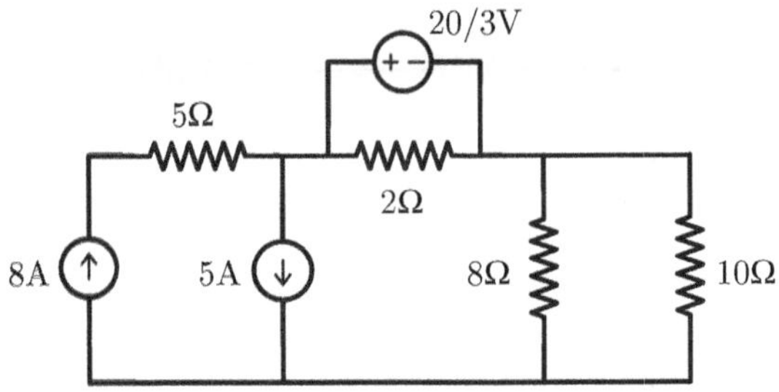

Exercise 69: In the following circuit, find the power of the voltage source.

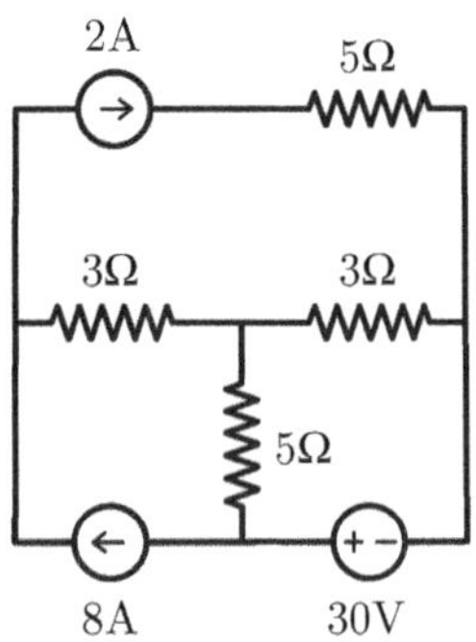

Exercise 70: In the following circuit, previously studied using nodal analysis (see Example 35), find the power of the 15 V source.

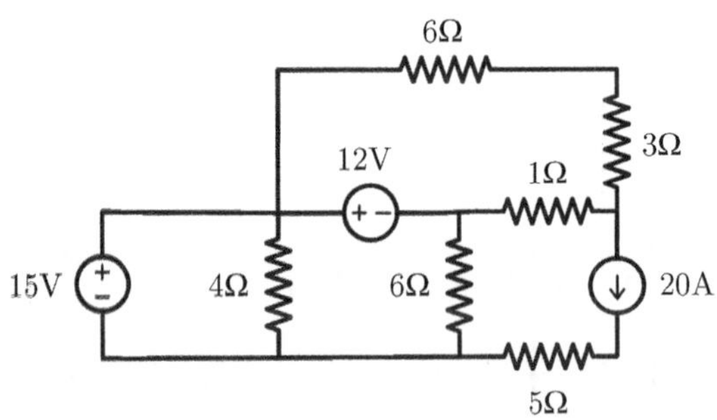

4.5 When Things Go Wrong with Mesh Analysis

We again consider some common issues and sources of confusion when dealing with circuits, now in the context of mesh analysis. One of the most common mistakes in mesh analysis is to assume zero voltage for a current source. This happens particularly when trying to apply KVL in a mesh containing a current source. As an example, consider the following circuit, previously solved using nodal analysis (see Example 37), but now to be analyzed via mesh analysis.

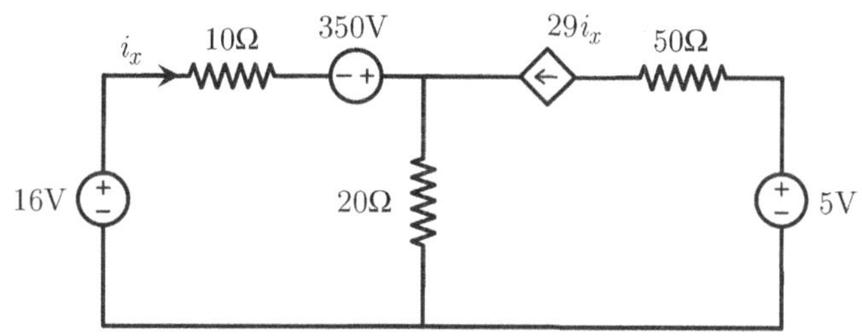

As usual, we can define two mesh currents as follows.

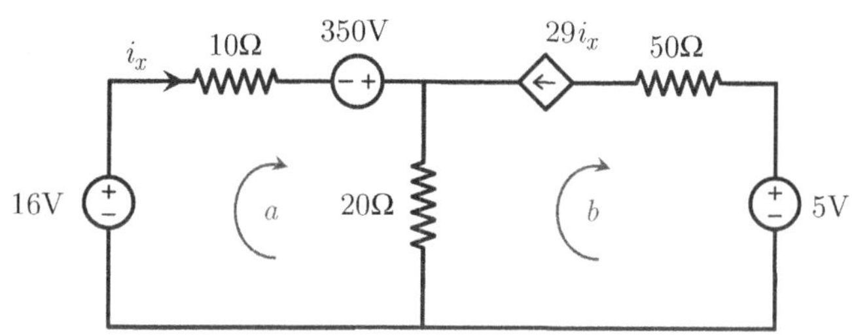

Then we have

- KVL(a): $-16 + 10i_a - 350 + 20(i_a - i_b) = 0 \longrightarrow 15i_a - 10i_b = 183$

as one of the equations to solve the mesh currents. However, the following equation is incorrect:

- KVL (b): $20(i_b - i_a) + 50i_b + 5 = 0$.

The correct version of KVL in mesh b is

- KVL(b): $20(i_b - i_a) - v_c + 50i_b + 5 = 0$,

where v_c is the voltage of the current-dependent current source in accordance with the sign convention. Unfortunately, this equation is not useful since it includes a new unknown v_c. Instead, one should deduce the link between the mesh currents by using the definition of the current source as

$$i_b = -29i_x = -29i_a.$$

Using this equation in the first one, we obtain $i_a = 3/5$ A and $i_b = -87/5$ A. Now, if needed, we can use this information to obtain the voltage of the dependent source as

- KVL(b): $20(i_b - i_a) - v_c + 50i_b + 5 = 0 \longrightarrow v_c = -280 - 870 + 5 = -1145$ V.

Obviously, it is not zero.

Confusion often occurs in mesh analysis when current sources are connected in parallel to components. Consider the following circuit.

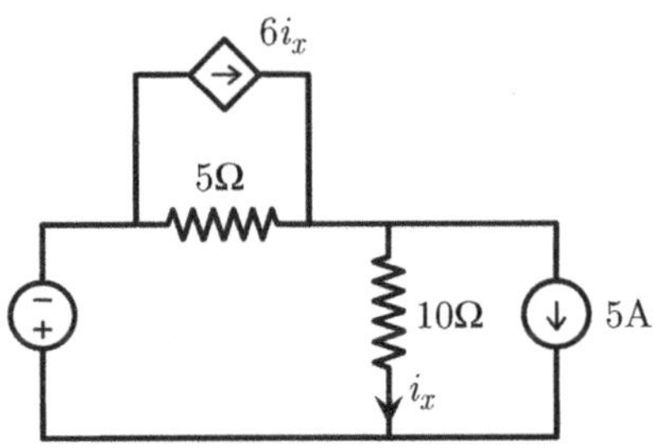

Our aim is to find the power delivered by the independent voltage source, if the power of the independent current source is given as 250 W. Formally, we must define three mesh currents as follows.

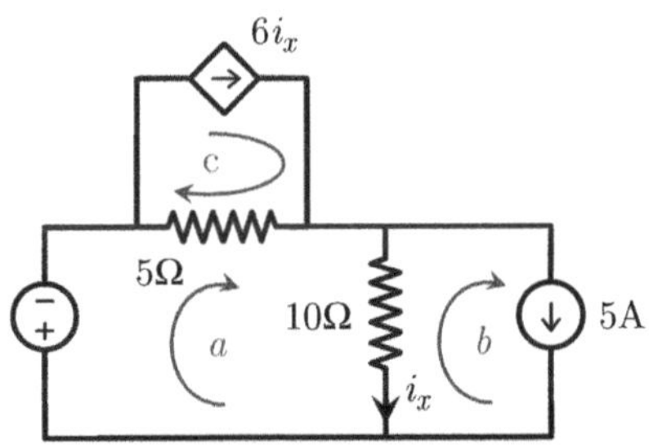

At this stage, it is not useful to apply a supermesh in this circuit. This is because none of the combinations *a*&*b*, *a*&*c*, or *a*&*b*&*c* contains a current source as an inner component. Applying KVL in any of these supermesh requires the voltage across a current source, which must be avoided. On the other hand, considering that the power of the independent current source is given as 250 W, we already have

$$i_x = 50/10 = 5\text{ A}.$$

In addition, considering the relationship between true and mesh currents, we obtain $i_b = 5$ A and $i_c = 6i_x = 30$ A. Finally, we have $i_a - i_b = i_x$, leading to $i_a = 35$ A. Therefore, we obtain the values of all mesh currents without using KVL. At this stage, KVL in mesh a can be used to find the value of the voltage source,

- KVL(a): $v_s + 5(i_a - i_c) + 10(i_a - i_b) = 0 \longrightarrow v_s = -325$ V.

Then the power of this source can be obtained as

$$p_s = v_s \times i_a = -325 \times 35 = 11\,375 \text{ W} = 11.375 \text{ kW}.$$

4.6 What You Need to Know before You Continue

Similar to nodal analysis, mesh analysis is a systematic way of analyzing circuits. Before moving on to the next chapter, we emphasize the following points.

- Equations in mesh analysis: Only KVL is used in mesh analysis, while KCL is usually not required. All equations are written by using mesh currents as unknowns, which are linearly independent. Once the analysis is completed, KCL may be applied to find the current through a component, if it is not already known.
- A supermesh, which involves a combination of multiple meshes and components between them, is often required to simplify mesh analysis. KVL is applied by considering the supermesh as a global loop with zero sum of voltages. The extra equation needed is obtained from the supermesh itself.

In the next chapter, we continue with a useful simplification method based on the application of the Thévenin and Norton equivalent circuits.

5

Black-Box Concept

Kirchhoff's laws (KVL and KCL) and their systematic applications in nodal and mesh analysis enable the solution of all circuits involving complex networks of electrical components. On the other hand, various simplification and transformation techniques may facilitate the analysis of a given circuit. For example, as discussed in Section 2.4, resistors that are connected in series and parallel can be combined, reducing the number of components, before the circuit is further analyzed. In this chapter, we discuss the black-box concept for the simplification of complex and crowded circuits into simple and neat ones. We particularly consider Thévenin and Norton equivalent circuits, which are suitable for simplifying linear circuits. These transformations are useful in various applications, such as finding load resistor values for maximum power transfer.

5.1 Thévenin and Norton Equivalent Circuits

Consider a complex circuit involving many components and two nodes (e.g., terminals). These terminals can be thought as input/output of the circuit to be used for a connection to another circuit or simply to a component. The circuit seen from these terminals can be simplified into two different circuits.

- Thévenin equivalent circuit: A circuit involving a series connection of a voltage source v_{th} and a resistor R_{th}.
- Norton equivalent circuit: A circuit involving a parallel connection of a current source i_{no} and a resistor R_{no}.

In the following, we find ways to derive the values of v_{th}, R_{th}, i_{no}, and R_{no}. Thus, instead of using the whole circuit, we represent it as a simple one, either a Thévenin circuit or a Norton circuit. After such a transformation and finding the equivalent simple circuit, we do not need to reconsider the actual components in the full circuit; hence, we consider it as a black box.

In order to use the Thévenin or Norton equivalence, the circuit to be simplified must be linear, that is, it must contain linear elements. A linear element has electrical properties that do not change with the applied voltage and/or current. Hence, a resistor or a source (independent or dependent) with a given fixed value can be considered as a linear element. In the next chapters, we also consider capacitors and inductors as linear elements with fixed capacitance and inductance values. In real life, all elements change behavior with the applied voltage and current values. However, these

Introduction to Electrical Circuit Analysis, First Edition. Özgür Ergül.

Companion Website: www.wiley.com/go/ergul4412

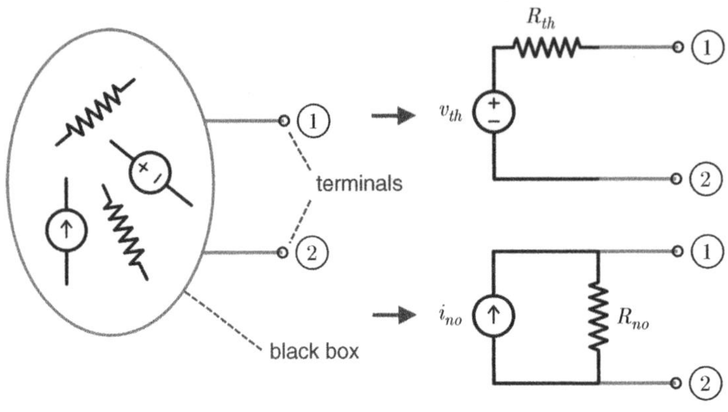

Figure 5.1 Thévenin and Norton equivalent circuits.

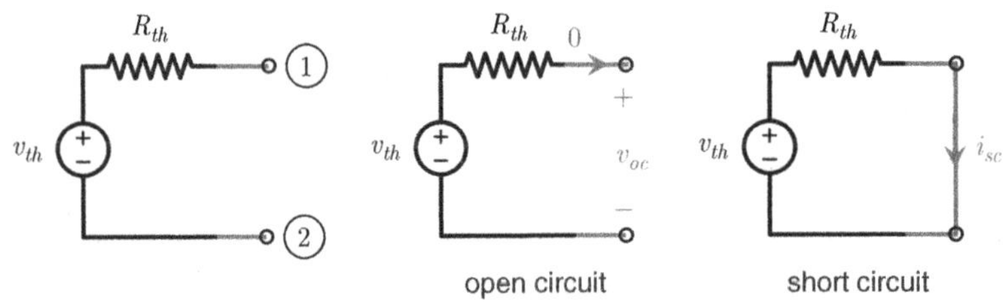

Figure 5.2 Open-circuit and short-circuit cases to find the values in a Thévenin equivalent circuit.

nonlinearities can be neglected in most cases. Some electrical components, such as transistors, demonstrate nonlinear behaviors that cannot be neglected. Even for such elements, linear approximations are commonly used depending on given conditions, for example, for small signal values.

We now discuss how to find the components in the equivalent circuits. Consider a Thévenin equivalent circuit, where the values of v_{th} and R_{th} need to be found. We can design two different experiments.

- Open circuit: We can measure the open-circuit voltage v_{oc}. Specifically, by making the output open circuit, no current passes through the resistor R_{th}. Therefore, the measured voltage v_{oc} must be equal to v_{th},

$$v_{\text{th}} = v_{\text{oc}}.$$

- Short circuit: We can measure the short-circuit current. Specifically, by making the output short circuit, a current i_{sc} passes through the closed loop,

$$i_{\text{sc}} = v_{\text{th}}/R_{\text{th}}.$$

Hence, the value of R_{th} can be obtained as

$$R_{\text{th}} = v_{\text{th}}/i_{\text{sc}} = v_{\text{oc}}/i_{\text{sc}}.$$

From a practical point of view, these two experiments can be performed in real life to simplify a circuit, even when there is no direct access inside the circuit (i.e., when it is a real black box). In circuit analysis, however, we perform these experiments

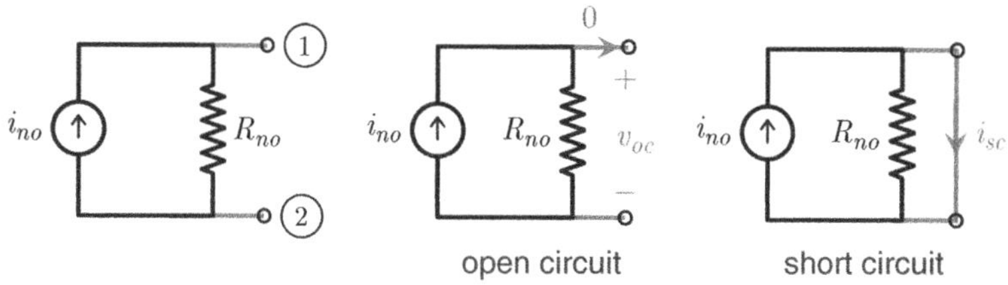

Figure 5.3 Open-circuit and short-circuit cases to find the values in a Norton equivalent circuit.

numerically, by solving the real circuit twice—once for the open-circuit case and once for the short-circuit case.

Next, we consider the derivation of the Norton circuit elements. In order to find the values of i_{no} and R_{no}, we again need two tests.

- Open circuit: Making the output open circuit, all current passes through the resistor R_{no}. Therefore, the measured voltage v_{oc} can be derived as

$$v_{oc} = i_{no}R_{no}.$$

- Short circuit: Making the output short circuit, no current passes through the resistor R_{no}. Then the measured current i_{sc} should be the same as i_{no},

$$i_{no} = i_{sc}.$$

Therefore, the value of R_{no} can be obtained as

$$R_{no} = v_{oc}/i_{no} = v_{oc}/i_{sc},$$

which is the same as R_{th}.

It can be observed that both Thévenin and Norton equivalent circuits can be obtained via two experiments, that is, solutions to two different problems: one with open-circuit output and another with short-circuit output. These can be considered as using test resistors with extreme values (0 and ∞). From these tests, one can find v_{oc}, i_{sc}, and $R_{th} = R_{no} = v_{oc}/i_{sc}$ which can be used to simplify the circuit for further analysis.

Example 79: Consider the following problem, with a large circuit involving various components and a resistor R_o.

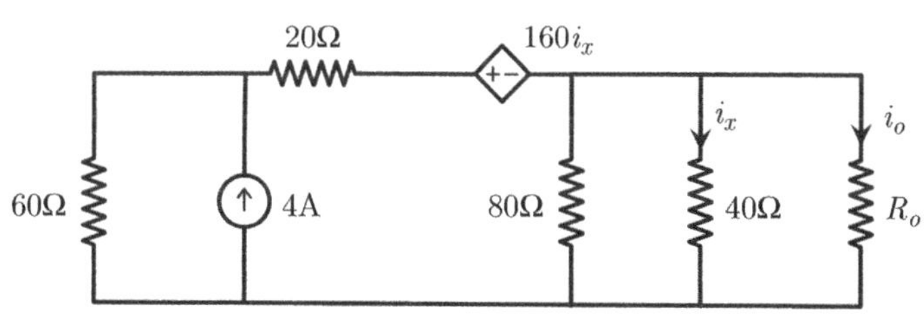

Find the current across R_o for different values of R_o, particularly for $R_o = \{5, 10, 20, 80, 290\}$ Ω.

Solution: Using a conventional approach, the overall circuit must be solved several (here five) times, where each solution corresponds to a given value of R_o. Obviously, if the large circuit (except R_o) could be represented by a simple one, that would be very helpful to solve the problems subsequently. In fact, Thévenin and Norton equivalence theorems enable this.

In order to solve the problem, we consider the circuit without R_o as follows.

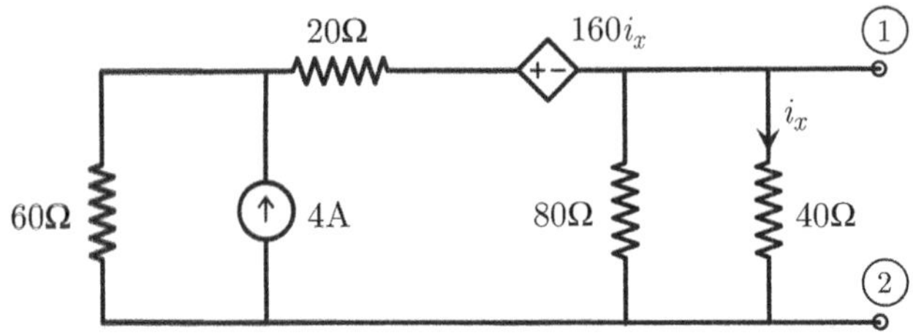

We can now find the open-circuit voltage by assuming an open circuit at the terminals.

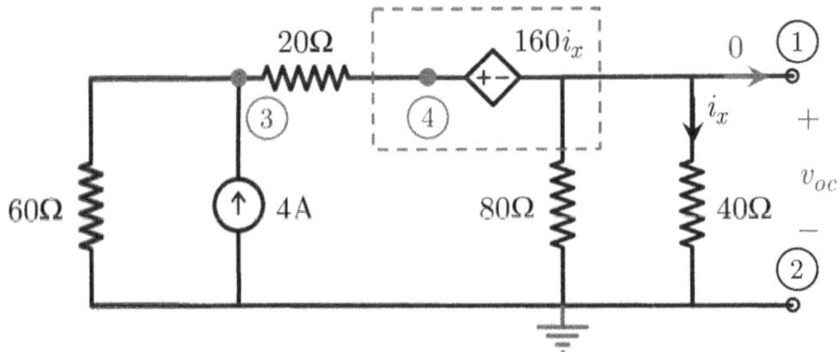

In the circuit above, which is solved via nodal analysis, we have $i_x = v_1/40$. Furthermore, applying KCL at node 3 and at the supernode, we derive

- KCL(3): $-v_3/60 + 4 - (v_3 - v_4)/20 = 0 \longrightarrow 4v_3 - 3v_4 = 240$,
- KCL(1&4): $(v_3 - v_4)/20 - v_1/80 - v_1/40 = 0 \longrightarrow 3v_1 - 4v_3 + 4v_4 = 0$.

Using $v_4 = 160i_x + v_1 = 5v_1$, one obtains $v_1 = 30$ V, $v_3 = 345/2$ V, $v_4 = 150$ V, and $v_{oc} = 30$ V.

Next, we consider the case when the circuit is terminated with a short circuit.

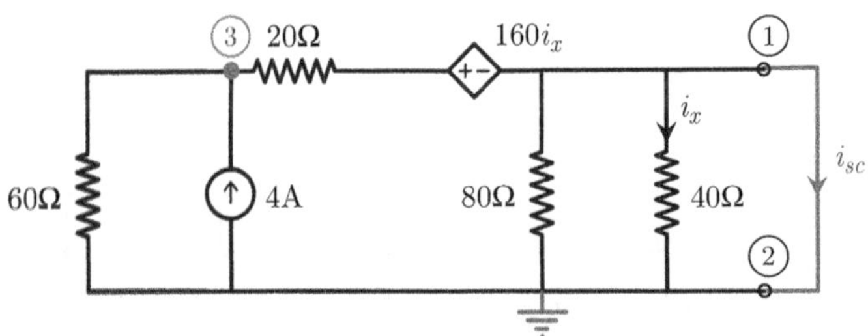

It must be well understood that this short-circuit version is a completely new circuit that must be analyzed separately. Specifically, currents and voltages across components may not be the same as those in the open-circuit case. Of course, there may be similarities if the same technique is used to solve circuits, but possible differences do not indicate an error in the solution. In the above, we have $v_1 = v_2 = 0$ V and $i_x = 0$ A, leading to

- KCL (3): $-v_3/60 + 4 - v_3/20 = 0 \longrightarrow v_3 = 60$ V.

Then we obtain $i_{sc} = 60/20 = 3$ A.

Finally, in order to obtain the Thévenin or Norton equivalent circuit, we calculate $R_{th} = R_{no}$ as

$$R_{th} = R_{no} = v_{oc}/i_{sc} = 10\ \Omega.$$

Then the Thévenin and Norton equivalent circuits are obtained as follows.

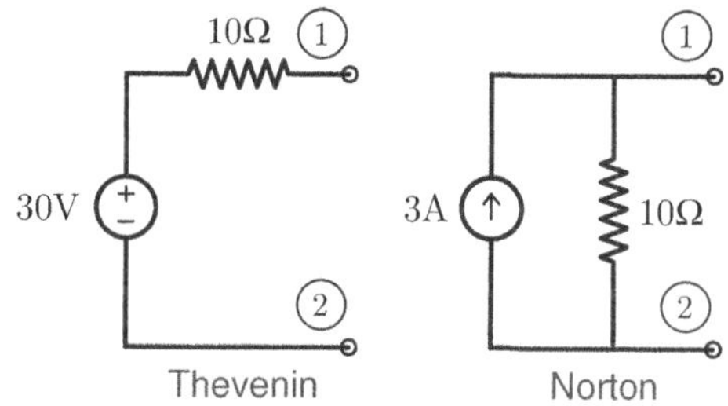

After representing the circuit with an equivalent one, we can be go back to the original scenario with R_o connected to the terminals, at which the equivalent circuits are found. A simplified representation of the circuit excluding R_o is called the equivalent circuit seen by R_o. Using the Thévenin and Norton equivalents, we have the following simplified circuits.

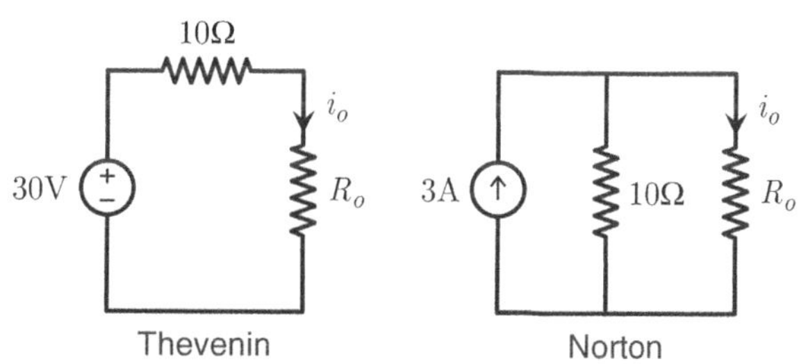

Both of these circuits can be used to solve the problem. Using the Thévenin equivalent circuit, we can find i_o for different values of R_o,

$$i_o = \frac{v_{th}}{R_{th} + R_o} = \frac{30}{10 + R_o}.$$

Alternatively, we can use the Norton equivalent circuit, leading to

$$i_o = i_{no}\frac{R_{no}}{R_{no} + R_o} = 3 \times \frac{10}{10 + R_o} = \frac{30}{10 + R_o},$$

which gives exactly the same expression.

Finally, we can list the current and power values for different resistance values as follows.

R_o (Ω)	5	10	20	80	290
$i_o = 30/(10 + R_o)$ (A)	2	3/2	1	1/3	1/10
$p_o = R_o i_o^2$ (W)	20	45/2	20	80/9	29/10

As shown above, the power of the load resistor R_o is maximized when its value is the same as $R_{th} = R_{no} = 10\ \Omega$. This is not a coincidence, as we discuss in detail under the rubric of maximum power transfer below.

Example 80: Consider the following circuit involving a large part connected to R_o.

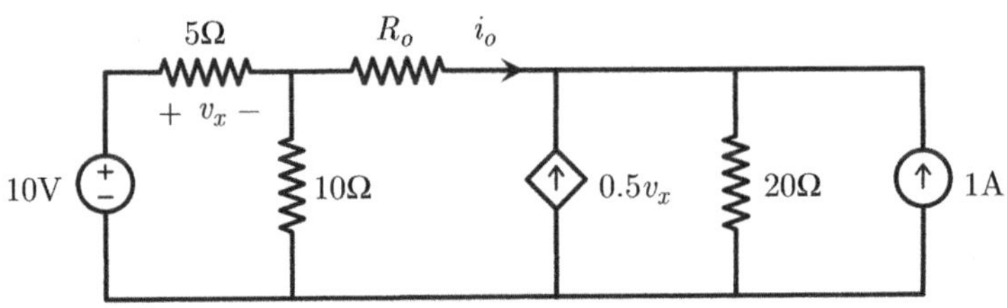

Find the Thévenin equivalent circuit seen by R_o.

Solution: In order to find the open-circuit voltage, we can use nodal analysis.

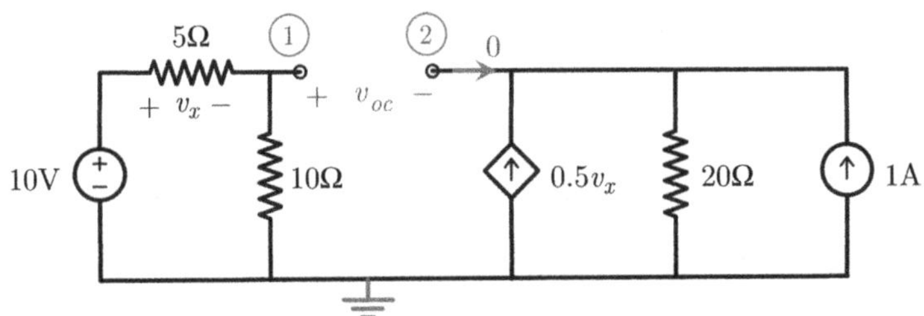

Since the circuit is already divided into two parts, we place the ground at the bottom and find one of the node voltages to be

$$v_1 = 10 \times 10/(10 + 5) = 20/3\ \text{V}.$$

In addition, $v_x = 10 - 20/3 = 10/3$ V. Using KCL at node 2, we get

- KCL(2): $0.5v_x - v_2/20 + 1 = 0 \longrightarrow v_2 = 160/3$ V.

Therefore,

$$v_{oc} = v_1 - v_2 = 20/3 - 160/3 = -140/3 \text{ V}.$$

Next, we consider the short-circuit case as follows.

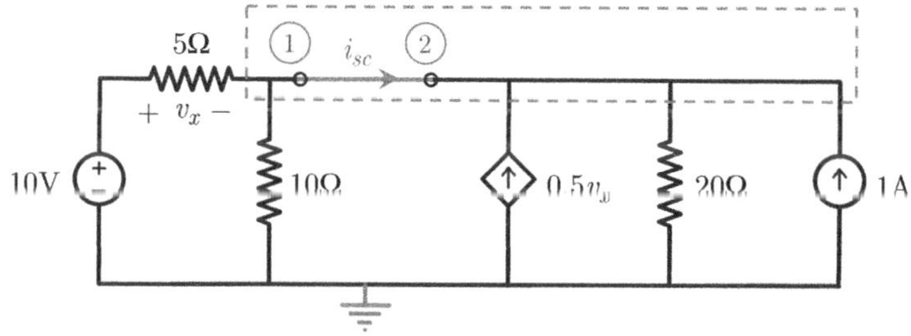

In this case, v_1 cannot be found easily as the nodes 1 and 2 are connected. If nodal analysis is used, we need to apply KCL at both nodes 1 and 2 to derive

- KCL(1&2): $(10 - v_1)/5 - v_1/10 + 0.5v_x - v_2/20 + 1 = 0$.

In addition, we have $v_1 = v_2$ and $v_x = 10 - v_1$, leading to

$$2 - v_1/5 - v_1/10 + 5 - v_1/2 - v_1/20 + 1 = 0 \longrightarrow 17v_1/20 = 8$$

or $v_1 = 160/17$ V. In order to find i_{sc}, we further apply KCL at node 1 to give

- KCL(1): $(10 - v_1)/5 - v_1/10 - i_{sc} = 0 \longrightarrow i_{sc} = 2 - 3v_1/10$,

leading to

$$i_{sc} = -14/17 \text{ A}.$$

Therefore, the Thévenin equivalent circuit involves

$$v_{th} = v_{oc} = -140/3 \text{ V},$$
$$R_{th} = v_{oc}/i_{sc} = 170/3 \ \Omega.$$

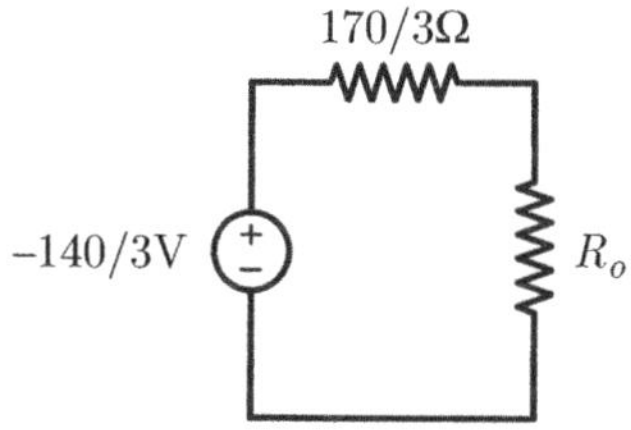

It should be emphasized that v_{oc} and i_{sc} should always follow the sign convention, which leads to positive R_{th} and R_{no}. Depending on the selection, however, v_{oc} and i_{sc} may or may not be positive, while their signs should be consistent.

Example 81: Consider the following circuit involving a large part connected to R_o.

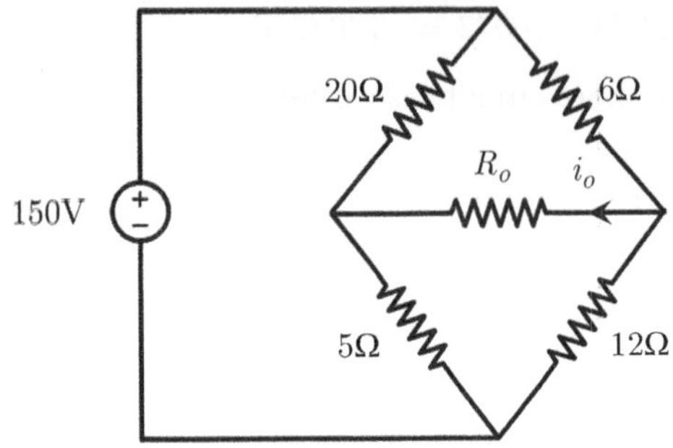

Find the Thévenin equivalent circuit seen by R_o.

Solution: In order to find the open-circuit voltage, we again prefer nodal analysis.

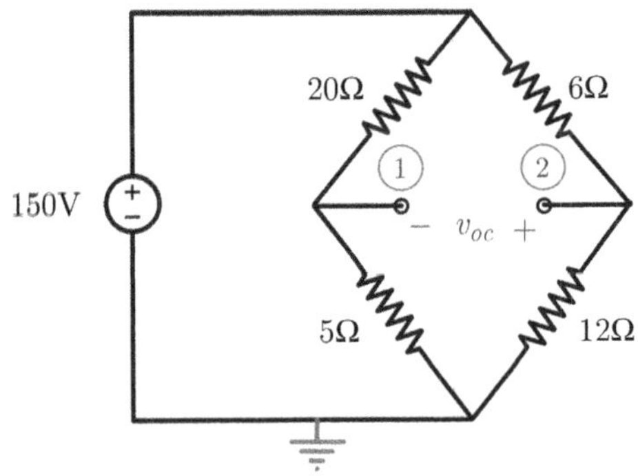

Interestingly, we obtain node voltages without recourse to KCL, but by simply considering voltage division,

$$v_1 = 150 \times 5/(5+20) = 30 \text{ V},$$

$$v_2 = 150 \times 12/(12+6) = 100 \text{ V}.$$

Therefore, we obtain the open-circuit voltage as

$$v_{\text{oc}} = v_2 - v_1 = 70 \text{ V}.$$

On the other hand, the short-circuit case is a bit more challenging.

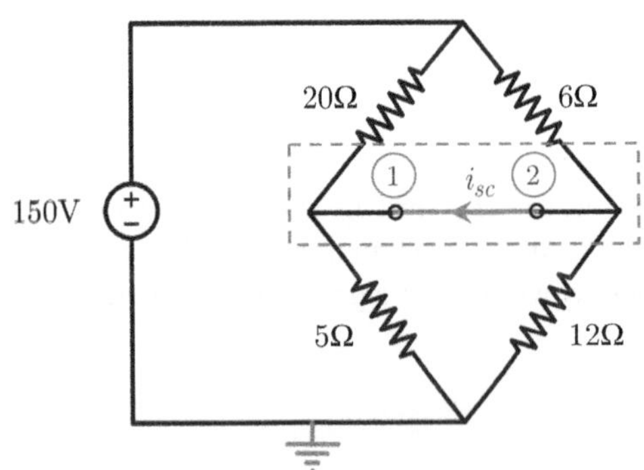

Connecting nodes 1 and 2 leads to equal voltages at these nodes, but KCL is required at the supernode. We have

- KCL(1&2): $(150 - v_1)/20 + (150 - v_2)/6 - v_1/5 - v_2/12 = 0$,

leading to

$$15/2 - v_1/20 + 25 - v_1/6 - v_1/5 - v_1/12 = 0 \longrightarrow v_1 = v_2 = 65 \text{ V}.$$

The current i_{sc} can only be found by using an extra KCL either at node 1 or node 2. Using node 2, we derive

- KCL(2): $(150 - v_2)/6 - i_{sc} - v_2/12 = 0$,

leading to

$$i_{sc} = 85/6 - 65/12 = 105/12 = 35/4 \text{ A}.$$

Therefore, we obtain

$$R_{th} = v_{oc}/i_{sc} = 8\ \Omega,$$

leading to the following simplified circuit.

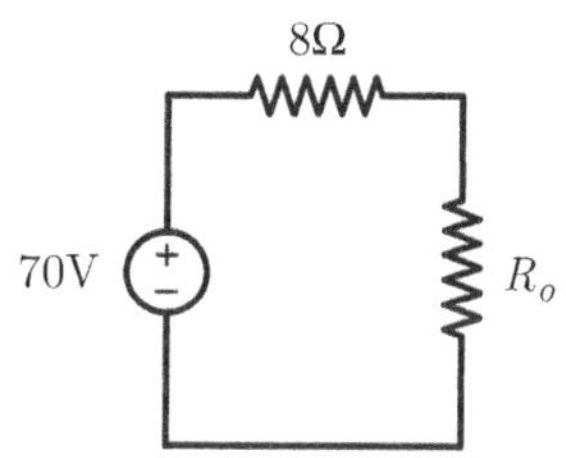

Example 82: Consider the following circuit involving a large part connected to R_o.

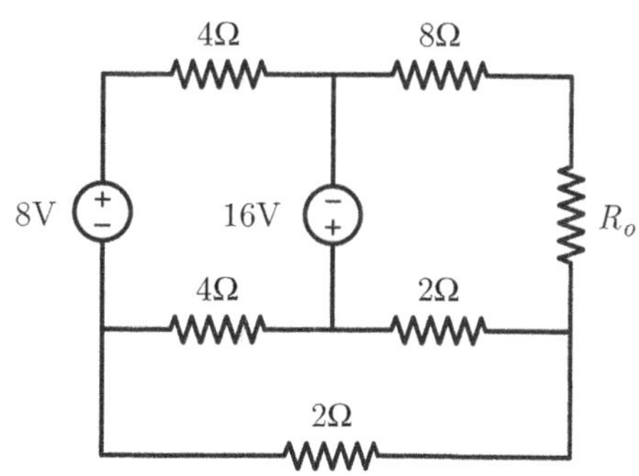

Find the Thévenin equivalent circuit seen by R_o.

Solution: First, we can find the open-circuit voltage using mesh analysis.

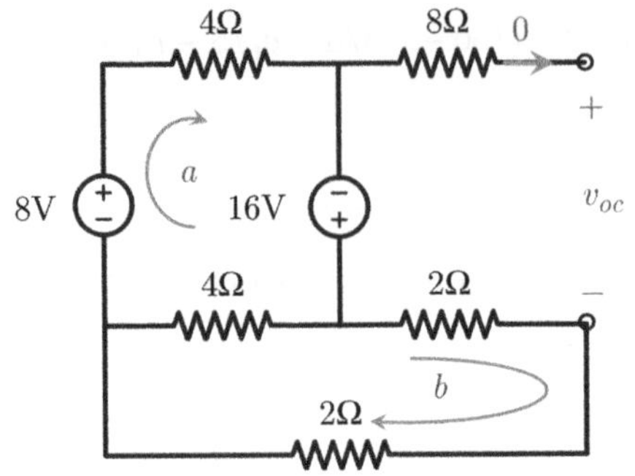

Applying KVL in meshes *a* and *b*, we obtain

- KVL(a): $-8 + 4i_a - 16 + 4(i_a - i_b) = 0 \longrightarrow 2i_a - i_b = 6$,
- KVL(b): $4(i_b - i_a) + 2i_b + 2i_b = 0 \longrightarrow i_a = 2i_b$.

Solving the equations, we get $i_a = 4\,\text{A}$, $i_b = 2\,\text{A}$, and $v_{\text{oc}} = -16 + 2i_b = -12\,\text{V}$.

Next, we consider the short-circuit case.

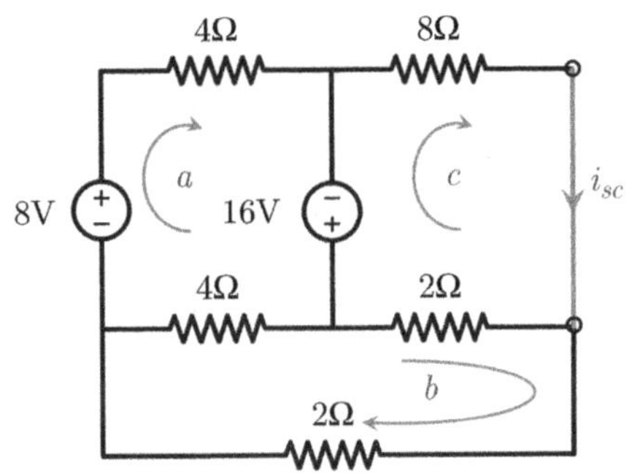

Applying KVL in meshes *a*, *b*, and *c*, we have

- KVL(a): $-8 + 4i_a - 16 + 4(i_a - i_b) = 0 \longrightarrow 2i_a - i_b = 6$,
- KVL(b): $4(i_b - i_a) + 2(i_b - i_c) + 2i_b = 0 \longrightarrow 2i_a + i_c - 4i_b = 0$,
- KVL(c): $16 + 8i_c + 2(i_c - i_b) = 0 \longrightarrow i_b - 5i_c = 8$.

We note that the equation for mesh *a* does not change in comparison to the open-circuit case, while the equation for mesh *b* does. Using substitution, we eliminate i_a, obtaining $i_c - 3i_b = -6$. Then, solving the equations, we obtain $i_b = 11/7\,\text{A}$, $i_c = -9/7\,\text{A}$, and $i_{\text{sc}} = i_c = -9/7\,\text{A}$. Overall, we have

$$v_{\text{th}} = v_{\text{oc}} = -12\text{ V},$$

$$R_{\text{th}} = v_{\text{oc}}/i_{\text{sc}} = 28/3\ \Omega.$$

The Thévenin equivalent seen by R_o, when combined with R_o, is as follows.

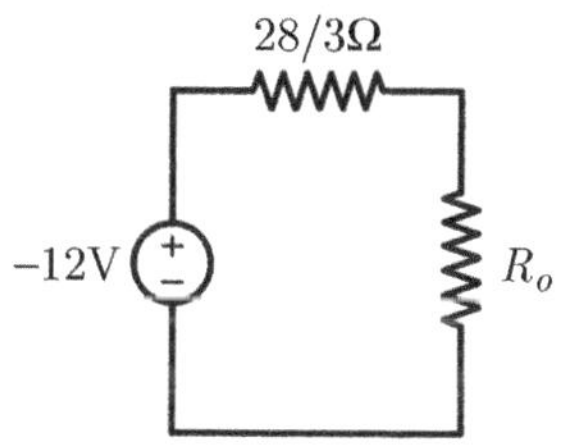

Example 83: Consider the following circuit.

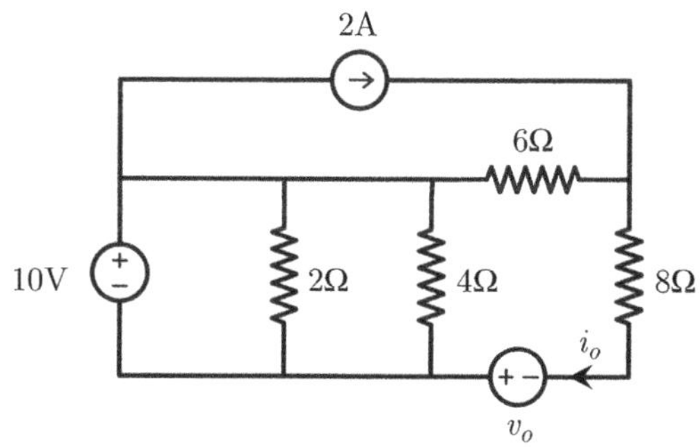

Find i_o if $5 \leq v_o \leq 15$.

Solution: We can solve this problem easily by finding the Thévenin equivalent circuit seen by the voltage source. First, we find the open-circuit voltage as follows.

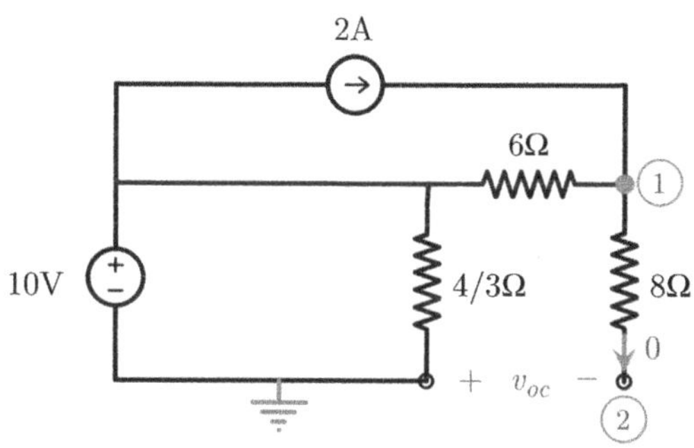

In the above, $v_1 = 10 + 6 \times 2 = 22$ V, since 2 A current passes through the 6 Ω resistor. Therefore, $v_{oc} = -v_2 = -v_1 = -22$ V.

For the short-circuit case, we have the following.

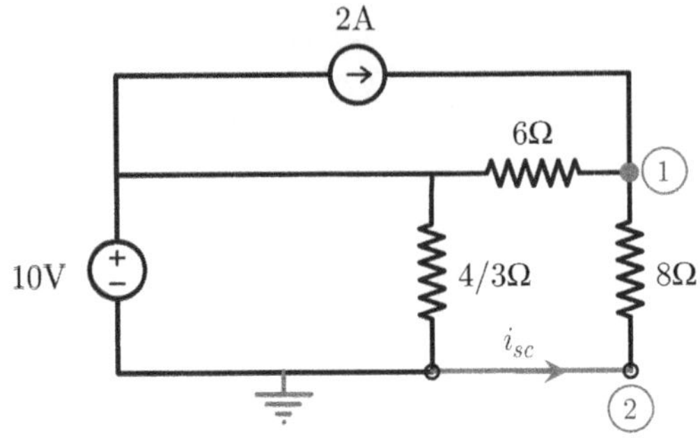

In this case, applying KCL at node 1, we have

- KCL(1): $2 + (10 - v_1)/6 - v_1/8 = 0 \longrightarrow v_1 = 88/7\,\text{V}$.

Considering that $v_2 = 0$ (it is now grounded), we find that

$$i_{sc} = (v_2 - v_1)/8 = -v_1/8 = -11/7\ \text{A}.$$

To sum up, we have

$$v_{\text{th}} = v_{\text{oc}} = -22\ \text{V},$$
$$R_{\text{th}} = v_{\text{oc}}/i_{\text{sc}} = 14\ \Omega.$$

Hence, using the Thévenin equivalence, we have the following simple circuit.

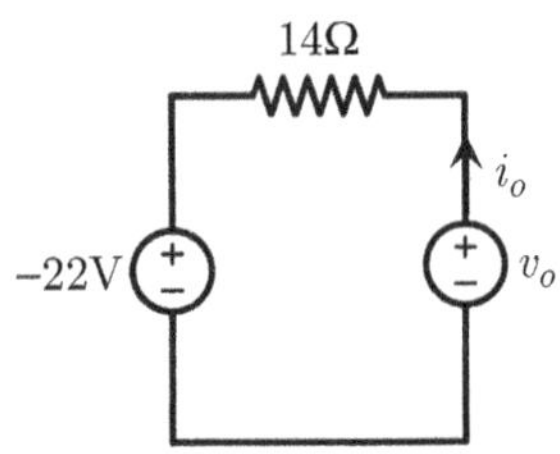

Then we obtain

$$i_o = (v_o + 22)/14$$

and the range of i_o is found to be

$$27/14 \leq i_o \leq 37/14.$$

Example 84: Consider the following circuit.

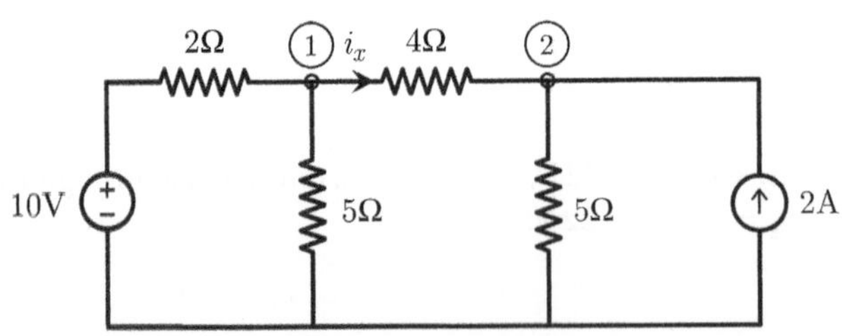

Simplify the circuit by finding the Thévenin equivalent seen by the 4 Ω resistor.

Solution: For the open-circuit case, we have the following circuit.

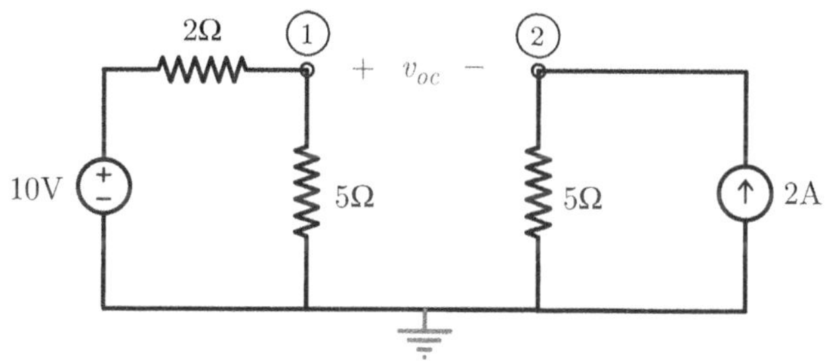

Applying nodal analysis, we derive

- $v_1 = 10 \times 5/7 = 50/7\,\text{V}$,
- $v_2 = 2 \times 5 = 10\,\text{V}$,

and $v_{oc} = 50/7 - 10 = -20/7\,\text{V}$.

Next, we consider the following short-circuit case.

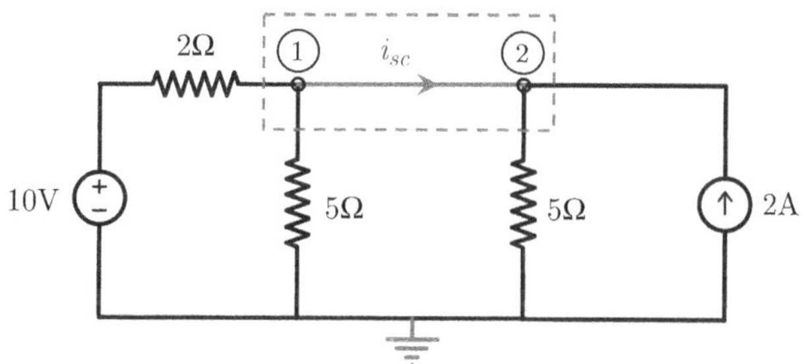

In this case, we have $v_1 = v_2$, and

- KCL(1&2): $(10 - v_1)/2 - v_1/5 - v_2/5 + 2 = 0 \longrightarrow v_1 = v_2 = 70/9\,\text{V}$,
- KCL(1): $(10 - 70/9)/2 - (70/9)/5 - i_{sc} = 0 \longrightarrow i_{sc} = -4/9\,\text{A}$.

Consequently, the equivalent resistance is found to be $R_{th} = 45/7\ \Omega$, leading to the following simplified circuit.

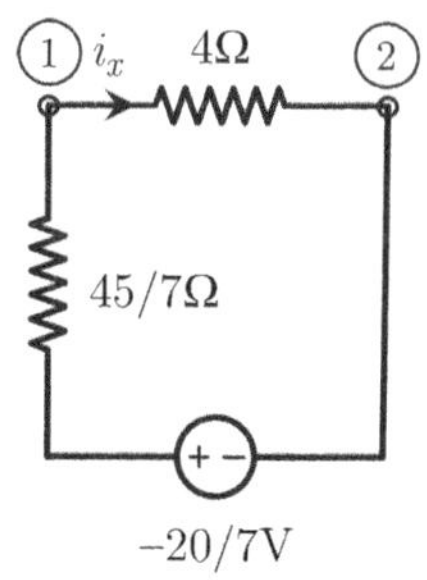

Example 85: Consider the following circuit.

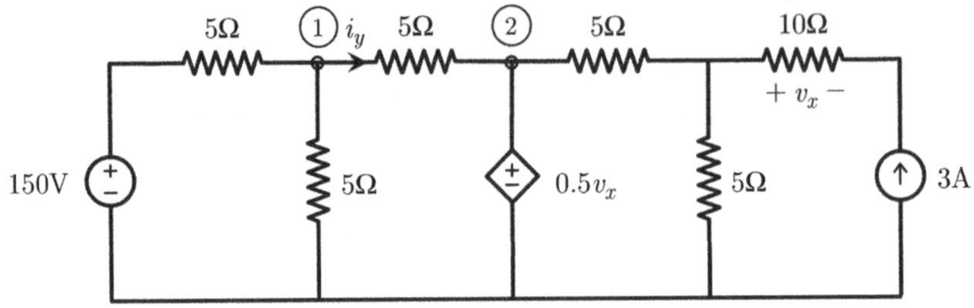

Simplify the circuit by finding the Thévenin equivalent seen by the 5 Ω resistor between nodes 1 and 2.

Solution: We first consider the open-circuit case as follows.

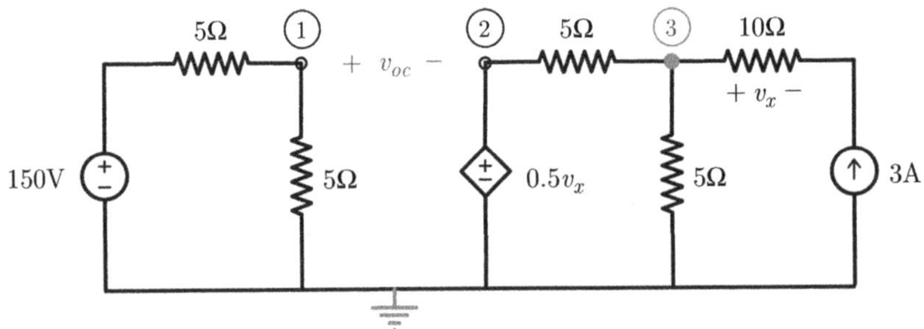

In the above, we note that $v_x = -3 \times 10 = -30$ V. Moreover, using nodal analysis, we have $v_1 = 150 \times 5/10 = 75$ V and $v_2 = 0.5v_x = -15$ V, leading to $v_{\text{oc}} = v_1 - v_2 = 90$ V. Then we consider the short-circuit case as follows.

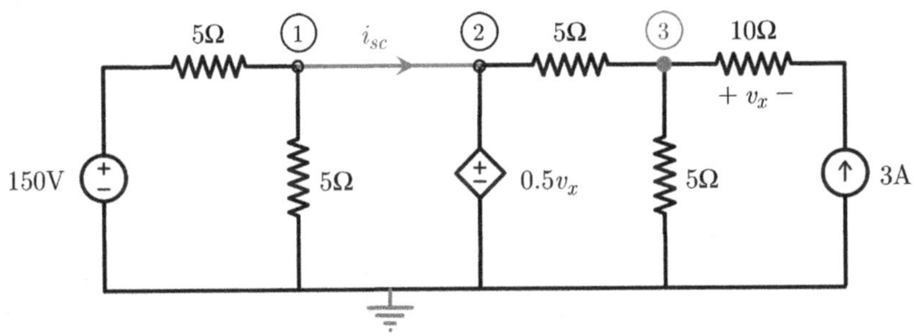

In this case, we still have $v_x = -3 \times 10 = -30$ V and $v_2 = 0.5v_x = -15$ V. On the other hand, due to the short circuit, $v_1 = v_2 = -15$ V and

- KCL(1): $(150 + 15)/5 + 15/5 - i_{\text{sc}} = 0 \longrightarrow i_{\text{sc}} = 36$ A.

Therefore, $R_{th} = 90/36 = 5/2\ \Omega$, and we obtain the following simple circuit.

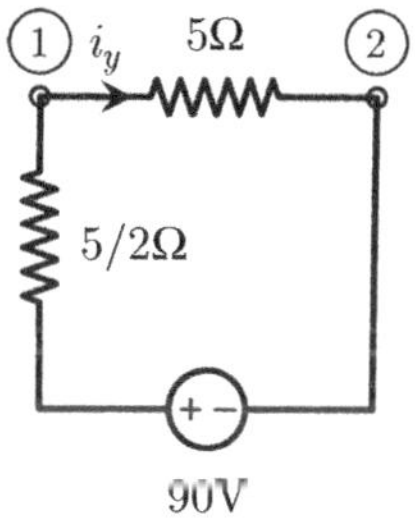

Example 86: Consider the following circuit.

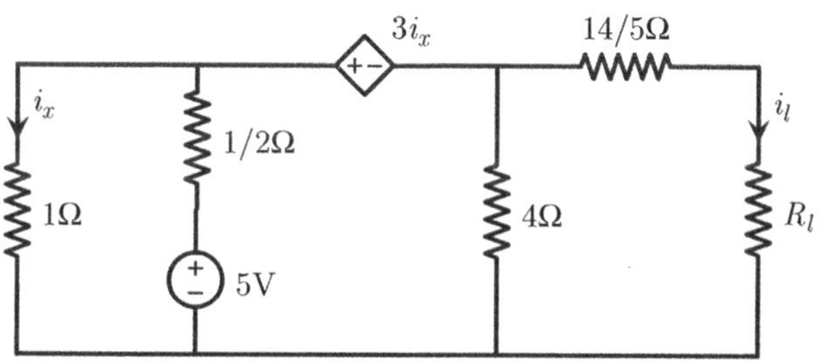

Find the power of R_l for its different values: 0, 1, 2, 4, 6, and 8 Ω.

Solution: Instead of solving the circuit six times or trying to find a general expression for the power of R_l, we can find the Thévenin equivalent circuit seen by this resistor. As usual, we consider the open-circuit case as follows.

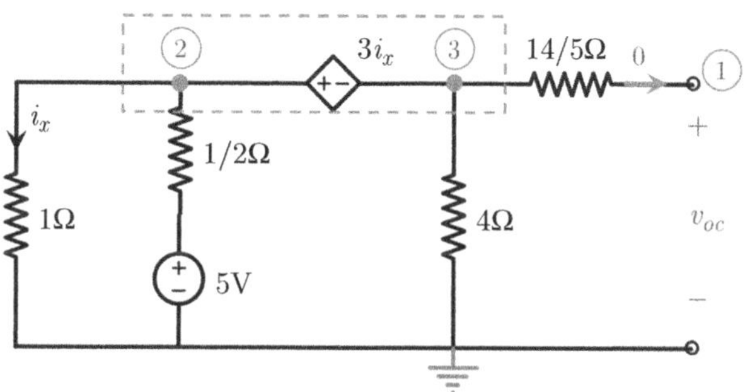

Applying nodal analysis, we derive $i_x = v_2$ and $v_2 - v_3 = 3i_x = 3v_2$, leading to $v_3 = -2v_2$. In addition, we derive

- KCL(2&3): $-v_2 - 2(v_2 - 5) - v_3/4 = 0 \longrightarrow 12v_2 + v_3 = 40$,

and we obtain $v_2 = 4$ V and $v_{oc} = v_3 = -8$ V.

Next, we consider the short-circuit case as follows.

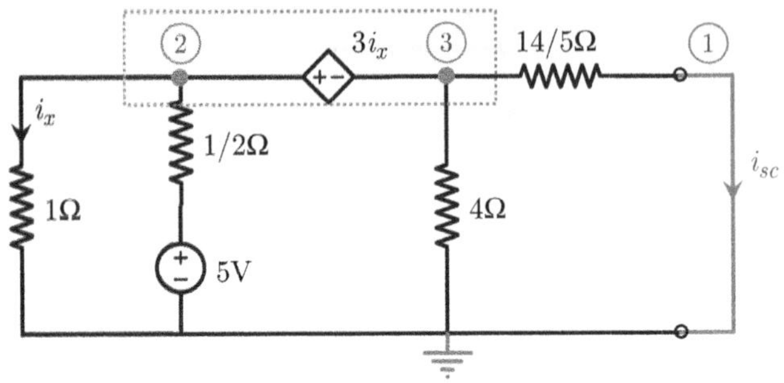

We still have $i_x = v_2$ and $v_3 = -2v_2$. However, in this case, $v_1 = 0$ V and

- KCL(2&3): $-v_2 - 2(v_2 - 5) - v_3/4 - v_3/(14/5) = 0 \longrightarrow 84v_2 + 17v_3 = 280.$

Hence, we find that $v_2 = 28/5$ V, $v_3 = -56/5$ V, and $i_{sc} = (-56/5)/(14/5) = -4$ A. This means that $R_{th} = 2\ \Omega$, and the simplified circuit can be drawn as follows.

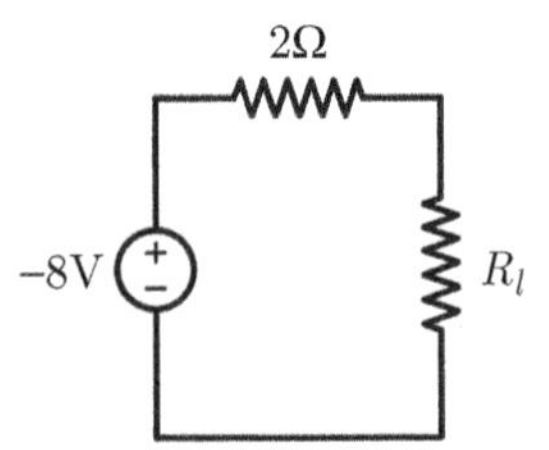

Finally, the power of the resistor can be listed as follows.

R_l (Ω)	0	1	2	4	6	8
i_l (A)	−4	−8/3	−2	−4/3	−1	−4/5
p_l (W)	0	64/9	8	64/9	6	128/25

Example 87: Consider the following circuit.

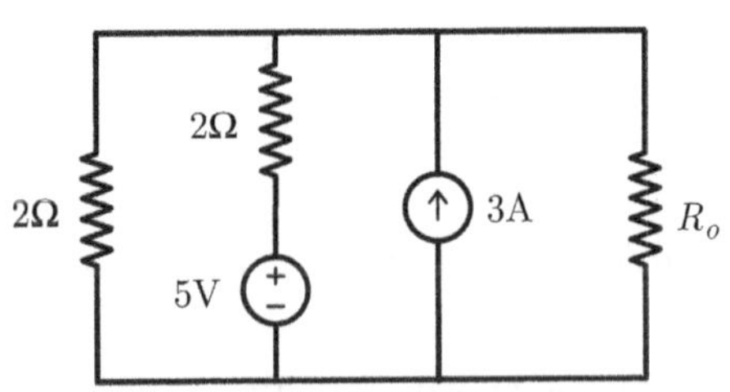

Find the power of R_o as a function of its resistance.

Solution: In this case, we can again use the Thévenin theorem to simplify the circuit. The open-circuit case is as follows.

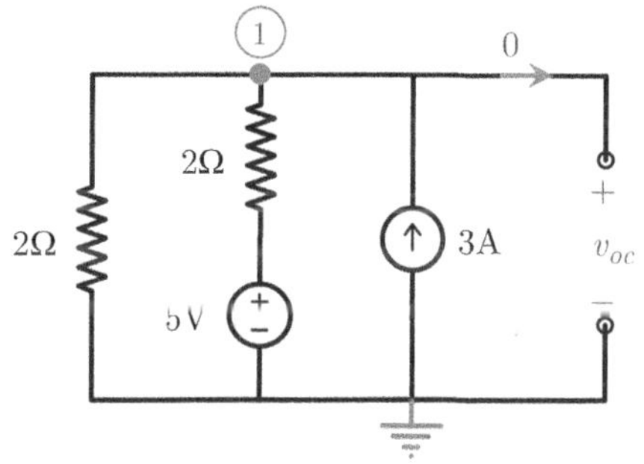

We easily derive

- KCL(1): $-v_1/2 + (5 - v_1)/2 + 3 = 0 \longrightarrow v_1 = 11/2\,\text{V}$

and $v_{oc} = v_1 = 11/2\,\text{V}$.

Next, the short-circuit case is as follows.

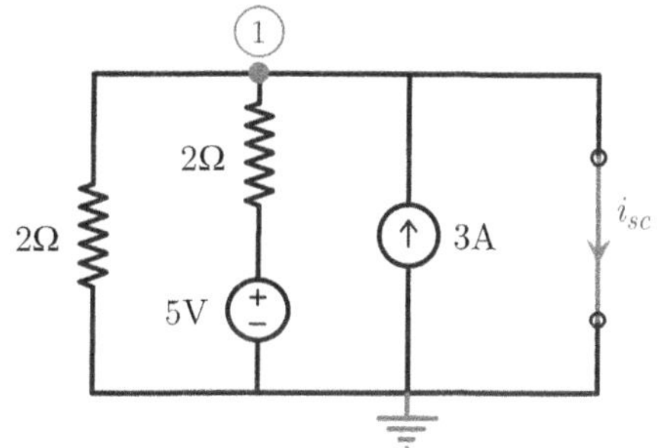

Now, we have $v_1 = 0\,\text{V}$ and

- KCL(1): $5/2 + 3 - i_{sc} = 0 \longrightarrow i_{sc} = 11/2\,\text{A}$.

Then, finding $R_{th} = 1\ \Omega$, we have the following simplified circuit.

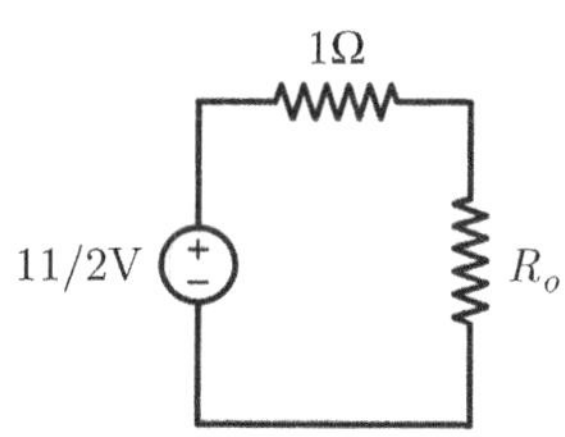

Obviously, the current through the circuit is found to be $i_o = 11/(2 + 2R_o)$. Therefore, we obtain

$$p_o = R_o i_o^2 = \frac{121R_o}{4(1 + R_o)^2}.$$

Exercise 71: In the following circuit, find the Thévenin equivalent circuit seen from terminals 1 and 2.

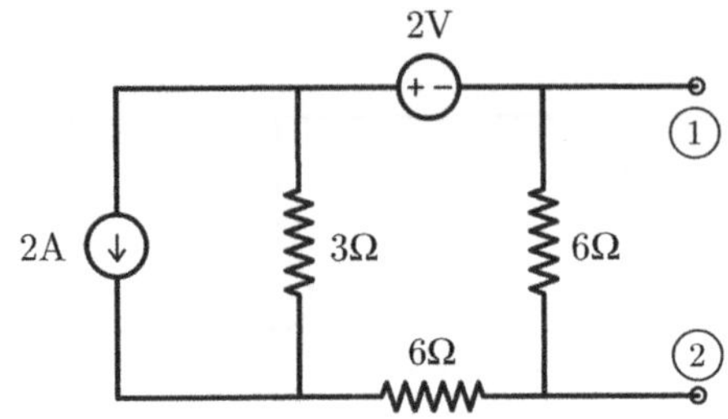

Exercise 72: In the following circuit, find the Thévenin equivalent circuit seen from terminals 1 and 2.

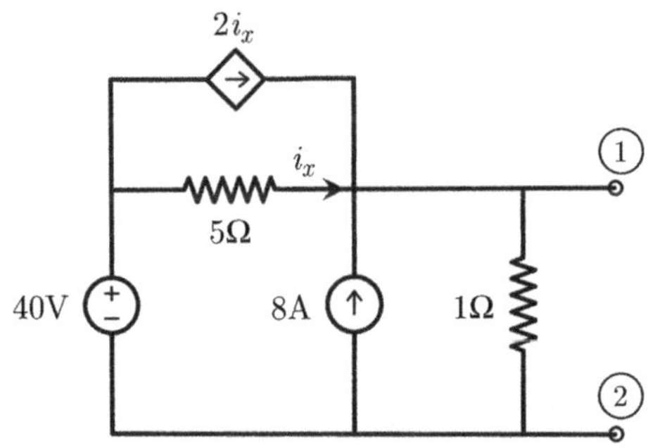

Exercise 73: Simplify the following circuit by finding the Thévenin equivalent seen by the 12 Ω resistor.

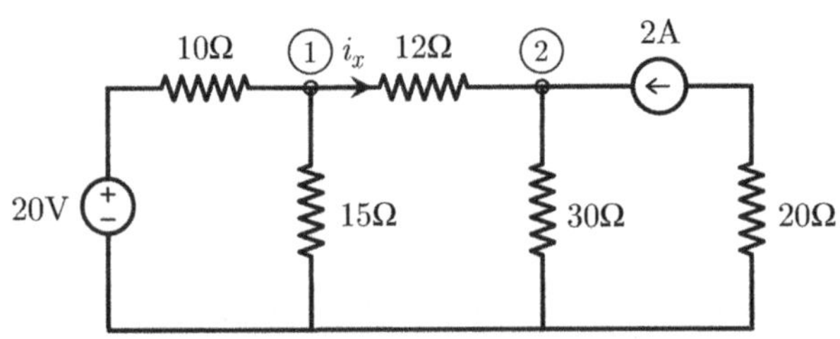

Exercise 74: In the following circuit, find the Thévenin equivalent circuit seen from terminals 1 and 2.

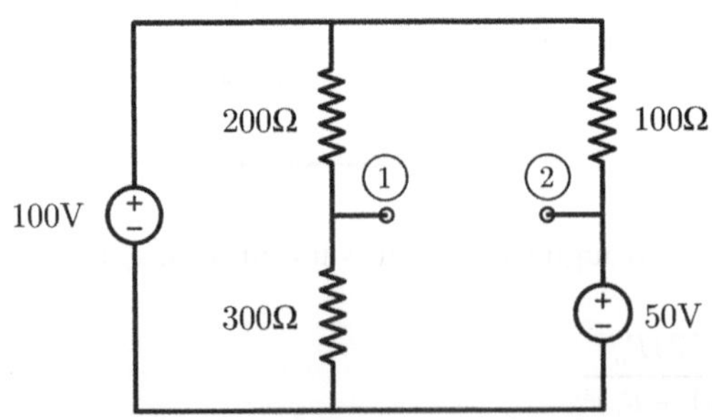

Exercise 75: Simplify the following circuit by finding the Thévenin equivalent seen by the 10 Ω resistor.

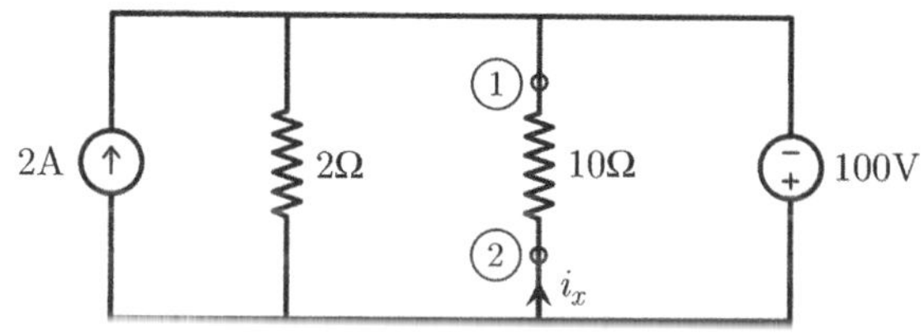

Exercise 76: Simplify the following circuit by finding the Thévenin equivalent seen by the 8 Ω resistor.

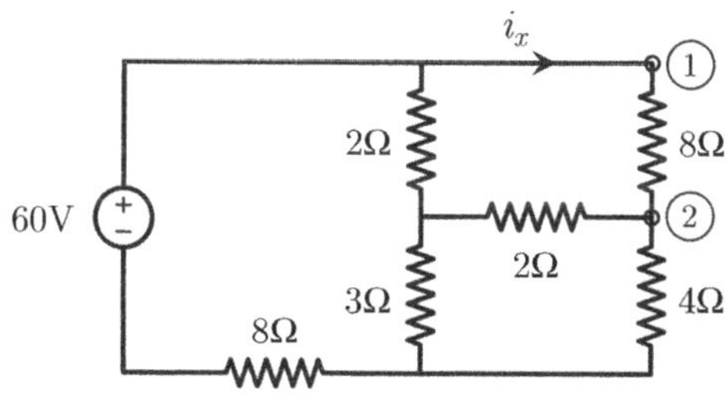

Exercise 77: In the following circuit, find the Thévenin equivalent circuit seen by the voltage source v_o.

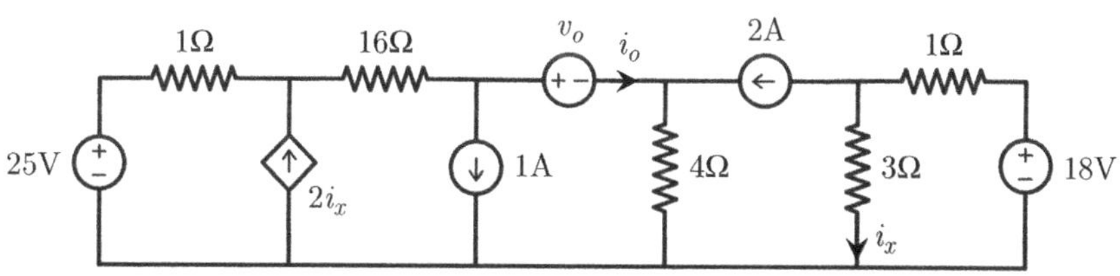

Exercise 78: In the following circuit, find the Thévenin equivalent circuit seen by R_l.

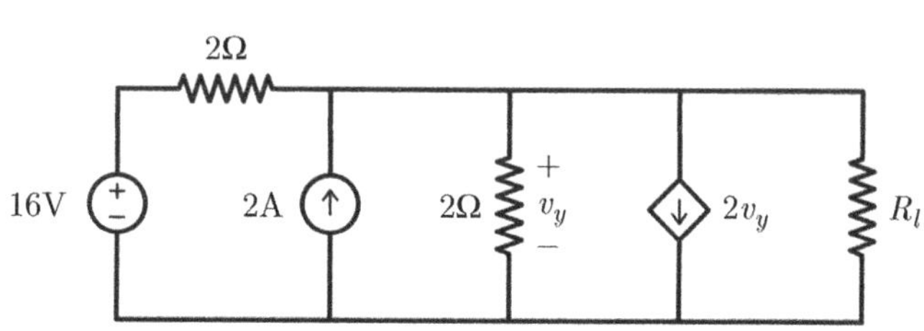

Exercise 79: In the following circuit, find the Thévenin equivalent circuit seen by R_o.

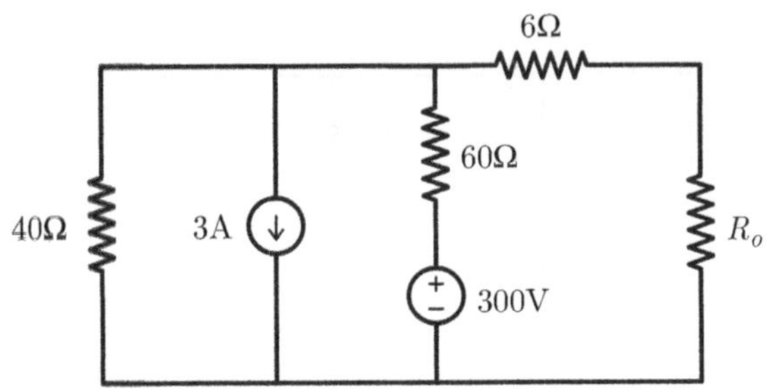

Exercise 80: In the following circuit, find the Thévenin equivalent circuit seen by R_l.

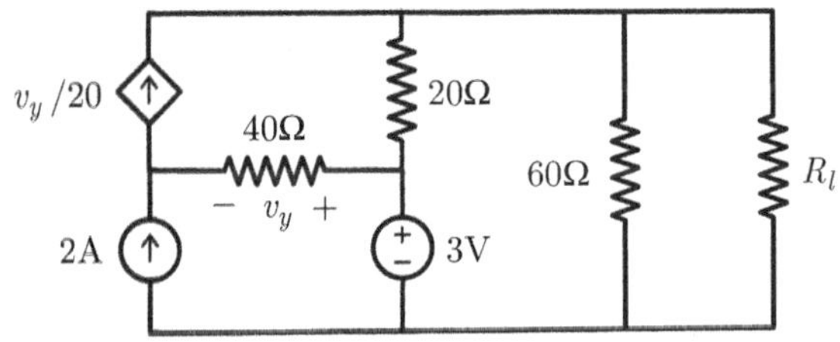

5.2 Maximum Power Transfer

In electrical and electronic engineering, it is often required to combine different circuits. While such combination seems trivial (just connect the nodes that need to be connected!), a good connection needs to be a good match between the circuits. In most cases, a good match is one that provides the maximum power transfer from one circuit to another. As an example, consider a simple case involving a voltage source and resistor (circuit 1) that are connected in series to another resistor (circuit 2).

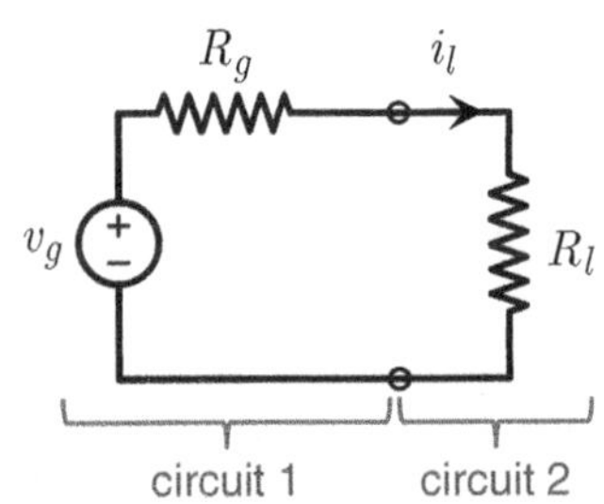

We would like to match circuits 1 and 2. Specifically, we would like to find the value of R_l such that the power consumed by this resistor is maximized. This can be interpreted as the maximum power transfer from circuit 1 to circuit 2. Obviously, both $R_l = 0$ (short circuit) and $R_l = \infty$ (open circuit) lead to $p_l = 0$.

We start with a general expression for the current,

$$i_l = \frac{v_g}{R_g + R_l}.$$

Therefore, the power of the resistor can be written as

$$p_l = R_l i_l^2 = \frac{R_l v_g^2}{(R_g + R_l)^2}.$$

Taking the derivative with respect to R_l and making it equal to zero,

$$\frac{\partial p_l}{\partial R_l} = \frac{v_g^2}{(R_g + R_l)^2} - \frac{2R_l v_g^2}{(R_g + R_l)^3} = \frac{v_g^2(R_g - R_l)}{(R_g + R_l)^3} = 0,$$

we obtain

$$R_l = R_g.$$

Hence, the best selection occurs by setting R_l equal to R_g, leading to

$$p_l^{\max} = \frac{R_g v_g^2}{4R_g^2} = \frac{v_g^2}{4R_g}.$$

We note that, in this matched case, R_g consumes the same amount of power.

In practice, more complex circuits (compared to the example above) need to be matched. Nevertheless, if the circuits are linear, they can be converted into Thévenin equivalent models such that matching corresponds to selecting the equivalent resistors equally. In the following examples, we consider various cases involving a resistor to be matched to a complex circuit.

Example 88: Consider the following circuit, which has previously been studied (see Example 86).

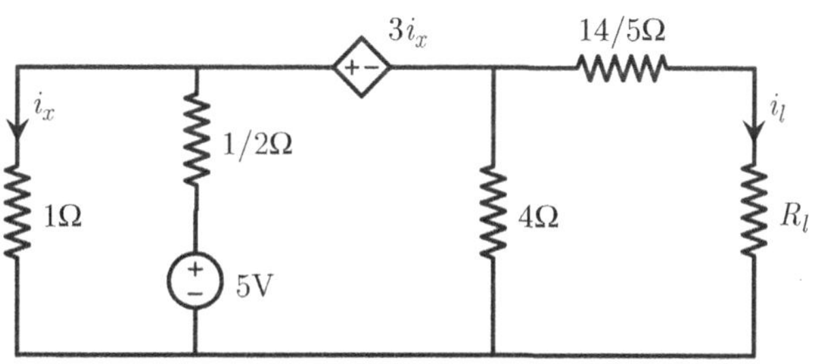

The power of R_l for different values was found and listed. The Thévenin equivalent circuit seen by R_l is also found as follows.

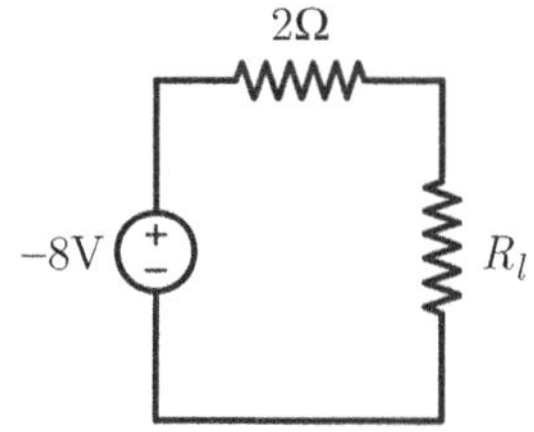

Obviously, the selection $R_l = 2\ \Omega$ maximizes the power transferred to this load. This power was found to be 8 W, the maximum among all trials.

Example 89: Consider the following circuit involving a resistor R_l.

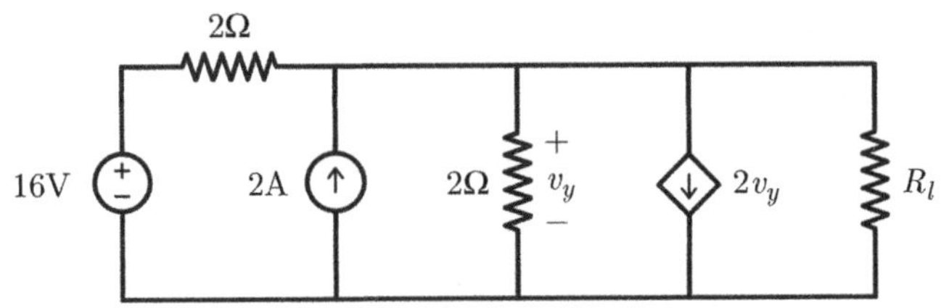

The Thévenin equivalent circuit seen by R_l was previously found to be as follows (see the solution to Exercise 78).

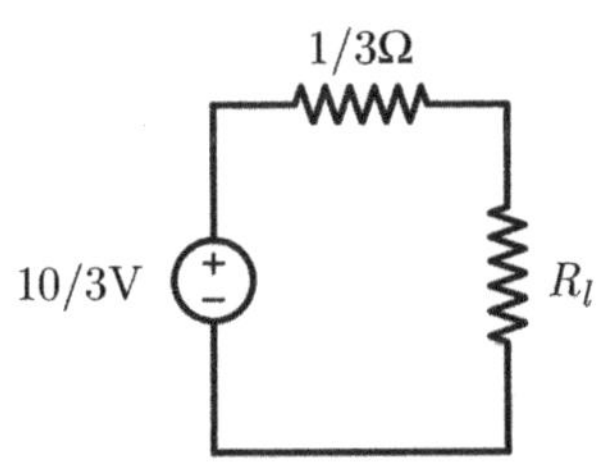

Therefore, in order to maximize the power transfer, one needs to select $R_l = 1/3\ \Omega$. In this case, the voltage $v_{oc} = 10/3$ V is divided equally among the resistors so that

$$v_l = v_{oc}/2 = 5/3\ \text{V}$$

and

$$p_l^{\max} = \frac{(5/3)^2}{1/3} = \frac{25}{3}\ \text{W}.$$

Example 90: In the following circuit, R_o again needs to be selected to maximize the power transfer.

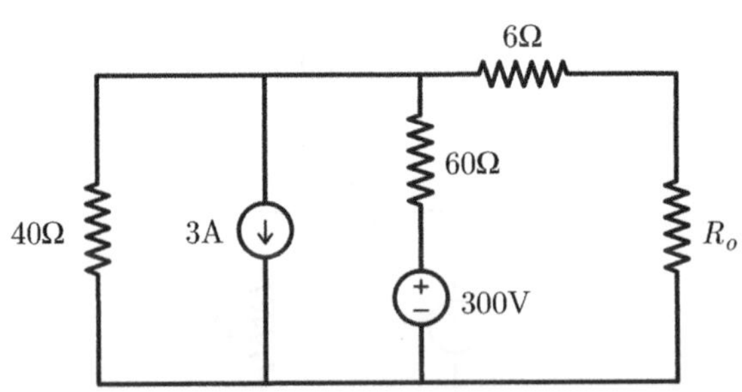

The Thévenin equivalent circuit seen by R_o was previously found to be as follows (see the solution to Exercise 79).

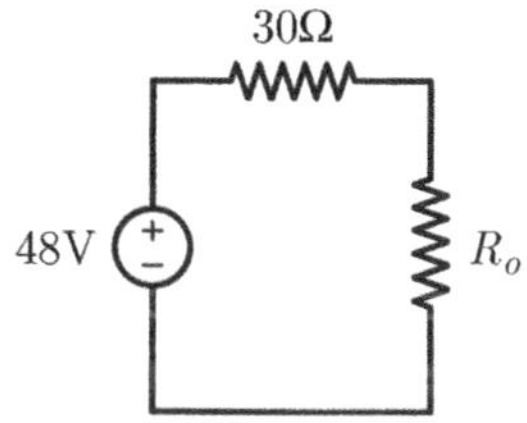

Then one must select $R_o = 30\ \Omega$, leading to

$$v_o = v_{oc}/2 = 24\ \text{V}$$

and

$$p_o^{\max} = \frac{24^2}{30} = \frac{96}{5}\ \text{W}.$$

One can check these voltage and power values as follows.

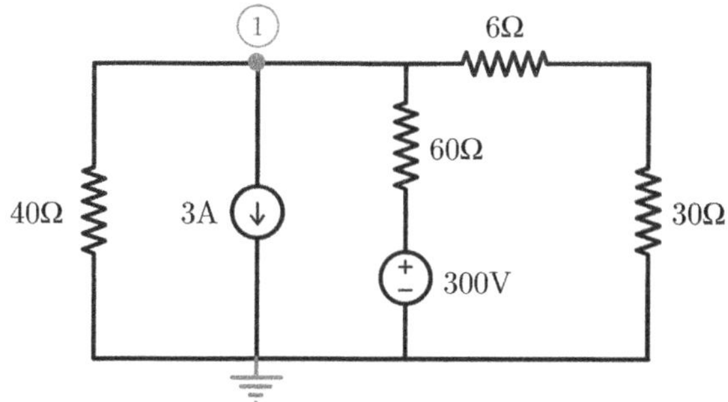

Applying KCL at node 1, we have

- KCL (1): $-v_1/40 - 3 - (v_1 - 300)/60 - v_1/36 = 0$,

leading to $v_1 = 144/5$ V. Then the voltage across the 30 Ω load resistor can be derived as

$$v_o = \frac{30}{36}v_1 = \frac{5}{6} \times \frac{144}{5} = 24\ \text{V},$$

as expected.

Example 91: Consider the following circuit.

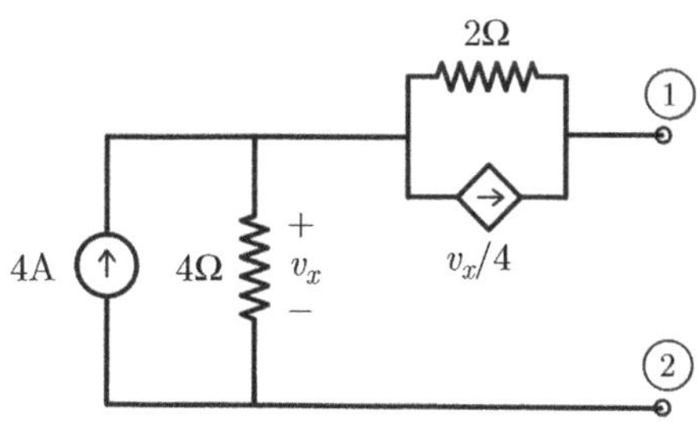

Find the value of the resistor to be connected between 1 and 2 for maximum power transfer.

Solution: In order to find the value of the resistor, we can first find the Thévenin equivalent circuit seen from 1 and 2. First, we consider the open-circuit case as follows.

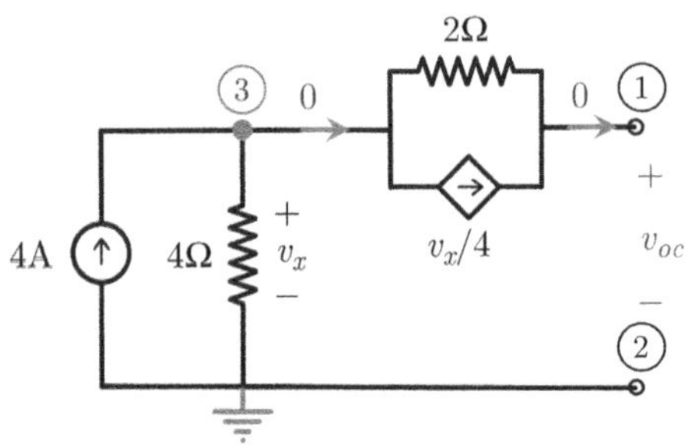

Using nodal analysis, we easily obtain $v_3 = v_x = 4 \times 4 = 16$ V. Then

$$v_{oc} = v_1 = v_3 + 2 \times v_x/4 = 24 \text{ V}.$$

It is interesting that $v_x/4$ current flows through the 2 Ω resistor as a closed loop.
Next, we consider the short-circuit case.

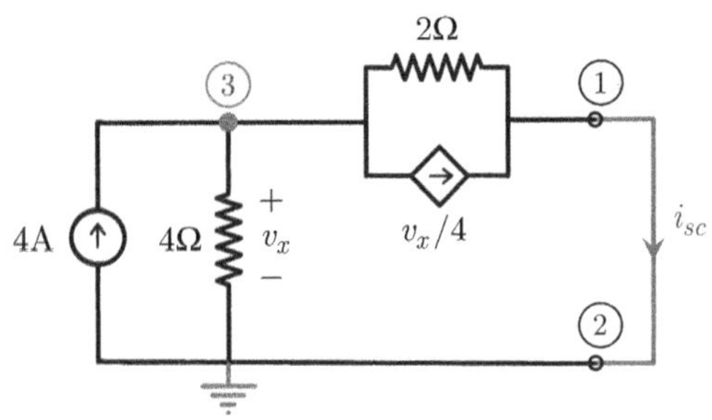

In this case, we have $v_1 = v_2 = 0$ and $v_3 = v_x$. Applying KCL at node 3, we get

- KCL (3): $4 - v_3/4 - v_3/4 - v_3/2 = 0 \longrightarrow v_3 = 4$ V.

Then $i_{sc} = v_3/4 + v_3/2 = 3$ A. Consequently, we obtain

$$v_{th} = v_{oc} = 24 \text{ V},$$

$$R_{th} = v_{oc}/i_{sc} = 24/3 = 8\ \Omega,$$

and the Thévenin equivalent circuit can be drawn as follows.

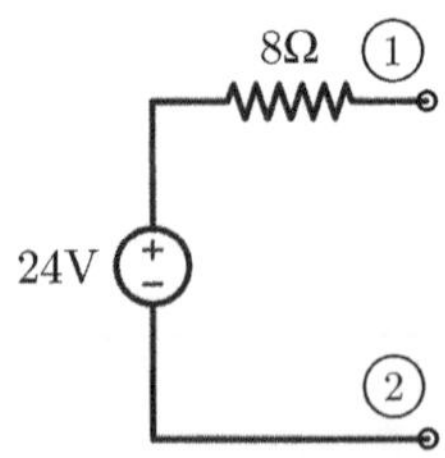

For maximum power transfer, the resistor to be connected to this circuit must be equal to $R_l = 8\ \Omega$. And, in this case, we have

$$p_l^{\max} = \frac{v_l^2}{R_l} = \frac{(v_{\text{th}}/2)^2}{R_l} = \frac{144}{8} = 18\ \text{W}.$$

Example 92: Consider the following circuit.

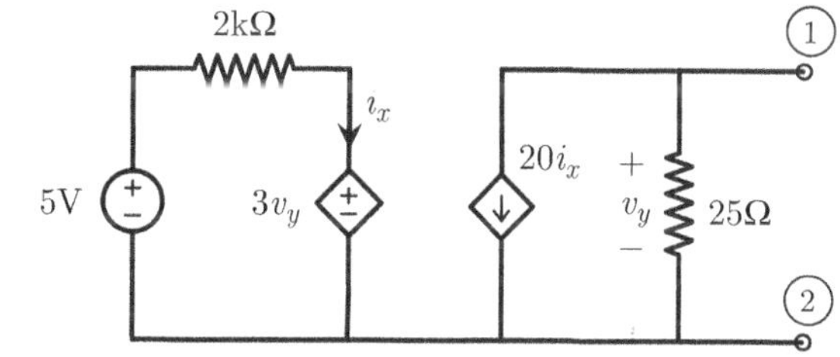

Find the value of the resistor to be connected between 1 and 2 for maximum power transfer.

Solution: In order to find the value of the resistor, one can find the Thévenin equivalent circuit seen from 1 and 2. The open-circuit case is considered as follows.

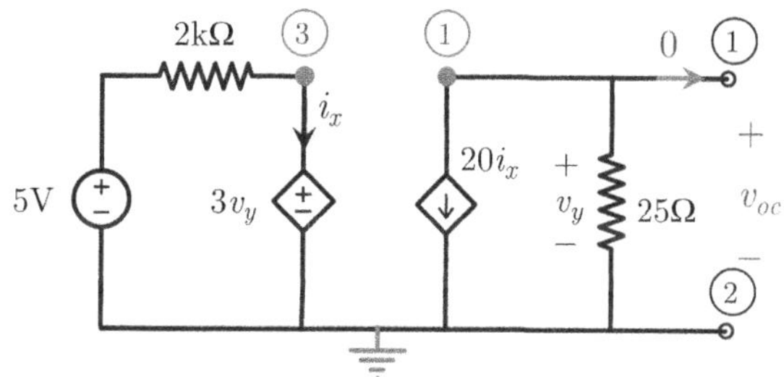

Using nodal analysis, we have $v_y = v_1$. Applying KCL at node 1 and node 3, we get

- KCL (1): $-20i_x - v_1/25 = 0 \longrightarrow v_1 = -500i_x$,
- KCL (3): $(5 - 3v_y)/2000 - i_x = 0 \longrightarrow 2000i_x + 3v_1 = 5$.

Then we obtain $i_x = 10\,\text{mA}$ and $v_{\text{oc}} = v_y = -5\,\text{V}$.

Next, we consider the short-circuit case.

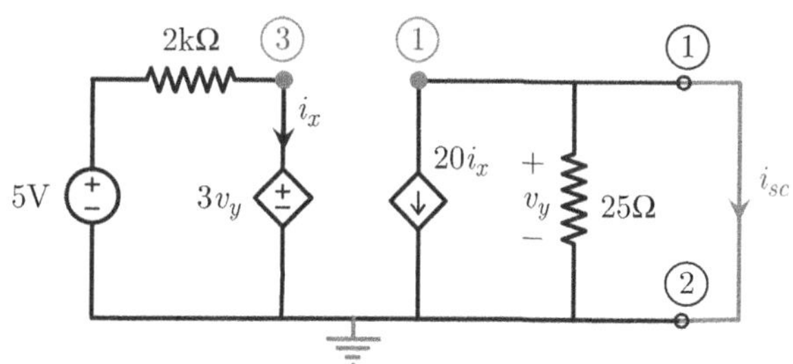

In this case, we have $v_y = v_1 = 0$ and $i_x = 5/2\,\text{mA}$. Then $i_{sc} = -20i_x = -50\,\text{mA}$. In summary, we obtain

$$v_{th} = v_{oc} = -5\text{ V},$$
$$R_{th} = v_{oc}/i_{sc} = -5/(-0.05) = 100\ \Omega,$$

which can be represented as follows.

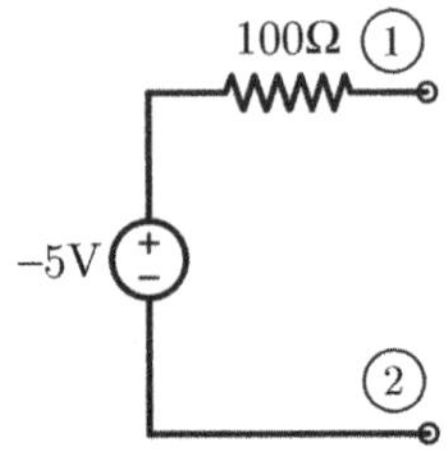

Maximum power transfer occurs when the resistor to be connected to this circuit is $R_l = 100\ \Omega$. In this case, we have

$$p_l^{max} = \frac{v_l^2}{R_l} = \frac{(v_{th}/2)^2}{R_l} = \frac{25}{400} = \frac{5}{80}\text{ W.}$$

Example 93: Consider the following circuit.

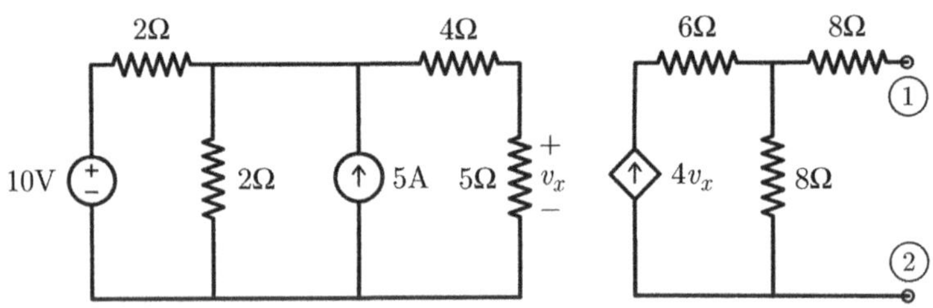

Find the value of the resistor to be connected between 1 and 2 for maximum power transfer.

Solution: In order to find the Thévenin equivalent circuit seen from 1 and 2, the open-circuit case is considered as follows.

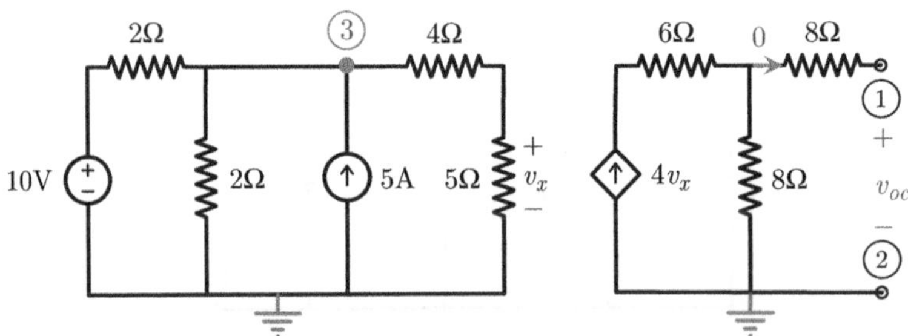

Using nodal analysis, we apply KCL at node 3 to obtain v_3 as

- KCL(3): $(10 - v_3)/2 - v_3/2 + 5 - v_3/9 = 0 \longrightarrow v_3 = 9\,\text{V}$.

Therefore, $v_x = 5\,\text{V}$ and $v_{oc} = 4v_x \times 8 = 32v_x = 160\,\text{V}$.

Next, we consider the short-circuit case.

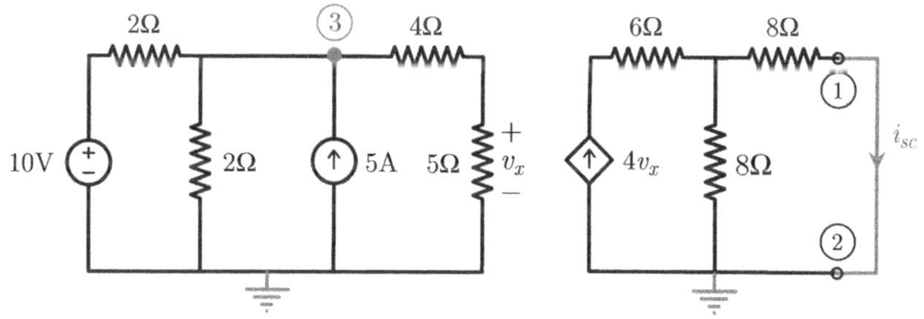

Interestingly, v_x does not change, equaling 5 V as in the open-circuit case. A current division occurs on the right-hand side of the circuit, that is, $i_{sc} = 4v_x/2 = 10\,\text{A}$. Consequently, we have

$$v_{th} = v_{oc} = 160\ \text{V},$$
$$R_{th} = v_{oc}/i_{sc} = 16\ \Omega.$$

The Thévenin equivalent circuit can be drawn as follows.

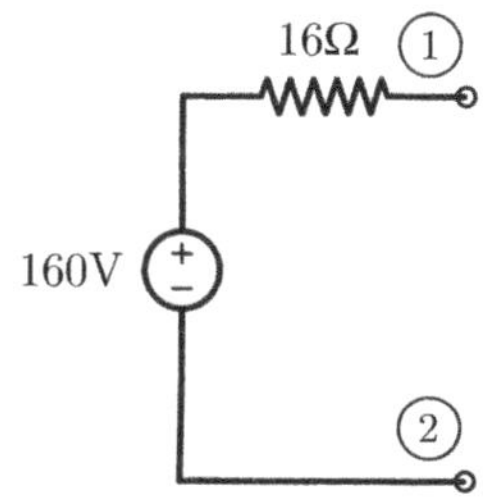

Maximum power transfer occurs when $R_l = 16\ \Omega$ is connected to the terminals, leading to

$$p_l^{\max} = \frac{v_l^2}{R_l} = \frac{(v_{th}/2)^2}{R_l} = \frac{80 \times 80}{16} = 400\ \text{W}.$$

Example 94: Consider the following circuit.

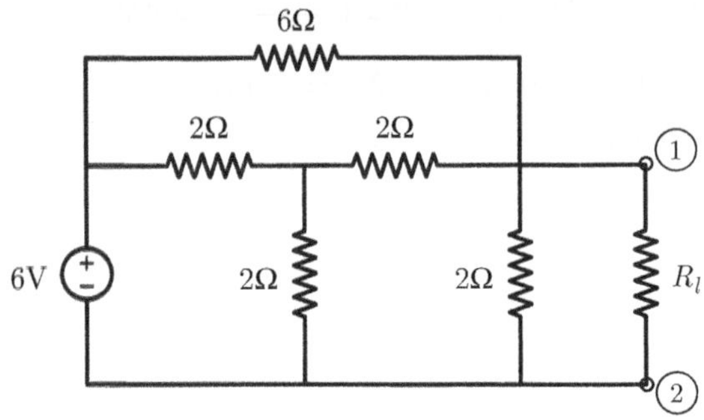

Find the value of the resistor R_l for maximum power transfer to this load.

Solution: First, we consider the open-circuit case as follows.

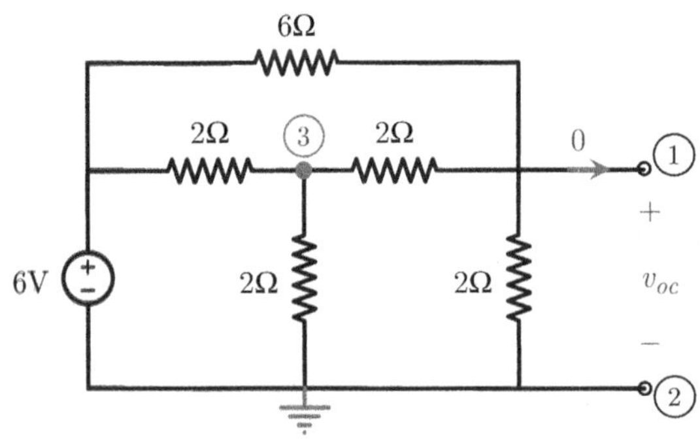

Using nodal analysis, we apply KCL at nodes 1 and 3 to derive

- KCL(1): $(v_3 - v_1)/2 + (6 - v_1)/6 - v_1/2 = 0 \longrightarrow 7v_1 - 3v_3 = 6$,
- KCL(3): $(6 - v_3)/2 - v_3/2 - (v_3 - v_1)/2 = 0 \longrightarrow v_1 - 3v_3 = -6$,

leading to $v_{oc} = v_1 = 2$ V.

Next, we can consider the short-circuit case as follows.

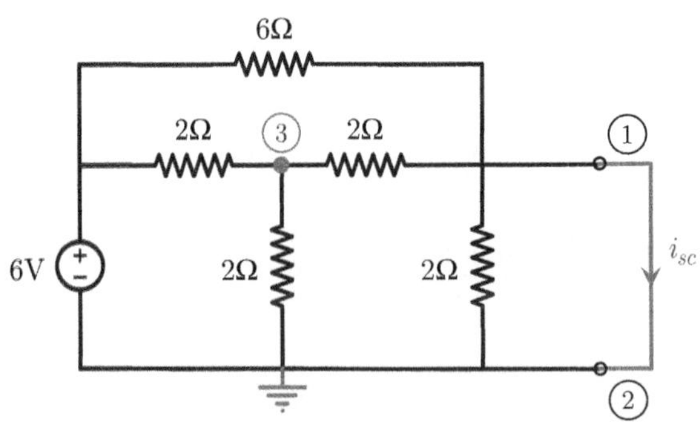

In this case, we have $v_1 = v_2 = 0$ V. Therefore, applying KCL at the nodes, we find

- KCL(3): $(6 - v_3)/2 - v_3/2 - v_3/2 = 0 \longrightarrow v_3 = 2$ V,
- KCL(1): $6/6 + 2/2 - i_{sc} = 0 \longrightarrow i_{sc} = 2$ V.

Therefore, $R_{th} = 1\ \Omega$, and the simplified circuit is as follows.

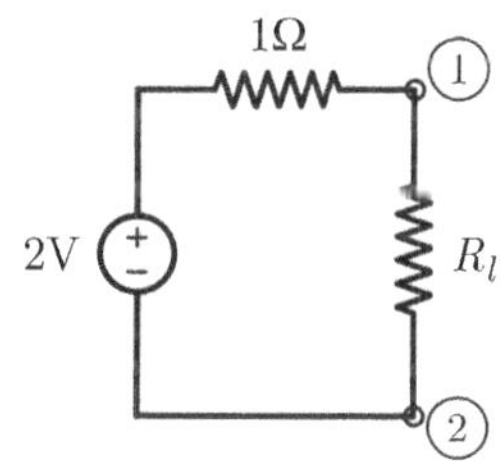

For maximum power transfer, one must select $R_l = 1\ \Omega$, leading to

$$p_l^{max} = \frac{v_l^2}{R_l} = 1 \text{ W}.$$

Example 95: Consider the following circuit.

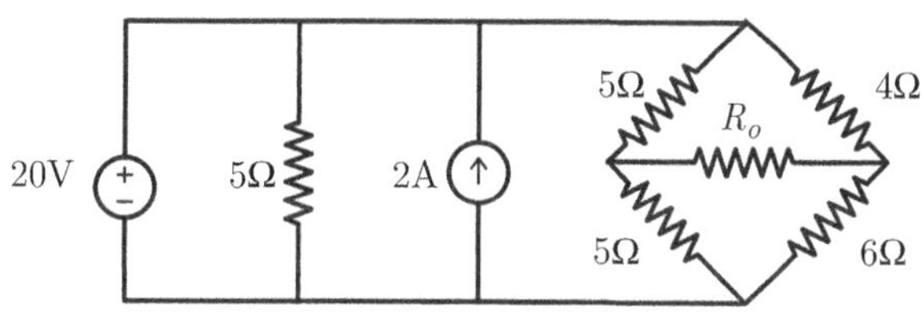

Find the value of the resistor R_o for maximum power transfer to this load.

Solution: We again solve this problem by using the Thévenin theorem. The open-circuit case is as follows.

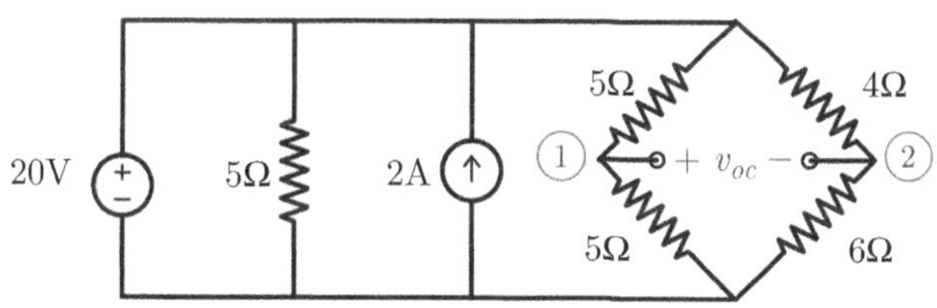

Using simple voltage division, we have

- $v_1 = 20 \times (5/10) = 10$ V,
- $v_2 = 20 \times (6/10) = 12$ V,

and $v_{oc} = v_1 - v_2 = -2$ V.

Similarly, the short-circuit case is as follows.

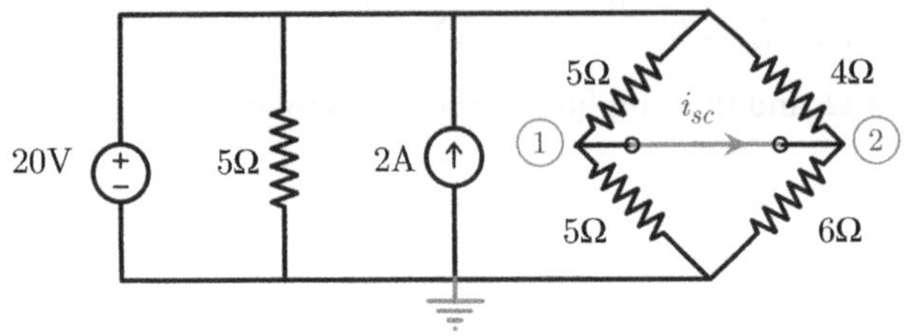

For this circuit, one finds the equivalent resistor for the combination of the four resistors to be

$$R_{\text{eq}} = 5 \parallel 4 + 5 \parallel 6 = \frac{20}{9} + \frac{30}{11} = \frac{490}{99}\ \Omega.$$

Then the current through the 5 Ω resistors can be found via current division to be

$$i_{5\ \Omega,\ \text{upper}} = \frac{4}{9}\frac{20}{R_{\text{eq}}} = \frac{80}{9}\frac{99}{490} = \frac{88}{49}\ \text{A},$$

$$i_{5\ \Omega,\ \text{lower}} = \frac{6}{11}\frac{20}{R_{\text{eq}}} = \frac{120}{11}\frac{99}{490} = \frac{108}{49}\ \text{A}.$$

Next, applying KCL at node 1, we derive

- KCL(1): $i_{\text{sc}} = (-108 + 88)/49 = -20/49$ A.

Therefore, $R_{\text{th}} = 49/10\ \Omega$ and one must select $R_o = 49/10\ \Omega$ for maximum power transfer.

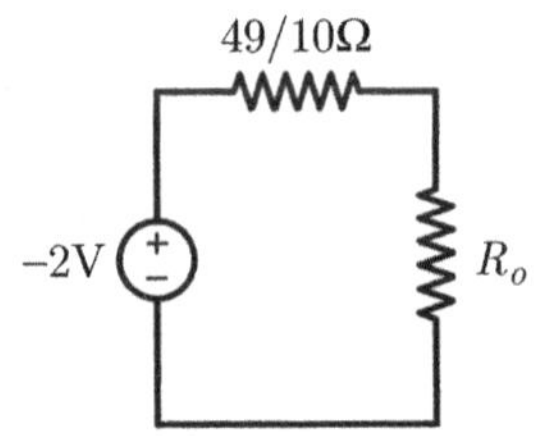

The value of the transferred power can be obtained as

$$p_o = \frac{1}{49/10} = \frac{10}{49}\ \text{W}.$$

Example 96: Consider the following circuit.

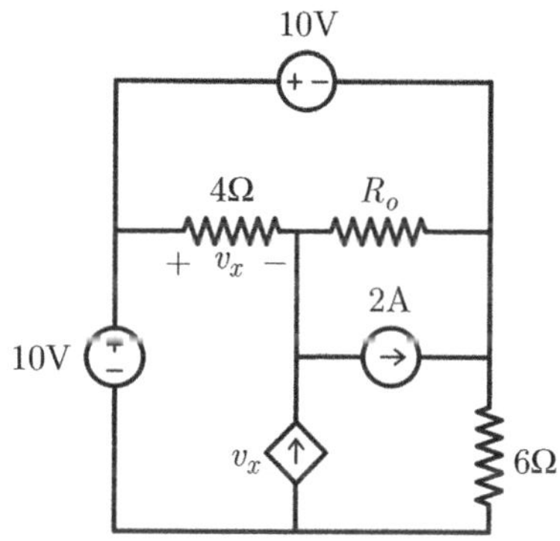

Find the value of the resistor R_o for maximum power transfer to this load.

Solution: We first consider the open-circuit case as follows.

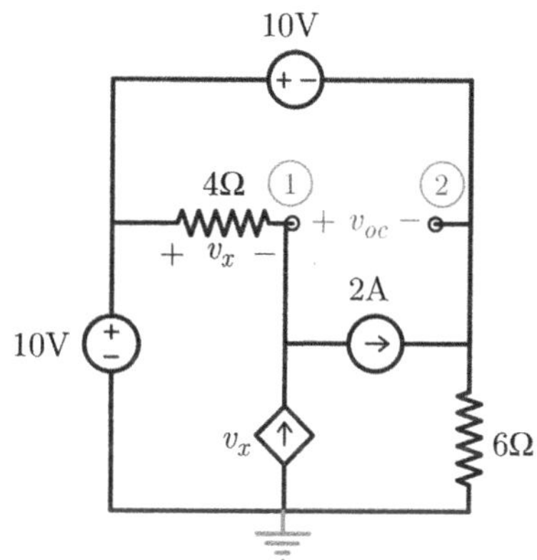

Using nodal analysis, one can derive $v_x = 10 - v_1$ and $v_2 = 10 - 10 = 0$ with a proper choice of the ground. Then, applying KCL at node 1, we have

- KCL(1): $(10 - v_1)/4 + v_x - 2 = 0 \longrightarrow v_1 = 42/5\,\text{V}$

leading to $v_{oc} = 42/5\,\text{V}$.

Next, we consider short-circuit case as follows.

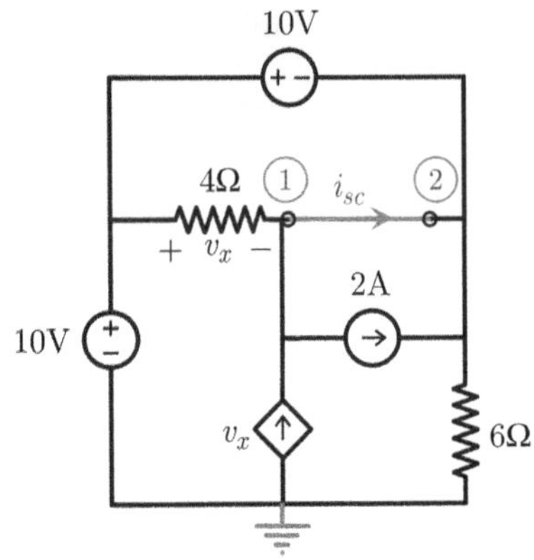

In this case, $v_1 = v_2 = 0$ V and $v_x = 10$ V, again using nodal analysis. Then, applying KCL at node 1, we derive

- KCL(1): $10/4 + 10 - 2 - i_{sc} = 0$,

leading to $i_{sc} = 21/2$ A and $R_{th} = 4/5\ \Omega$. The simplified circuit is as follows.

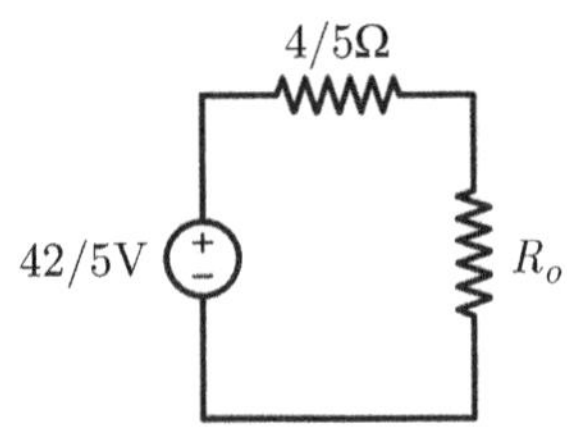

For maximum power transfer, one should choose $R_o = 4/5\ \Omega$. Then we obtain

$$p_o = \left(\frac{21}{5}\right)^2 \times \frac{5}{4} = \frac{441}{20}\ \text{W}$$

as the transferred power.

Example 97: Consider the following circuit.

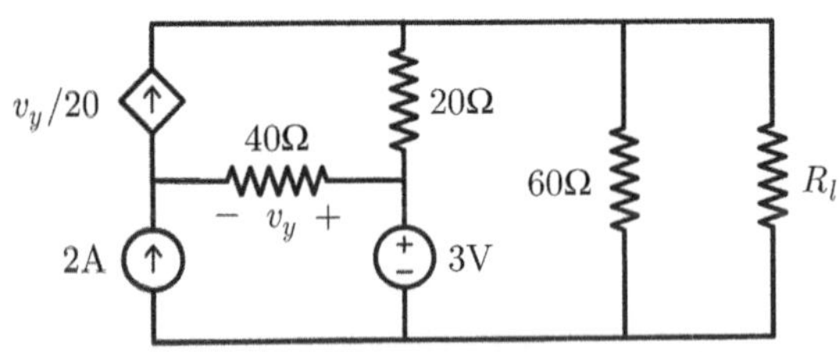

The Thévenin equivalent circuit seen by R_l was previously found to be as follows (see the solution to Exercise 80).

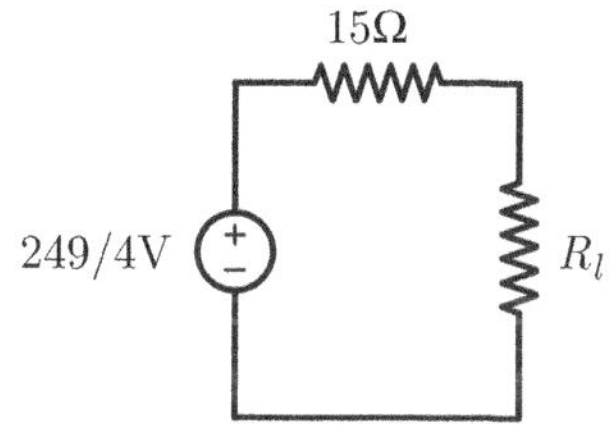

Hence, when $R_l = 15\ \Omega$, the transferred power is maximized and the corresponding voltage across R_l can be found to be $v_l = v_{oc}/2 = 249/8$ V.

Exercise 81: In the following circuit, find the maximum power that can be transferred to R_o.

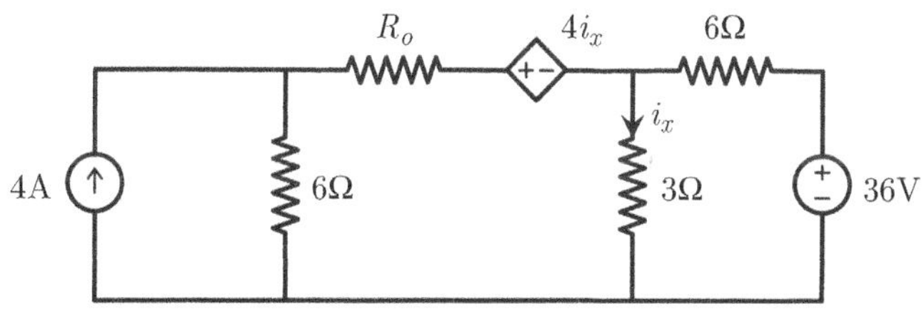

Exercise 82: In the following circuit, find the maximum power that can be transferred to R_o.

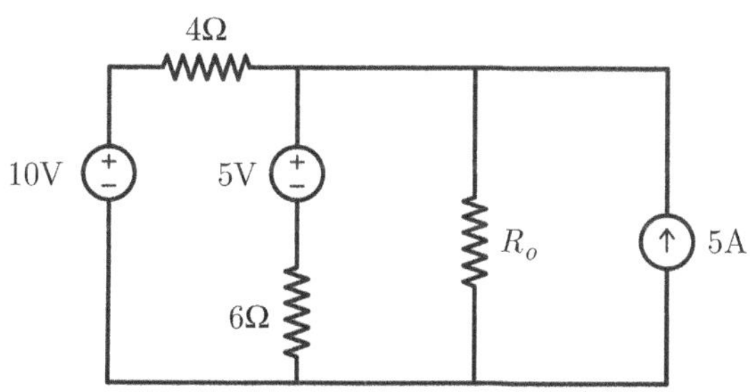

Exercise 83: In the following circuit, find the maximum power that can be transferred to R_o.

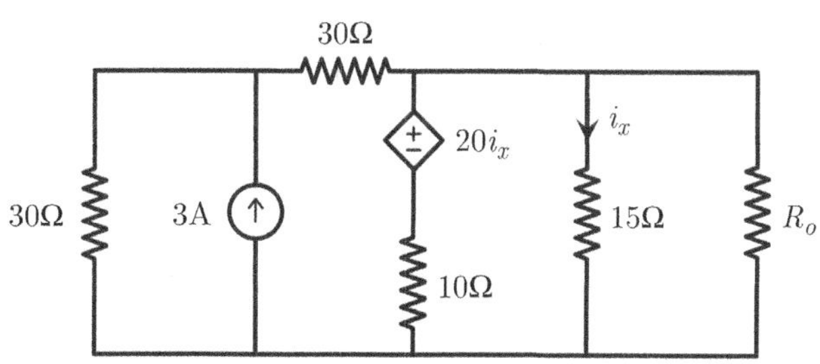

Exercise 84: In the following circuit, find the maximum power that can be transferred to R_o.

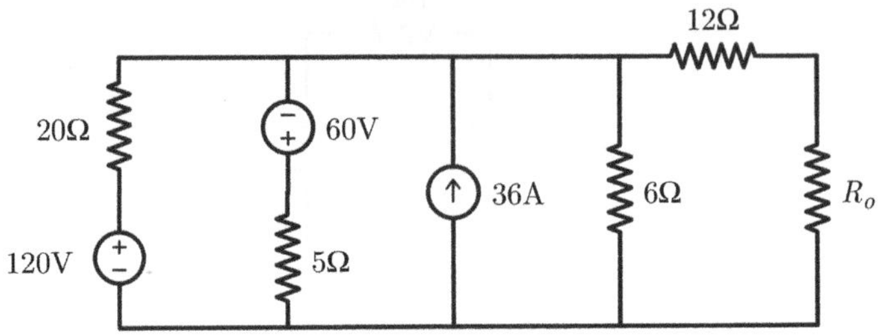

Exercise 85: In the following circuit, find the maximum power that can be transferred to R_o.

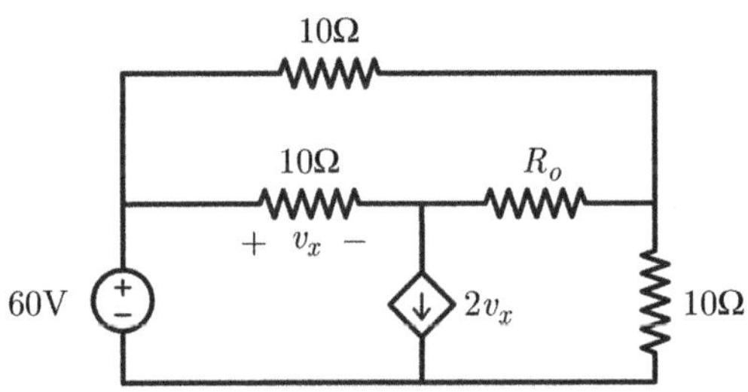

Exercise 86: In the following circuit, find the maximum power that can be transferred to R_o.

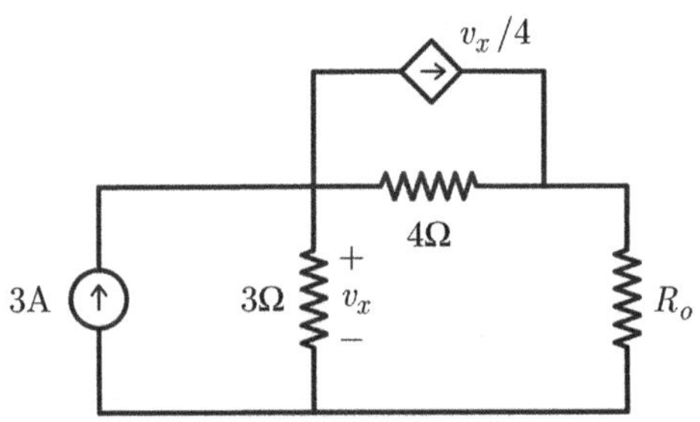

Exercise 87: In the following circuit, find the maximum power that can be transferred to R_o.

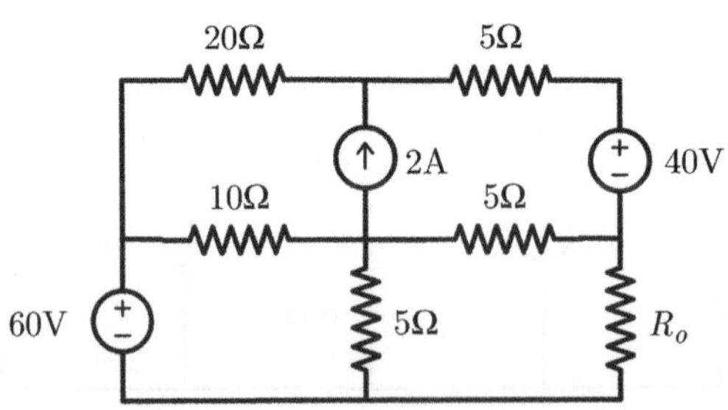

Exercise 88: In the following circuit, find the maximum power that can be transferred to R_o.

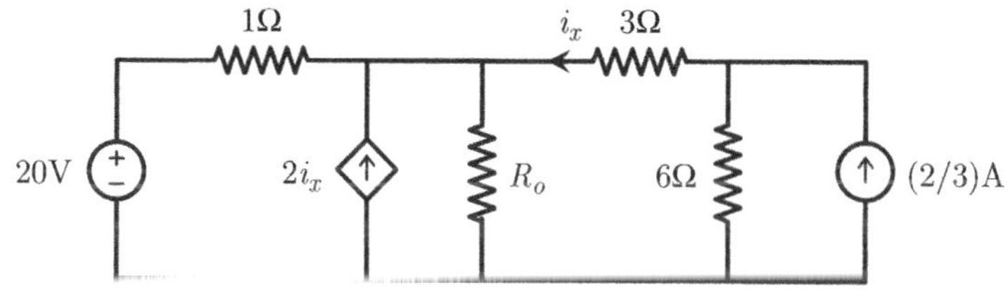

Exercise 89: In the following circuit, find the expression for the maximum power that can transferred to R_o.

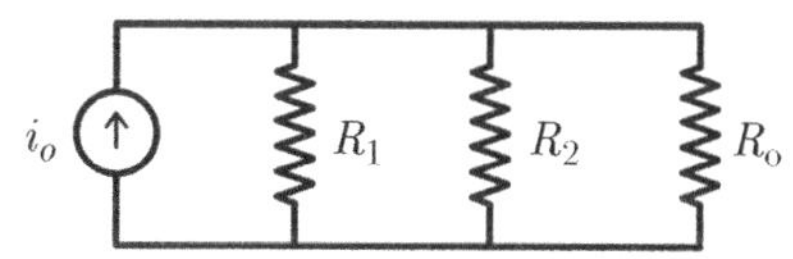

5.3 Shortcuts in Equivalent Circuits

As shown in the previous section, maximum power transfer is achieved when the equivalent resistances of two circuits (e.g., a large circuit and a single resistor as a small circuit) are equal. On the other hand, in order to find the Thévenin/Norton equivalent resistance, we need to find both the open-circuit voltage and short-circuit current. In this section, we discuss a shortcut to find the equivalent resistance.

In order to derive a shortcut to find the equivalent resistance for a given circuit, we reconsider the Thévenin and Norton equivalent circuits in Figure 5.1. We already know that $R_{\text{th}} = R_{\text{no}} = v_{\text{th}}/i_{\text{no}}$. On the other hand, if we replace v_{th} with a short circuit, the resistance measured in the Thévenin equivalent circuit is the same as R_{th}. Similarly, if i_{no} is replaced with an open circuit, R_{no} is directly seen from the terminals of the Norton circuit. Consequently, given a complex circuit, one can perform the following steps to find the equivalent resistance.

- Replace all independent voltage sources with short circuits.
- Replace all independent current sources with open circuits.
- Calculate the overall resistance seen from the terminals.

This shortcut is particularly useful when replacing the independent sources with short/open circuits leads only to resistors such that series and parallel connections of resistors can easily be calculated. When dependent sources (that cannot be replaced) are involved, however, the equivalent resistance may not be trivial to find.

As an example, we reconsider the following circuit, where the resistance seen by v_o needs to be found.

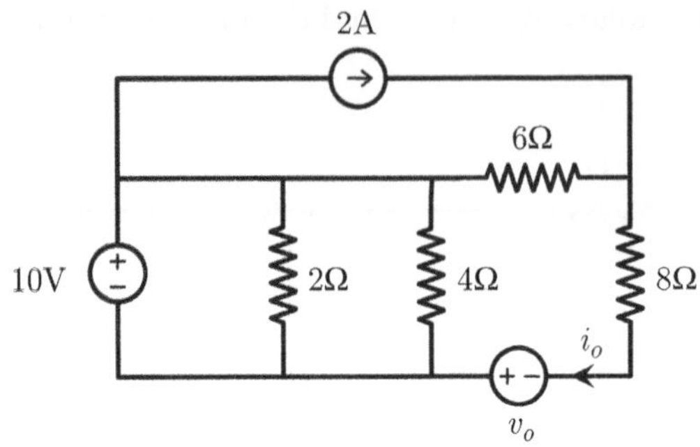

When the independent sources (other than v_o) are replaced with short and open circuits, we have the following circuit.

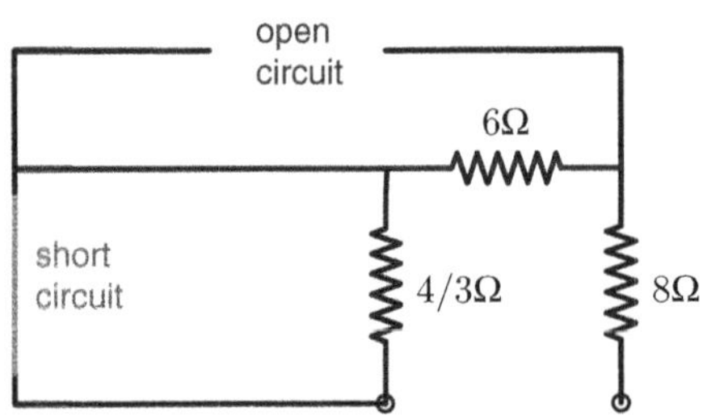

Rearranging the resistors, the circuit becomes clearer as follows.

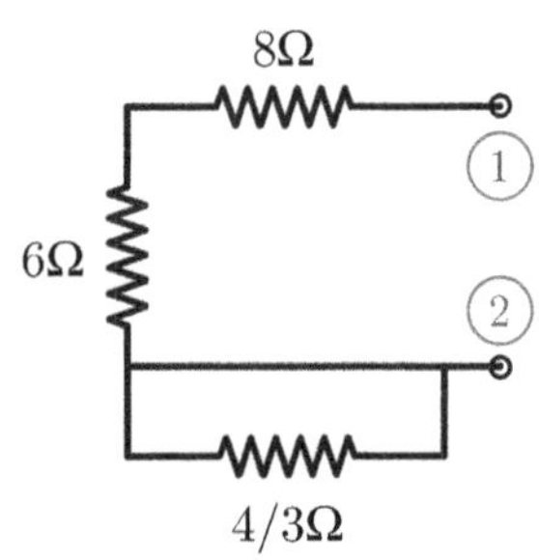

Obviously, the 4/3 Ω resistor is short-circuited, and the overall resistance seen from the terminals can easily be found to be

$$R_{eq} = R_{th} = R_{no} = 6 + 8 = 14\ \Omega.$$

Equivalence of Thévenin and Norton circuits provides other types of interesting shortcuts when solving circuits. Specifically, we know that a series connection of a voltage source v_{th} and a resistor R_{eq} is equivalent to a parallel connection of a current source $i_{no} = v_{th}/R_{eq}$ and the resistor R_{eq}. Similarly, a parallel connection of a current source i_{no} and a resistor R_{eq} can be replaced with a series connection of a voltage source $v_{th} = R_{eq}i_{no}$

and the same resistor R_{eq}. Such substitutions can facilitate the analysis of circuits with many connections of resistors.

As an example, consider the following circuit involving five resistors and two independent voltage sources.

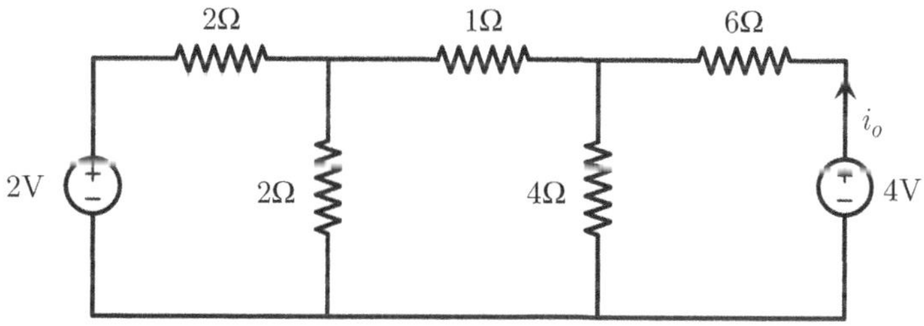

We would like to find the value of the current i_o, which can be done in many different ways. In this case, however, we consider the equivalence of circuits. First, a series connection of the 2 V voltage source and the 2 Ω resistor on the left can be replaced with a $2/2 = 1$ A current source and a 2 Ω resistor as follows.

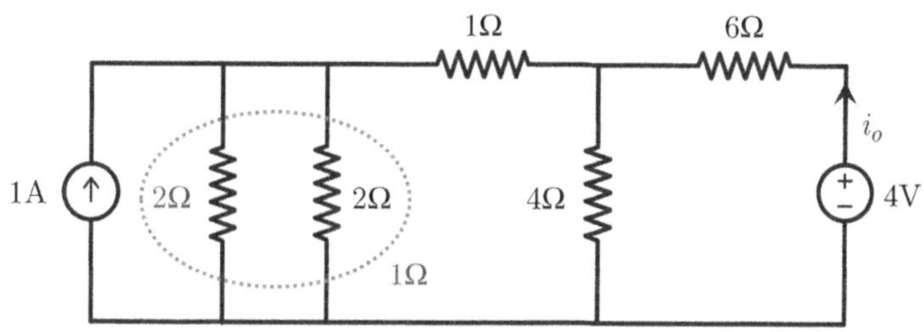

In the above, the new 2 Ω resistor is parallel to another 2 Ω resistor, leading to 1 Ω overall. Hence, we have a 1 A current source that is connected in parallel to a 1 Ω resistor in the updated circuit. Using the equivalence again, this parallel connection can be replaced as follows.

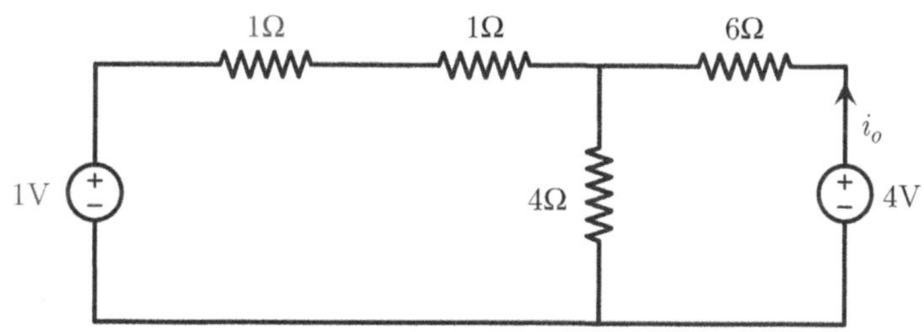

In the updated circuit, we observe that the new 1 Ω resistor is connected in series to another 1 Ω resistor. Therefore, we have a series connection of a 1 V voltage source and a 2 Ω resistor, which can be replaced again as follows.

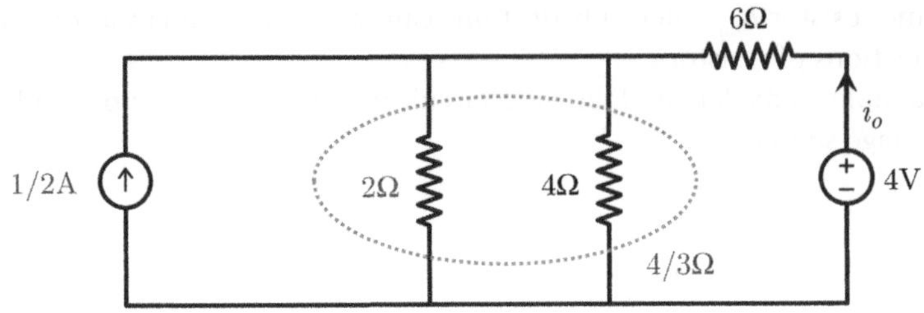

Considering the new parallel connection of two resistors, and using the equivalence theorem one more time, we obtain the following circuit.

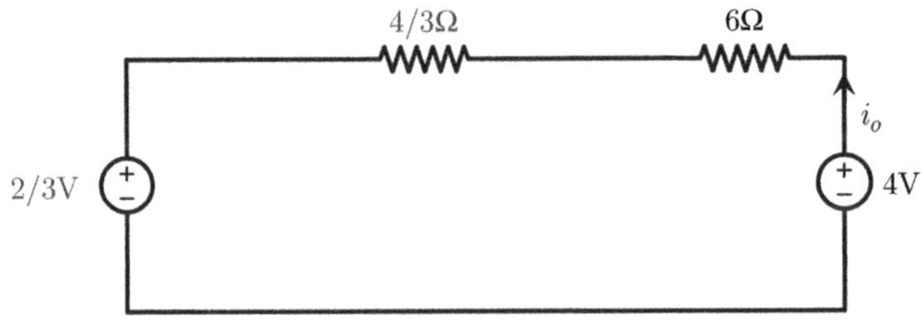

In this final form, the value of i_o can easily be found to be

$$i_o = \frac{4 - 2/3}{6 + 4/3} = \frac{10}{22} = 5/11 \text{ A}.$$

5.4 When Things Go Wrong with Equivalent Circuits

As in the previous chapters, we now discuss some possible mistakes when working with equivalent circuits. A fatal error in using Thévenin or Norton equivalent circuits is mixing two analyses (for v_{oc} and i_{sc}). Consider the following example, where the range of i_o needs to be found for given $-8 \leq v_o \leq 4$ V.

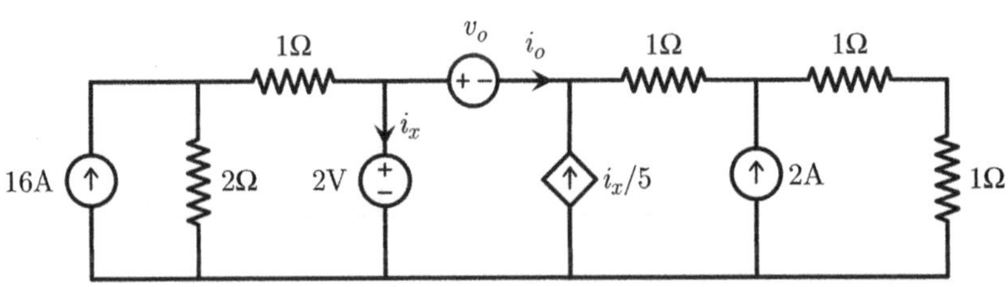

For the open-circuit case, we have the following scenario.

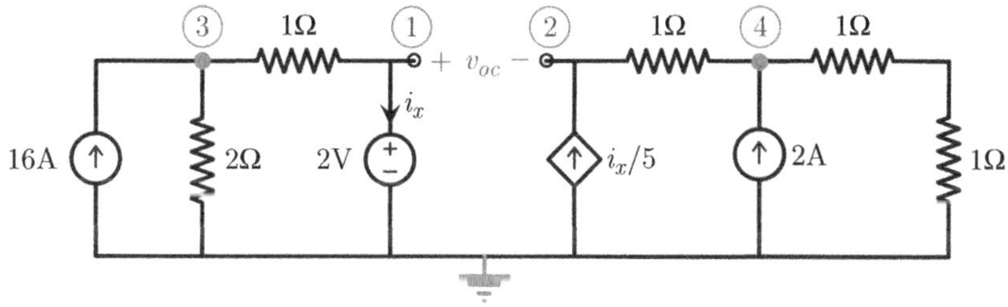

Using nodal analysis, we have $v_1 = 2$ V and

- KCL(3): $16 - v_3/2 - (v_3 - v_1)/1 = 0 \longrightarrow v_3 = 12$ V.

Then we obtain $i_x = (v_3 - v_1)/1 = 10$ A and

- KCL(4): $(v_2 - v_4)/1 + 2 - v_4/2 = i_x/5 + 2 - v_4/2 = 0 \longrightarrow v_4 = 8$ V.

Hence, v_2 and the open-circuit voltage are found to be

$$v_2 = i_x/5 + v_4 = 10 \text{ A},$$
$$v_{oc} = v_1 - v_2 = -8 \text{ V}.$$

In the short-circuit case, one can again use nodal analysis.

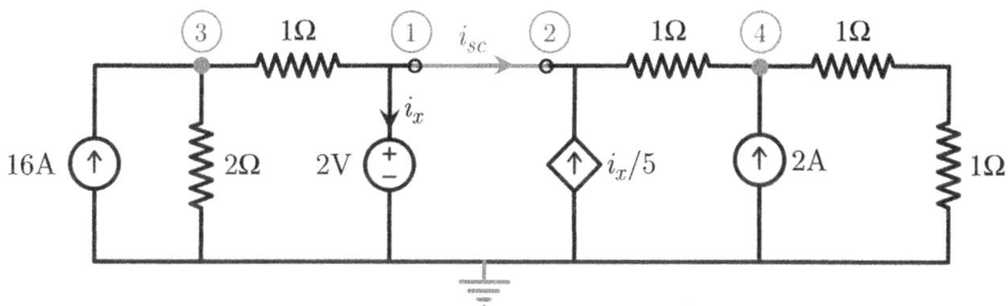

Some of the equations and values remain the same as in the open-circuit case. For example, $v_1 = 2$ V, and at node 3, we have

- KCL(3): $16 - v_3/2 - (v_3 - v_1)/1 = 0 \longrightarrow v_3 = 12$ V.

In addition, KCL at node 4 can be written as

- KCL(4): $(v_2 - v_4)/1 + 2 - v_4/2 = 0$,

again the same as before. On the other hand, the following flow is incorrect.

- $i_x = (v_3 - v_1)/1 = 10 \longrightarrow v_4 = 8$ V.

Instead, KCL at node 1 reveals that

- KCL(1): $i_x + i_{sc} = (v_3 - v_1)/1 = 10$,

which is valid but not useful as the equation involves a new unknown i_{sc}. For a correct solution, one may use $v_1 = v_2 = 2$ V, and find $v_4 = 8/3$ V by using the equation derived from KCL at node 4. Then the correct value of i_x can be obtained by applying KCL at the supernode involving nodes 1 and 2,

- KCL(1&2): $(v_3 - v_1)/1 - i_x + i_x/5 - (v_2 - v_4)/1 = 0 \longrightarrow i_x = 40/3$ A.

Finally, using KCL at node 1, we have $i_{sc} = 10 - i_x = -10/3$ A. To complete the solution, one can consider the Thévenin equivalent seen by the voltage source v_o as follows.

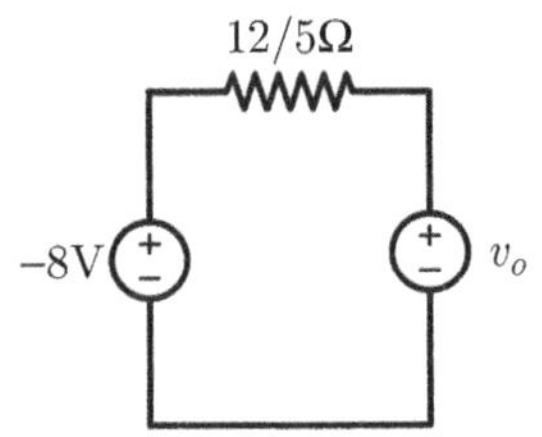

In the above, the value of the resistor is found to be $R_{th} = v_{oc}/i_{sc} = -8/(-10/3) = 12/5\ \Omega$. Then we derive

$$v_o = -8 - (12/5)i_o \longrightarrow -8 \leq [-8 - (12/5)i_o] \leq 4$$

leading to

$$0 \leq -(12/5)i_o \leq 12$$

or

$$-5 \leq i_o \leq 0 \text{ A.}$$

5.5 What You Need to Know before You Continue

As discussed in this chapter, Thévenin and Norton equivalences allow us to simplify circuits before analyzing them in some scenarios, such as maximum power transfer. Before proceeding to the next chapter, we emphasize a few important points.

- Equivalence: Thévenin and Norton equivalent circuits can be used to represent any linear circuit involving linear elements.
- Maximum power transfer: Circuits are matched best when their equivalent resistance values are the same. Considering a voltage source and a resistor, a matched resistor draws the maximum power that can be extracted.

- The equivalence of Thévenin and Norton circuits enables the simplification of complex resistive networks by applying the equivalence consecutively on series and parallel connections of sources and resistors.

In the next chapter, we start with two new types of components, namely, capacitors and inductors, which are commonly used in real-life circuits. Unlike with resistors, time is an important parameter in the analysis of capacitors and inductors, since these components present delayed responses (that cannot be neglected) to the change in voltage and current values.

6

Transient Analysis

As discussed at various points in Chapter 1, time is an important parameter in electrical circuits. On the other hand, in subsequent chapters, we neglected the time concept in the circuit analysis because

- all sources have been DC sources that deliver fixed voltage and current values,
- we have considered only resistors, in addition to dependent and independent sources, but not energy-storage elements.

Basically, there are two fundamental types of energy-storage elements, namely, capacitors and inductors. Similar to resistance, capacitance and inductance are natural properties of all structures, while capacitors and inductors appear as individual components in many circuits. Unlike resistors, the response of a capacitor and inductor to an applied voltage and current is not instant. Specifically, for DC circuits involving energy-storage elements (and resistors), it takes time for these components to store and release energy, leading to a transient state that does not occur in resistor-only circuits. This chapter is devoted to the analysis of such transient states. In Chapter 7 we focus on AC circuits, where the energy-storage elements add phases between current and voltage values that are oscillatory.

6.1 Capacitance and Capacitors

In general terms, the capacitance of a structure is its ability to store charge for a given voltage. Therefore, it is defined as

$$C = \frac{q}{v}.$$

The unit of capacitance is the farad (F), where 1 farad is 1 coulomb per volt. Typically, a capacitance is defined for a structure with two parts that are separated by a distance, while a voltage is applied between them. For such structures, the capacitance is usually proportional to the sizes (e.g., areas) of the parts and inversely proportional to distance between them. Capacitance also depends on (and is proportional to) the electrical properties (specifically, the electric permittivity) of the medium in which the structure is located. Capacitance can also be defined for a single body when the voltage is defined with respect to a reference (usually infinity).

Capacitors (see Figure 11.2) are two-port devices that have fixed capacitance values. While the capacitance may actually depend on the frequency, applied voltage, and

Introduction to Electrical Circuit Analysis, First Edition. Özgür Ergül.

Companion Website: www.wiley.com/go/ergul4412

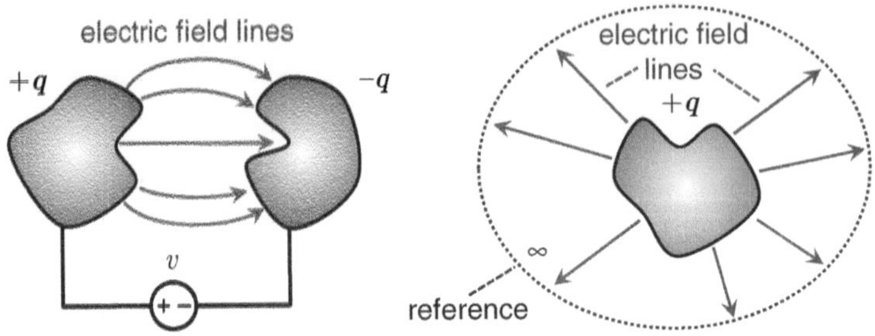

Figure 6.1 The capacitance between two bodies is proportional to the amount of charge (q) in the positively charged body for a given voltage difference between them. The capacitance of a single body can also be defined when the voltage difference is defined with respect to infinity.

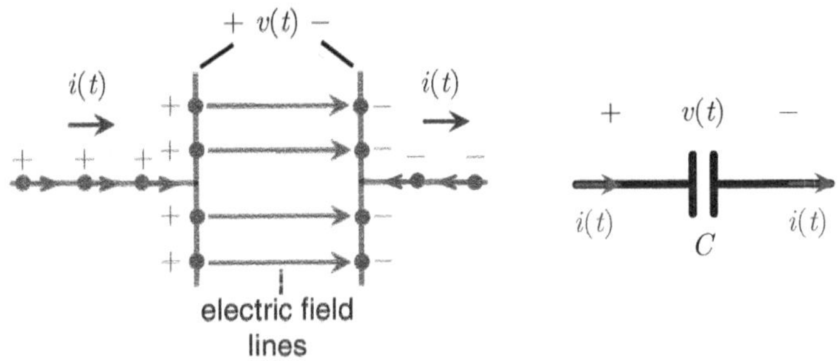

Figure 6.2 When a nonzero current exists through a capacitor involving two bodies (e.g., conducting plates), leading to the accumulation of positive and negative charges, it builds up electric field, hence a voltage difference, between the bodies. In circuit analysis, a capacitor is a two-port device that stores and releases energy.

current values (leading to nonlinear behavior), as well as on outer conditions, we consider only ideal capacitors with constant capacitance values. In order to derive the voltage/current response of a capacitor, we use the definition of the capacitance and take the time-derivative of the charge,

$$\frac{dq(t)}{dt} = \frac{d}{dt}[Cv(t)] = C\frac{dv(t)}{dt},$$

leading to

$$i(t) = C\frac{dv(t)}{dt},$$

using the definition of the current. Therefore, the current through a capacitor is proportional to the time derivative of its voltage. This can be interpreted as follows. An increase in the voltage of a capacitor corresponds to an increase in the amount of charge accumulated in the capacitor. The flow of these charges also corresponds to a positive current passing through the capacitor, as formulated in the equation above. Obviously, the larger the capacitance, larger the value of the flow of the charges, that is, the current.

Given a current with respect to time, the voltage across a capacitor is given by

$$v(t) = \frac{1}{C}\int_{t_0}^{t} i(t')dt' + v_0,$$

where $v_0 = v(t = t_0)$ is the initial voltage. If v_0 is not given, it is common to select $v_0 = 0$ and $t_0 = 0$. In the above, the integral term can be interpreted as the accumulated voltage during $t' \in [t_0, t]$. Hence, it explains how a voltage is produced as a results of charges that are flowing toward the capacitor. If there is a current through a capacitor, it builds up voltage.

At this stage, we can derive the power of a capacitor as

$$p(t) = v(t)i(t) = Cv(t)\frac{dv(t)}{dt}.$$

Therefore, the power of a capacitor can be positive or negative, depending on the value of the voltage and its time derivative. For example, suppose that the voltage is positive. Then if the time derivative is also positive, indicating that the voltage of the capacitor is further increasing, the power of the capacitor is positive. Therefore, similar to resistors, such a capacitor absorbs energy. On the other hand, unlike resistors, the energy absorbed by the capacitor is not consumed (e.g., not converted into heat) but stored. This is the main reason why capacitors are called energy-storage elements. Depending on the circuit, the stored energy may be released at a particular time (i.e., when $p(t) < 0$). In this case, if the voltage is positive, the time derivative of the capacitor becomes negative, indicating a decrease in the voltage. We emphasize that the energy released from a capacitor is not produced by the capacitor itself but is provided from the energy stored in the capacitor at an earlier time.

Finally, we can derive the energy stored in a capacitor by taking the time derivative of the power. Assuming $v(t_0) = 0$ for a given initial time t_0, we have

$$\begin{aligned} w(t) &= \int_{t_0}^{t} p(t')dt' = \int_{t_0}^{t} Cv(t')\frac{dv(t)'}{dt'}dt' \\ &= \frac{1}{2}C[v(t)]^2 \end{aligned}$$

as the energy of the capacitor. Therefore, in order to find the energy of a capacitor, it is sufficient to know the voltage across it. Obviously, there is no energy stored if the voltage is zero.

Example 98: Find the current i_C across a 1 F capacitor as a function of time, if the voltage is given as follows.

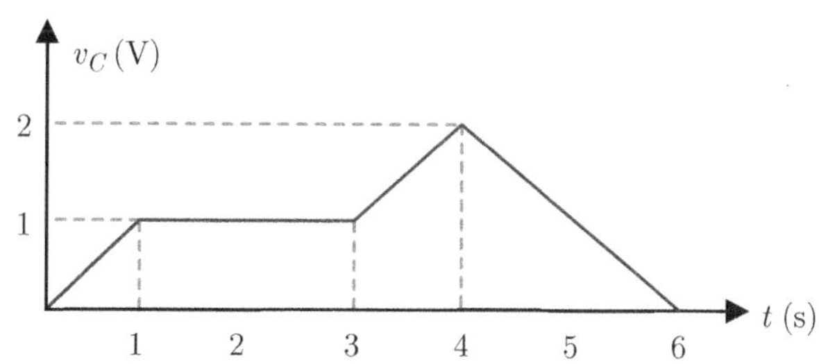

Also find the power and energy of the capacitor with respect to time.

Solution: Using the expression for the current of a capacitor, we list the values in different time ranges as

- $0 < t < 1\,\text{s}: i_C(t) = \dfrac{dv_C(t)}{dt} = \dfrac{dt}{dt} = 1\,\text{A},$
- $1 < t < 3\,\text{s}: i_C(t) = \dfrac{dv_C(t)}{dt} = \dfrac{d1}{dt} = 0\,\text{A},$
- $3 < t < 4\,\text{s}: i_C(t) = \dfrac{dv_C(t)}{dt} = \dfrac{d}{dt}(t-2) = 1\,\text{A},$
- $4 < t < 6\,\text{s}: i_C(t) = \dfrac{dv_C(t)}{dt} = \dfrac{d}{dt}(-t+6) = -1\,\text{A}.$

Therefore, the current through the capacitor can be drawn with respect to time as follows.

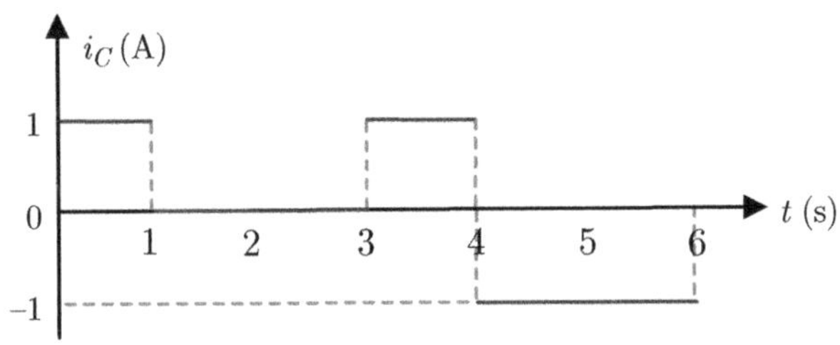

We note that the current through a capacitor can be discontinuous. Specifically, in this example, the current values at exactly $t = 0$, $t = 1$ s, $t = 3$ s, and $t = 6$ s are not well defined. Such an instant change in the polarization of a current is due to the sudden change in the slope of the corresponding voltage of the capacitor. This is a characteristic of ideal capacitors when they are modeled as purely capacitive. In real life, where each capacitor involves a small amount of resistance and inductance, it would be impossible to change its current as well as the slope of its voltage instantaneously.

Even when they are modeled ideally, capacitors that are connected to resistors cannot have discontinuous voltages. As is revealed later, it takes time for a capacitor (connected to some resistors) to store and release energy, that is, there cannot be a jump in its energy. Since the energy of a capacitor is proportional to the square of its voltage, there also cannot be a jump in the voltage, that is, it must be continuous. We note that a discontinuous voltage at a particular time would also need an infinite amount of current at the same time since it is the derivative of the voltage.

The power of the capacitor above can be obtained from $p_C(t) = v_C(t)i_C(t)$ as

- $0 < t < 1\,\text{s}: p_C(t) = (t)\,\text{W},$
- $1 < t < 3\,\text{s}: p_C(t) = 0,$
- $3 < t < 4\,\text{s}: p_C(t) = (t-2)\,\text{W},$
- $4 < t < 6\,\text{s}: p_C(t) = (t-6)\,\text{W}.$

Then we have the following plot.

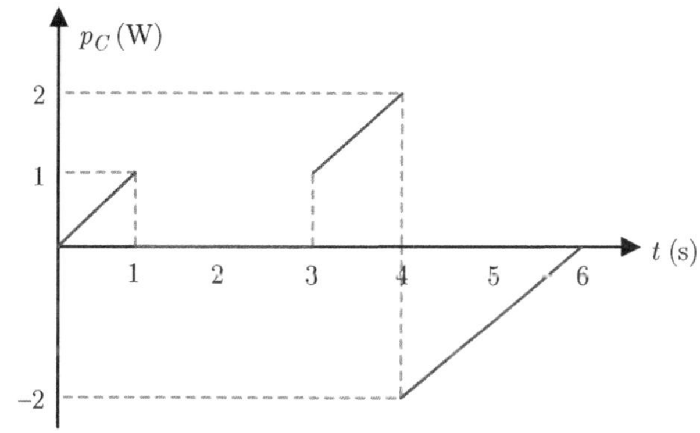

Obviously, the capacitor absorbs (stores) energy in the time intervals [0, 1] s and [3, 4] s, whereas it releases (delivers) energy in the time interval [4, 6] s. These are consistent with the increasing and decreasing voltage values of the capacitor. In the time interval [1, 3] s, where the voltage of the capacitor is constant, the power is zero, indicating no change in the energy.

Finally, the energy of the capacitor can be plotted as follows.

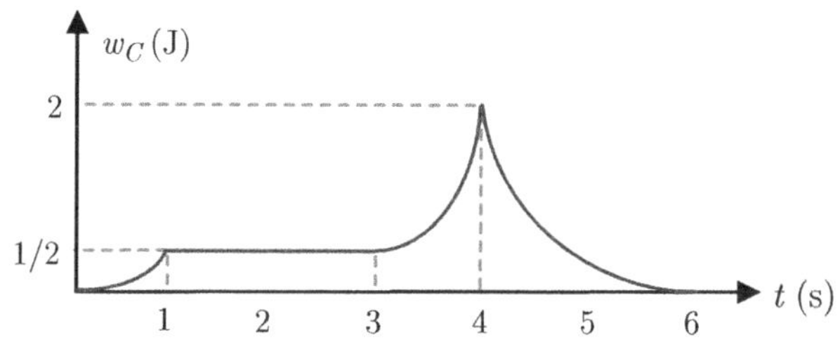

In order to obtain the energy as a function of time, the time integral of the power can be evaluated. Alternatively, one can use the relationship between the voltage and energy, $w_C(t) = 0.5C[v_C(t)]^2$. It can be observed that an energy peak occurs at $t = 4$ s, but then the energy drops to zero at $t = 6$ s.

Example 99: Consider a 1 F capacitor. The current across the capacitor is given by

$$i_C(t) = \begin{cases} 1, & 0 < t < 1 \text{ s} \\ 0, & \text{otherwise} \end{cases} \quad \text{(A)}.$$

Find and plot the voltage across the capacitor if $v_C(0) = 0$.

Solution: First, we note that $v_C(t) = 0$ for $t < 0$, since there is no current in this period. In addition, for $0 < t < 1$ s, we have

$$v_C(t) = v_C(0) + \int_0^t 1dt' = (t) \text{ V},$$

leading to $v_C(1) = 1$ V. Finally, for $t > 1$ s, we get

$$v_C(t) = v_C(1) + \int_1^t 0 dt' = 1 \text{ V}.$$

Overall, the voltage across the capacitor can be written as

$$v_C(t) = \begin{cases} 0, & t < 0 \\ t, & 0 < t < 1 \text{ s} \\ 1, & t > 1 \text{ s} \end{cases} \quad (\text{V}).$$

The current and voltage across the capacitor can be plotted as follows.

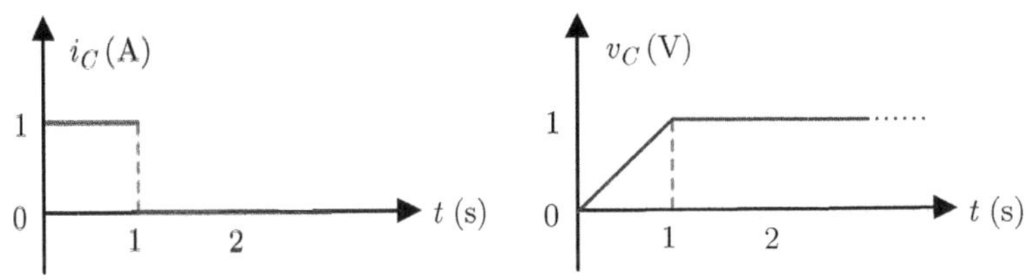

Once again, we note that the current across an ideal capacitor can be discontinuous, while the voltage must be continuous for finite values of the current.

Example 100: Consider a time-dependent current across a 1 F capacitor given by

$$i_C(t) = \begin{cases} 1, & 0 < t < 1 \text{ s} \\ -1, & 1 < t < 3 \text{ s} \\ 0, & \text{otherwise.} \end{cases} \quad (\text{A}).$$

Assuming $v_C(0) = 0$, find and plot the voltage across the capacitor, as well as the power of the capacitor and the stored energy.

Solution: We again have $v_C(t) = 0$ for $t < 0$. In addition,

$$v_C(t) = (t) \text{ V}$$

for $0 < t < 1$ s and $v_C(1) = 1$V, as in the previous example. Then, for $1 < t < 3$ s, we have

$$v_C(t) = v_C(1) + \int_1^t (-1) dt' = (-t + 2) \text{ V},$$

leading to $v_C(3) = -1$ V and $v_C(t) = -1$ V for $t > 3$ s. Overall, we have

$$v_C(t) = \begin{cases} 0, & t < 0 \\ t, & 0 < t < 1 \text{ s} \\ -t + 2, & 1 < t < 3 \text{ s} \\ -1, & t > 3 \text{ s} \end{cases} \quad (\text{V}).$$

Consequently, the current and voltage across the capacitor can be plotted as follows.

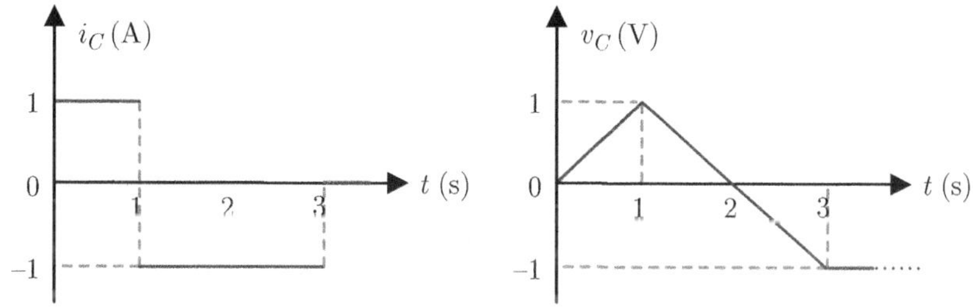

Using the expressions for the current and voltage, the power of the capacitor can be found as

$$p_C(t) = v_C(t)i_C(t) = \begin{cases} 0, & t < 0 \\ t, & 0 < t < 1 \text{ s} \\ t-2, & 1 < t < 3 \text{ s} \\ 0, & t > 3 \text{ s} \end{cases} \quad (\text{W}).$$

Considering the expression above, it can be deduced that the capacitor stores energy (positive power) in the time periods $0 < t < 1$ s and $2 < t < 3$ s, whereas it releases energy (negative power) for $1 < t < 2$ s. Finally, the energy of the capacitor can be found to be

$$w_C(t) = \frac{1}{2}C[v_C(t)]^2 = \begin{cases} 0, & t < 0 \\ t^2/2, & 0 < t < 1 \text{ s} \\ (t^2 - 4t + 4)/2, & 1 < t < 3 \text{ s} \\ 1/2, & t > 3 \text{ s} \end{cases} \quad (\text{J}).$$

The plots of the power and energy of the capacitor are as follows.

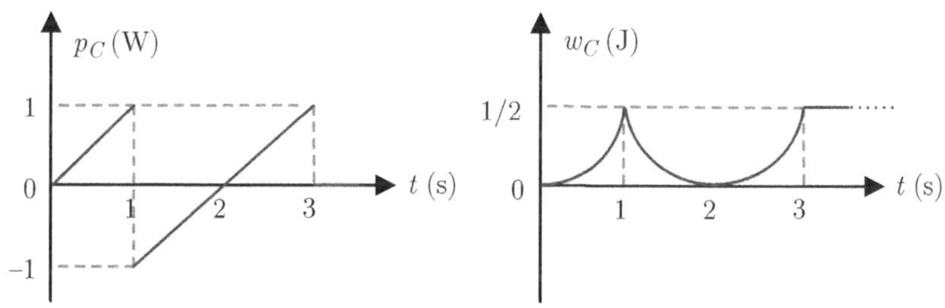

Example 101: Consider a time-dependent current across a 2 μF capacitor given as

$$i_C(t) = \begin{cases} t, & 0 < t < 1 \text{ s} \\ 1, & 1 < t < 3 \text{ s} \\ -t+4, & 3 < t < 4 \text{ s} \\ 0, & \text{otherwise}. \end{cases} \quad (\mu\text{A}).$$

Assuming $v_C(0) = 1$ V, find and plot the voltage across the capacitor. In addition, find the power during $1 < t < 3$ s, as well as the energy stored or released in this time period.

Solution: We have $v_C(t) = 1$ V for $t < 0$. For $0 < t < 1$ s, we derive

$$v_C(t) = v_C(0) + \frac{1}{2\times 10^{-6}}\int_0^t 10^{-6}t'dt' = \left(\frac{t^2}{2}+1\right)\text{ V},$$

leading to $v_C(1) = 3/2$ V. Then, for $1 < t < 3$ s, we have

$$v_C(t) = v_C(1) + \frac{1}{2\times 10^{-6}}\int_1^t 10^{-6}dt' = \frac{3}{2}+\frac{1}{2}(t-1) = \left(\frac{t}{2}+1\right)\text{ V}$$

and $v_C(3) = 5/2$ V. Moreover, when $3 < t < 4$ s, we get

$$v_C(t) = v_C(3) + \frac{1}{2\times 10^{-6}}\int_3^t 10^{-6}(4-t')dt' = \frac{5}{2}+\frac{1}{2}\left[4t' - \frac{(t')^2}{2}\right]_3^t$$
$$= \frac{5}{2}+\frac{1}{2}\left(4t-\frac{t^2}{2}\right)-\frac{1}{2}(12-9/2) = \left(-\frac{t^2}{4}+2t-\frac{5}{4}\right)\text{ V}$$

and $v_C(4) = 11/4$ V. Overall, the voltage across the capacitor can be written as

$$v_C(t) = \begin{cases} 1, & t<0 \\ t^2/2+1, & 0<t<1\text{ s} \\ t/2+1, & 1<t<3\text{ s} \\ -t^2/4+2t-5/4, & 3<t<4\text{ s} \\ 11/4, & t>4\text{ s} \end{cases} \quad (\text{V}).$$

Hence, the current and voltage across the capacitor can be plotted as follows.

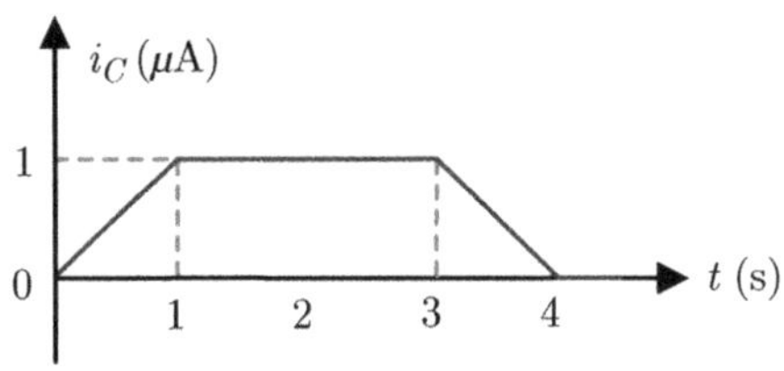

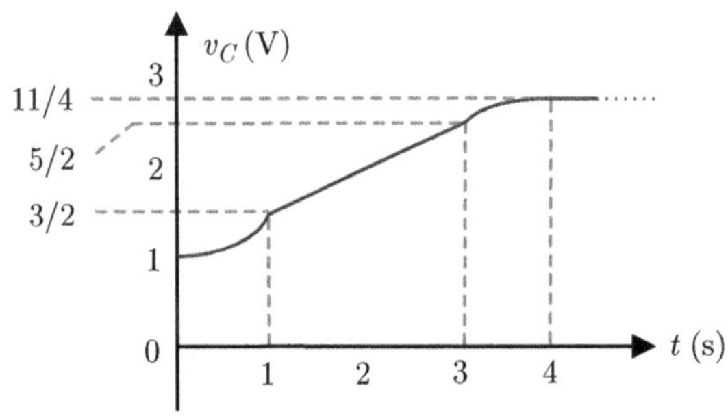

Next, the power during $1 < t < 3$ s can be found as

$$p_C(t) = v_C(t)i_C(t) = \left(\frac{t}{2}+1\right)\ \mu\text{W}.$$

Hence, the energy change in this time period can be obtained via integration as

$$w_C(3) - w_C(1) = \int_1^3 (t'/2 + 1)dt' = \left[\frac{t^2}{4} + t\right]_1^3 = 4\ \mu\text{J}.$$

Alternatively, we can use

$$w_C(3) - w_C(1) = \frac{1}{2}C[v_C(3)]^2 - \frac{1}{2}C[v_C(1)]^2 = \frac{25}{4} - \frac{9}{4} = 4\ \mu\text{J}.$$

Example 102: Find the voltage v_C across a capacitor of 1 F as a function of time, if $v_C(0) = 2$ V and the current is given as follows.

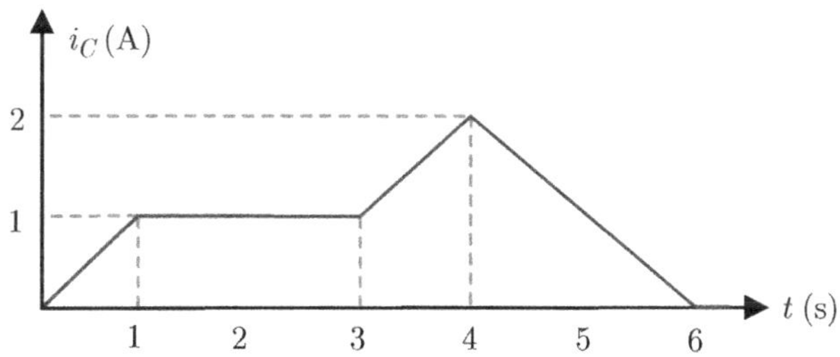

Also, find the energy stored in the capacitor at $t = 6$.

Solution: Using the expression for the voltage of a capacitor, we list the values in different time ranges as follows.

- $0 \le t \le 1$ s: $v_C(t) = 2 + \int_0^t t'dt' = \left(\frac{t^2}{2} + 2\right)$ V,
- $v_C(1) = 5/2$ V,
- $1 \le t \le 3$ s: $v_C(t) = 5/2 + \int_1^t 1dt' = \left(t + \frac{3}{2}\right)$ V,
- $v_C(3) = 9/2$ V,
- $3 \le t \le 4$ s: $v_C(t) = 9/2 + \int_3^t (t' - 2)dt' = \left(\frac{t^2}{2} - 2t + 6\right)$ V,
- $v_C(4) = 6$ V,
- $4 \le t \le 6$ s: $v_C(t) = 6 + \int_4^t (-t' + 6)dt' = \left(-\frac{t^2}{2} + 6t - 10\right)$ V,
- $v_C(6) = 8$ V,
- $t \ge 6$ s: $v_C(t) = 8$ V.

In addition, we obtain

$$w_C(6) = \frac{1}{2}C[v_C(6)]^2 = 32\ \text{J}.$$

Exercise 90: Find the voltage v_C across a capacitor of 1 F as a function of time, if $v_C(0) = 0$ V and the current is given as follows.

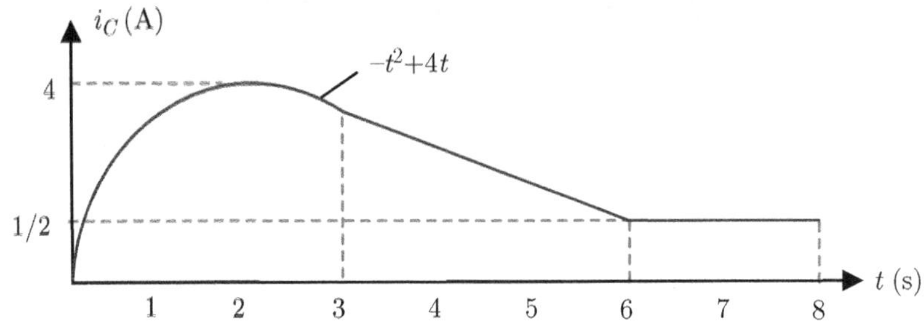

Exercise 91: Find the current i_C across a capacitor of 1 F as a function of time, if the voltage is given as

$$v_C(t) = \begin{cases} t^2+1, & 0 \le t \le 2 \text{ s} \\ t+3, & 2 \le t \le 5 \text{ s} \\ -t^2+12t-27, & 5 \le t \le 6 \text{ s} \\ 9, & t \ge 6 \text{ s} \end{cases} \quad \text{(V)}.$$

Exercise 92: Find the power p_C of a capacitor of 1 F as a function of time for $t \ge 0$, if $v_C(0) = 1$ V and the current is given as follows.

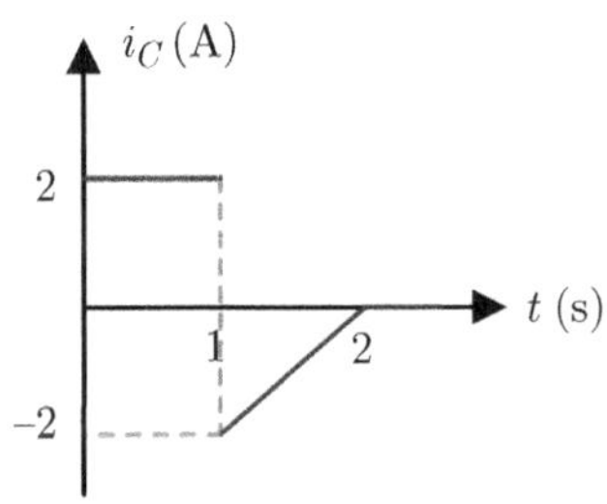

Exercise 93: The current through a 2 F capacitor is shown below. Find the energy of the capacitor for $t \ge 3$ s, if its voltage is zero is at $t = 0$.

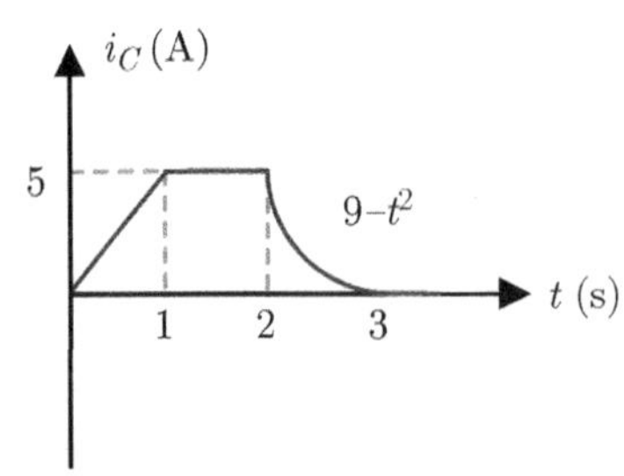

Exercise 94: The current across a capacitor of 1 F is shown below. Given that the voltage is zero at $t = 0$, derive the energy of the capacitor with respect to time.

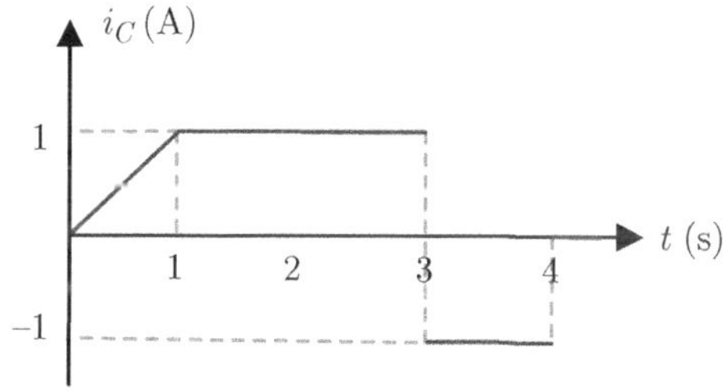

6.2 Inductance and Inductors

Inductance is another electrical property of objects, particularly conducting structures and especially wires. Similarly to capacitance, we often consider two structures separated by a distance, leading to the definition of mutual inductance. First of all, we can construct the following analogies.

- Similarly to charges that attract or repel each other, currents affect each other. For example, wires that carry currents apply force to each other.
- The electric force can be represented as an electric field created by a charge that affects another charge located inside the field. Similarly, the magnetic force can be represented by a magnetic field created by a current that affects another current located inside the field.

Consider two conducting wires with currents i_1 and i_2. Then the mutual inductance can be defined as

$$L = \frac{\Lambda_{12}}{i_1} = \frac{\Lambda_{21}}{i_2},$$

where Λ_{ab} is the magnetic flux linkage created by the wire a on the wire b. The unit of inductance is the henry (H), where 1 henry is 1 weber per ampere. Similarly to capacitance which represents the ability to collect charges (hence establish an electric field and store electric energy) for given a voltage, inductance represents the ability to establish magnetic flux (hence store magnetic energy) for a given current. And again similarly to capacitance, inductance can be defined for a single-body structure, which is called self-inductance.

Similar to capacitors, inductors (see Figure 11.3) are two-port devices that can store energy. Once again, we only consider ideal inductors, where the inductance does not depend on frequency, outer conditions, and applied voltage and current values. The voltage–current relationship of an inductor can be derived by considering the definition of inductance and taking the time derivative,

$$\frac{d\Lambda}{dt} = \frac{d}{dt}[Li(t)] = L\frac{di(t)}{dt}.$$

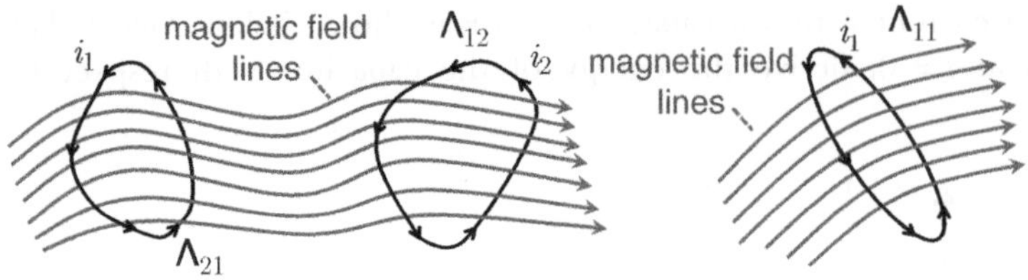

Figure 6.3 The inductance between two wires is proportional to the amount of magnetic flux created by one of the wires on the other, for a given current of the former, Λ_{12}/i_1 or Λ_{21}/i_2. The self-inductance of a single wire can be defined as the flux created by the wire on itself divided by its current.

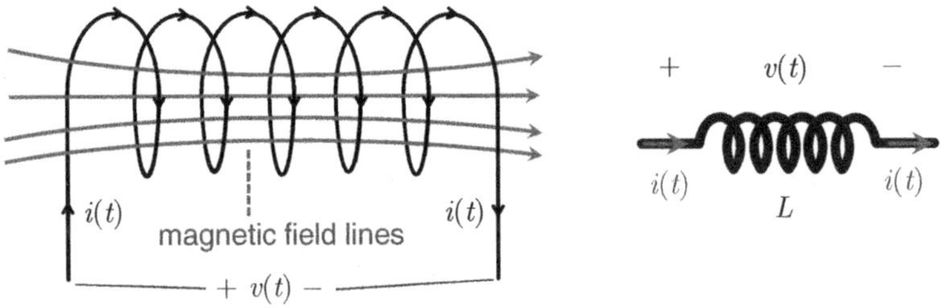

Figure 6.4 When the current through a wire changes with respect to time, leading to a change in the magnetic flux, a voltage difference is created between the terminals, even when the wire is perfectly conducting. In circuit analysis, an inductor is a two-port device that stores and releases energy.

According to Faraday's law of induction, a change in the magnetic flux corresponds to the electromotive force (voltage) which appears across the inductor, that is,

$$v(t) = L\frac{di(t)}{dt}.$$

Therefore, the voltage across an inductor is proportional to the time derivative of its current. Any increase in the current leads to a voltage drop as a response of the inductor to the changing magnetic field. In addition, the magnitude of the voltage is proportional to the inductance; the higher the inductance, the higher the voltage produced.

Now, very similarly to the capacitor case, one can derive the inverse relationship between the current and voltage. We have

$$i(t) = \frac{1}{L}\int_{t_0}^{t} v(t')dt' + i_0,$$

where $i_0 = i(t = t_0)$ is the initial current. If i_0 is not given, it is common to assume that $i_0 = 0$ and $t_0 = 0$. The equation above provides another insight for interpreting the working principles of inductors. Having a nonzero voltage across an inductor builds up current through it, and the integral term in the above can be interpreted as the accumulated current during $t' \in [t_0, t]$.

Next, we derive the power of an inductor as

$$p(t) = v(t)i(t) = Li(t)\frac{di(t)}{dt}.$$

Since the expression above can be positive or negative, an inductor may absorb or deliver energy at a particular time. For example, when the current is positive, a further increase in the current means that the inductor absorbs energy. The absorbed energy is stored in the inductor, and can be retrieved later. The energy stored can be derived as

$$w(t) = \int_{t_0}^{t} p(t')dt' = \int_{t_0}^{t} Li(t')\frac{di(t)'}{dt'}dt'$$
$$= \frac{1}{2}L[i(t)]^2,$$

assuming that $i(t_0) = 0$. Consequently, in order to find the energy stored by an inductor, it is sufficient to know its current, while zero current means zero energy.

Example 103: Consider a 1 H inductor. The current across the inductor is given by

$$i_L(t) = \begin{cases} t, & 0 \le t \le 2 \text{ s} \\ 5/2 - t/4, & 2 \le t \le 10 \text{ s} \\ 0, & t \ge 10 \text{ s} \end{cases} \quad \text{(A)}.$$

Find and plot the voltage across the inductor, as well as its power and energy for $t > 0$.

Solution: Using the expression for the voltage of an inductor, we have

$$v_L(t) = L\frac{di_L(t)}{dt} = \begin{cases} 1, & 0 < t < 2 \text{ s} \\ -1/4, & 2 < t < 10 \text{ s} \\ 0, & t > 10 \text{ s} \end{cases} \quad \text{(V)}.$$

Then the current and voltage across the inductor can be plotted as follows.

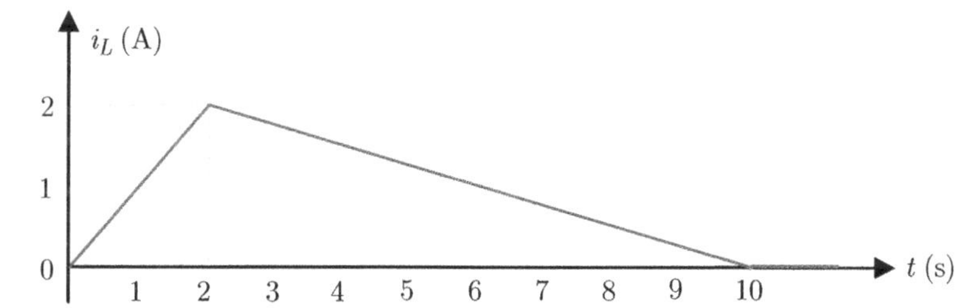

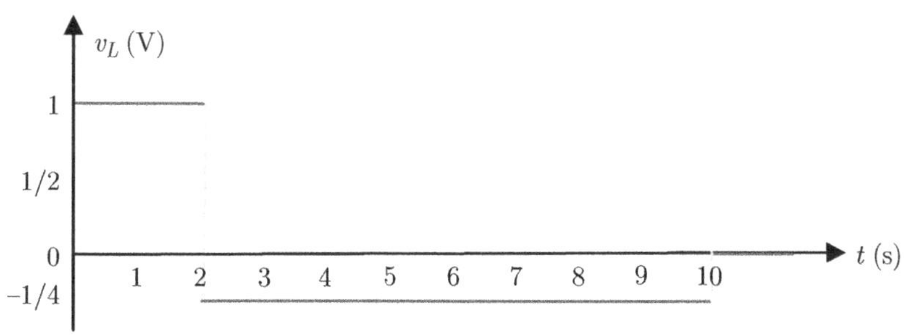

Even when they are modeled ideally, inductors in resistive circuits cannot have discontinuous currents. Specifically, it takes time for an inductor (connected to some resistors)

to store and release energy. Since the energy of an inductor is proportional to the square of its current, a jump in the current of an inductor requires a jump (instant change) in its energy that may not be physically possible. Such a discontinuous current at a particular time also needs an infinite voltage across the inductor.

Next, for the given voltage and current values above, the power of the inductor can be obtained as

$$p_L(t) = v_L(t)i_L(t) = \begin{cases} t, & 0 < t < 2\text{ s} \\ t/16 - 5/8, & 2 < t < 10\text{ s} \\ 0, & t > 10\text{ s} \end{cases} \quad \text{(W)}.$$

While the energy can be found by integrating the power, it can also be obtained as

$$w_L(t) = \frac{1}{2}L[i_L(t)]^2 = \begin{cases} t^2/2, & 0 \le t \le 2\text{ s} \\ 25/8 - 5t/8 + t^2/32, & 2 \le t \le 10\text{ s} \\ 0, & t \ge 10\text{ s} \end{cases} \quad \text{(J)}.$$

We note that, for $0 < t < 2$ s, the inductor absorbs energy, whereas it delivers energy for $2 < t < 10$ s. The power and energy of the inductor can be plotted with respect to time as follows.

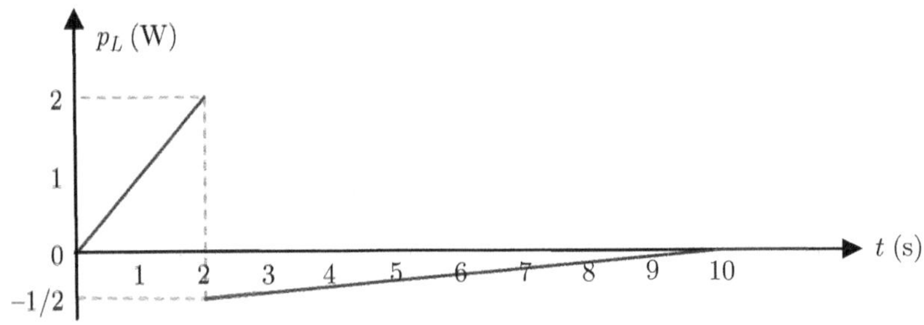

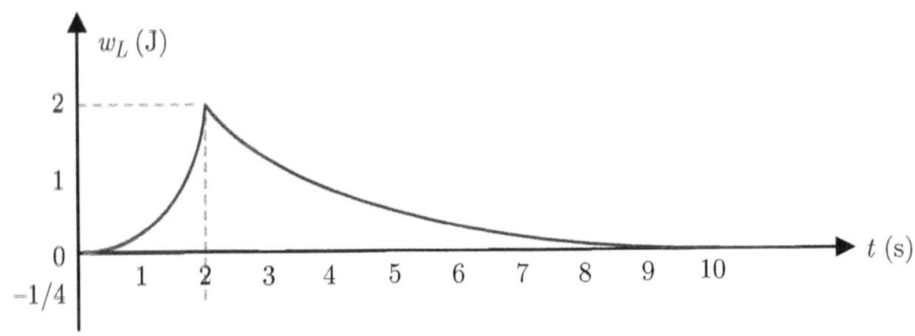

Exercise 95: The current through a 10 H inductor is given by

$$i_L(t) = \begin{cases} 40t, & 0 \le t \le 4\text{ s} \\ 140 + 5t, & 4 \le t \le 8\text{ s} \\ 180, & t \ge 8\text{ s} \end{cases} \quad \text{(A)}.$$

Find the voltage across the inductor, as well as its power and the stored energy.

6.3 Time-Dependent Analysis of Circuits in Transient State

Until now, we have considered capacitors and inductors as individual components. Given voltage/current across a capacitor or inductor and necessary initial conditions, one can find the current/voltage as well as the power and energy of the component. In addition, we claim that the voltage/current of a capacitor/inductor that is connected to resistors cannot have a discontinuity, since any jump in these values corresponds to a jump in the energy. In this section, we proceed further by considering simple circuits involving capacitors, inductors, and resistors connected to DC sources. We focus on the electrical variables with respect to time and study how circuits behave in the transient state.

6.3.1 Time-Dependent Analysis of RC Circuits

Consider the following circuit involving a capacitor connected in series to a resistor. Since it contains a resistor (R) and a capacitor (C), it is commonly called an RC circuit in the literature.

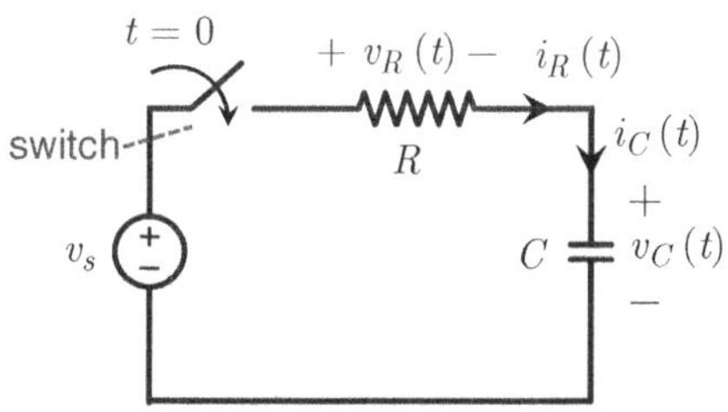

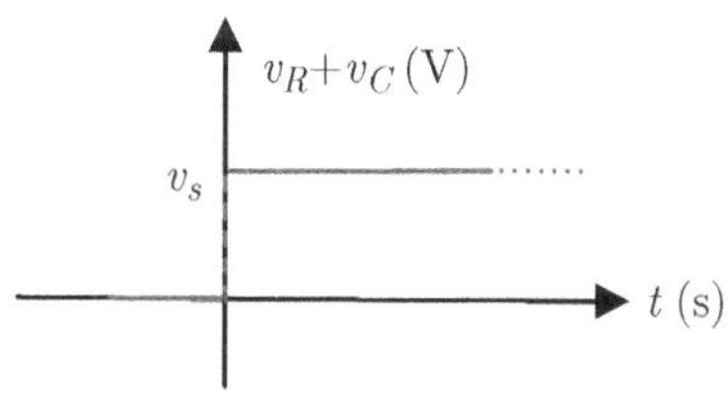

It is also given that $v_C(t) = 0$ for $t < 0$, which indicates that the capacitor is empty with zero energy for $t < 0$. However, at $t = 0$, a switch is closed and a voltage source with value v_s becomes connected to the capacitor and resistor. We assume that the switch is perfect, so that it closes exactly at $t = 0$. Therefore, using KVL, the voltage provided by the source must appear across the combination of the resistor and capacitor. Specifically, for $t > 0$, we have

$$v_R(t) + v_C(t) = v_s,$$

while v_R and v_C depend on time. Our aim is to find these voltages, as well as the value of the current through the resistor and capacitor with respect to time.

In order to solve the problem, we note that

$$i_R(t) = i_C(t) = \frac{v_R(t)}{R} = C\frac{dv_C(t)}{dt}$$

due to the voltage–current characteristics of the resistor and capacitor. In addition, $v_R(t) = v_s - v_C(t)$, leading to

$$C\frac{dv_C(t)}{dt} + \frac{v_C(t)}{R} = \frac{v_s}{R}$$

or

$$\frac{dv_C(t)}{dt} + \frac{v_C(t)}{RC} = \frac{v_s}{RC}.$$

The equation above can be solved by substituting

$$v_C(t) = b\exp(at) - b,$$

where a and b are constants. Using the expression for the voltage in the main equation, we get

$$ab\exp(at) + \frac{b}{RC}\exp(at) - \frac{b}{RC} = \frac{v_s}{RC},$$

leading to

$$a = -\frac{1}{RC}$$
$$b = -v_s.$$

Therefore, we obtain

$$v_C(t) = -v_s\exp(-t/RC) + v_s = v_s[1 - \exp(-t/RC)].$$

In addition, we can derive

$$v_R(t) = v_s - v_C(t) = v_s\exp(-t/RC),$$
$$i_R(t) = i_C(t) = \frac{v_s}{R}\exp(-t/RC).$$

The voltage and current values above provide interesting information on the behavior of the RC circuit. First, for a better understanding, we plot these values with respect to time as follows.

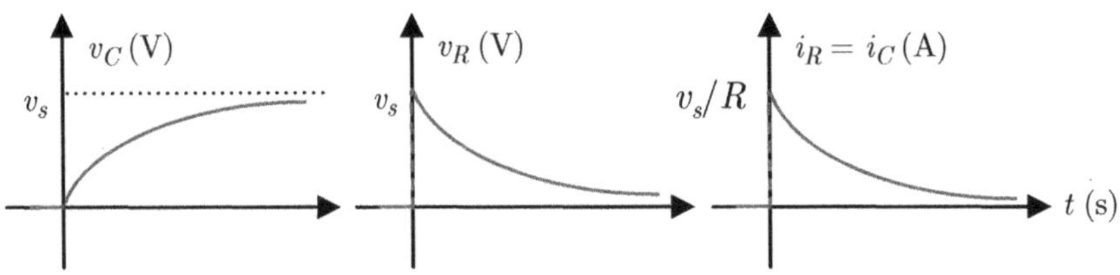

We have the following observations.

- The voltage of the capacitor increases with respect to time from 0 toward v_s. As it increases, however, the rate of change decreases continuously. Specifically, the voltage (hence the energy) of the capacitor increases more and more slowly with time.
- The voltage of the resistor has a jump at $t = 0$ from 0 to v_s. Then it decreases toward 0 as a function of time, since the sum of the voltages across the resistor and capacitor must be fixed.
- The current through the resistor and capacitor has a jump at $t = 0$ from 0 to v_s/R. Then it also decreases toward 0 as time proceeds. Larger values of the current at the beginning mean that the capacitor is filled (with electric energy) faster. But, as time progresses, the filling process slows down and the current value drops to zero.

Basically, in this transient analysis, we witness how the capacitor is filled with energy provided by the voltage source. Obviously, there is a limit to the energy that can be stored, given by

$$w_C^{\max} = \frac{1}{2}C[v_s]^2,$$

since v_s is the maximum voltage that the capacitor can have. Theoretically, the voltage of the capacitor can never reach v_s, unless we consider the limit case as

$$v_C(t = \infty) \to v_s,$$
$$v_R(t = \infty) \to 0,$$
$$i_R(t = \infty) = i_C(t = \infty) \to 0.$$

These final values represent the steady state of the circuit. As defined before, a steady state is an equilibrium state, where the circuit is not affected by outer effects (a switch in this case). For a circuit involving only DC sources, the steady state corresponds to the case where the voltage and current values do not change. For the above circuit, the steady state occurs when $v_C = v_s$ and

$$i_C = C\frac{dv_C}{dt} = 0.$$

In general, it can be deduced that the current though any capacitor must be zero in steady state when only DC sources are involved. This is because when the voltage of a capacitor becomes constant, its derivative (hence the current across it) must be zero. In the literature, this is often interpreted as the capacitor becoming open circuit in steady state, since its voltage can be anything (depending on the rest of the circuit) while its current must be zero in the DC equilibrium.

As also discussed before, except for very special and ideal cases, infinite time is required to pass from transient state to equilibrium. In the expressions for the voltage and current above, the time variance is $\exp(-t/RC)$, which describes how the values change with respect to time. In this exponential function, $\tau = RC$ is a special quantity called the time constant. It represents how fast the circuit approaches the steady state, that is, the state when electrical variables (voltages and currents) become constant (due to DC sources). Specifically, if RC is large, it takes a longer time for the variables to come close to steady state. If RC is small, however, the circuit quickly approaches the steady state. We note that the unit of RC is volt/ampere × coulomb/volt = coulomb/ampere = second. While, theoretically, it takes infinite time to completely enter steady state, most circuits are assumed to reach equilibrium after a sufficient period. For example, when $t = 5\tau$ and $t = 10\tau$, the voltage of the capacitor can be evaluated as

$$v_C(t = 5\tau) \approx 0.9932621 v_s$$
$$v_C(t = 10\tau) \approx 0.9999546 v_s.$$

Obviously, the smaller the value of τ, the shorter the time to steady state that one can assume. The following figure illustrates how the value of v_C changes with respect to time for different values of τ.

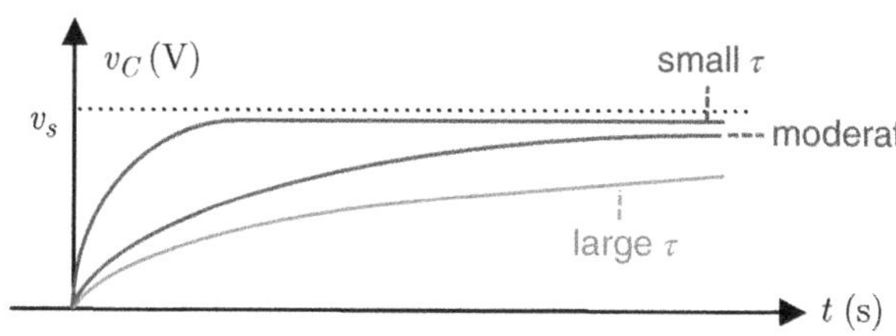

At this stage, it is interesting to consider two special cases, namely, when $R = 0$ and $C = 0$ in the RC circuit above. We have the following circuits.

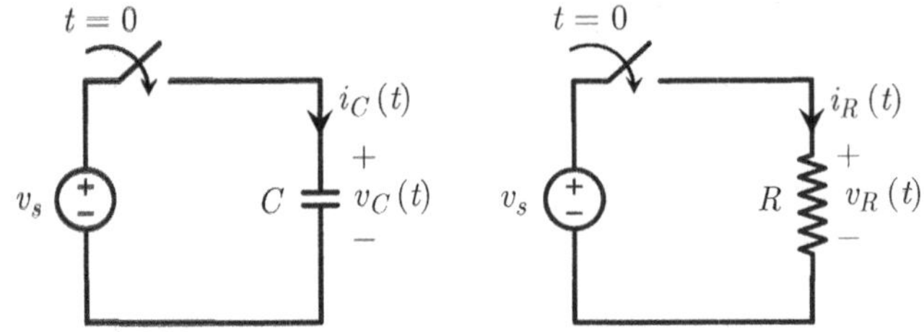

In both circuits, we first note that $\tau = RC = 0$, that is, the time constant is zero. Then the exponential function $\exp(-t/RC)$ is also zero for $t > 0$. This means that, in both cases, it takes zero time to reach the steady state. In the case of the resistor, the total voltage applied by the source appears across the resistor immediately when the switch is closed. In general, any circuit involving only DC sources and resistors has this kind of ideal behavior. This is indeed the reason why we simply do not consider the transient state in the previous chapters. In the case of the capacitor, the total voltage again appears across the capacitor once the switch is closed. In this ideal (zero-resistance) case, it takes zero time for the capacitor to collect all charges that are required to set up the voltage v_s so that the current becomes immediately zero for $t > 0$. But then how are these charges are collected in the capacitor? If we consider that the voltage is discontinuous, changing from zero to v_s at $t = 0$, its derivative must be infinite at $t = 0$. This super-ideal current that has an infinite value at a single time is responsible for the setup of the capacitor voltage.

In the above, we studied a series connection of a resistor and capacitor in detail. Now, we consider another circuit involving a parallel connection of a resistor and capacitor.

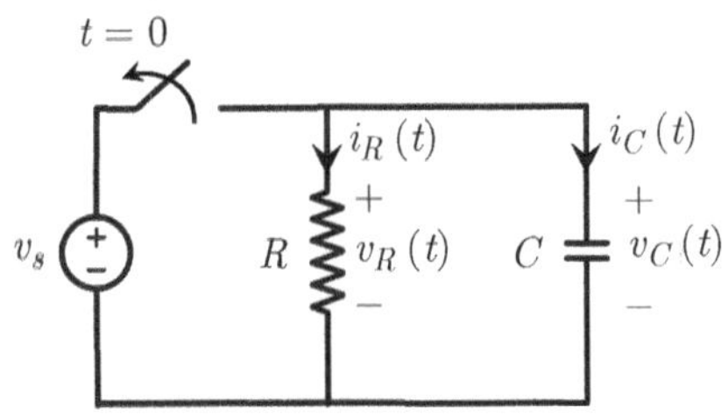

After a long time, the switch is opened at $t = 0$. Therefore, it can be assumed that the circuit is in steady state before the switch is opened. This means that the capacitor is full at $t = 0$ and $v_C(0) = v_s$. After the switch is opened, the voltage source becomes disconnected from the RC part (resistor and capacitor). In the RC part, we have

$$v_R(t) = v_C(t),$$
$$i_R(t) = -i_C(t),$$

or

$$\frac{v_R(t)}{R} = -C\frac{dv_C(t)}{dt}.$$

Therefore,

$$\frac{dv_C(t)}{dt} + \frac{v_C(t)}{RC} = 0.$$

The equation above can be solved by using

$$v_C(t) = v_s \exp(at),$$

where a is a constant. Substituting the voltage, we further get

$$av_s \exp(at) + \frac{v_s}{RC} \exp(at) = 0,$$

leading to

$$a = -\frac{1}{RC}.$$

Hence, we get

$$v_C(t) = v_R(t) = v_s \exp(-t/RC),$$
$$i_R(t) = \frac{v_s}{R} \exp(-t/RC) = -i_C(t).$$

Once again, we reach voltage and current values that are changing with a time constant RC. The capacitor voltage and current can be plotted as follows.

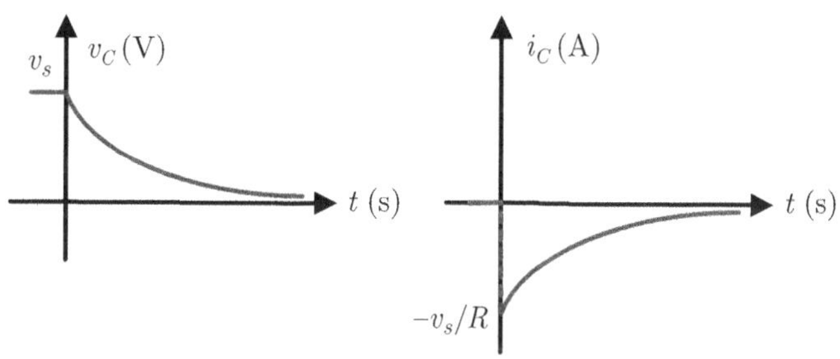

Furthermore, in steady state, we have

$$v_C(t = \infty) \to 0,$$
$$v_R(t = \infty) \to 0,$$
$$i_R(t = \infty) = -i_C(t = \infty) \to 0.$$

But what about the power of the capacitor in the circuit above? Using the expressions for the voltage and current, we derive

$$p_C(t) = -\frac{v_s^2}{R} \exp(-2t/RC).$$

Therefore, the power is negative for $t > 0$, and it is larger in magnitude for small values of t. In fact, when the switch is opened, the capacitor starts to supply energy (with decreasing rate), which is consumed by the resistor. This can be verified by finding the power of the resistor, $p_R(t) = -p_C(t)$ for $t > 0$.

Example 104: Consider the following circuit involving a capacitor of 0.5 μF.

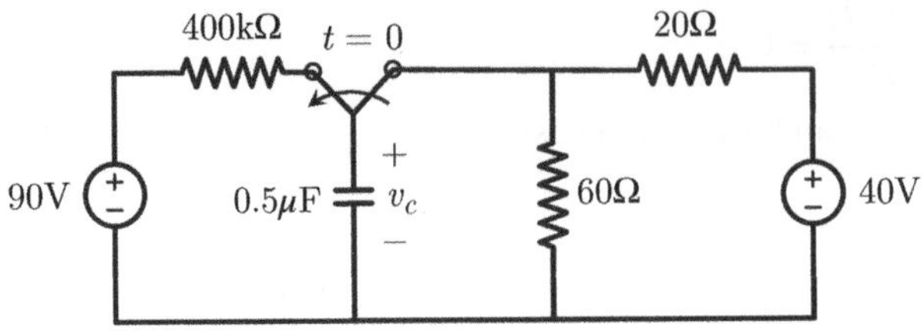

The switch stays in position a for a long time. Then, at $t = 0$, it is changed to position b. Find the voltage of the capacitor as a function of time for $t > 0$.

Solution: Before the switch is changed, the capacitor behaves like an open circuit (after a long time). Therefore, 40 V voltage is divided between 20 Ω and 60 Ω resistors that are simply connected in series. We have

$$v_C(0) = \frac{60}{60+20} \times 40 = 30 \text{ V}.$$

When the switch is changed to position b, the capacitor becomes connected to 400 kΩ and 90 V. The circuit in this case can be considered as follows.

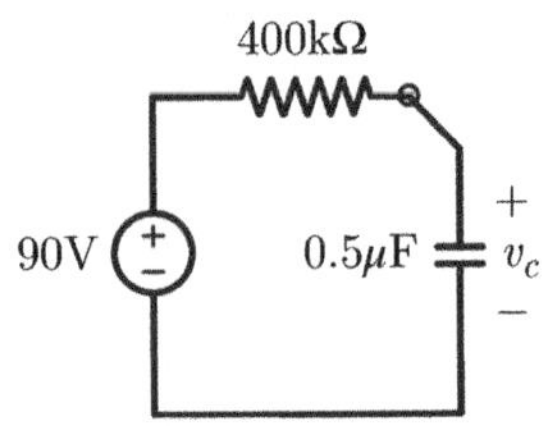

This is simply an RC circuit, and the time constant can be found to be

$$RC = 400 \times 10^3 \times 0.5 \times 10^{-6} = 200 \times 10^{-3} = 0.2 \text{ s}.$$

Then the voltage of the capacitor can be written in general as

$$v_C(t) = b\exp(-t/RC) + d,$$

where

$$v_C(0) = b + d = 30 \text{ V}.$$

In addition, we have the final (steady) state as

$$v_C(\infty) = d = 90 \text{ V}.$$

Therefore, we find $b = 30 - d = -60$ V, and

$$v_C(t) = -60\exp(-t/0.2) + 90 \text{ V}.$$

Using the final expression above, we can find the value of the voltage at any time $t > 0$. For example, at $t = 5$ s, we have

$$\begin{aligned} v_C(5) &= -60\exp(-5/0.2) + 90 \\ &= -60\exp(-25) + 90 \approx 89.9999999992 \text{ V}. \end{aligned}$$

Since the given time (5 s) is much larger than the time constant (0.2 s), the value of $v_C(5)$ is very close to the final state 90 V. We can find the time at which the voltage of the capacitor equals 89 V:

$$v_C(t) = -60 \exp(-t/0.2) + 90 = 89 \longrightarrow \exp(-t/0.2) = 1/60,$$

leading to

$$t = 0.2 \ln(60) \approx 0.82 \text{ s}.$$

Example 105: Consider the following circuit, which is very similar to the previous one, except for the reversed voltage source on the right.

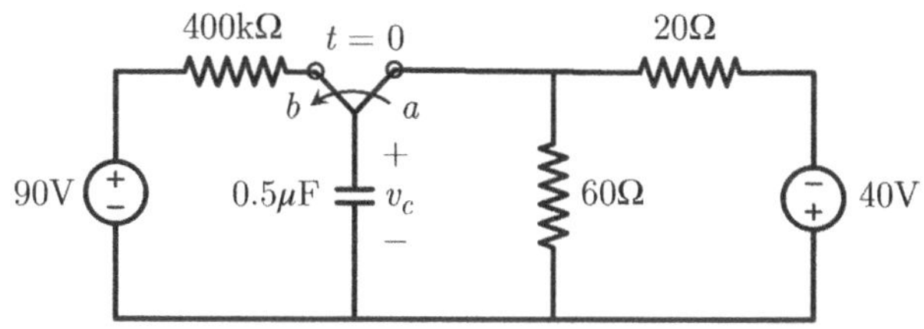

In this case, we have

$$v_C(0) = \frac{60}{60 + 20} \times (-40) = -30 \text{ V}$$

as the initial voltage of the capacitor. After the switch is changed, the circuit again becomes an RC circuit, with time constant 0.2 s. Using the general expression

$$v_C(t) = b \exp(-t/0.2) + d,$$

we have $d = 90$ due to the final state $v_C(\infty) = 90$ V and $b = -120$ V to satisfy $v_C(0) = -30$ V. Therefore, the voltage of the capacitor can be written as

$$v_C(t) = -120 \exp(-t/0.2) + 90 \text{ V}.$$

Example 106: Consider the following circuit involving a 2 F capacitor connected to resistors and a voltage source.

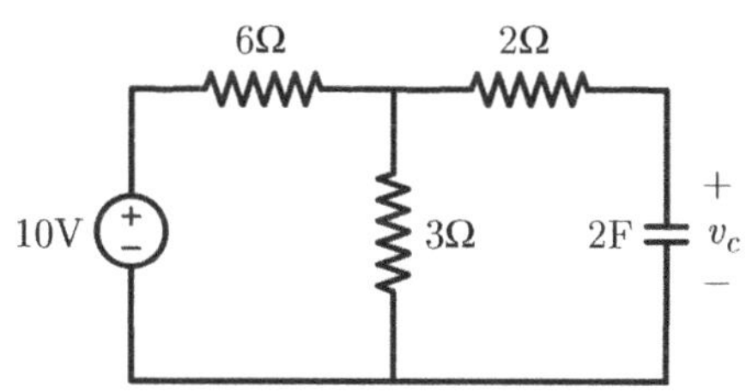

Given that $v_C(0) = 2$ V, find the voltage of the capacitor as a function of time for $t > 0$.

Solution: This circuit is not trivial to analyze as it contains parallel and series connections. On the other hand, one can find the Thévenin equivalent seen by the capacitor so that the overall structure reduces to an RC circuit. First, we have

$$v_{oc} = \frac{3}{6+3} \times 10 = 10/3 \text{ V},$$

considering an open circuit instead of the capacitor. In addition, one can derive

$$i_{sc} = \frac{3}{3+2} \times \frac{10}{6+2 \parallel 3} = \frac{3}{5} \times \frac{10}{6+6/5} = \frac{3}{5} \times \frac{50}{36} = \frac{5}{6} \text{ A}.$$

Therefore, the Thévenin equivalent resistor can be found to be

$$R_{th} = \frac{v_{oc}}{i_{sc}} = \frac{10/3}{5/6} = 4\ \Omega.$$

Consequently, the following RC circuit should be considered, where $\tau = RC = 8$ s.

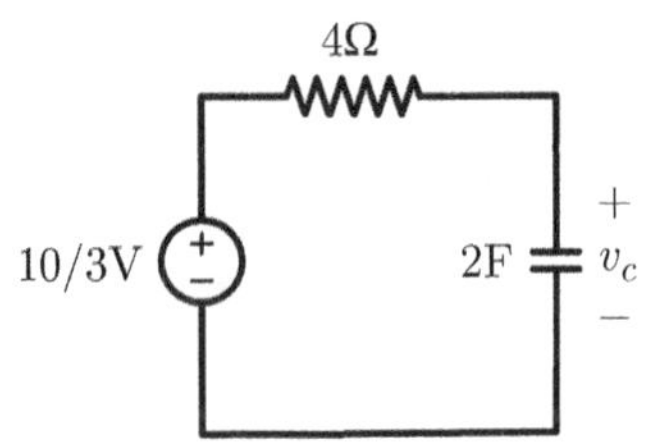

Using $v_C(0) = 2$ V and $v_C(\infty) = 10/3$ V, we obtain

$$v_C(t) = -\frac{4}{3} \exp(-t/8) + \frac{10}{3} \text{ V}.$$

Example 107: Consider the following circuit.

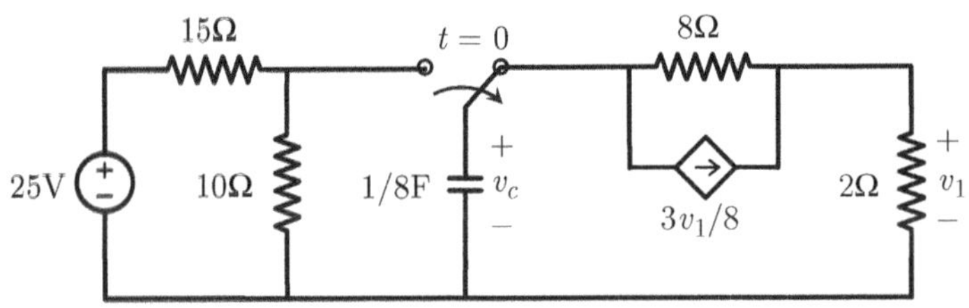

After a long time, the switch changes position. Find $v_C(t)$ as a function of time for $t > 0$.

Solution: For $t = 0^-$, we again have a voltage division, leading to

$$v_C(0^-) = 25 \times 10/(10+15) = 10 \text{ V}.$$

After the switch is changed, the circuit becomes as follows.

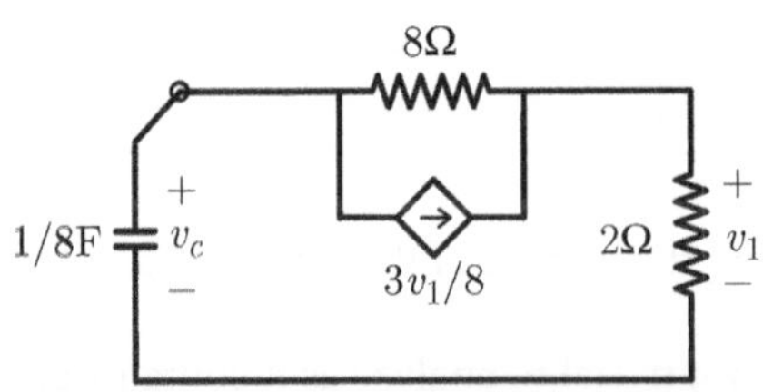

This circuit cannot be solved by considering the standard Thévenin equivalent seen by the capacitor. However, we can proceed by considering that

$$v_1(t) = -2i_C(t).$$

In addition, applying KCL, we have

$$i_C(t) = \frac{v_C(t) - v_1(t)}{8} + \frac{3v_1(t)}{8},$$

leading to

$$v_1(t) = 4i_C(t) - \frac{v_C(t)}{2}.$$

Combining the equations, we derive

$$i_C(t) = \frac{v_C(t)}{4}.$$

Finally, considering the equation for the capacitor (with $C = 1/8$ F), we have

$$i_C(t) = \frac{v_C(t)}{4} = -\frac{1}{8}\frac{dv_C(t)}{dt}$$

or

$$\frac{dv_C(t)}{dt} + \frac{v_C(t)}{1/2} = 0.$$

Obviously, this equation is the same as an RC circuit with a time constant of 1/2 s. This means that, if the Thévenin circuit seen by the capacitor were found, the equivalent resistor value would be 4 Ω. The solution can be written as

$$v_C(t) = b\exp(-2t) + d,$$

where $d = v_C(\infty) = 0$ and $b = v_C(0) - d = 10$. Therefore, we have

$$v_C(t) = 10\exp(-2t)$$

as the capacitor voltage for $t > 0$.

Exercise 96: For the following circuit, find the expression for $v_C(t)$ as a function of time, given that $v_C(0) = 3/8$ V.

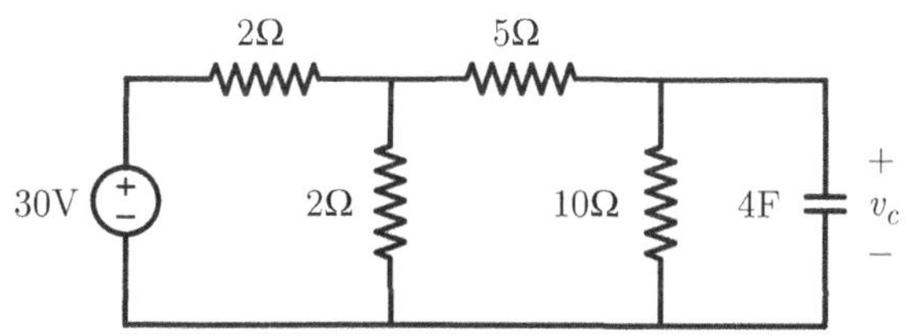

Exercise 97: Consider the following circuit, where it is given than $v_C(0) = 10$ V.

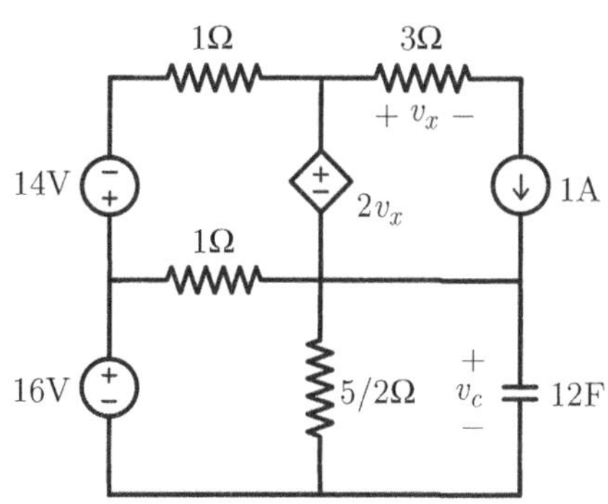

- Find the expression for the capacitor voltage $v_C(t)$ as a function of time for $t > 0$.
- Find the expression for the capacitor current $i_C(t)$ as a function of time for $t > 0$.
- Find the energy stored in the capacitor at $t = 5\ln(2)$ s.

6.3.2 Time-Dependent Analysis of RL Circuits

Circuits involving inductors and resistors also demonstrate transient behaviors that can be formulated with time constants. Consider the following circuit involving an inductor connected in parallel to a resistor.

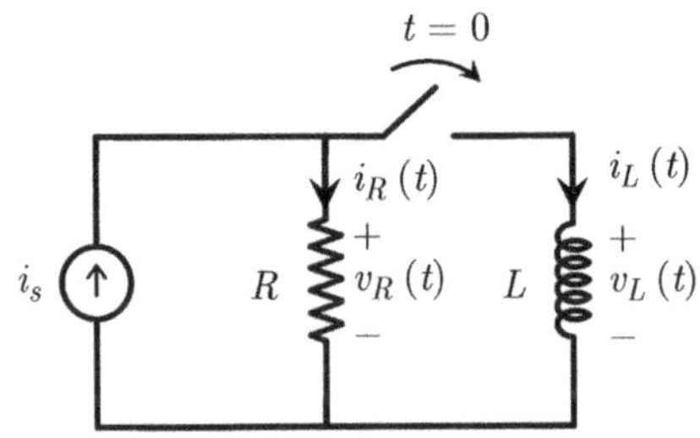

At $t = 0$, the switch is closed and a current source becomes connected to both inductor and resistor. In addition, we note that $i_L(0) = 0$ since the inductor is not connected to any closed loop before the switch is closed. After the switch is closed, we have

$$v_R(t) = v_L(t) = L\frac{di_L(t)}{dt},$$

where

$$v_R(t) = Ri_R(t) = Ri_s - Ri_L(t)$$

since $i_R(t) = i_s - i_L(t)$. Therefore,

$$L\frac{di_L(t)}{dt} + Ri_L(t) = Ri_s$$

or

$$\frac{di_L(t)}{dt} + \frac{R}{L}i_L(t) = \frac{R}{L}i_s.$$

The final equation can be solved by substituting

$$i_L(t) = b\exp(at) - b,$$

where a and b are constants. We get

$$ab\exp(at) + \frac{Rb}{L}\exp(at) - \frac{Rb}{L} = \frac{R}{L}i_s,$$

leading to

$$a = -\frac{R}{L},$$
$$b = -i_s.$$

Hence, the current of the inductor can be rewritten as

$$i_L(t) = -i_s \exp(-tR/L) + i_s = i_s[1 - \exp(-tR/L)].$$

Furthermore, we have

$$i_R(t) = i_s - i_L(t) = i_s \exp(-tR/L),$$
$$v_R(t) = v_L(t) = Ri_s \exp(-tR/L),$$

for $t > 0$. The current and voltage values can be plotted with respect to time as follows.

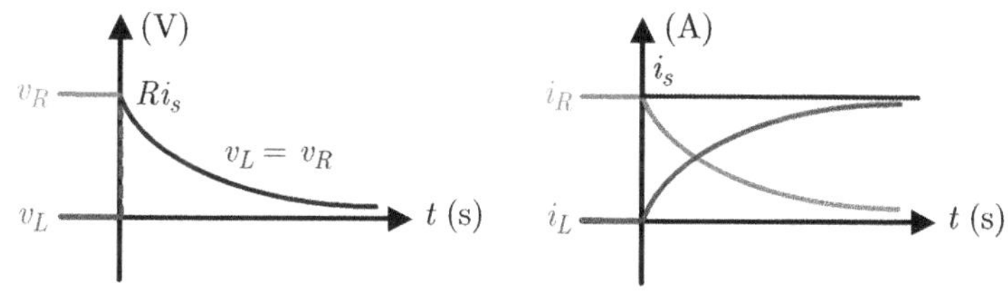

It can be seen that the RL circuit above demonstrates transient characteristics similar to RC circuits. Once the switch is closed, the current of the inductor starts to increase with a time constant of $\tau = L/R$. While the limit value is i_s, the current through the inductor never reaches this value for finite values of t. Specifically, the rate of change in the inductor current decreases as time passes. With the increase in the current value, the energy stored in the inductor also increases and approaches a limit value

$$w_L^{\max} = \frac{1}{2}L[i_s]^2.$$

When the voltage across the components is investigated, it can be seen that it decays from a maximum value of Ri_s to zero, while the rate of change again decreases for large values of t. In the steady state, we have

$$i_L(t = \infty) \to i_s,$$
$$i_R(t = \infty) \to 0,$$
$$v_R(t = \infty) = v_L(t = \infty) \to 0.$$

In practice, RL circuits are also assumed to enter steady state for sufficiently large values of t (e.g., for $t > 5\tau$).

Obviously, the inductor in the circuit above has a zero voltage in steady state. This is because the inductor energy is saturated (the current through the inductor is maximized) and it cannot store any further energy. Since the current through the inductor can be arbitrary depending on the rest of the circuit (the current source in this case), the inductor can be thought of as a short circuit after a sufficient time. In fact, for DC circuits, any inductor can be modeled as a short circuit, very similarly to the open-circuit characteristics of the capacitors, in steady state. This is particularly because a constant current i_L leads to

$$v_L(t) = L\frac{di_L(t)}{dt} = 0,$$

no matter what the value of i_L is.

Similarly to RC circuits, there are different versions of RL circuits. For example, we consider the following scenario involving an inductor connected in series to a resistor and voltage source when the switch is closed at $t = 0$.

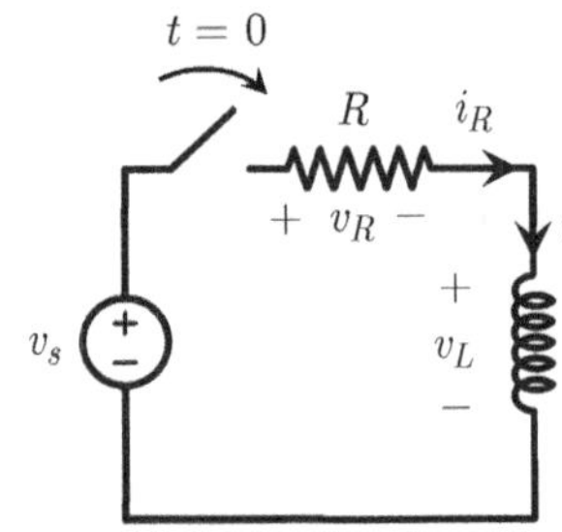

After the switch is closed, we have $i_R(t) = i_L(t)$, while

$$v_R(t) + v_L(t) = v_s$$

for all $t > 0$. Using the voltage–current relationships for both resistor and inductor, we derive

$$Ri_R(t) + L\frac{di_L(t)}{dt} = v_s$$

or

$$\frac{di_L(t)}{dt} + \frac{R}{L}i_R(t) = v_s.$$

The solution to this equation, which is similar to the previous one, can be found to be

$$i_L(t) = -\frac{v_s}{R}\exp\left(-tR/L\right) + \frac{v_s}{R} = \frac{v_s}{R}[1 - \exp\left(-tR/L\right)],$$

considering that $i_L(t = \infty) = v_s/R$ and $i_L(t = 0) = 0$. We note that, in steady state, the inductor again becomes short circuit and all voltage provided by the source appears across the resistor. This leads to a steady current as

$$i_R(t = \infty) = i_L(t = \infty) = v_s/R.$$

The energy stored in the inductor in steady state can be found to be

$$w_L^{\max} = \frac{1}{2}L[i_L]^2 = \frac{1}{2}L\frac{v_s^2}{R^2},$$

which is actually the maximum energy that the inductor can have.

Exercise 98: Consider the following circuit, where it is given than $i_L(0) = 8\,\text{V}$.

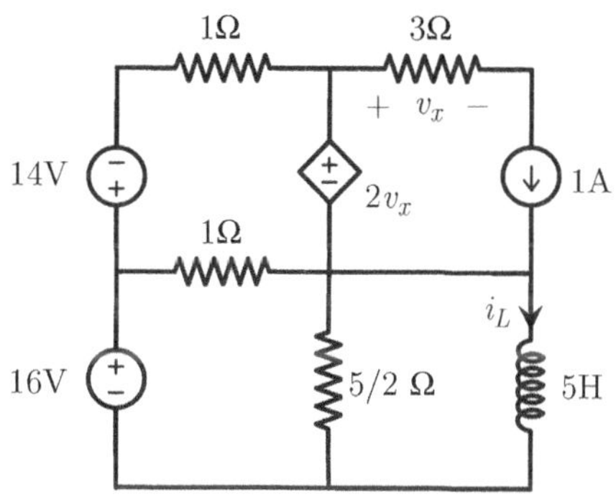

Find the power of the inductor at $t = 12\,\text{s}$.

6.3.3 Impossible Cases

As discussed in Section 1.5, some connections of voltage and current sources, as well as open and short circuits, are not allowed if ideal components are assumed. Similar impossible connections exist when ideal capacitors and inductors are considered, since these components become open and short circuits in steady state. As an example, we consider the following case.

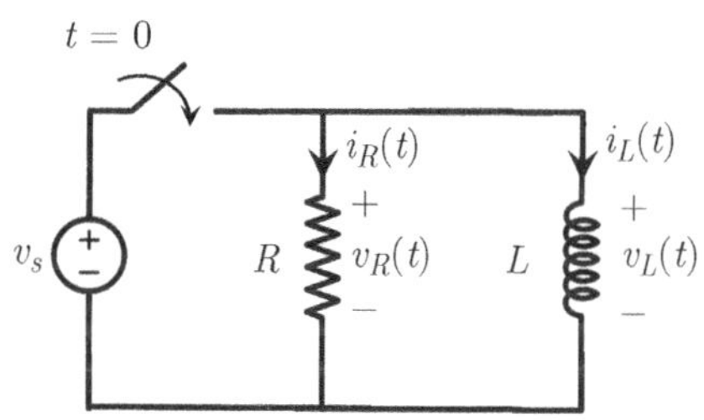

This circuit is theoretically impossible. In order to understand this, one can use the definition of the inductor as

$$v_L(t) = L\frac{di_L(t)}{dt},$$

which must be equal to v_s once the switch is closed. This means that $di_L(t)/dt$ is constant for all $t > 0$. Therefore, the circuit never reaches a steady state; the current and energy stored in the inductor increase to infinity as time passes. In practice, the inductor has a small internal resistance, which makes the circuit reach steady state with a large but almost constant current passing through the inductor (obviously this can be a dangerous experiment).

Another impossible scenario is a simple connection of a current source and a capacitor.

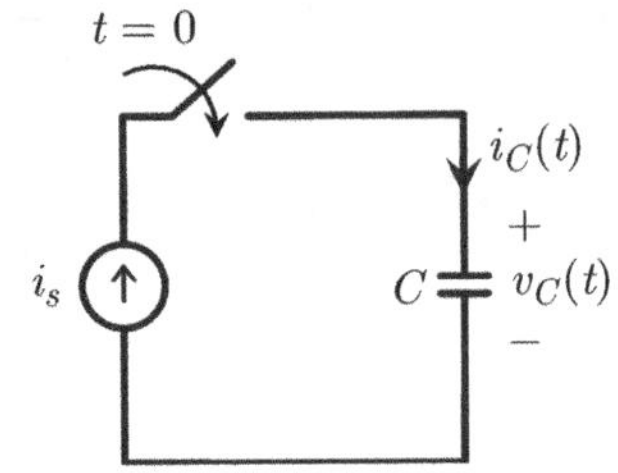

This circuit also does not have a steady state. With ideal components, the voltage and energy of the capacitor increase without bound. In practice, a series internal resistance of the capacitor inhibits infinite voltage accumulation, while the voltage of the capacitor can be extremely large.

6.4 Switching and Fixed-Time Analysis

As discussed in the previous sections, we can analyze simple circuits involving combinations of resistors and capacitors/inductors by using the voltage/current characteristics of the components. In fact, more complex circuits involving resistors, capacitors, and inductors can be analyzed with respect to time, from transient state to steady state, while this can be challenging when multiple capacitors and inductors are involved. As shown in Section 6.5, parallel and series connections of capacitors and inductors can be simplified. In addition, when a circuit involves large numbers of resistors and sources, it can be investigated by using the Thévenin/Norton equivalent circuits seen by a capacitor/inductor.

In some cases, it may be sufficient to analyze circuits at a fixed time when switches are opened and/or closed. In this type of analysis, one can enforce the continuity of the voltage and current values across capacitors and inductors, respectively, considering that these quantities cannot jump if they are connected to resistors. This section is devoted to such fixed-time analysis of circuits.

Example 108: Consider the following circuit involving a capacitor connected in parallel to a resistor $R_2 = 4\ \Omega$, and both are connected in series to another resistor $R_1 = 4\ \Omega$.

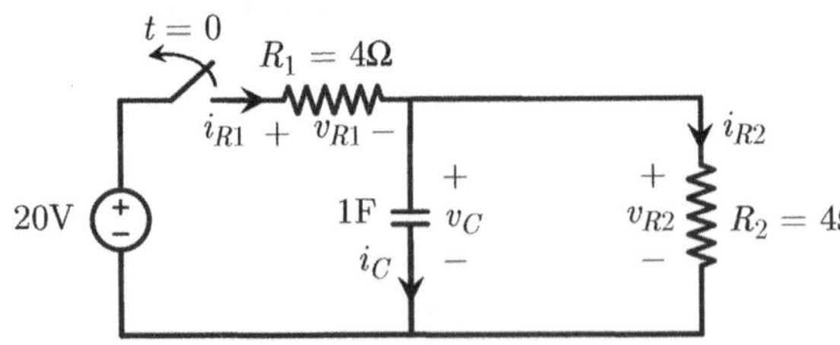

A switch is opened at $t = 0$. Just before the switch is opened, the current across R_2 given by $i_{R2}(0^-) = 2$ A. Analyze the circuit just after the switch is opened.

Solution: This problem can easily be solved by drawing up a table and filling it by inspecting the circuit both before and after the switch is changed, at $t = 0^-$ and $t = 0^+$, respectively. We start by listing the variables as follows.

	i_{R1}	i_{R2}	i_C	v_{R1}	v_{R2}	v_C
$t = 0^-$		2 A				
$t = 0^+$						

Then, using Ohm's law, we have

$$v_{R2}(0^-) = 4i_{R2}(0^-) = 8 \text{ V}.$$

In addition, we derive

$$v_C(0^-) = v_{R2}(0^-) = 8 \text{ V},$$
$$v_{R1}(0^-) = 20 - v_{R2}(0^-) = 12 \text{ V},$$
$$i_{R1}(0^-) = v_{R1}(0^-)/4 = 3 \text{ A}$$

by inspecting the circuit. We update the table as follows.

	i_{R1}	i_{R2}	i_C	v_{R1}	v_{R2}	v_C
$t = 0^-$	3 A	2 A		12 V	8 V	8 V
$t = 0^+$						

In order to complete the first row, we use KCL to get

$$i_C(0^-) = i_{R1}(0^-) - i_{R2}(0^-) = 3 - 2 = 1 \text{ A}.$$

After the switch is opened, the voltage of the capacitor should not change, $v_C(0^+) = v_C(0^-) = 8$ V. This is the key to analyzing the circuit just after the switch is opened. Now, due to the parallel connection, we also have $v_{R2}(0^+) = v_C(0^+) = 8$ V and $i_{R2}(0^+) = v_{R2}(0^+)/4 = 2$ A. Hence, the table can be updated as follows.

	i_{R1}	i_{R2}	i_C	v_{R1}	v_{R2}	v_C
$t = 0^-$	3 A	2 A	1 A	12 V	8 V	8 V
$t = 0^+$		2 A			8 V	8 V

Finally, due to the open circuit, $i_{R1}(0^+) = 0$, $v_{R1}(0^+) = 0$, and

$$i_C(0^+) = i_{R1}(0^+) - i_{R2}(0^+) = 0 - 2 = -2 \text{ A}.$$

As a result, we can complete our table as shown below.

	i_{R1}	i_{R2}	i_C	v_{R1}	v_{R2}	v_C
$t = 0^-$	3 A	2 A	1 A	12 V	8 V	8 V
$t = 0^+$	0	2 A	−2 A	0	8 V	8 V

Several points must be emphasized at this stage.

- The current across a capacitor can be discontinuous while its voltage must be continuous, if it is connected to resistors.
- Continuity of the capacitor voltage is inconsistent with the fact that the energy stored in a capacitor needs time to be released if there are resistors in the circuit.
- The power of a capacitor can be discontinuous. In the example above, we have $p_C(0^-) = 8$ W and $p_C(0^+) = -16$ W. Hence, the capacitor is absorbing (storing) energy before the switch is changed. When the switch is turned off, it starts to deliver (release) energy, which is obviously consumed by R_2.
- The fixed-time analysis above does not give information for $t > 0$.

Example 109: Consider the following circuit.

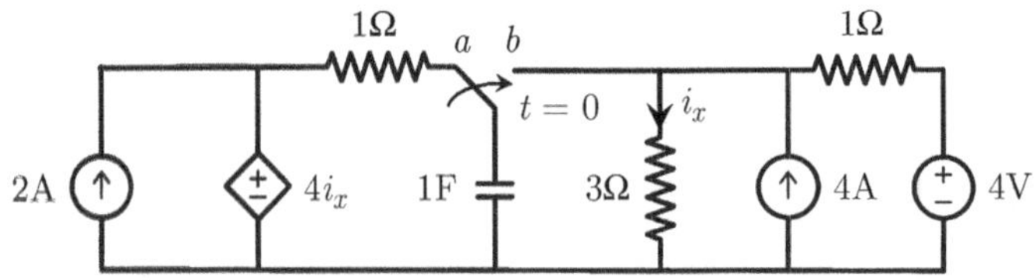

The switch remains in position a for a long time. Then, at $t = 0$, it is changed to position b. Find the power of the 2 A current source at $t = 0^+$, just after the switch is changed.

Solution: When the switch is in position a, we have the following circuit.

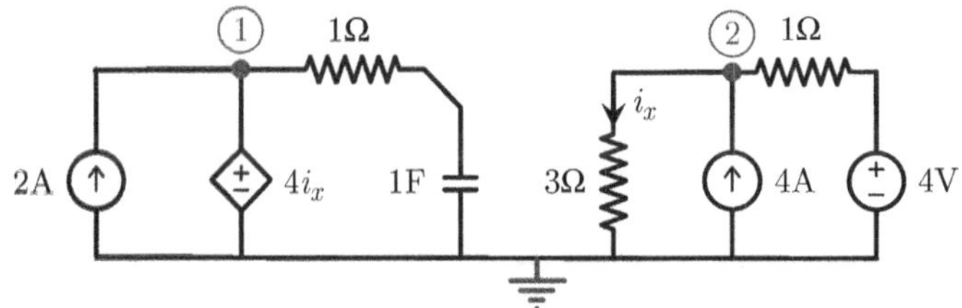

Applying KCL at node 2, we obtain

- KCL(2): $4 - v_2/3 - (v_2 - 4)/1 = 0 \longrightarrow v_2(0^-) = 6$ V.

Therefore, we have $i_x(0^-) = v_2(0^-)/3 = 2$ A, $v_1(0^-) = 4i_x(0^-) = 8$ V, and $v_C(0^-) = v_1(0^-) = 8$ V since the capacitor acts like an open circuit.

Then, when the switch is changed to position b, the circuit becomes as follows.

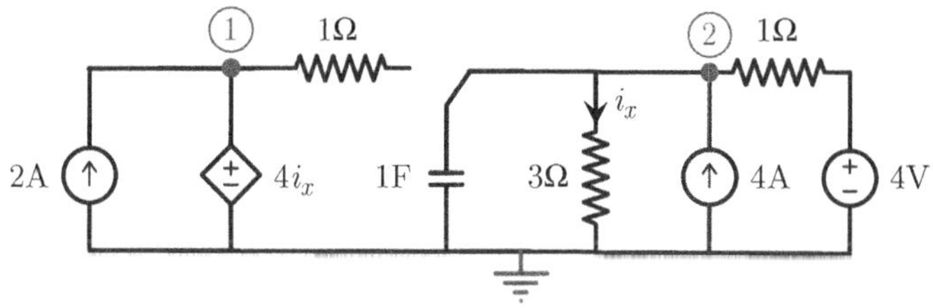

Since the voltage of the capacitor does not jump, we still have $v_C(0^+) = 8$ V. Therefore, the value of i_x can be found to be

$$i_x(0^+) = v_C(0^+)/3 = 8/3 \text{ A},$$

leading to

$$v_1(0^+) = 4i_x(0^+) = 32/3 \text{ V}$$

and

$$p_{2\text{ A}}(0^+) = -v_1(0^+) \times 2 = -64/3 \text{ W}.$$

This power value indicates that the 2 A current source delivers power at $t = 0^+$.

Example 110: Consider the following circuit involving an inductor connected to resistors and a current source.

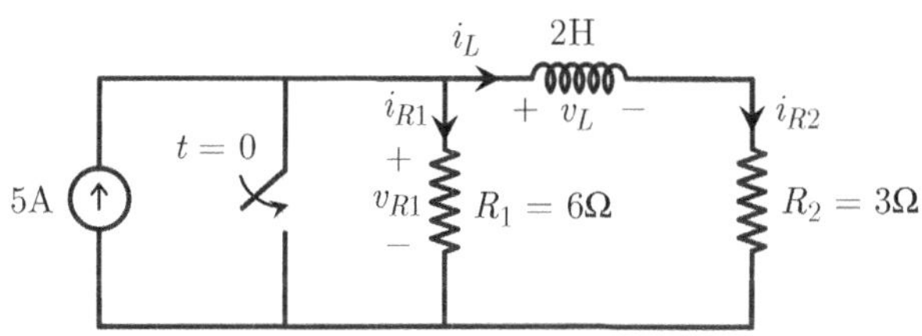

A switch is closed at $t = 0$, making the current source short circuit. Just before the switch is closed, the current across R_1 is given by $i_{R1}(0^-) = 2$ A. Analyze the circuit just after the switch is closed.

Solution: For a fixed-time analysis, we again draw up a table and fill it by inspection of the circuit both before and after the switch is changed. We start by listing the variables as follows.

	i_{R1}	i_{R2}	i_L	v_{R1}	v_{R2}	v_L
$t = 0^-$	2 A					
$t = 0^+$						

We have

$$i_{R2}(0^-) = i_L(0^-) = 5 - 2 = 3 \text{ A}$$

using KCL, and

$$v_{R2}(0^-) = 3i_{R2}(0^-) = 9 \text{ V}$$

using Ohm's law. In addition, $v_{R1}(0^-) = 6i_{R1}(0^-) = 12$ V, and

$$v_L(0^-) = v_{R1}(0^-) - v_{R2}(0^-) = 12 - 9 = 3 \text{ V}.$$

We update the first row of the table as follows.

	i_{R1}	i_{R2}	i_L	v_{R1}	v_{R2}	v_L
$t = 0^-$	2 A	3 A	3 A	12 V	9 V	3 V
$t = 0^+$						

After the switch is closed, we have $v_{R1}(0^+) = 0$ and $i_{R1}(0^+) = 0$ since this resistor becomes short-circuited. Furthermore, the current of the inductor must be continuous; hence, $i_L(0^+) = i_{R2}(0^+) = 3$ A. Then, using Ohm's law, we obtain $v_{R2}(0^+) = 3i_{R2}(0^+) = 9$ V. Finally, applying KVL, we have

$$v_L(0^+) = -v_{R2}(0^+) = -9 \text{ V}.$$

The final form of the table is as follows.

	i_{R1}	i_{R2}	i_L	v_{R1}	v_{R2}	v_L
$t = 0^-$	2 A	3 A	3 A	12 V	9 V	3 V
$t = 0^+$	0	3 A	3 A	0	9 V	−9 V

We note that the voltage of an inductor can be discontinuous. In the example above, v_L jumps from 3 V to −9 V, while i_L stays at 3 A when the switch is changed. Hence, the power of the inductor changes from 9 W to −27 W, that is, the inductor absorbs energy at $t = 0^-$ (just before the switch is closed) whereas it delivers energy at $t = 0^+$ (just after the switch is closed). Another interesting quantity is the rate of change of the current of the inductor at a particular time (e.g., at $t = 0^-$ and $t = 0^+$ in this case). Using the properties of inductors, we have

$$\left.\frac{di_L(t)}{dt}\right|_{t=0^-} = \frac{v_L(0^-)}{L} = \frac{3}{2} \text{ A/s},$$

$$\left.\frac{di_L(t)}{dt}\right|_{t=0^+} = \frac{v_L(0^+)}{L} = -\frac{9}{2} \text{ A/s}.$$

Hence, after the switch is closed, the current of the inductor starts to decrease; however, a fixed-time analysis provides no information on the circuit as time proceeds.

Example 111: Consider the following circuit involving both an inductor and a capacitor, in addition to resistors and a voltage source.

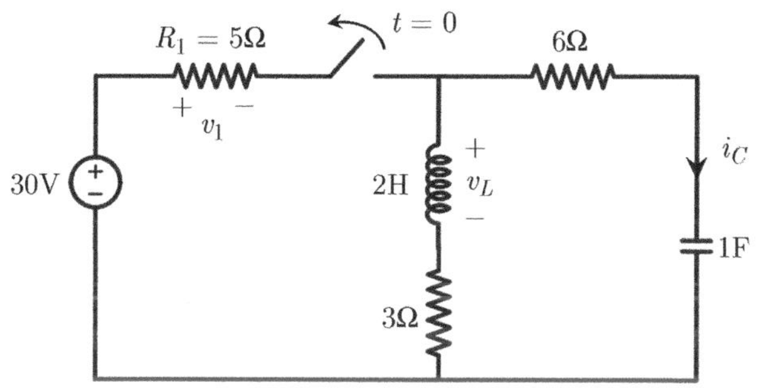

A switch is opened at $t = 0$, disconnecting the voltage source from the inductor and capacitor. Just before the switch is opened, the voltage across R_1 is given by $v_{R1}(0^-) = 18$ V, while the current across the capacitor is given by $i_C(0^-) = 1$ A. Find $i_C(0^+)$, $v_L(0^+)$, and $di_C(t)/dt$ at $t = 0^+$, just after the switch is opened.

Solution: For the solution, we analyze the circuit twice, at $t = 0^-$ and $t = 0^+$. Just before the switch is opened, we have the following circuit.

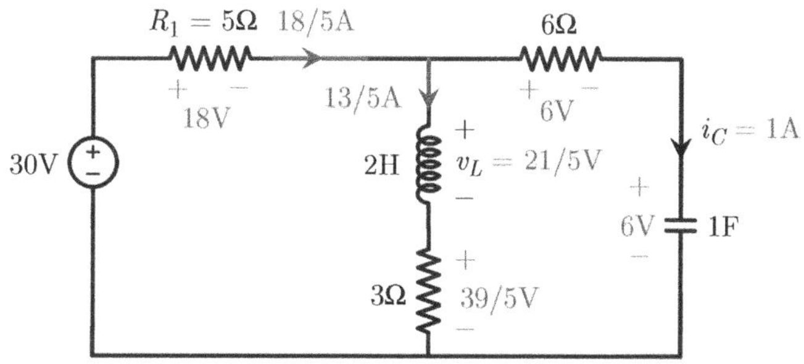

One can obtain important voltages and currents as follows:

$$i_{R1}(0^-) = v_{R1}(0^-)/5 = 18/5 \text{ A} \qquad \text{(Ohm's law)},$$
$$v_C(0^-) = 30 - v_{R1}(0^-) - 6i_C(0^-) = 6 \text{ V} \qquad \text{(KVL)},$$
$$i_L(0^-) = i_{R1}(0^-) - 1 = 13/5 \text{ A} \qquad \text{(KCL)}.$$

Just after the switch is opened, the capacitor voltage and inductor current must remain the same, $v_C(0^+) = v_C(0^-) = 6$ V and $i_L(0^+) = i_L(0^-) = 13/5$ A. Hence, we have the following circuit.

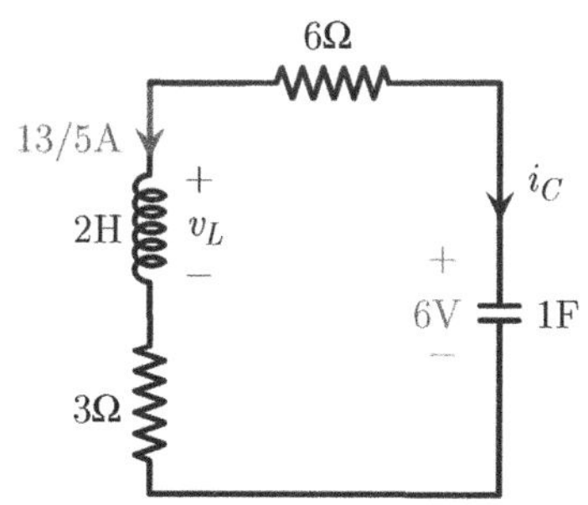

In this case, one obtains

$$i_C(0^+) = -i_L(0^+) = -13/5 \text{ A} \qquad \text{(KCL)},$$
$$v_L(0^+) = v_C(0^+) - 9i_L(0^+) = -87/5 \text{ V} \qquad \text{(KVL)}.$$

In order to find $di_C(t)/dt$ at $t = 0^+$, we use the relation between the current of the capacitor and the current of the inductor,

$$\left.\frac{di_C(t)}{dt}\right|_{t=0^+} = -\left.\frac{di_L(t)}{dt}\right|_{t=0^+} = -\frac{v_L(0^+)}{L} = \frac{87}{10} \text{ A/s}.$$

Example 112: Consider the following circuit, where the switch is closed at $t = 0$ after a long time.

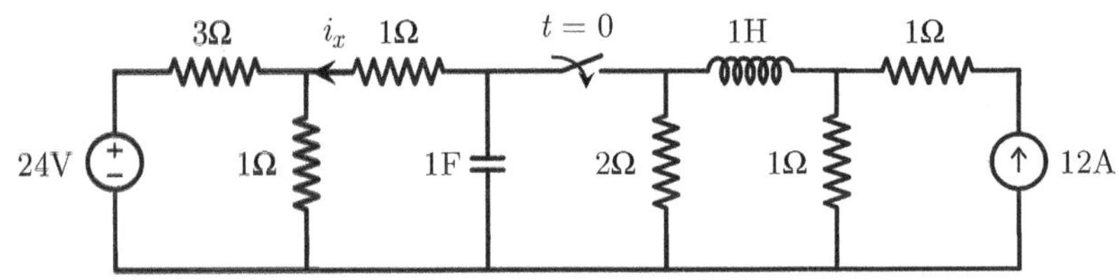

Find the power of the capacitor just after the switch is closed.

Solution: Before the switch is closed, the capacitor that acts like an open circuit ($i_x = 0$) has voltage

$$v_C(0^-) = 24 \times \frac{1}{1+3} = 6 \text{ V},$$

due to the voltage division between 3 Ω and 1 Ω. Similarly, the inductor acts like a short circuit and the current through it can be found to be

$$i_L(0^-) = 12 \times \frac{1}{1+2} = 4 \text{ A},$$

due to the current division between 2 Ω and 1 Ω. Hence, after the switch is closed, the circuit is as follows.

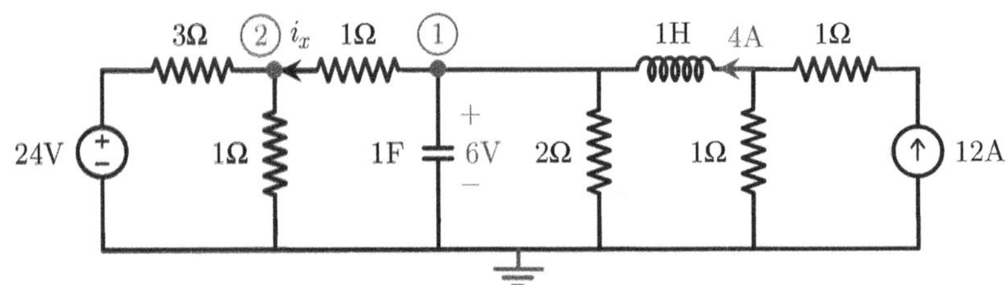

Applying KCL at nodes 1 and 2, we have

- KCL(2): $(24 - v_2)/3 - v_2/1 - (v_2 - 6)/1 = 0 \longrightarrow v_2(0^+) = 6\,\text{V}$,
- KCL(1): $(v_2 - 6)/1 - i_C/1 + 4 - 6/2 = 0 \longrightarrow i_C(0^+) = 1\,\text{A}$.

Therefore, the power of the capacitor can be obtained as

$$p_C(0^+) = 6 \times 1 = 6\ \text{W},$$

indicating that the capacitor absorbs energy at $t = 0^+$.

Example 113: Consider the following circuit.

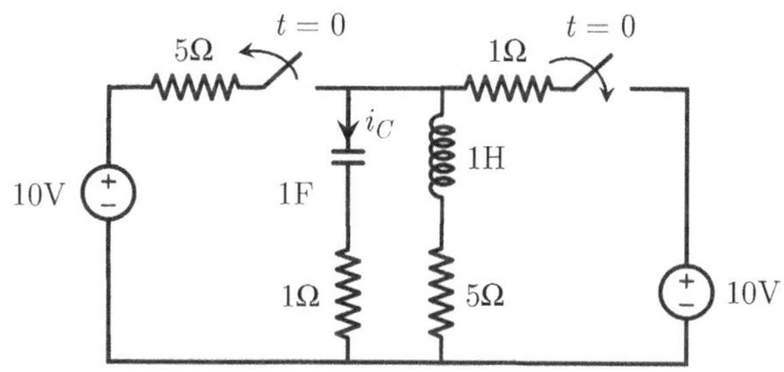

After a long time, the switch on the left is opened while the switch on the right is closed simultaneously at $t = 0$. Find the powers of the voltage source on the right, the capacitor, and the inductor, as well as di_C/dt at $t = 0^+$.

Solution: Before the switches are changed, we have the following circuit, where the capacitor and inductor act like open and short circuits, respectively, due to the long elapsed time.

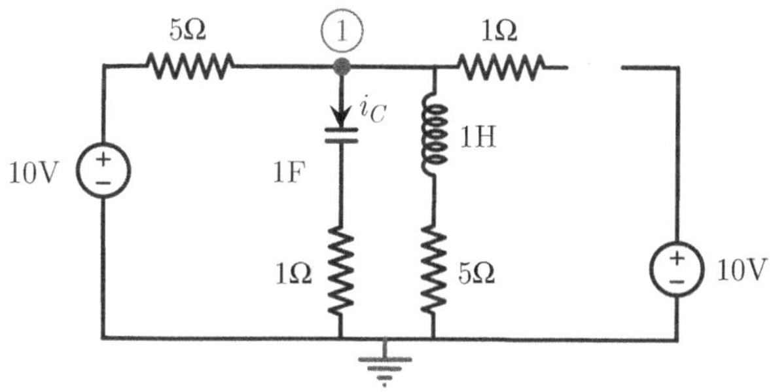

Then, using simple voltage division, we have

$$v_1(0^-) = 10 \times \frac{5}{5+5} = 5\ \text{V}.$$

In addition, we obtain $v_C(0^-) = 5\,\text{V}$ (no current flows through the resistor that is in serieswith the capacitor) and $i_L(0^-) = 1\,\text{A}$ (no voltage drop occurs across the inductor).

When the switches are changed, we have the following case at $t = 0^+$.

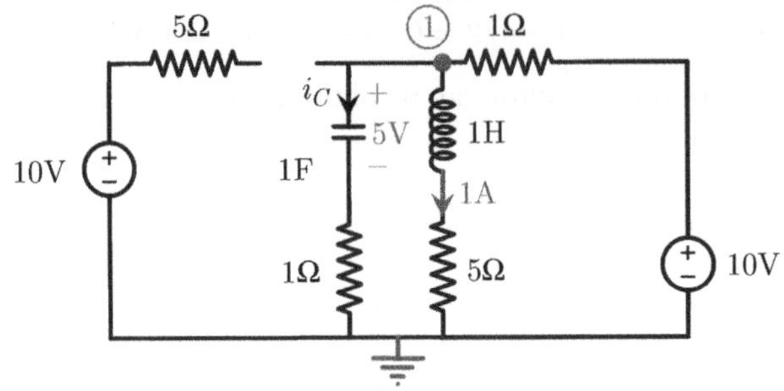

Once again the voltage of the capacitor and the current of the inductor do not change, as time does not pass. Applying KCL at node 1, we derive

- KCL(1): $-(v_1 - 5)/1 - 1 - (v_1 - 10)/1 = 0$,

leading to $v_1(0^+) = 7$ V. Consequently, the power values are found to be

$$p_{10\text{ V}}(0^+) = 10 \times \frac{v_1(0^+) - 10}{1} = -30 \text{ W},$$
$$p_C(0^+) = 5 \times \frac{v_1(0^+) - 5}{1} = 10 \text{ W},$$
$$p_L(0^+) = (v_1(0^+) - 5 \times 1) \times 1 = 2 \text{ W},$$

indicating that the voltage source provides energy while both capacitor and inductor store energy at $t = 0^+$.

In order to find di_C/dt at $t = 0^+$, we note that KCL at node 1 provides

$$-i_C - i_L + (10 - v_C - 1 \times i_C) = 0,$$

or

$$2i_C + i_L + v_C = 10,$$

where i_L is the current through the inductor with a selected direction from node 1 to the ground. Similarly, v_C is the voltage of the capacitor in accordance with the sign convention. We note that $v_C(0) = 5$ V and $i_L = 1$ A, while these quantities are now written as variables in the equation above. This is because we would like to find the rate of change. Taking the derivative of both sides and considering the voltage–current dependencies for the capacitor and inductor, we further obtain

$$2\frac{di_C}{dt} + \frac{di_L}{dt} + \frac{dv_C}{dt} = 0$$
$$2\frac{di_C}{dt} + \frac{v_L}{L} + \frac{i_C}{C} = 0,$$

leading to

$$\frac{di_C}{dt} = -\frac{v_L}{2L} - \frac{i_C}{2C}.$$

Finally, using $C = 1\,\text{F}$, $L = 1\,\text{H}$, $v_L(0^+) = 2\,\text{V}$, and $i_C(0^+) = 2\,\text{A}$, we have

$$\left.\frac{di_C}{dt}\right|_{t=0^+} = -1 - 1 = -2\ \text{A/s}.$$

This derivative value indicates that the current of the capacitor starts to decrease just after the switches are changed.

Exercise 99: In the following circuit, the switch is opened after a long time. Find the voltage of the 12 Ω resistor at $t = 0^+$, just after the switch is opened.

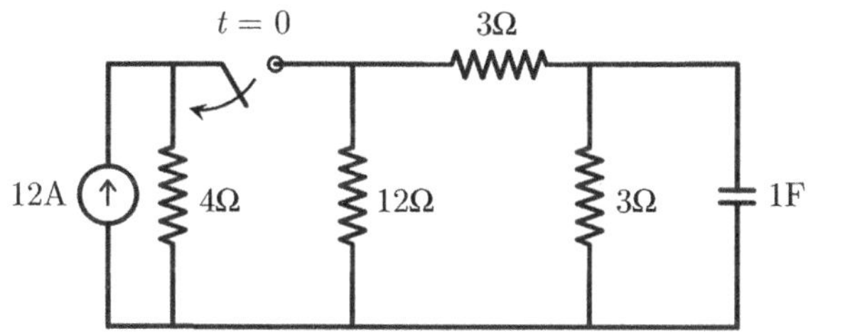

Exercise 100: Consider the following circuit, where the switch is closed after a long time. Find the power of the 6 V voltage source at $t = 0^+$, just after the switch is closed.

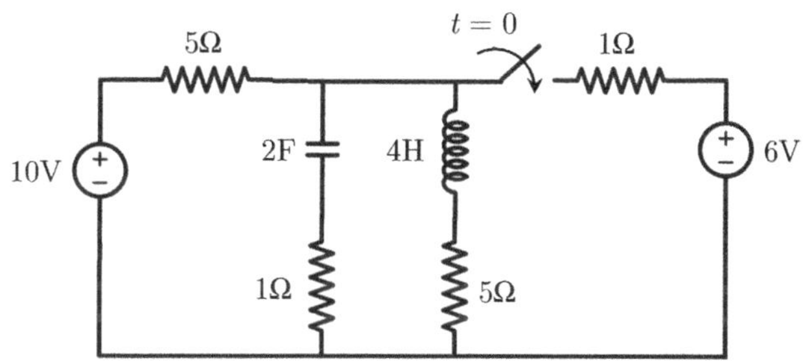

Exercise 101: Consider the following circuit, where the switch is opened at $t = 0$ after a long time. Find the power of the inductor and the 1 F capacitor just after the switch is opened.

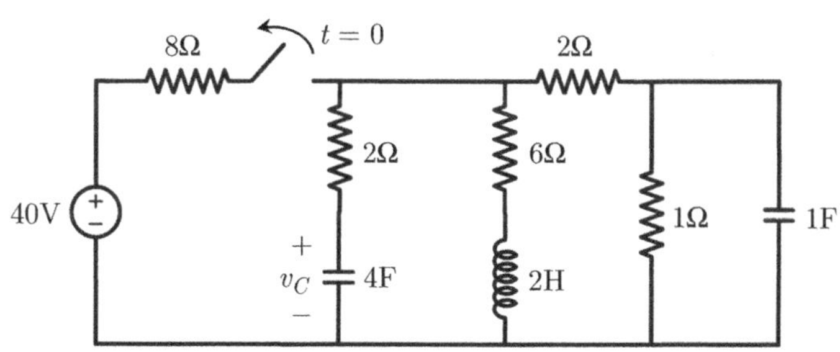

Also find the the value of dv_C/dt at $t = 0^+$.

Exercise 102: In the following circuit, the switch is opened at $t = 0$. Given that $i_C(0^-) = 2$ A and $v_1(0^-) = 20$ V, find $v_{L1}(0^+)$, just after the switch is changed.

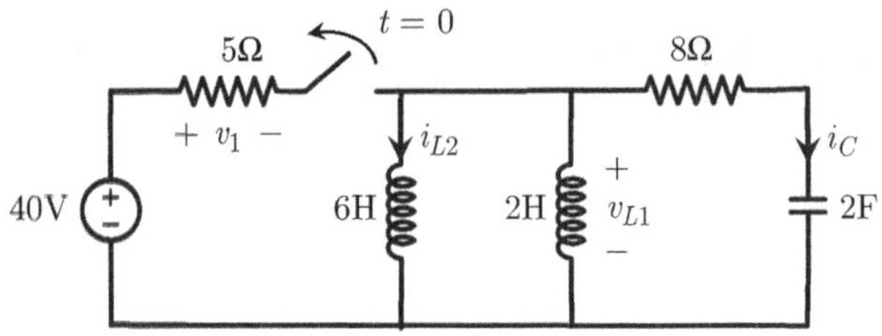

Exercise 103: Consider the following circuit, where the switch remains at position a for a long time. Then, at $t = 0$, it is changed to position b. Find the power of the capacitor and inductor at $t = 0^+$, just after the switch is changed.

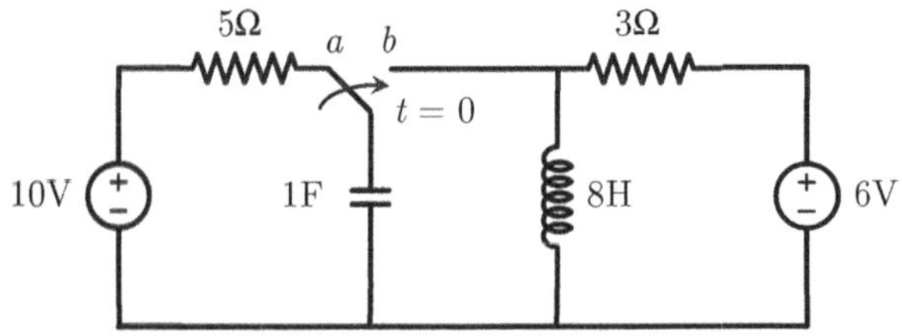

6.5 Parallel and Series Connections of Capacitors and Inductors

Similarly to resistors, capacitors and inductors that are connected in series and in parallel can be analyzed easily by combining the capacitance and inductance values appropriately. In this section, we discuss how the voltage and current values are shared between alternative connections of capacitors and inductors.

6.5.1 Connections of Capacitors

First, we consider the series connection of two capacitors as follows.

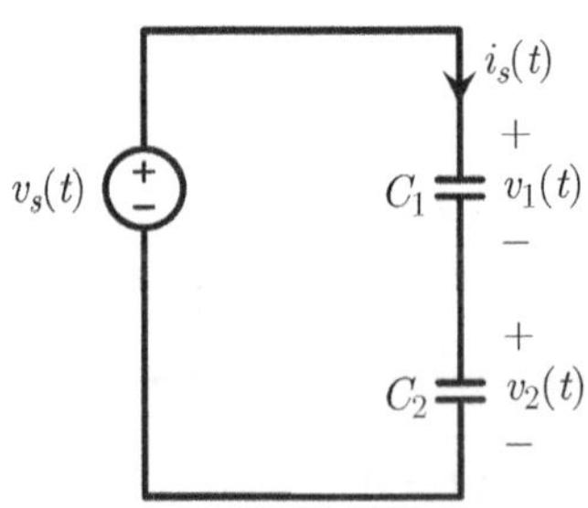

Note that, in order to obtain the overall response, we assume v_s also depends on time. We have

$$v_s(t) = v_1(t) + v_2(t),$$

where

$$v_1(t) = \frac{1}{C_1}\int_{t_0}^{t} i_s(t')dt' + v_1(t_0),$$

$$v_2(t) = \frac{1}{C_2}\int_{t_0}^{t} i_s(t')dt' + v_2(t_0),$$

according to the definition of the capacitors. Therefore, the relationship between the overall voltage and the current is found to be

$$v_s(t) = \left(\frac{1}{C_1} + \frac{1}{C_2}\right)\int_{t_0}^{t} i_s(t')dt' + v_1(t_0) + v_2(t_0)$$

or

$$v_s(t) = \left(\frac{1}{C_1} + \frac{1}{C_2}\right)\int_{t_0}^{t} i_s(t')dt' + v_s(t_0)$$

since $v_s(t_0) = v_1(t_0) + v_2(t_0)$. Finally, we obtain

$$v_s(t) = \frac{1}{C_{\text{eq}}}\int_{t_0}^{t} i_s(t')dt' + v_s(t_0),$$

where

$$\frac{1}{C_{\text{eq}}} = \frac{1}{C_1} + \frac{1}{C_2}.$$

As a result, the equivalent capacitance corresponding to the series connection of C_1 and C_2 is

$$C_{\text{eq}} = \frac{C_1C_2}{C_1 + C_2}.$$

At first glance, the expression above may seem strange since $C_{\text{eq}} \leq C_1$ and $C_{\text{eq}} \leq C_2$, meaning that the ability to store charges for a given voltage decreases when two capacitors are connected in series. This can be understood if we consider the fact that the negative terminal of one capacitor is actually connected to the positive terminal of the other, if we fix the current flowing between them in a single direction. Therefore, these capacitances work opposite to each other, making the overall combination have less capacitance. In the limit as $C_2 \rightarrow 0$, we have

$$C_{\text{eq}} \rightarrow 0,$$

since this corresponds to the case where an open circuit exists instead of C_2. In the limit as $C_2 \rightarrow \infty$, corresponding to the short-circuited C_2, one can derive

$$C_{\text{eq}} \rightarrow C_1$$

as expected.

We now consider a parallel connection of two capacitors and a time-varying current source.

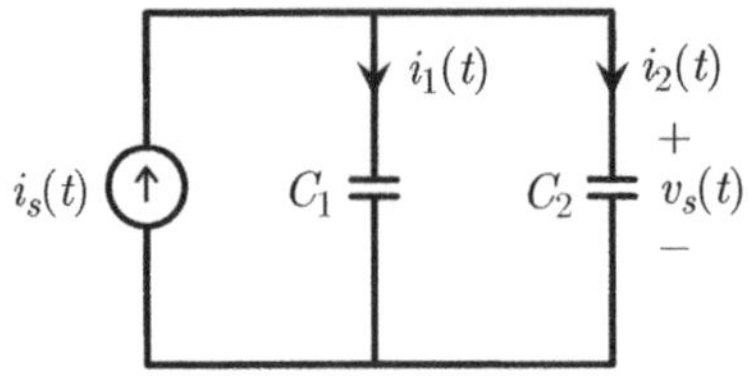

In this case, KCL can be used to write

$$i_s(t) = i_1(t) + i_2(t),$$

where

$$i_1(t) = C_1 \frac{dv_s(t)}{dt},$$

$$i_2(t) = C_2 \frac{dv_s(t)}{dt}.$$

Hence, the overall current and voltage are related by

$$i_s(t) = C_1 \frac{dv_s(t)}{dt} + C_2 \frac{dv_s(t)}{dt} = (C_1 + C_2)\frac{dv_s(t)}{dt}$$

or

$$i_s(t) = C_{eq} \frac{dv_s(t)}{dt},$$

where

$$C_{eq} = C_1 + C_2$$

is the equivalent capacitance for a parallel connection of two capacitors. Obviously, when two the capacitors are connected in parallel, their combination is able to store more charge for a given voltage. Therefore, the overall capacitance grows in parallel connections of capacitors.

6.5.2 Connections of Inductors

We now discuss the series and parallel connections of inductors. When two inductors are connected in series, we have the following.

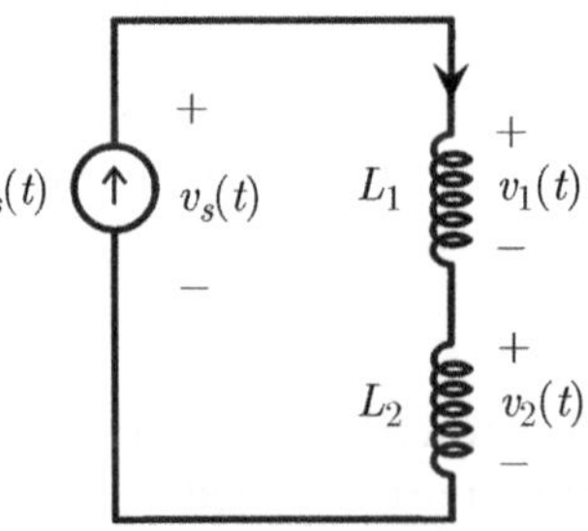

We note that the choice of source does not change the result of the analysis, while we often select the one that leads to easier manipulations. Using the inductor definition, we derive

$$v_1(t) = L_1 \frac{di_s(t)}{dt},$$
$$v_2(t) = L_2 \frac{di_s(t)}{dt},$$

leading to

$$v_s(t) = v_1(t) + v_2(t) = L_1 \frac{di_s(t)}{dt} + L_2 \frac{di_s(t)}{dt}$$

or

$$v_s(t) = (L_1 + L_2) \frac{di_s(t)}{dt}.$$

Therefore, the overall inductance can be written as

$$L_{\text{eq}} = L_1 + L_2.$$

Next, we consider a parallel connection of two inductors as follows.

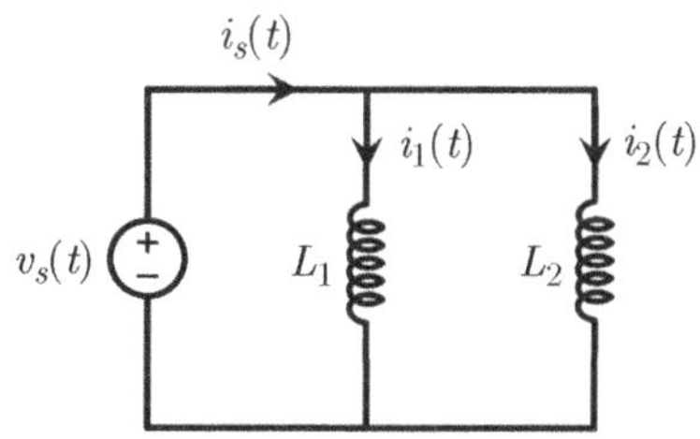

We derive

$$i_s(t) = i_1(t) + i_2(t),$$

where

$$i_1(t) = \frac{1}{L_1} \int_{t_0}^{t} v_s(t')dt' + i_1(t_0),$$
$$i_2(t) = \frac{1}{L_2} \int_{t_0}^{t} v_s(t')dt' + i_2(t_0).$$

Therefore, the relationship between the overall voltage and the current is found to be

$$i_s(t) = \left(\frac{1}{L_1} + \frac{1}{L_2}\right) \int_{t_0}^{t} v_s(t')dt' + i_1(t_0) + i_2(t_0)$$

or

$$i_s(t) = \left(\frac{1}{L_1} + \frac{1}{L_2}\right) \int_{t_0}^{t} v_s(t')dt' + i_s(t_0),$$

using $i_s(t_0) = i_1(t_0) + i_2(t_0)$. Therefore, we obtain

$$i_s(t) = \frac{1}{L_{\text{eq}}} \int_{t_0}^{t} v_s(t')dt' + i(t_0),$$

where

$$\frac{1}{L_{\text{eq}}} = \frac{1}{L_1} + \frac{1}{L_2}.$$

The equation above can be rewritten as

$$L_{\text{eq}} = \frac{L_1 L_2}{L_1 + L_2}$$

as the equivalent inductance of a parallel connection of two inductors. It can be observed that inductors behave similarly to resistors when they are combined: the inductance increases/decreases if they are connected in series/parallel.

6.6 When Things Go Wrong in Transient Analysis

As shown in this chapter, RC and RL circuits involving a resistor and capacitor/inductor demonstrate transient behaviors with time constants, such as $\tau = RC$ and $\tau = L/R$. In the absence of a resistor, that is, when a capacitor or inductor is directly connected to a source, either the circuit immediately enters steady state without a transient state or it is an impossible case. But what happens when a capacitor and an inductor are connected to each other?

Consider the following case, where a capacitor and an inductor are connected in series to a DC voltage source when a switch is closed at $t = 0$.

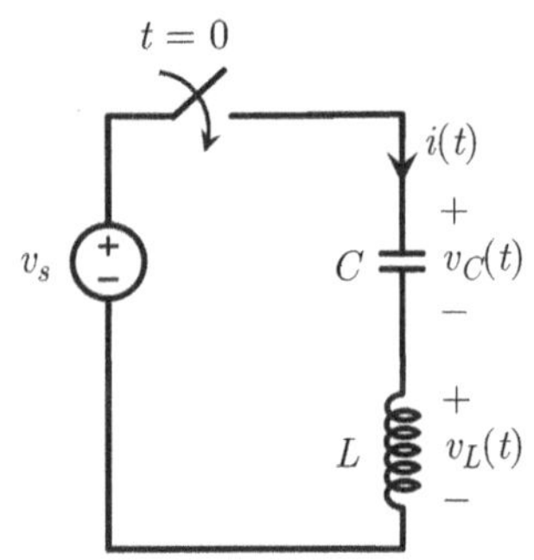

Using the definitions of capacitors and inductors, we have

$$i(t) = C\frac{dv_C(t)}{dt},$$

$$v_L(t) = L\frac{di(t)}{dt},$$

leading to

$$v_L(t) = L\frac{di(t)}{dt} = LC\frac{d^2v_C(t)}{dt^2}.$$

Then, using $v_C(t) + v_L(t) = v_s$, we derive

$$\frac{d^2v_C(t)}{dt^2} + \frac{1}{LC}v_C(t) = \frac{1}{LC}v_s.$$

A general solution to this equation can be written as

$$v_C(t) = a\cos(\omega_o t) + b\sin(\omega_o t) + v_s,$$

where a and b can be found by checking the initial conditions and $\omega_o = 1/\sqrt{LC}$. Then the current through the loop is found to be

$$i(t) = C\frac{dv_C(t)}{dt} = -Ca\omega_o\sin(\omega_o t) + Cb\omega_o\cos(\omega_o t).$$

Now, choosing $v_C(t=0) = 0$ and $i(t=0) = 0$, one obtains

$$a = -v_s,$$
$$b = 0.$$

Consequently, the voltage and current values for these initial conditions are found to be

$$v_C(t) = v_s[1 - \cos(\omega_o t)],$$
$$v_L(t) = v_s - v_C(t) = v_s\cos(\omega_o t),$$
$$i(t) = v_s\omega_o C\sin(\omega_o t) = v_s\sqrt{\frac{C}{L}}\sin(\omega_o t).$$

But what does this mean, considering that the cos and sin functions represent oscillations for all values of t?

The expressions above indicate that the voltage and current values oscillate infinitely, without reaching constant values. On the other hand, the oscillations fit into a repetitive behavior, starting from $t = 0$. In fact, despite the fact that voltage and current values vary, this circuit does not have any transient state; it immediately enters steady state once the switch is closed. The oscillation rate, ω_o, is often called the natural angular frequency of the circuit since it represents how fast the voltage and current values change. In the next chapter, we consider time-harmonic sources, where the oscillation of the voltage and current values are enforced by the sources, where they also vary with respect to time in steady state.

For the circuit above, an interesting case occurs when there is no voltage source ($v_s = 0$), but $v_C(t=0) = v_{C0} \neq 0$, that is, the capacitor is charged before the switched is closed. Once the connection is established between the capacitor and inductor, we have

$$v_C(t) = v_{C0}\cos(\omega_o t),$$
$$i(t) = -v_{C0}\sqrt{\frac{C}{L}}\sin(\omega_o t).$$

This circuit is called an LC tank, since it involves an oscillatory (repetitive) transfer of energy between the capacitor and inductor. Specifically, these energies can be calculated as

$$w_C(t) = \frac{1}{2}Cv_{C0}^2\cos^2(\omega_o t),$$
$$w_L(t) = \frac{1}{2}Cv_{C0}^2\sin^2(\omega_o t),$$

and

$$w_C(t) + w_L(t) = \frac{1}{2}Cv_{C0}^2 = w_C(t=0).$$

Hence, the total energy of the capacitor and inductor is constant and equal to the initially stored energy in the capacitor, while this energy oscillates back and forth between the capacitor and inductor with respect to time. This circuit also does not have any transient state.

As discussed in this chapter, there cannot be a jump in the voltage/current of a capacitor/inductor (hence its energy) if it is connected to a resistor. We note that such a resistor can be a part of the circuit, while they can also be internal resistances of the capacitor/inductor in real life. Considering ideal components, this claim may be misinterpreted as saying that any circuit involving a resistor must have continuous energy values. As a (counter)example, the following scenario involves two sources, a capacitor, and a resistor. After a long time, the switch is closed at $t = 0$.

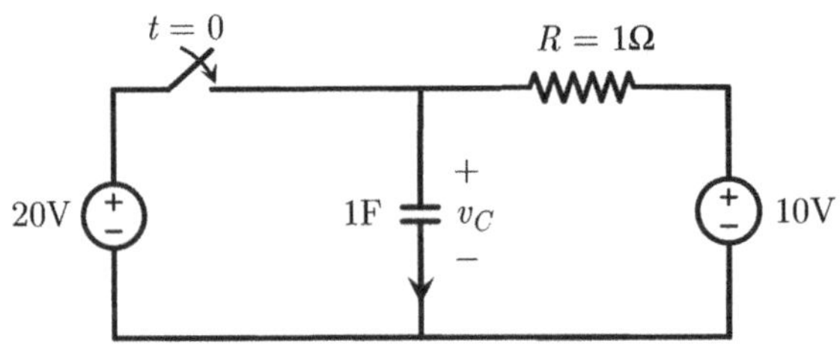

Before the switch is closed, the voltage of the capacitor can be written as

$$v_C(0^-) = 10 \text{ V},$$

considering that the capacitor acts as an open circuit. When the switch is closed, however, the voltage of the capacitor must jump to 20 V,

$$v_C(0^+) = 20 \text{ V},$$

as a result of the voltage source on the left. Interestingly, the circuit is still valid after the switch is closed. In this case, 20 V voltage of the capacitor leads to 10 V across the resistor. On the other hand, there are jumps in the voltage and energy of the capacitor at $t = 0$ and this circuit immediately enters steady state without any transient behavior.

6.7 What You Need to Know before You Continue

As discussed in this chapter, transient analysis is essential to understand the behavior of circuits involving energy storage elements in addition to resistors. We emphasize a few important points, before proceeding to the next chapter.

- Voltage/current values of energy storage elements (capacitors and inductors) are related to each other via the time derivative. Therefore, their responses depend on time, leading to transient states.
- While theoretically a steady state never occurs, it is assumed to be reached after a sufficiently long time such that the unwanted variations in the voltage and current values diminish below some threshold values.
- When only DC sources are involved in a circuit, capacitors and inductors become open and short circuits, respectively, in steady state. This happens when the capacitors and inductors are saturated and cannot store any more energy.

- When connected directly to resistors, the voltage/current value of a capacitor/inductor cannot have a jump since this would require a jump in their energy. Such jumps can occur in some ideal cases, for example, when a capacitor is connected to a voltage source without a resistor.

In the next chapter, we consider the steady state of AC circuits, where sources provide time-varying voltage and current values. We focus in particular on circuits driven by sinusoidal sources, which can be analyzed in the phasor domain.

7

Steady-State Analysis of Time-Harmonic Circuits

As discussed in Chapter 6, there are only rare and ideal cases (e.g., when a capacitor and inductor are connected to each other) leading to oscillatory behavior of DC circuits in steady state. Otherwise, a steady state corresponds to constant voltage and current values, if only DC sources are involved in a circuit. But what happens if AC sources are used such that voltage and current values are forced to change with respect to time? In fact, as mentioned in Section 1.1.6, AC sources and signals are commonly used in modern circuits due to their advantages. In this chapter, we consider steady-state analysis of such time-harmonic circuits, which involve sinusoidal sources with given frequencies. In order to handle the time-dependent behaviors of voltage and current values, we take advantage of phasor notation that enables the transformation of signals and components into the phasor (complex) domain. Once transformed, we analyze each circuit as usual, for exmaple, using KVL, KCL, nodal or mesh analysis, as well as Ohm's law.

In the phasor domain, where complex numbers are used to represent voltages and currents, the power has also a general form with possibly complex values. This generalized definition of the power allows for the categorization of the energy as dissipated (e.g., spent as heat, light, etc.) and stored.

7.1 Steady-State Concept

In general, other than sources, we have three different basic components, namely, resistors, capacitors, and inductors, with voltage–current relationships given by

$$v_R(t) = Ri_R(t), \qquad i_C(t) = C\frac{dv_C(t)}{dt}, \qquad v_L(t) = L\frac{di_L(t)}{dt},$$

respectively. When only DC sources exist in a circuit, a steady state corresponds to constant voltage and current values. Hence, the time derivatives in the equations above must be zero, leading to $i_C = 0$ and $v_L = 0$ for capacitors and inductors, respectively, independent of the values of v_C and i_L. As depicted in Figure 7.1, these conditions for the capacitors and inductors can be interpreted as open and short circuits. We also note that an ideal resistor with resistance R always has the same voltage/current behavior whether it is in transient state or in steady state.

When AC sources are involved in a circuit, one can expect voltage and current values to oscillate as dictated by the sources. For example, when a capacitor is connected to an AC voltage source, the voltage of the capacitor increases and decreases with respect to

Introduction to Electrical Circuit Analysis, First Edition. Özgür Ergül.

Companion Website: www.wiley.com/go/ergul4412

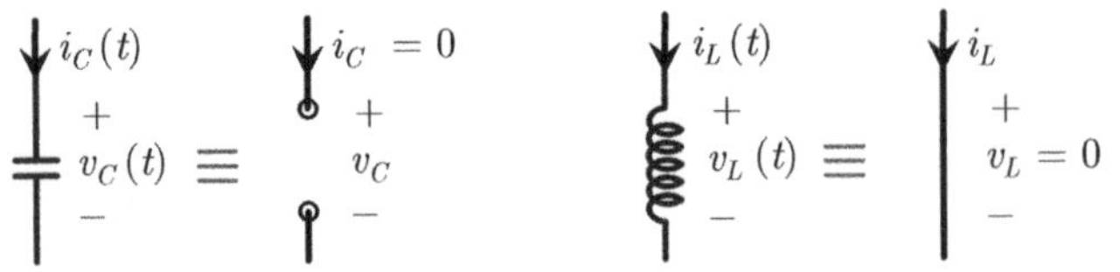

Figure 7.1 Steady-state equivalents of capacitors and inductors when only DC sources are involved.

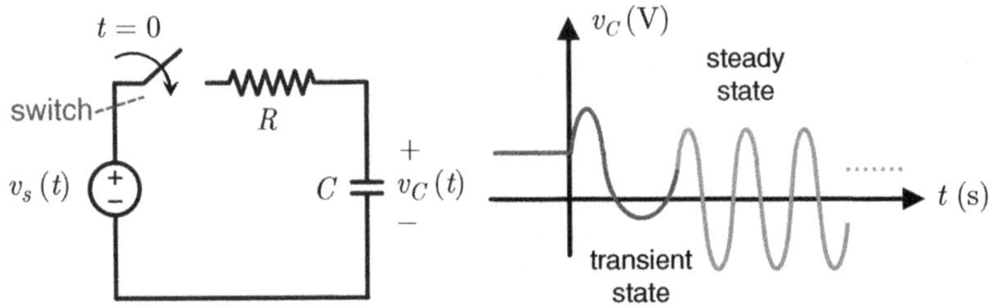

Figure 7.2 A representation of passage from transient to steady state in an AC circuit.

time. When the voltage of the capacitor increases, its energy also increases, which can be interpreted as the capacitor storing energy provided by the source. When the voltage of the capacitor decreases, however, the capacitor delivers energy to the source. This oscillatory behavior continues infinitely (in theory) or until the capacitor is disconnected from the source (in practice). Obviously, the steady state of such a circuit involves oscillatory voltage and current values, which change with respect to time, unlike to those in DC circuits.

But what about transient states of AC circuits? Very similarly to the DC case, AC circuits have transient states in which the voltage and current values behave differently than desired, before they settle. The transient behaviors can be dominated by temporary responses that usually decay exponentially such that the circuit enters steady state after a sufficient time. The required time is again dependent on time constants, such as in the form of RC or L/R.

7.2 Time-Harmonic Circuits with Sinusoidal Sources

The most common form of AC signals is sinusoidal, and we consider only sinusoidal sources in the rest of this chapter. A sinusoidal voltage source provides a voltage value that depends on time as

$$v(t) = v_m \cos(\omega t)$$

or

$$v(t) = v_m \sin(\omega t) = v_m \cos(\omega t - \pi/2),$$

where ω is defined as the angular frequency. The unit of ω is radians per second, and it is further defined as

$$\omega = 2\pi f,$$

Figure 7.3 Sinusoidal voltage and current sources.

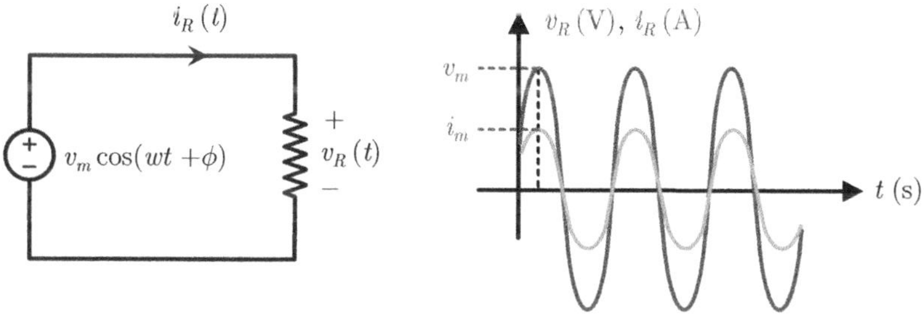

Figure 7.4 A resistor connected to a sinusoidal voltage source.

where the value of the frequency is set to 50–60 Hz in domestic usage. The voltage expressions above are periodic, that is,

$$v(t + nT) = v(t)$$

for any integer n, where $T = 2\pi/\omega = 1/f$ is the period. More generally, sinusoidal voltage expressions may have phases such that

$$v(t) = v_m \cos(\omega t + \phi),$$
$$v(t) = v_m \sin(\omega t + \phi) = v_m \cos(\omega t - \pi/2 + \phi),$$

where ϕ represents the phase in radians or degree ($180^\circ = \pi$ rad).

Very similarly to sinusoidal voltage sources, there can be sinusoidal current sources that can provide current expressed as

$$i(t) = i_m \cos(\omega t + \phi),$$
$$i(t) = i_m \sin(\omega t + \phi) = i_m \cos(\omega t - \pi/2 + \phi).$$

7.2.1 Resistors Connected to Sinusoidal Sources

We now consider a voltage source connected to a single resistor R. The voltage of the resistor is given by

$$v_R(t) = v_m \cos(\omega t + \phi).$$

Then the current through the resistor has the same sinusoidal form,

$$i_R(t) = i_m \cos(\omega t + \phi) = \frac{v_m}{R} \cos(\omega t + \phi).$$

Specifically, the voltage and current across a resistor are in phase, that is, they do not have any phase shift with respect to each other.

7.2.2 Capacitors Connected to Sinusoidal Sources

Next, we consider another simple circuit involving a sinusoidal voltage source and a capacitor. Using

$$v_C(t) = v_m \cos(\omega t + \phi),$$

the current through the capacitor is found to be

$$\begin{aligned} i_C(t) &= C\frac{dv_C(t)}{dt} = -\omega C v_m \sin(\omega t + \phi) \\ &= -\omega C v_m \cos(\omega t + \phi - \pi/2). \end{aligned}$$

The expressions above have two interesting properties, as follows.

- There is a $\pi/2$ rad $= 90^\circ$ phase difference between the voltage and current of a capacitor. For the sake of clarity, we consider further the case $\phi = 0$, leading to

  $$\begin{aligned} v_C(t) &= v_m \cos(\omega t), \\ i_C(t) &= -\omega C v_m \cos(\omega t - \pi/2) = \omega C v_m \cos(\omega t + \pi/2). \end{aligned}$$

  As also illustrated in Figure 7.5, the current of the capacitor leads its voltage by 90°; for example, the peak of the current occurs earlier than the peak of the voltage by an amount $\Delta t = (\pi/2)/\omega$.
- The amplitude of the capacitor current is ωC times the amplitude of its voltage. Obviously, when C is large, more charge must be collected in the capacitor to set up a given voltage. This corresponds to more current flow through the capacitor. But what about the effect of the frequency? According to the expressions above, the amount of current increases with the frequency. In the limits, we have

  $$\begin{aligned} &\lim_{\omega \to 0} i_C(t) = 0, \\ &\lim_{\omega \to \infty} i_C(t) = \infty, \end{aligned}$$

  for a finite value of v_m. We are already familiar with the first case, when $\omega \to 0$, which corresponds to the DC response of a capacitor (open circuit) in steady state. In the second case, when $\omega \to \infty$, the current is infinite for a finite value of the voltage. This is often interpreted as saying that the capacitor becomes like a short circuit at high frequencies.

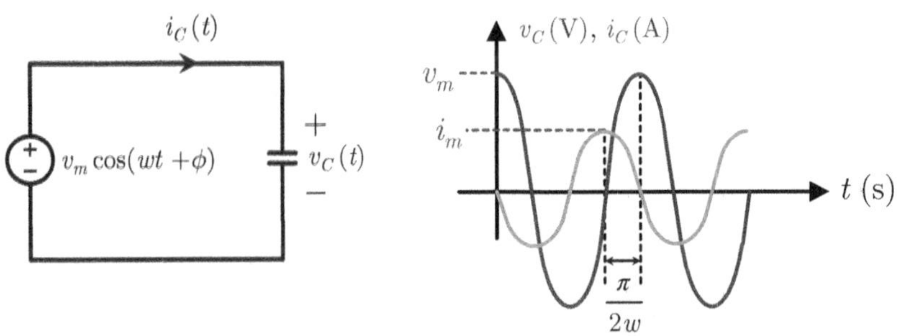

Figure 7.5 A capacitor connected to a sinusoidal voltage source.

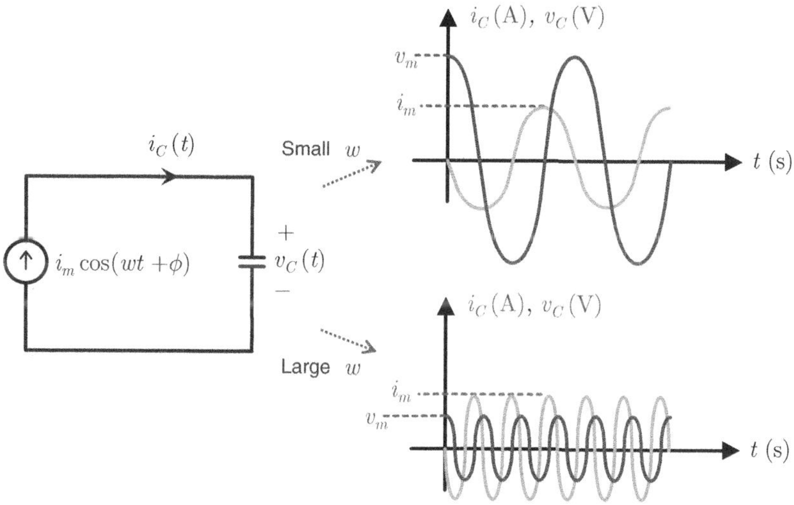

Figure 7.6 A capacitor connected to a sinusoidal current source at different frequencies.

In order to understand the behavior of capacitors at high frequencies, one can consider the case where a capacitor is connected to a current source. Such a circuit is easier to interpret as ω goes to infinity. If the current is given by

$$i_C(t) = i_m \cos(\omega t + \phi),$$

the voltage of the capacitor can be derived as

$$v_C(t) = \frac{i_m}{\omega C} \sin(\omega t + \phi),$$

assuming that $v_C(t = 0) = 0$. Obviously, when $\omega \to \infty$, we have

$$\lim_{\omega \to \infty} v_C(t) = 0,$$

for a finite value of i_m. Hence, the voltage across the capacitor becomes zero (it becomes a short circuit), despite a finite value of the current. This can be understood if we consider that current through a capacitor corresponds to the accumulation and depletion of charge. When the capacitor is connected to an AC source, such accumulation and depletion processes occur consecutively in cycles. If the oscillation is very fast (the frequency is very high) such that the direction of the current flow changes quickly, only a small amount of charge *is able* to accumulate to set up a voltage across the capacitor. In the theoretical limit of $\omega \to \infty$, the capacitor cannot store any charge.

7.2.3 Inductors Connected to Sinusoidal Sources

Consider an inductor connected to a sinusoidal current source, leading to

$$i_L(t) = i_m \cos(\omega t + \phi).$$

In this case, we have

$$\begin{aligned} v_L(t) &= L\frac{di_L(t)}{dt} = -\omega L i_m \sin(\omega t + \phi) \\ &= -\omega L i_m \cos(\omega t + \phi - \pi/2). \end{aligned}$$

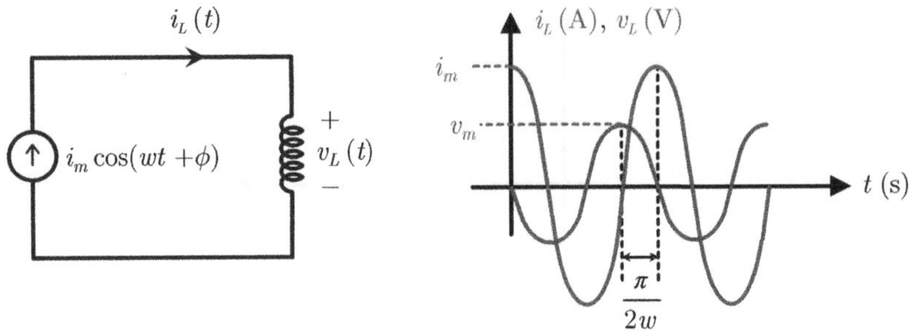

Figure 7.7 An inductor connected to a sinusoidal current source.

We again note two interesting properties.

- There is $\pi/2$ rad $= 90°$ phase difference between the voltage and current of an inductor. Considering the particular case of $\phi = 0$, we have

$$i_L(t) = i_m \cos(\omega t),$$
$$v_L(t) = -\omega L i_m \cos(\omega t - \pi/2) = \omega L i_m \cos(\omega t + \pi/2).$$

  Therefore, as illustrated in Figure 7.7, the voltage of the inductor leads its current by 90° or the current of the inductor lags its voltage by 90°.
- The amplitude of the inductor voltage is ωL times the amplitude of its current. Naturally, for large values of L, the inductor can create more magnetic flux, hence voltage across its terminals, for a given current. Similarly to capacitors, there is also a frequency dependence in the voltage and current values of inductors. Specifically, for a fixed current value, the voltage across an inductor increases with frequency. Two limit cases of particular interest are

$$\lim_{\omega \to 0} v_L(t) = 0,$$
$$\lim_{\omega \to \infty} v_L(t) = \infty,$$

  for a finite value of i_m. Hence, we verify that an inductor becomes a short circuit for the DC case. On the other hand, as the frequency increases, an inductor becomes an open circuit.

To sum up, capacitors and inductors that behave as open and short circuits, respectively, in DC circuits, become short and open circuits as the frequency increases. The reason why an inductor turns into an open circuit at high frequencies can be explained similarly to the short-circuit behavior of capacitors. When the voltage across an inductors oscillates rapidly, the current across its terminals is reduced. At the infinite frequency limit, no current can be produced anymore, for a finite value of the voltage.

7.2.4 Root-Mean-Square Concept

Obviously, the value of a sinusoidal voltage or current depends on time. In fact, considering both positive and negative values, the overall average of a sinusoidal function is zero. Therefore, we need a new quantity to measure the strength of a voltage or current in AC circuits. A useful quantity, which is called the root-mean-square (RMS) value (corresponding to applications of root, mean and square operations) of a function $f(t)$, is defined as

$$f_{\text{RMS}} = \left\{ \frac{1}{T} \int_0^T [f(t)]^2 dt \right\}^{1/2},$$

where T is the period. Now, considering a sinusoidal function

$$f(t) = A_f \cos(\omega t + \phi_f),$$

where A_f and ϕ_f represent the amplitude and phase, respectively, we have

$$\begin{aligned} f_{\text{RMS}} &= \left\{ \frac{\omega}{2\pi} \int_0^{2\pi/\omega} A_f^2 \cos^2(\omega t + \phi_f) dt \right\}^{1/2} \\ &= \left\{ \frac{\omega A_f^2}{2\pi} \int_0^{2\pi/\omega} \left[\frac{1}{2} + \frac{1}{2}\cos(2\omega t + 2\phi_f) \right] dt \right\}^{1/2} \\ &= \left\{ \frac{\omega A_f^2}{2\pi} \frac{\pi}{\omega} \right\}^{1/2} = \frac{A_f}{\sqrt{2}}. \end{aligned}$$

Therefore for a sinusoidal function, the RMS value is simply the amplitude divided by the square root of 2. In AC lines and circuits, the RMS value of the voltage (e.g., 220 V) is often indicated to describe this strength.

Example 114: Plot the sinusoidal voltages

$$v_1(t) = 2\sin(2t),$$
$$v_2(t) = 3\sin(2t + \pi/4),$$

with respect to time t, labeling all critical values.

Solution: The given voltages can be plotted as follows.

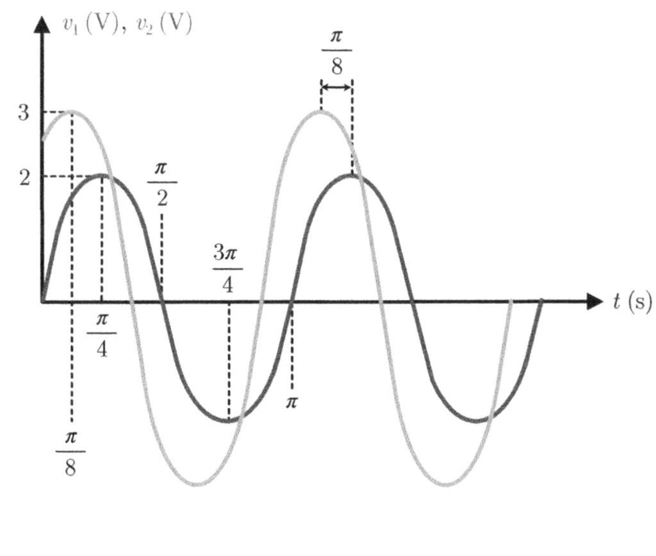

In these plots, it can be seen that the closest peaks of the voltages are separated by $\pi/8$. If the plots are generated with respect to ωt, we have the following graph.

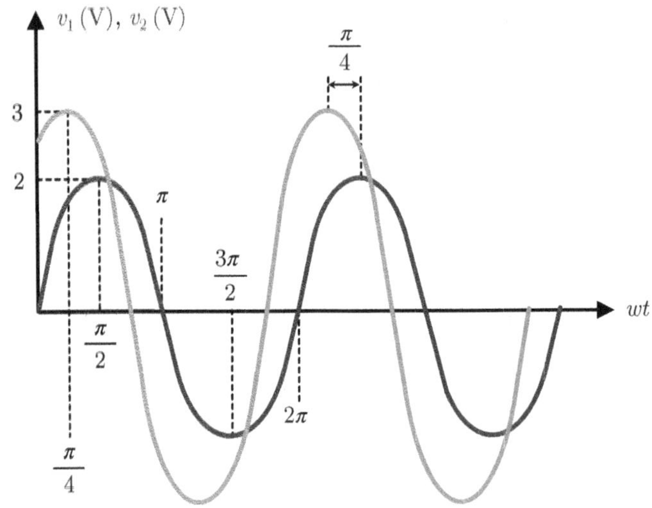

Using ωt as the variable, the peaks are separated by $\pi/4$. Since the peak of v_2 appears before the peak of v_1, the graph above can be interpreted as v_2 leading v_1 by $\pi/4$ or v_1 lagging v_2 by $\pi/4$. The RMS values for v_1 and v_2 are $2/\sqrt{2} = \sqrt{2}$ V and $3/\sqrt{2} = 3\sqrt{2}/2$ V, respectively.

7.3 Concept of Phasor Domain and Component Transformation

Circuits with sinusoidal sources and components, such as resistors, capacitors, and inductors, are difficult to analyze in the time domain. However, given that a circuit is linear and all sources have the same frequency ω, it can be derived that all voltages and currents have the form $f(t) = A_f \cos(\omega t + \phi_f)$. Therefore, one can actually drop the frequency parameter, making the analysis significantly easier. Such quantities after the frequency is dropped are called phasors.

In order to derive the phasor of a time-harmonic function $f(t)$, we note that

$$f(t) = A_f \cos(\omega t + \phi_f) = A_f \mathrm{Re}\{\exp(j\omega t + j\phi_f)\},$$

where A_f and ϕ_f uniquely define the function $f(t)$, given that the angular frequency is ω. Therefore, a complex quantity

$$f_c = A_f \exp(j\phi_f) = A_f(\cos\phi_f + j\sin\phi_f),$$

which does not depend on time, provides all the information regarding $f(t)$. Specifically, given f_c, we have

$$f(t) = \mathrm{Re}\{f_c \exp(j\omega t)\}.$$

In circuit analysis, it is useful to define a phasor quantity as

$$\underline{f} = \frac{A_f}{\sqrt{2}} \exp(j\phi_f),$$

leading to

$$f(t) = \sqrt{2}\text{Re}\{\underline{f} \exp(j\omega t)\}.$$

where $\underline{f}$ is the phasor of $f(t)$. In the above, $A_f/\sqrt{2}$ corresponds to the RMS of the sinusoidal function, that is,

$$\underline{f} = f_{\text{RMS}} \exp(j\phi_f).$$

Therefore, given the phasor, the RMS of the function is given by

$$f_{\text{RMS}} = |\underline{f}|.$$

Example 115: Find the phasors of the voltage values

$$v_1(t) = 10\cos(4t + \pi/6),$$
$$v_2(t) = 20\sin(3t + 45^\circ).$$

Solution: From the definition of the phasor, we have

$$\underline{v}_1 = \frac{10}{\sqrt{2}} \exp(j\pi/6)$$
$$= 5\sqrt{2}(\cos(\pi/6) + j\sin(\pi/6)) = 5\sqrt{2}\left(\frac{\sqrt{3}}{2} + \frac{j}{2}\right)$$
$$= \frac{5\sqrt{6}}{2} + j\frac{5\sqrt{2}}{2}$$

for the first voltage. For the second one, we first convert it into cosine form as

$$v_2(t) = 20\cos(3t - 45^\circ).$$

Then we derive

$$\underline{v}_2 = \frac{20}{\sqrt{2}} \exp(-j\pi/4)$$
$$= \frac{20}{\sqrt{2}}(\cos(\pi/4) - j\sin(\pi/4)) = \frac{20}{\sqrt{2}}\left(\frac{1}{\sqrt{2}} - \frac{j}{\sqrt{2}}\right)$$
$$= 10 - j10 = 10(1 - j).$$

We note that, without the angular frequency, the phasor of a function does not provide full information. For example, given only $\underline{v}_2 = 10(1 - j)$, one cannot find $v_2(t)$ without knowing the corresponding angular frequency (i.e., $\omega = 3$ rad/s).

A major advantage of using the phasor domain appears when derivatives of functions are used. In steady-state analysis of circuits in the phasor domain, resistors are not affected while capacitors and inductors are converted into imaginary resistors.

7.3.1 Resistors in Phasor Domain

In a circuit involving sinusoidal sources, the voltage and current of a resistor R take the form

$$v(t) = v_m \cos(\omega t + \phi),$$
$$i(t) = \frac{v(t)}{R} = \frac{v_m}{R} \cos(\omega t + \phi).$$

Therefore, in the phasor domain, we have

$$\underline{v} = \frac{v_m}{\sqrt{2}} \exp(j\phi) = \frac{v_m}{\sqrt{2}} \angle \phi,$$
$$\underline{i} = \frac{v_m}{R\sqrt{2}} \exp(j\phi) = \frac{v_m}{R\sqrt{2}} \angle \phi,$$

and

$$\underline{v} = R\underline{i}.$$

Hence, the voltage and current of a resistor are in phase, that is, they have the same phase.

7.3.2 Capacitors in Phasor Domain

Consider a capacitor in a circuit involving sinusoidal sources and its voltage in the form of

$$v(t) = v_m \cos(\omega t + \phi).$$

As discussed before, the current of the capacitor is given by

$$i(t) = C\frac{dv(t)}{dt} = -\omega C v_m \sin(\omega t + \phi)$$
$$= -\omega C v_m \cos(\omega t + \phi - \pi/2).$$

Therefore, we have

$$\underline{v} = \frac{v_m}{\sqrt{2}} \exp(j\phi) = \frac{v_m}{\sqrt{2}} \angle \phi,$$
$$\underline{i} = -\omega C \frac{v_m}{\sqrt{2}} \angle \phi - \pi/2 = j\omega C \frac{v_m}{\sqrt{2}} \angle \phi.$$

Consequently, in the phasor domain, the current and voltage of a capacitor are related by

$$\underline{v} = \frac{1}{j\omega C}\underline{i}.$$

This can be interpreted as a capacitor acting like an imaginary resistor with a resistance of $1/(j\omega C)$.

As also discussed before, the voltage of a capacitor lags its current by $\pi/2$. In order to see this in the phasor domain, we note that, if $\underline{i} = |i| \angle \phi_i$, then

$$\underline{v} = \frac{1}{j\omega C} |i| \exp(j\phi_i) = \frac{|i|}{\omega C} \exp(j\phi_i - j\pi/2) = \frac{|i|}{\omega C} \angle \phi_i - \pi/2.$$

Hence, we have the following representative graph of the complex current and voltage of a capacitor.

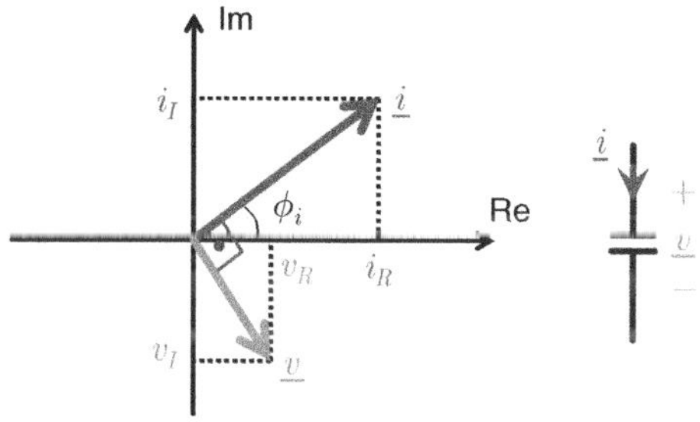

In this plot, the $\pi/2$ angle between the current and voltage phasors can clearly be seen.

7.3.3 Inductors in Phasor Domain

Consider an inductor in a circuit involving sinusoidal sources and its current in the form of

$$i(t) = i_m \cos(\omega t + \phi).$$

In this case, we have

$$\begin{aligned} v(t) &= L\frac{di(t)}{dt} = -\omega L i_m \sin(\omega t + \phi) \\ &= -\omega L i_m \cos(\omega t + \phi - \pi/2). \end{aligned}$$

Therefore, in the phasor domain, we get

$$\underline{i} = \frac{i_m}{\sqrt{2}} \exp(j\phi) = \frac{i_m}{\sqrt{2}} \underline{/\phi}$$

$$\underline{v} = -\omega L \frac{i_m}{\sqrt{2}} \underline{/\phi - \pi/2} = j\omega L \frac{i_m}{\sqrt{2}} \underline{/\phi},$$

and

$$\underline{v} = j\omega L \underline{i}.$$

Consequently, an inductor acts like an imaginary resistor with a resistance of $j\omega L$. There is again a $\pi/2$ phase difference between the voltage and current. However, unlike the capacitor, the voltage of an inductor leads its current by $\pi/2$. This relationship can be visualized as follows.

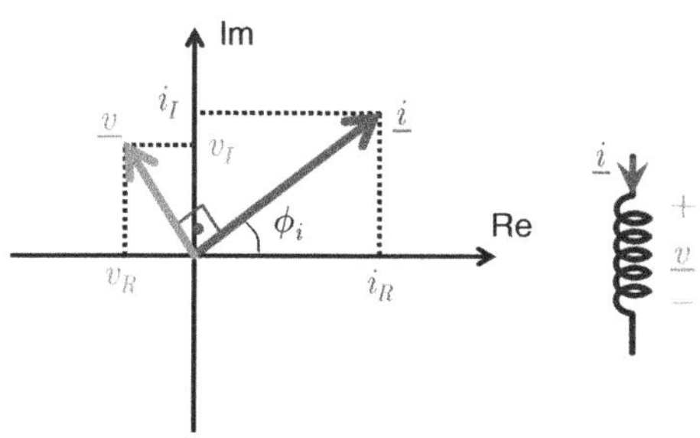

7.3.4 Impedance Concept

In general, AC circuits involve resistors, capacitors, and inductors, in addition to AC sources. In a steady-state analysis of circuits with time-harmonic sources, Ohm's law can be used for all of these components by treating capacitors and inductors as imaginary resistors. A combination of resistors, capacitors, and inductors in part of a circuit may lead to a complex resistance value, which is called impedance, in the form of

$$Z = R + jX,$$

where R is the normal resistance (due to resistors) and X is the reactance (due to capacitors and/or inductors). Such a combination can be considered as a single component for easier analysis. For a component with impedance Z, its voltage and current are related by

$$\underline{v} = Z\underline{i},$$

where complex values of Z introduce phase differences between $v(t)$ and $i(t)$.

Example 116: Consider the following circuit involving two capacitors and three inductors that are connected to a time-harmonic voltage source.

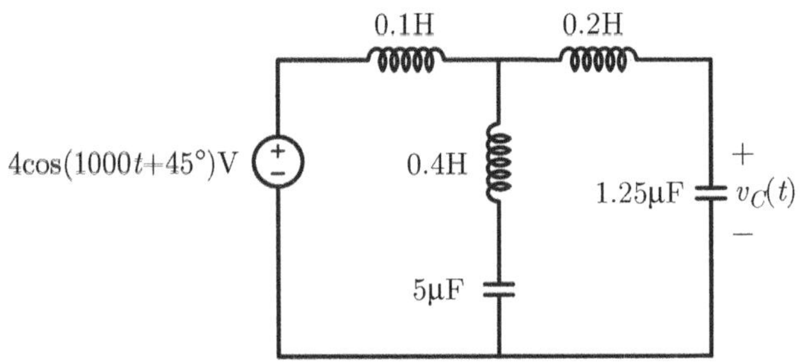

Find $v_C(t)$ in steady state.

Solution: In order to solve this problem, we need to convert the time-domain circuit into a phasor-domain circuit. Considering that $\omega = 1000$ rad/s, we have the following conversions:

- $4\cos(1000t + 45°)$ V $\longrightarrow \dfrac{4}{\sqrt{2}}\angle\pi/4$ V,
- 0.1 H $\longrightarrow j \times 1000 \times 0.1 = j100\ \Omega$,
- 0.2 H $\longrightarrow j \times 1000 \times 0.2 = j200\ \Omega$,
- 0.4 H $\longrightarrow j \times 1000 \times 0.4 = j400\ \Omega$,
- 1.25 μF $\longrightarrow 1/(j \times 1000 \times 1.25 \times 10^{-6}) = -j800\ \Omega$,
- 5 μF $\longrightarrow 1/(j \times 1000 \times 5 \times 10^{-6}) = -j200\ \Omega$.

Consequently, in the phasor domain, we have the following circuit.

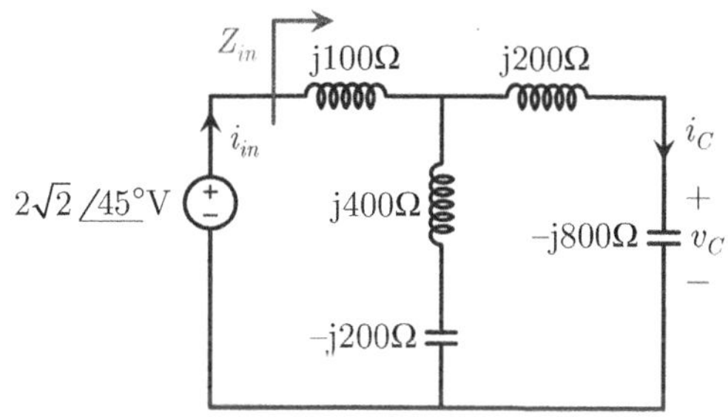

Now the circuit can be solved via any method: nodal analysis, mesh analysis, or by directly using KCL and KVL. Since the circuit is quite simple, one can also use series and parallel connections of impedances to find the overall input impedance seen by the voltage source:

$$Z_{\text{in}} = j100 + (j400 - j200) \parallel (j200 - j800) = j100 + \frac{j200 \times (-j600)}{j200 - j600}$$

$$= j100 + \frac{j200 \times (-j600)}{-j400} = j100 + j300 = j400\ \Omega.$$

We note that the input impedance is purely reactive and does not have any real part due to lack of resistance and power dissipation. The current flowing through the voltage source can be found to be

$$\underline{i}_{\text{in}} = \frac{\frac{4}{\sqrt{2}}\angle \pi/4}{j400}\ \text{A}.$$

Since the numerator and denominator of the fraction above have different forms, one needs to manipulate one of them in order to evaluate the expression. For example, modifying the numerator, we have

$$\underline{i}_{\text{in}} = \frac{\frac{4}{\sqrt{2}}(\cos(\pi/4) + j\sin(\pi/4))}{j400}$$

$$= \frac{2 + j2}{j400} = \frac{1 + j}{j200} = \frac{1 - j}{200} = \frac{\sqrt{2}}{200}\angle(-\pi/4)\ \text{A}.$$

Alternatively, one can find

$$\underline{i}_{\text{in}} = \frac{\frac{4}{\sqrt{2}}\angle \pi/4}{400\angle \pi/2} = \frac{\sqrt{2}}{200}\angle \pi/4 - \pi/2 = \frac{\sqrt{2}}{200}\angle(-\pi/4)\ \text{A}.$$

In order to find the current through the 1.25 μF capacitor, we can use the rule of current division:

$$\underline{i}_C = \frac{j200}{j200 - j600} \times \underline{i}_{\text{in}} = -\frac{1}{2}\underline{i}_{\text{in}} = \frac{\sqrt{2}}{400}\angle(-\pi/4 + \pi) = \frac{\sqrt{2}}{400}\angle 3\pi/4\ \text{A}.$$

Finally, the voltage across the capacitor can be found to be

$$\underline{v}_C = Z_C \underline{i}_C = -j800 \times \frac{\sqrt{2}}{400}\angle 3\pi/4 \text{ A} = 2\sqrt{2}\angle 3\pi/4 - \pi/2$$
$$= 2\sqrt{2}\angle \pi/4 \text{ V}.$$

Converting the final expression into the time domain, we have

$$v_C(t) = \sqrt{2}\mathrm{Re}\{\underline{v}_C \exp(j\omega t)\} = 4\cos(1000t + \pi/4)$$
$$= 4\cos(1000t + 45^\circ) \text{ V}.$$

We note that, as a coincidence, the voltage of the capacitor is in phase with the voltage source, while this could not be understood directly without analyzing the circuit.

Example 117: Consider the following circuit.

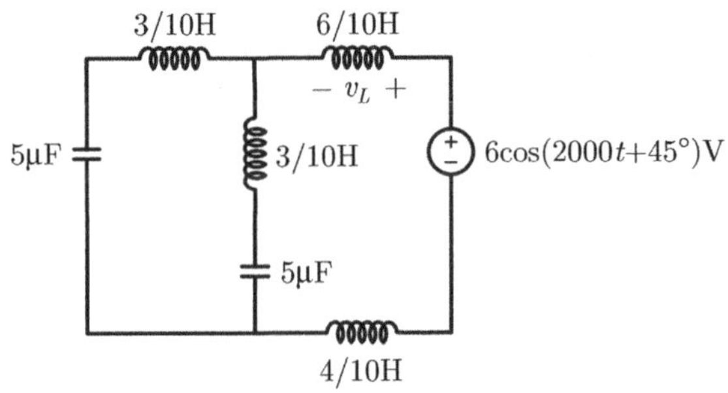

Find $v_L(t)$ in steady state.

Solution: We again consider the following conversions:

- $6\cos(2000t + 45^\circ)$ V $\longrightarrow \frac{3}{\sqrt{2}}\angle \pi/4$ V,
- 3/10 H $\longrightarrow j \times 2000 \times 3/10 = j600\ \Omega$,
- 4/10 H $\longrightarrow j \times 2000 \times 4/10 = j800\ \Omega$,
- 6/10 H $\longrightarrow j \times 2000 \times 6/10 = j1200\ \Omega$,
- 5 μF $\longrightarrow 1/(j \times 2000 \times 5 \times 10^{-6}) = -j100\ \Omega$,

Therefore, in the phasor domain, the circuit can be represented as follows.

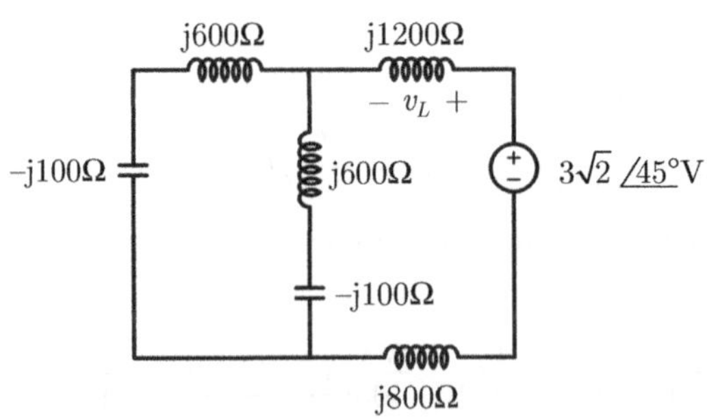

The input impedance seen by the source can be found to be

$$Z_{\text{in}} = (j600 - j100) \parallel (j600 - j100) + j1200 + j800 = j2250\ \Omega.$$

Then, since

$$3\sqrt{2}\underline{/45^\circ} = 3\sqrt{2}(\cos(\pi/4) + j\sin(\pi/4)) = 3(1+j)\ \text{V},$$

the current through the $j1200\ \Omega$ inductor can be obtained as

$$\underline{i}_L = \frac{3+j3}{j2250} = \frac{1-j}{750}\ \text{A}.$$

Finally, the voltage across the $j1200\ \Omega$ inductor can be derived as

$$\underline{v}_L = j1200 \times \underline{i}_L = \frac{8+j8}{5}\ \text{V}.$$

Considering that the above expression can also written as

$$\underline{v}_L = j1200 \times \underline{i}_L = \frac{8\sqrt{2}}{5}\underline{/\pi/4}\ \text{V},$$

we further have

$$v_L(t) = \frac{16}{5}\cos(2000t + \pi/4) = \frac{16}{5}\cos(2000t + 45^\circ)\ \text{V}$$

in the time domain.

Example 118: Consider the following circuit.

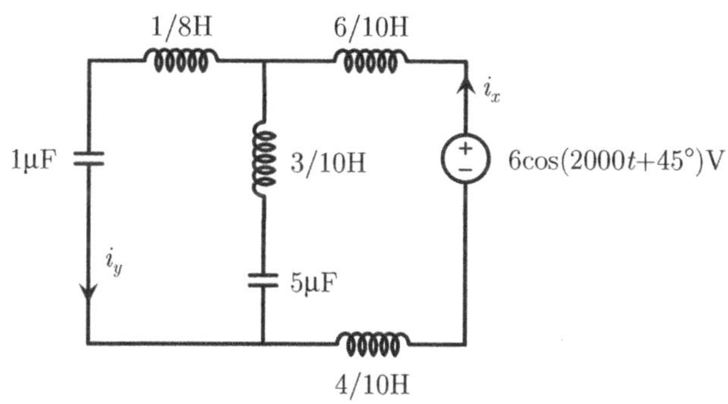

Find $i_y(t)$ as a function of time in steady state.

Solution: This circuit is very similar to the previous one, except

- $1/8\ \text{H} \longrightarrow j \times 2000 \times 1/8 = j250\ \Omega$
- $1\ \mu\text{F} \longrightarrow 1/(j \times 2000 \times 1 \times 10^{-6}) = -j500\ \Omega.$

Therefore, the input impedance seen by the source should be revised as

$$Z_{\text{in}} = (j250 - j500) \parallel (j600 - j100) + j1200 + j800 = j1500\ \Omega.$$

Therefore, we obtain $\underline{i}_x$ as

$$\underline{i}_x = \frac{3+j3}{j1500} = \frac{1-j}{500}\ \text{A}.$$

Then $\underline{i}_y$ can be obtained via current division as

$$\underline{i}_y = \frac{j500}{j250}\underline{i}_x = 2\underline{i}_x = \frac{1-j}{250}\ \text{A}.$$

Therefore, in the time domain, we have

$$i_y(t) = \frac{2}{250}\cos(2000t - \pi/4)\ \text{A}.$$

Exercise 104: In the following circuit, find $v_x(t)$ in steady state.

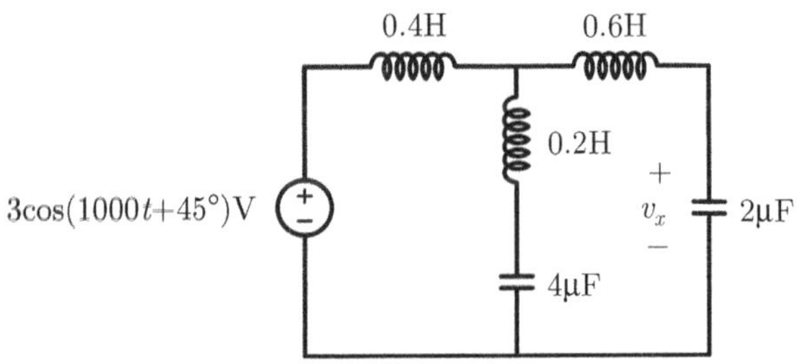

Exercise 105: Consider the following circuit.

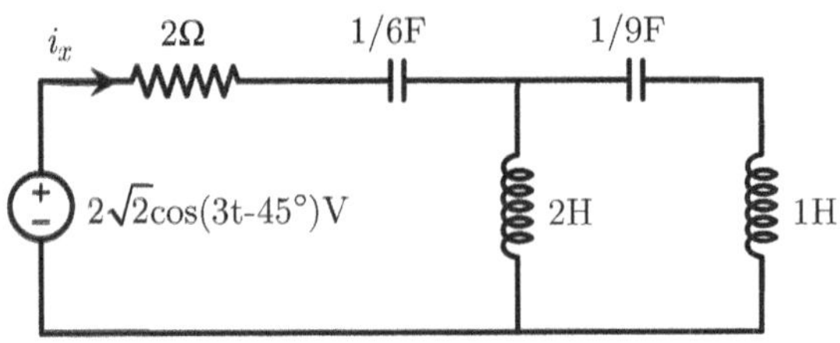

Find $i_x(t)$ in steady state.

Exercise 106: In the following circuit, find $v_L(t)$ in steady state.

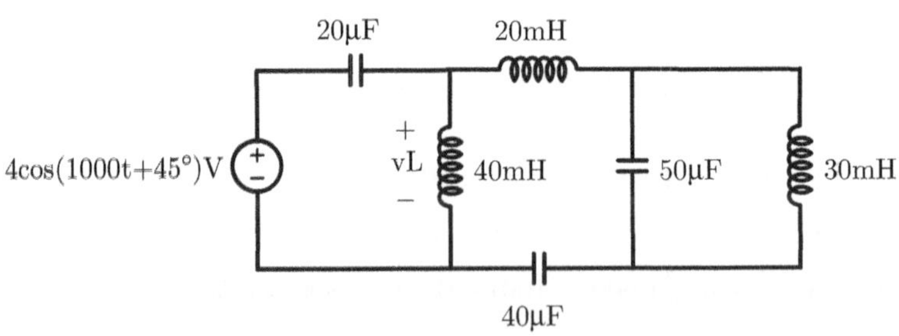

7.4 Special Circuits in Phasor Domain

Once converted into the phasor domain, circuits involving resistors, energy storage elements, and AC sources become easy to analyze in steady state. In this section, we consider some special circuits and their responses, especially with respect to frequency.

7.4.1 RC Circuits in Phasor Domain

Consider the following circuit involving a series connection of a resistor and capacitor to a sinusoidal voltage source $v_s = v_m \cos(\omega t)$.

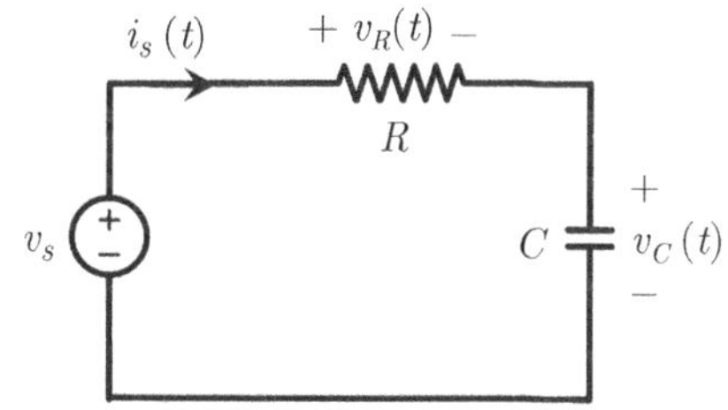

In the phasor domain, we have

$$\underline{i}_s = \frac{\underline{v}_s}{R + 1/(j\omega C)} = \frac{j\omega C \underline{v}_s}{1 + j\omega RC},$$

leading to

$$\underline{v}_C = \frac{1}{j\omega C} \frac{j\omega C \underline{v}_s}{1 + j\omega RC} = \frac{\underline{v}_s}{1 + j\omega RC}.$$

Furthermore, using $\underline{v}_s = \frac{v_m}{\sqrt{2}} \angle 0$, we derive

$$\underline{v}_C = \frac{v_m}{\sqrt{2}} \frac{1}{1 + j\omega RC} = \frac{v_m}{\sqrt{2}} \frac{1 - j\omega RC}{1 + \omega^2 R^2 C^2}$$
$$= \frac{v_m}{\sqrt{2}} \frac{1}{\sqrt{1 + \omega^2 R^2 C^2}} \angle \tan^{-1}(-\omega RC).$$

Two cases are of particular interest.

- For small values of ω, that is, for $\omega \ll 1/RC$, we have $\omega RC \approx 0$ and

  $$\underline{v}_C \approx \frac{v_m}{\sqrt{2}} \angle 0,$$

  leading to

  $$v_C(t) \approx v_m \cos(\omega t).$$

  Therefore, the capacitor voltage is approximately the same as the input voltage. This can be verified by considering the DC limit, that is, the capacitor becomes an open circuit so that no voltage drops across the resistor and all voltage appears on the capacitor.

- For large values of ω, that is, for $\omega \gg 1/RC$, we have

$$\underline{v}_C \approx \frac{v_m}{\sqrt{2}} \frac{1}{\omega RC} \angle(-\pi/2),$$

leading to

$$v_C(t) \approx \frac{v_m}{\omega RC} \cos(\omega t - \pi/2).$$

Hence, the amplitude of the capacitor voltage drops to small values, while it becomes 90° out of phase compared to the source. In the limit $\omega \to \infty$, we further have

$$v_C(t) \to 0$$

since the capacitor becomes a short circuit.

Considering the special cases above, as well as the behavior for different values of ω, the RC circuit can be considered as a low-pass filter. Specifically, if the capacitor voltage is selected as the output, low-frequency signals easily pass through the circuit, while high-frequency signals are filtered out (cannot appear across the capacitor). This property makes the RC circuit an important ingredient of many practical circuits. Obviously, the values of R and C in the circuit determine which frequencies are to be filtered for a given threshold. For example, in order to filter 90% of any signal above 100 MHz, one must select

$$\frac{1}{\sqrt{1 + \omega^2 R^2 C^2}} < 0.9 \to RC > 7.71 \times 10^{-10} \text{ s}.$$

Finally, we note that the same RC circuit can be considered as a high-pass circuit if the resistor voltage is selected as the output.

7.4.2 RL Circuits in Phasor Domain

Consider the following circuit involving a series connection of a resistor and inductor to a voltage source $v_s = v_m \cos(\omega t)$.

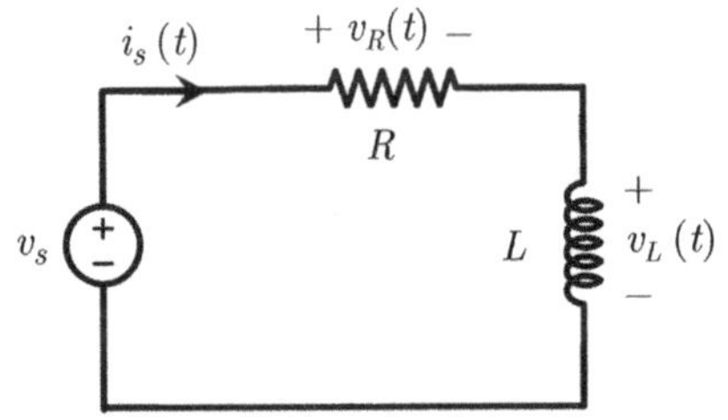

Investigating the circuit in the phasor domain, we derive

$$\underline{i}_s = \frac{\underline{v}_s}{R + j\omega L}$$

and

$$\underline{v}_L = j\omega L \frac{\underline{v}_s}{R + j\omega L} = \frac{\underline{v}_s}{1 - jR/(\omega L)}.$$

In addition, using $\underline{v}_s = \frac{v_m}{\sqrt{2}}\angle 0$, we have

$$\underline{v}_L = \frac{v_m}{\sqrt{2}} \frac{1}{1 - jR/(\omega L)} = \frac{v_m}{\sqrt{2}} \frac{1 + jR/(\omega L)}{1 + R^2/(\omega^2 L^2)}$$
$$= \frac{v_m}{\sqrt{2}} \frac{1}{\sqrt{1 + R^2/(\omega^2 L^2)}} \angle \tan^{-1}(R/(\omega L)).$$

Similarly to the RC circuit discussed above, two special cases are interesting for this RL circuit.

- For small values of ω, that is., for $\omega \ll R/L$, we have

  $$\underline{v}_L \approx \frac{v_m}{\sqrt{2}} \frac{\omega L}{R} \angle \pi/2,$$

  leading to

  $$v_L(t) \approx \frac{v_m \omega L}{R} \cos(\omega t - \pi/2).$$

  Therefore, as the value of ω drops to zero, the amplitude of the inductor voltage becomes zero. In the DC limit, that is, when $\omega \to 0$, we further have

  $$v_L(t) \to 0$$

  since the inductor becomes a short circuit.
- For large values of ω, that is, for $\omega \gg R/L$, we derive

  $$\underline{v}_L \approx \frac{v_m}{\sqrt{2}} \angle 0$$

  and

  $$v_C(t) \approx v_m \cos(\omega t).$$

  In such a case, most of the voltage provided by the source appears on the inductor. In the limit as $\omega \to \infty$, where the inductor can be considered as an open circuit, no current flows through the circuit and the inductor voltage becomes the same as the source voltage.

Considering the behavior of the RL circuit with the inductor voltage as the output, it can be called a high-pass filter since the low-frequency signals are not transferred and blocked. As the frequency increases, the voltage of the source (input) appears more across the inductor (output), that is, signals are allowed to pass.

7.4.3 RLC Circuits in Phasor Domain

Now, we consider the following circuit involving a series connection of a resistor R, a capacitor C, and an inductor L.

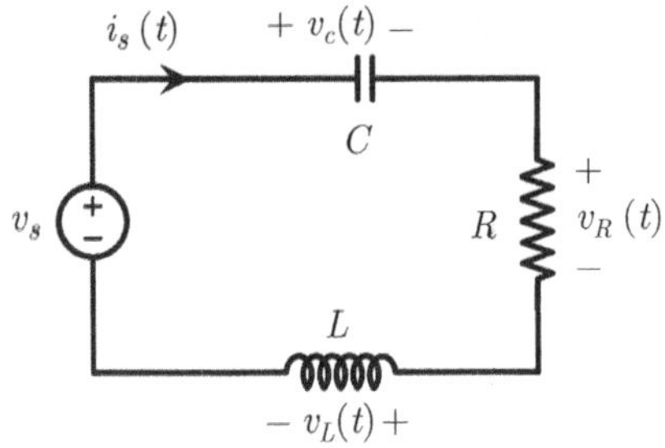

A sinusoidal voltage source with angular frequency ω is connected to the components, leading to

$$\underline{i}_s = \frac{\underline{v}_s}{R + j\omega L - j/(\omega C)} = \frac{\underline{v}_s}{R + j(\omega L - 1/(\omega C))}.$$

The RMS of the current can be found directly from the phasor as

$$i_{\text{RMS}} = |\underline{i}_s| = \frac{v_{\text{RMS}}}{|R + j(\omega L - 1/(\omega C))|} = \frac{v_{\text{RMS}}}{\sqrt{R^2 + (\omega L - 1/(\omega C))^2}}.$$

For the circuit above, the frequency at which the RMS of the current is maximum is particularly interesting. This corresponds to the case where the power of the resistor is maximized. Using the expression above, one can derive

$$\frac{\partial i_{\text{RMS}}}{\partial \omega} = v_{\text{RMS}} \frac{\partial}{\partial \omega}\left[R^2 + \left(\omega L - \frac{1}{\omega C}\right)^2\right]^{-1/2}$$
$$= -v_{\text{RMS}}\left[R^2 + \left(\omega L - \frac{1}{\omega C}\right)^2\right]^{-3/2}\left(\omega L - \frac{1}{\omega C}\right)\left(L + \frac{1}{\omega^2 C}\right).$$

Then, for $\partial i_{\text{RMS}}/\partial\omega = 0$, a positive value of the angular frequency can be found,

$$\omega = \frac{1}{\sqrt{LC}}.$$

For this value of ω, the overall impedance of the series connection becomes R, so that we have $\underline{i}_s = \underline{v}_s/R$ and $i_{\text{RMS}} = v_{\text{RMS}}/R$. This can be seen as a result of the cancelation of the impedances of the inductor and capacitor, that is,

$$Z_L = j\omega L = j\frac{1}{\sqrt{LC}}L = j\sqrt{\frac{L}{C}},$$
$$Z_C = \frac{1}{j\omega C} = -j\sqrt{LC}\frac{1}{C} = -j\sqrt{\frac{L}{C}},$$
$$Z_L + Z_C = 0.$$

Consequently, if the resistor is considered as the output of this circuit, frequencies at around $\omega = 1/\sqrt{LC}$ pass through the circuit, while the other signals (with smaller and larger values of ω) are filtered out due to the open-circuit behavior of the capacitor and inductor, respectively. Such a circuit can be called a band-pass filter.

Next, we consider a parallel connection of a resistor R, a capacitor C, and an inductor L to a current source with angular frequency ω.

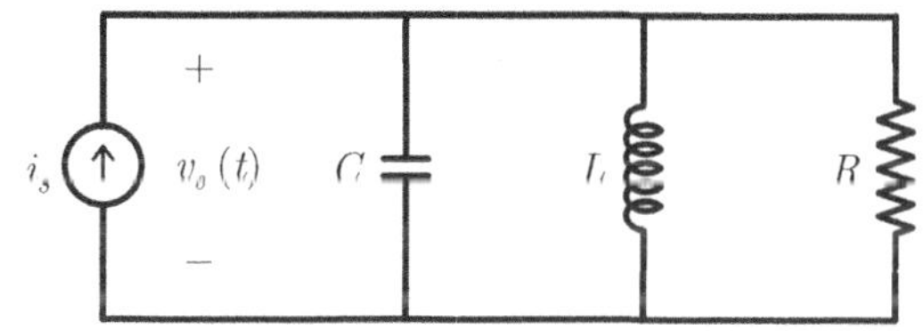

In this case, we have

$$\underline{i}_s = \frac{\underline{v}_s}{R} + \frac{\underline{v}_s}{j\omega L} + \frac{\underline{v}_s}{-j/(\omega C)} = \underline{v}_s \left[\frac{1}{R} + j\left(-\frac{1}{\omega L} + \omega C\right)\right]$$

and

$$i_{\mathrm{RMS}} = v_{\mathrm{RMS}} \sqrt{\frac{1}{R^2} + \left(\omega C - \frac{1}{\omega L}\right)^2}.$$

Therefore, in order to maximize the RMS of the voltage value, such that the power consumed by the resistor is again maximized, one can choose $\omega = 1/\sqrt{LC}$, leading to $v_{\mathrm{RMS}} = Ri_{\mathrm{RMS}}$. This can be considered as the open circuit created by the parallel connection of the capacitor and inductor as

$$Z_L \parallel Z_C = \frac{(j\sqrt{L/C}) \times (-j\sqrt{L/C})}{j\sqrt{L/C} - j\sqrt{L/C}} = \infty.$$

We note that neither capacitor nor inductor alone is an open circuit, while their combination creates the open-circuit effect. In the time domain, this can be explained as a perfect LC resonator, with oscillatory transfer of energy between capacitor and inductor, without affecting the resistor and its connection to the source.

Exercise 107: Consider the following circuit.

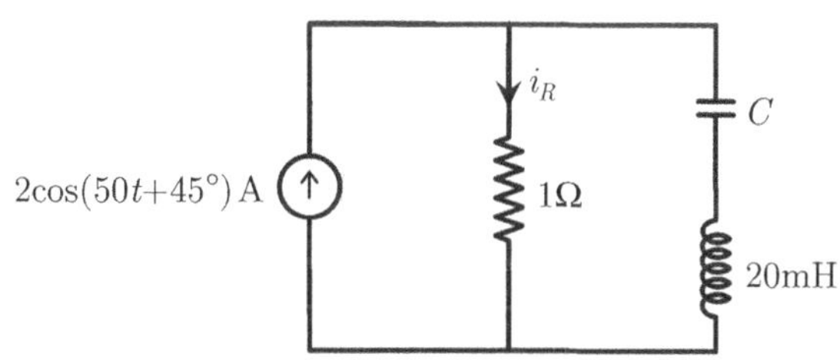

Find $i_R(t)$ in steady state, when (a) $C = 20$ mF and (b) $C = 10$ mF.

7.4.4 Other Combinations

In the above, we consider some of the popular combinations, including RC and RL circuits connected to a voltage source, as well as series and parallel RLC variations with voltage and current sources, respectively. However, considering different kinds of sources (voltage and current) and alternative connections (series and parallel), there can be many other basic combinations. For example, a parallel connection of a resistor and inductor can be connected in series to a capacitor, leading to a high-pass filter if the resistor is considered as output. All those combinations are frequently used in modern circuits, while their analysis can be performed as shown in this section.

7.5 Analysis of Complex Circuits at Fixed Frequencies

We now turn to the analysis of more complex circuits involving sinusoidal sources in steady state. Once again, we transform the circuits into phasor-domain versions such that they can be analyzed using the techniques that we discussed in the previous chapters.

Example 119: Consider the following circuit.

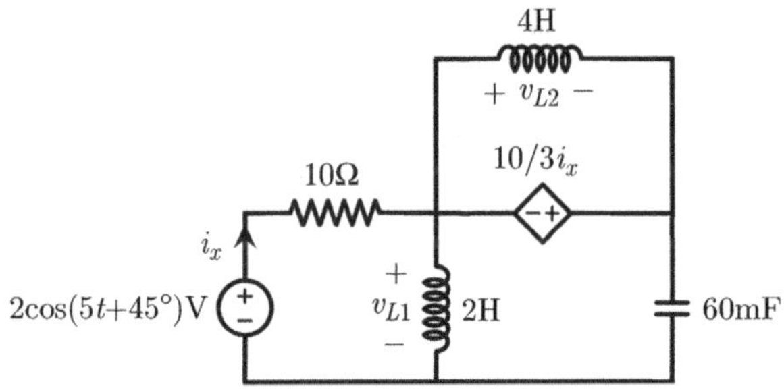

Find $v_{L1}(t)$ and $v_{L2}(t)$ in steady state.

Solution: First we convert the time-domain circuit into a phasor-domain version. Considering that $\omega = 5$ rad/s, we have the following conversions:

- $2\cos(5t + 45°)$ V $\longrightarrow \sqrt{2}\angle\pi/4$ V,
- 2 H $\longrightarrow j \times 5 \times 2 = j10\ \Omega$,
- 4 H $\longrightarrow j \times 5 \times 4 = j20\ \Omega$,
- 60 mF $\longrightarrow 1/(j \times 5 \times 60 \times 10^{-3}) = -j10/3\ \Omega$.

With these conversions, we have the following circuit in the phasor domain.

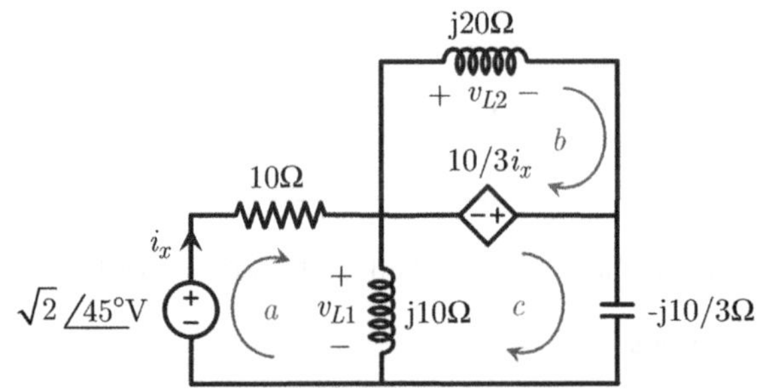

In order to solve the problem, we apply the mesh analysis with three mesh currents. Considering that $\underline{i}_x = \underline{i}_a$, we obtain

- KVL(a): $-\sqrt{2}\underline{/\pi/4} + 10\underline{i}_a + j10(\underline{i}_a - \underline{i}_c) = 0$,

leading to

$$(10 + j10)\underline{i}_a - j10\underline{i}_c = 1 + j.$$

Similarly, applying KVL in mesh b, we derive

- KVL(b): $(10/3)\underline{i}_x + j20\underline{i}_b = 0 \longrightarrow \underline{i}_b = (j/6)\underline{i}_a$.

Finally, considering mesh c, we have

- KVL(c): $-(10/3)\underline{i}_x - j(10/3)\underline{i}_c + j10(\underline{i}_c - \underline{i}_a) = 0$

or

$$j(20/3)\underline{i}_c = (10/3 + j10)\underline{i}_a.$$

The final equality can be used to find $\underline{i}_c$ in terms of $\underline{i}_a$:

$$\underline{i}_c = \frac{3}{j20} \times \left(\frac{10}{3} + j10\right)\underline{i}_a = (3 - j)\underline{i}_a/2.$$

Therefore, we have

$$(10 + j10)\underline{i}_a - j5(3 - j)\underline{i}_a = 1 + j$$

$$(5 - j5)\underline{i}_a = 1 + j \longrightarrow \underline{i}_a = \frac{1 + j}{5(1 - j)} = j/5 \text{ A},$$

and

$$\underline{i}_b = \frac{j}{6} \times \frac{j}{5} = -1/30 \text{ A},$$

$$\underline{i}_c = \frac{3 - j}{2} \times \frac{j}{5} = (1 + j3)/10 \text{ A}.$$

Then the voltage across the 2 H inductor can be found to be

$$\underline{v}_{L1} = j10 \times (\underline{i}_a - \underline{i}_c) = j10 \times \left(\frac{j}{5} - \frac{1 + j3}{10}\right) = -2 - j + 3 = 1 - j \text{ V}.$$

Similarly, we have

$$\underline{v}_{L2} = j20\underline{i}_b = -j2/3 \text{ V}.$$

Finally, we obtain the time-domain voltage values

$$\begin{aligned} v_{L1}(t) &= \sqrt{2}\text{Re}\{\underline{v}_{L1}\exp(j\omega t)\} \\ &= \sqrt{2}\text{Re}\{\sqrt{2}\underline{/(-\pi/4)}\exp(j5t)\} = 2\cos(5t - \pi/4) \text{ V} \end{aligned}$$

and

$$v_{L2}(t) = \sqrt{2}\text{Re}\{\underline{v}_{L2}\exp(j\omega t)\}$$
$$= \sqrt{2}\text{Re}\{(2/3)\underline{/(-\pi/2)}\exp(j5t)\} = \frac{2\sqrt{2}}{3}\cos(5t - \pi/2)\text{ V}.$$

Example 120: Consider the following circuit.

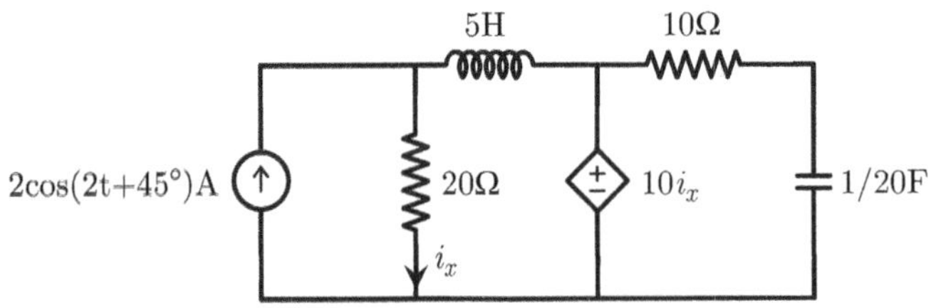

Find $i_x(t)$ and the voltage across the capacitor in steady state.

Solution: We convert the circuit into the phasor domain, considering that $\omega = 2$ rad/s, as follows.

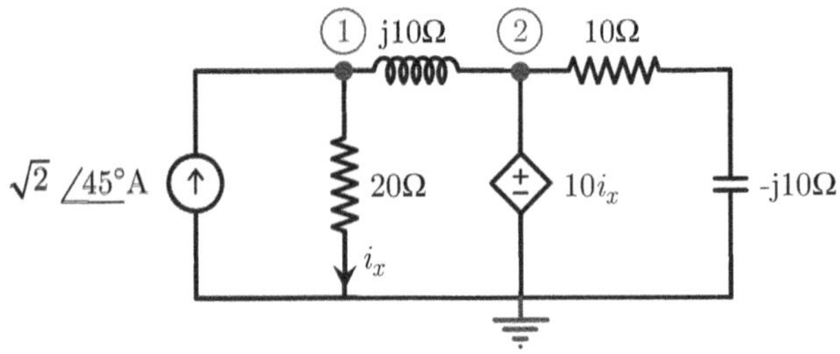

Using nodal analysis as also shown above, we have $\underline{i}_x = \underline{v}_1/20$ and $\underline{v}_2 = 10\underline{i}_x = \underline{v}_1/2$. Furthermore, applying KCL at node 1, we derive

- KCL(1): $(1 + j) - \underline{v}_1/20 - (\underline{v}_1 - \underline{v}_2)/j10 = 0$,

leading to $\underline{v}_1 = j20$ V. In the above, we note that

$$\sqrt{2}\underline{/45^\circ} = \sqrt{2}\cos(45^\circ) + j\sqrt{2}\sin(45^\circ) = 1 + j.$$

Using $\underline{v}_1$, one obtains $\underline{i}_x = j$ A, $\underline{v}_2 = j10$ V, and

$$\underline{v}_C = \frac{-j10}{10 - j10}\underline{v}_2 = \frac{10}{1 - j} = 5(1 + j)\text{ V}.$$

Therefore, in the time domain, we derive

$$v_C(t) = 10\cos(2t + \pi/4)\text{ V},$$
$$i_x(t) = \sqrt{2}\cos(2t + \pi/2) = -\sqrt{2}\sin(2t)\text{ A}.$$

Example 121: Consider the following circuit involving two sources at the same frequency.

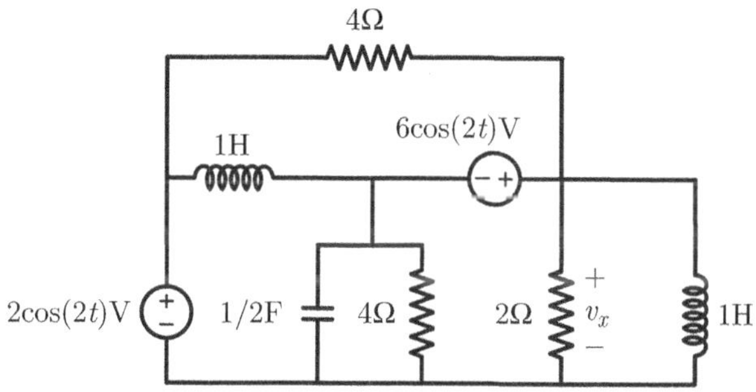

Find $v_x(t)$ in steady state.

Solution: Considering that $\omega = 2$ rad/s, we have the following conversions:

- $2\cos(2t)$ V $\longrightarrow$ $\sqrt{2}\angle 0$ V,
- $6\cos(2t)$ V $\longrightarrow$ $3\sqrt{2}\angle 0$ V,
- 1 H $\longrightarrow$ $j \times 2 \times 1 = j2\ \Omega$,
- 1/2 F $\longrightarrow$ $1/(j \times 2 \times 1/2) = -j\ \Omega$.

In addition, the parallel connections lead to

- $Z_1 = -j \parallel 4 = \dfrac{-j4}{4-j}\ \Omega$,
- $Z_2 = j2 \parallel 2 = \dfrac{j4}{2+j2} = \dfrac{j2}{1+j}\ \Omega$.

Consequently, we have the following phasor-domain circuit, which can be analyzed via nodal analysis.

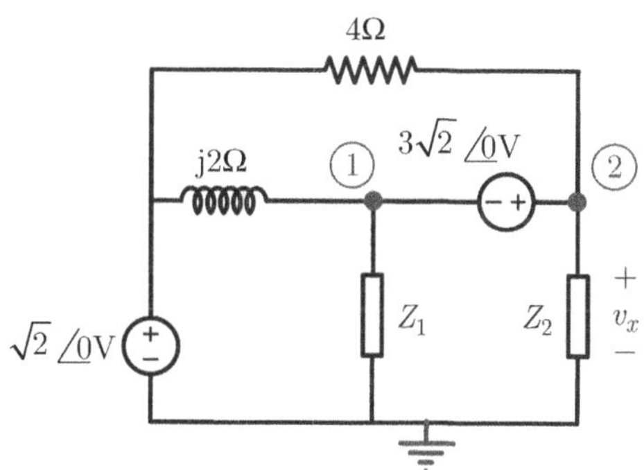

Applying KCL at nodes 1 and 2, we have

- KCL(1&2): $\dfrac{\sqrt{2} - \underline{v}_1}{j2} + \dfrac{\sqrt{2} - \underline{v}_2}{4} - \dfrac{\underline{v}_1}{-j4/(4-j)} - \dfrac{\underline{v}_2}{j2/(1+j)} = 0.$

Furthermore, using $\underline{v}_2 - \underline{v}_1 = 3\sqrt{2}$, we obtain

$$\frac{\sqrt{2}-\underline{v}_1}{j2} + \frac{-2\sqrt{2}-\underline{v}_1}{4} + \frac{(4-j)\underline{v}_1}{j4} - \frac{(1+j)\underline{v}_1}{j2} - \frac{(1+j)3\sqrt{2}}{j2} = 0$$

or

$$\frac{\sqrt{2}}{j2} - \frac{2\sqrt{2}}{4} - \frac{(1+j)3\sqrt{2}}{j2} = \frac{\underline{v}_1}{j2} + \frac{\underline{v}_1}{4} - \frac{(4-j)\underline{v}_1}{j4} + \frac{(1+j)\underline{v}_1}{j2}$$
$$= \underline{v}_1\left(\frac{1}{j2} + \frac{1}{4} - \frac{4-j}{4j} + \frac{1+j}{j2}\right).$$

Simplifying the combinations of complex numbers as

$$\frac{\sqrt{2}}{j2} - \frac{2\sqrt{2}}{4} - \frac{(1+j)3\sqrt{2}}{j2} = -\frac{\sqrt{2}}{2} - j\frac{\sqrt{2}}{2} + j\frac{3\sqrt{2}}{2} - \frac{3\sqrt{2}}{2} = -2\sqrt{2} + j\sqrt{2}$$
$$\frac{1}{j2} + \frac{1}{4} - \frac{4-j}{j4} + \frac{1+j}{j2} = -\frac{j}{2} + \frac{1}{4} + j + \frac{1}{4} - \frac{j}{2} + \frac{1}{2} = 1,$$

we derive

$$\underline{v}_1 = -\sqrt{2}(2-j) \text{ V}.$$

Then we obtain $\underline{v}_x = \underline{v}_2$ as

$$\underline{v}_x = 3\sqrt{2} + \underline{v}_1 = 3\sqrt{2} - \sqrt{2}(2-j) = \sqrt{2}(1+j) \text{ V}.$$

Finally, converting the final expression into the time domain, we have

$$v_x(t) = 2\sqrt{2}\cos(2t + \pi/4) \text{ V}.$$

Example 122: Consider the following circuit.

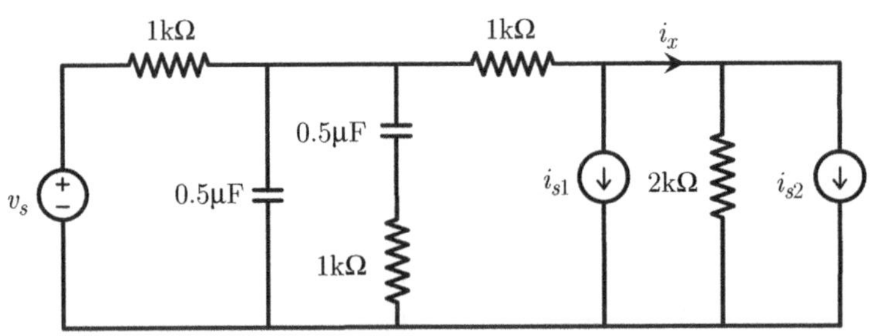

The source values are given as follows:

- $v_s = 8\cos(2000t)$ V,
- $i_{s1} = 18\sin(2000t)$ mA,
- $i_{s2} = 10\cos(2000t - 180°)$ mA.

Find $i_x(t)$ in steady state.

Solution: We again start with the conversions as follows:

- $8\cos(2000t)$ V $\longrightarrow 4\sqrt{2}\underline{/0} = 4\sqrt{2}$ V,
- $18\sin(2000t)$ mA $\longrightarrow 9\sqrt{2}\underline{/-\pi/2} = -j9\sqrt{2}$ mA,
- $10\cos(2000t - 180°)$ mA $\longrightarrow 5\sqrt{2}\underline{/-\pi} = -5\sqrt{2}$ mA,
- $0.5\ \mu$F $\longrightarrow 1/(j \times 2000 \times 0.5 \times 10^{-6}) = -j1000\ \Omega = -j$ kΩ.

Note that a zero phase can directly be omitted,

$$\underline{f} = f_{\text{RMS}}\underline{/0} = f_{\text{RMS}},$$

while keeping in mind that $\underline{f}$ is still a complex number with zero imaginary part. Hence, we have the following phasor-domain circuit.

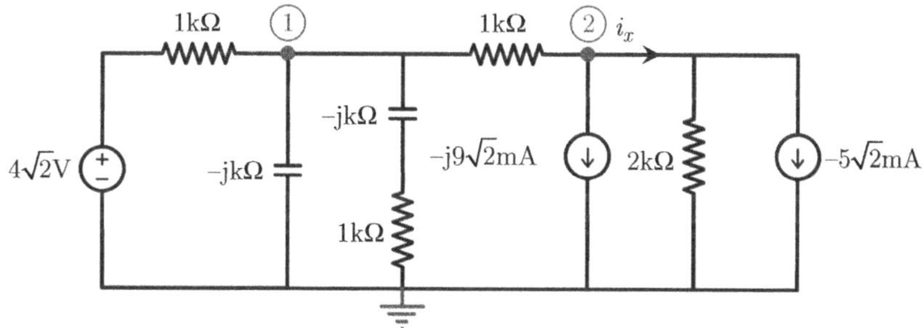

In the following, the current and impedance units are milliamperes and kilohms, respectively. Applying KCL at node 1, we derive

- KCL(1): $\dfrac{4\sqrt{2} - \underline{v}_1}{1} - \dfrac{\underline{v}_1}{-j} - \dfrac{\underline{v}_1}{1-j} - \dfrac{\underline{v}_1 - \underline{v}_2}{1} = 0,$

leading to

$$\underline{v}_1(5/2 + j3/2) - \underline{v}_2 = 4\sqrt{2}.$$

Then, applying KCL at node 2, we have

- KCL(2): $\dfrac{v_1 - v_2}{1} + j9\sqrt{2} - \dfrac{\underline{v}_2}{2} + 5\sqrt{2} = 0,$

leading to

$$\underline{v}_1 - 3\underline{v}_2/2 = -5\sqrt{2} - i9\sqrt{2}.$$

Solving the equations, we obtain

$$\underline{v}_1 = 4\sqrt{2}\text{ V},$$
$$\underline{v}_2 = 6\sqrt{2}(1 + j)\text{ V},$$

and

$$\underline{i}_x = (-2\sqrt{2} + j3\sqrt{2}) = \sqrt{26}\underline{/\pi - \tan^{-1}(3/2)}\text{ mA}.$$

Finally, the time-domain i_x can be written as

$$i_x(t) = 2\sqrt{13}\cos(2000t + \pi - \tan^{-1}(3/2))\ \text{mA}.$$

Example 123: Consider the following circuit.

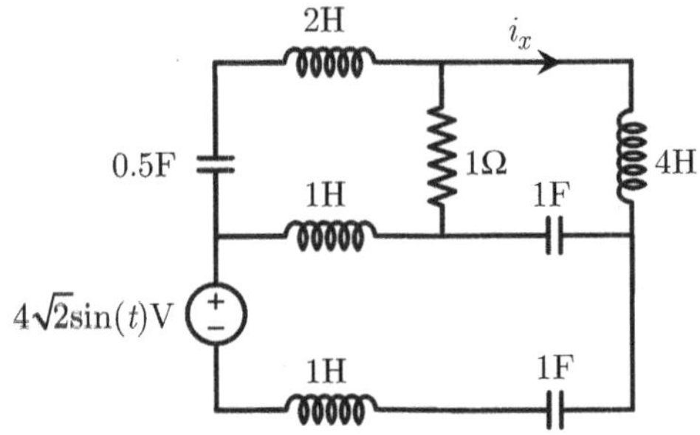

Find $i_x(t)$ in steady state.

Solution: Once again, considering that $\omega = 1$ rad/s, we have the following conversions:

- $4\sqrt{2}\sin(t)\ \text{V} \longrightarrow 4\angle(-\pi/2)\ \text{V}$,
- $1\ \text{H} \longrightarrow j \times 1 \times 1 = j\ \Omega$,
- $2\ \text{H} \longrightarrow j \times 1 \times 2 = j2\ \Omega$,
- $4\ \text{H} \longrightarrow j \times 1 \times 4 = j4\ \Omega$,
- $1\ \text{F} \longrightarrow 1/(j \times 1 \times 1) = -j\ \Omega$.

Consequently, the circuit in the phasor domain is as follows.

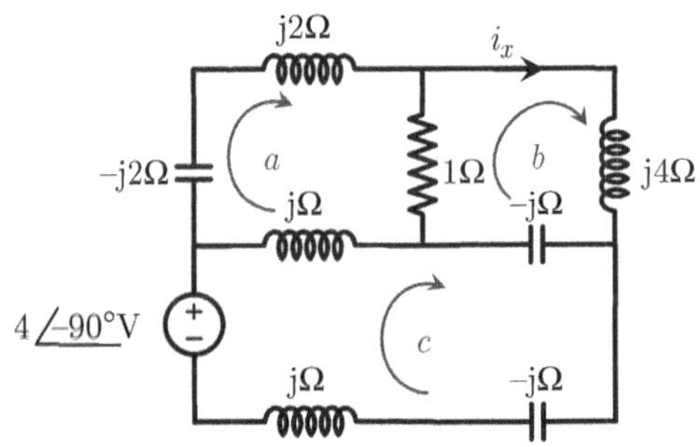

Using mesh analysis, we apply KVL in mesh a to derive

- KVL(a): $(-j2 + j2)\underline{i}_a + (\underline{i}_a - \underline{i}_b) + j(\underline{i}_a - \underline{i}_c) = 0 \longrightarrow \underline{i}_a(1 + j) - \underline{i}_b - j\underline{i}_c = 0.$

Similarly, using KVL in mesh b, we have

- KVL(b): $(\underline{i}_b - \underline{i}_a) + j4\underline{i}_b - j(\underline{i}_b - \underline{i}_c) = 0 \longrightarrow -\underline{i}_a + (1 + j3)\underline{i}_b + j\underline{i}_c = 0.$

Combining the two equations above, we derive

$$j\underline{i}_a + j3\underline{i}_b = 0,$$

leading to $\underline{i}_a = -3\underline{i}_b$. Finally, KVL in mesh c leads to

- KVL(c): $j4 + j(\underline{i}_c - \underline{i}_a) - j(\underline{i}_c - \underline{i}_b) = 0 \longrightarrow \underline{i}_a - \underline{i}_b = 4.$

Solving the equations, one obtains $\underline{i}_a = 3$ A and $\underline{i}_b = \underline{i}_x = -1$ A. Therefore, in the time domain, we have

$$\underline{i}_x(t) = -\sqrt{2}\cos(t) \text{ A}$$

or

$$\underline{i}_x(t) = \sqrt{2}\cos(t + \pi) = \sqrt{2}\cos(t + 180°) \text{ A}.$$

In this circuit, we note that the capacitor and inductor in the bottom branch actually cancel each other out in the phasor domain. Therefore, their combination can be replaced by a short circuit.

Example 124: Consider the following circuit, where $v_o(t) = 2\sqrt{2}\sin(2t)$.

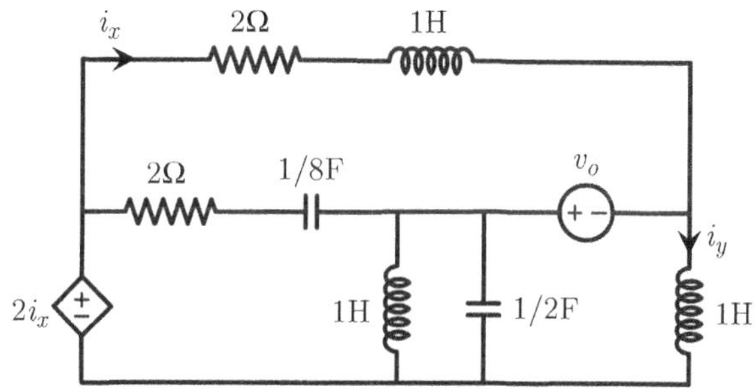

Find $i_x(t)$ in steady state.

Solution: Considering that $\omega = 2$ rad/s, the circuit in the phasor domain is as follows:

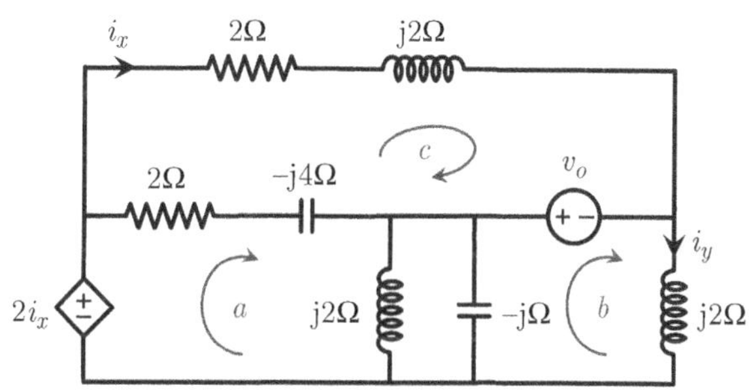

First, we note that

$$\underline{v}_o = 2\angle{-\pi/2} = -j2 \text{ V}$$

in the phasor domain. In addition, the parallel connection of the inductor and capacitor leads to

$$j2 \parallel (-j) = -j2\ \Omega.$$

Using mesh analysis, we further have $\underline{i}_x = \underline{i}_c$. Applying KCL in mesh b, we derive

- KVL(b): $-j2(\underline{i}_b - \underline{i}_a) - j2 + j2\underline{i}_b = 0$,

leading to $\underline{i}_a = 1$ A, without knowing $\underline{i}_b$. Then KCL in mesh c gives $\underline{i}_c$ as

- KVL(c): $(2+j2)\underline{i}_c + j2 + (2-j4)(\underline{i}_c - \underline{i}_a) = 0 \longrightarrow \underline{i}_c = (1-j)$ A.

Therefore, we have

$$\underline{i}_x = 1 - j = \sqrt{2}\underline{/(-\pi/4)} \text{ A}$$

and

$$i_x(t) = 2\cos(2t - \pi/4) = 2\cos(2t - 45^\circ) \text{ A}.$$

Example 125: Consider the following circuit.

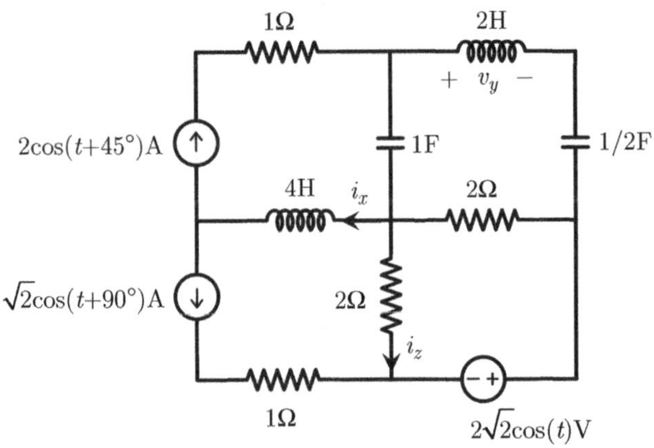

Find $v_y(t)$ in steady state.

Solution: In the phasor domain, the circuit is as follows.

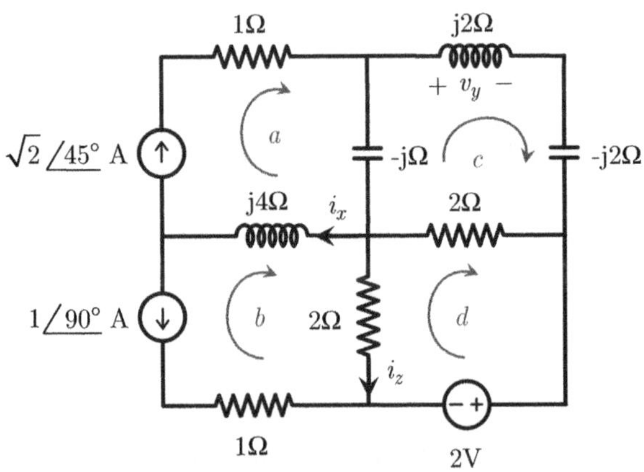

Using mesh analysis, we immediately have

$$\underline{i}_a = \sqrt{2}\underline{/\pi/4} = \sqrt{2}(\sqrt{2}/2 + j\sqrt{2}/2) = 1 + j \text{ A}$$

and

$$\underline{i}_b = -1\angle \pi/2 = -j \text{ A}.$$

Applying KVL in mesh c, we further have

- KVL(c): $-j(\underline{i}_c - \underline{i}_a) + 2(\underline{i}_c - \underline{i}_d) = 0$,

leading to

$$(2-j)\underline{i}_c - 2\underline{i}_d = 1 - j.$$

In addition, applying KVL in mesh d, we have

- KVL(d): $2 + 2(\underline{i}_d - \underline{i}_b) + 2(\underline{i}_d - \underline{i}_c) = 0$,

which can be simplified as

$$2\underline{i}_d - \underline{i}_c = -1 - j.$$

Combining the equations, we obtain

$$\underline{i}_c = (1-j) \text{ A},$$
$$\underline{i}_d = -j \text{ A},$$

and

$$\underline{v}_y = j2 \times \underline{i}_c = j2 \times (1-j) = 2 + j2 = 2\sqrt{2}\angle \pi/4 \text{ V}.$$

Finally, in the time domain, this voltage can be written as

$$v_y(t) = 4\cos(t + \pi/4) = 4\cos(t + 45^\circ) \text{ V}.$$

Exercise 108: In the following circuit, find $i_x(t)$ in steady state.

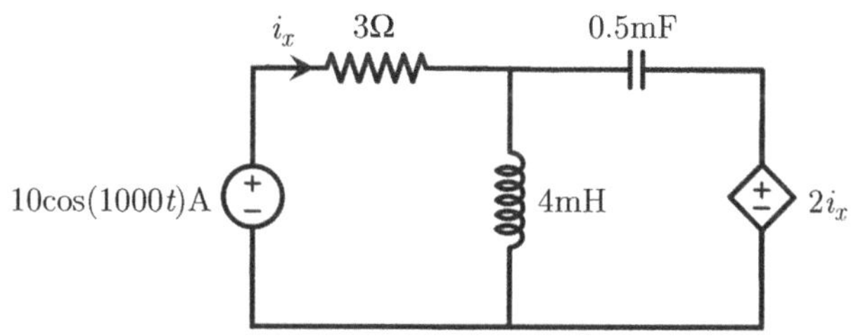

Exercise 109: In the following circuit, find $i_L(t)$ in steady state.

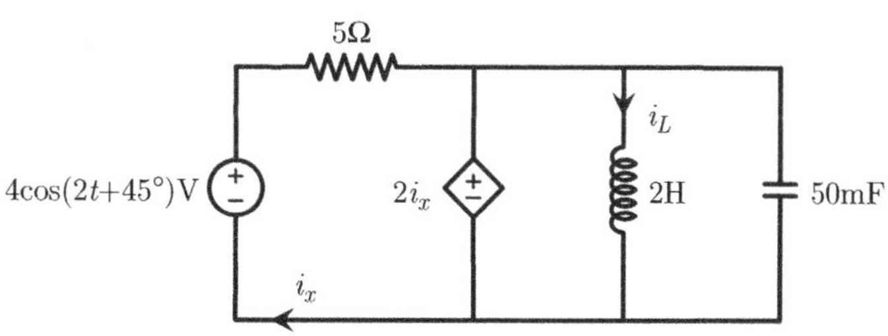

Exercise 110: In the following circuit, find $i_x(t)$ in steady state.

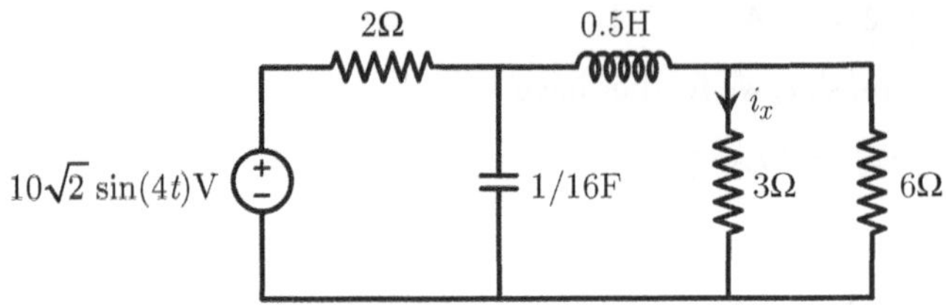

Exercise 111: In the following circuit, find $v_x(t)$ in steady state.

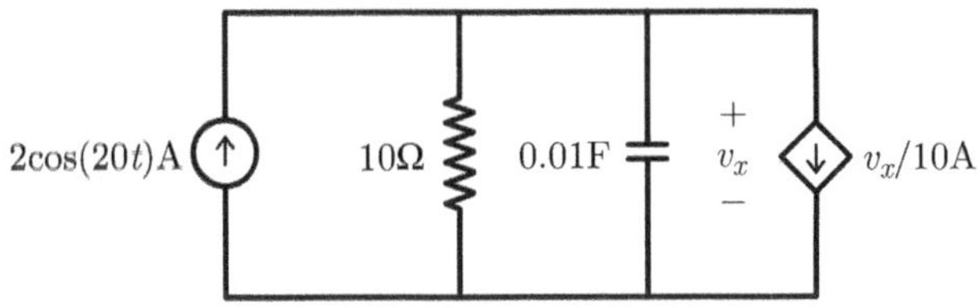

Exercise 112: Consider the following circuit in steady state.

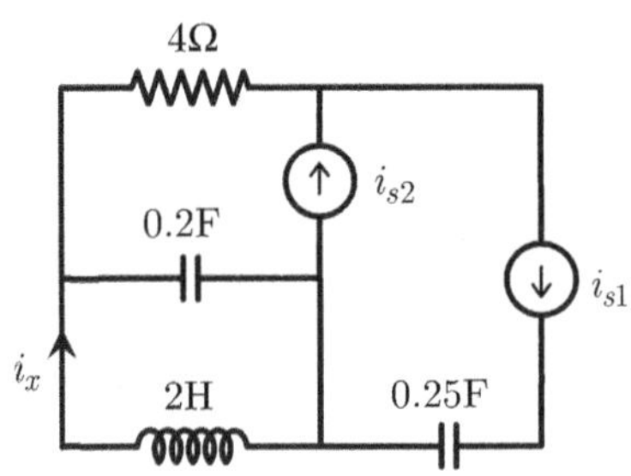

Find $i_x(t)$, given that

- $i_{s1}(t) = 2\cos(t - 45^\circ)$ A,
- $i_{s2}(t) = \sqrt{2}\cos(t)$ A.

Exercise 113: Consider the following circuit in steady state.

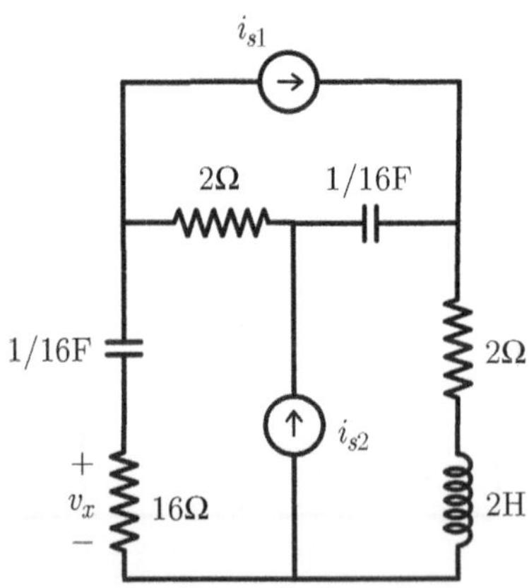

Find $v_x(t)$, given that

- $i_{s1} = 12\cos(4t)$ A,
- $i_{s2} = 8\cos(4t)$ A.

7.6 Power in Steady State

Considering a time-harmonic circuit with sinusoidal sources in steady state, the power of a component can be found as usual as

$$p(t) = v(t)i(t),$$

which depends on time. If the voltage and current are periodic with period T, we have

$$p(t + nT) = v(t + nT)i(t + nT) = v(t)i(t) = p(t),$$

indicating that $p(t)$ is also periodic with T. On the other hand, T may not be the period of $p(t)$. As an example, we consider a resistor with

$$v_R(t) = v_m \cos(\omega t),$$
$$i_R(t) = \frac{v_m}{R}\cos(\omega t).$$

Then the power (energy consumed by the resistor per unit time) can be found to be

$$p_R(t) = v_m \cos(\omega t)\frac{v_m}{R}\cos(\omega t) = \frac{v_m^2}{R}\cos^2(\omega t)$$
$$= \frac{v_m^2}{2R}[1 + \cos(2\omega t)].$$

Therefore, the period of the power is half that of the voltage and current,

$$T_p = \frac{\pi}{\omega} = \frac{T_v}{2} = \frac{T_i}{2}.$$

As shown below, this is also valid for capacitors and inductors that are connected to sinusoidal sources.

7.6.1 Instantaneous and Average Power

In general, the power of a component with respect to time is called its instantaneous power. In order to find a general expression for the instantaneous power of a component with sinusoidal signals, we consider arbitrary voltage and current values

$$v(t) = v_m \cos(\omega t + \phi),$$
$$i(t) = i_m \cos(\omega t + \phi - \theta),$$

leading to

$$p(t) = v_m i_m \cos(\omega t + \phi)\cos(\omega t + \phi - \theta).$$

The expression above can be used for any component, including resistors ($\theta = 0$), capacitors ($\theta = -\pi/2$), and inductors ($\theta = \pi/2$). Using trigonometric identities, the instantaneous power can be rewritten as

$$p(t) = \frac{v_m i_m}{2}[\cos(2\omega t + 2\phi - \theta) + \cos(\theta)].$$

Hence, the power has two components: an oscillating part

$$p_o(t) = \frac{v_m i_m}{2}\cos(2\omega t + 2\phi - \theta),$$

and a constant part (shift in the amplitude)

$$p_s(t) = \frac{v_m i_m}{2}\cos\theta$$

that exists only if the component has a resistive part ($\theta \neq \pm\pi/2$).

The instantaneous power provides information on the power of a component at any given time. Another important quantity is the average power, which can be obtained via integration as

$$p_{\text{avg}} = \frac{1}{T}\int_0^T p(t)dt = \frac{1}{T_p}\int_0^{T_p} p(t)dt,$$

where $T_p = \pi/\omega$. Using the expression above, we have

$$p_{\text{avg}} = \frac{\omega}{\pi}\frac{\pi}{\omega}\frac{v_m i_m}{2}\cos\theta = \frac{v_m i_m}{2}\cos\theta.$$

Conventionally, the average power is written in terms of RMS voltage and current values as

$$p_{\text{avg}} = \frac{v_m}{\sqrt{2}}\frac{i_m}{\sqrt{2}}\cos\theta = v_{\text{RMS}} i_{\text{RMS}}\cos\theta,$$

where θ is the phase difference between the voltage and current. Hence, for a single capacitor or inductor, the average power is zero. In fact, for a passive element, the average power represents the dissipated power; hence, it is nonzero only when the component involves a resistive part that consumes energy.

7.6.2 Complex Power

In phasor domain, the voltage and current of an arbitrary element can be written as

$$\underline{v} = v_{\text{RMS}}\underline{/\phi},$$
$$\underline{i} = i_{\text{RMS}}\underline{/\phi - \theta},$$

where

$$Z = \frac{\underline{v}}{\underline{i}} = \frac{v_{\text{RMS}}}{i_{\text{RMS}}}\underline{/\theta}$$

is the impedance of the element. The complex power of the element is defined as

$$\begin{aligned} s &= \underline{v} \times \underline{i}^* \\ &= v_{\text{RMS}}\underline{/\phi} \times i_{\text{RMS}}\underline{/\theta - \phi} \\ &= (v_{\text{RMS}} i_{\text{RMS}})\underline{/\theta}. \end{aligned}$$

Hence, the complex power can be written as

$$s = v_{\text{RMS}} i_{\text{RMS}}\cos\theta + j v_{\text{RMS}} i_{\text{RMS}}\sin\theta,$$

where the real part is the same as the average power in the time domain. Therefore, the real part of the complex power is related to the resistive part of the component.

On the other hand, the imaginary part, which is called the reactive power, measured in volt-amperes (V A), is related to the stored/released energy that is not dissipated.

When solving circuits in the phasor domain, the complex power of a component with a given impedance Z can be found as

$$s = \underline{v} \times \underline{i}^* = Z\underline{i} \times \underline{i}^* = Z|\underline{i}|^2$$

or

$$s = \underline{v} \times \frac{\underline{v}^*}{Z^*} = \frac{|\underline{v}|^2}{Z^*}.$$

Then the average power can be obtained simply as

$$p_{\text{avg}} = \text{Re}\{s\}$$

without going back to the time-domain solution.

Example 126: Consider the following RLC circuit involving series connections of a voltage source, a resistor, a capacitor, and an inductor in steady state.

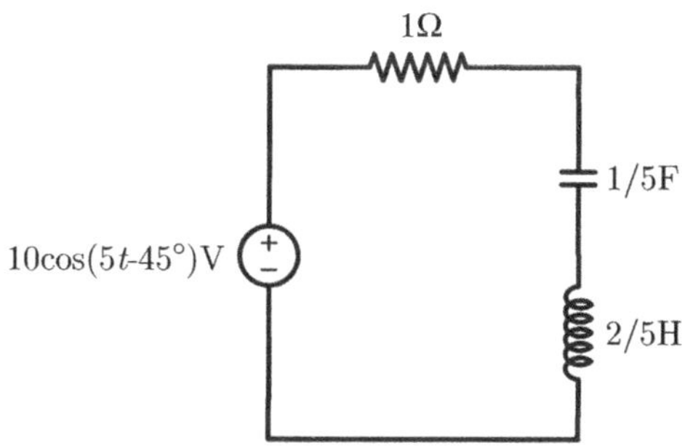

Find the complex powers of all elements.

Solution: Considering that $\omega = 5$ rad/s, we have the following conversions:

- $10\cos(5t - 45^\circ) \longrightarrow 5\sqrt{2}\angle(-\pi/4)$ V,
- $1/5\text{F} \longrightarrow -j\ \Omega$,
- $2/5\text{H} \longrightarrow j2\ \Omega$.

Then we obtain

$$\underline{i}_{\text{in}} = \frac{5\sqrt{2}\angle(-\pi/4)}{1 - j + j2} = \frac{5\sqrt{2}\angle(-\pi/4)}{\sqrt{2}\angle\pi/4}$$
$$= 5\angle(-\pi/2) = -j5 \text{ A}.$$

Hence, the complex powers can be obtained as

$$s_R = \underline{v}_R\underline{i}_R^* = Z_R|\underline{i}_R|^2 = 25 \text{ W},$$
$$s_C = \underline{v}_C\underline{i}_C^* = Z_C|\underline{i}_C|^2 = -j25 \text{ V A},$$
$$s_L = \underline{v}_L\underline{i}_L^* = Z_L|\underline{i}_L|^2 = j50 \text{ V A}.$$

As expected, the complex power of the resistor is purely real, while those of capacitor and inductor are purely imaginary. Furthermore, the power of the voltage source can be found to be

$$
\begin{aligned}
s_s &= \underline{v}_s \underline{i}_s^* = -\underline{v}_s \underline{i}_{\text{in}}^* \\
&= -5\sqrt{2}\angle(-\pi/4) \times 5\angle\pi/2 = -25\sqrt{2}\angle\pi/4 = 25\sqrt{2}\angle 5\pi/4 \\
&= -25(1 + j)\ \text{V A},
\end{aligned}
$$

which has both real and imaginary parts. Finally, we can check the conservation of energy:

$$s_R + s_C + s_L + s_s = 25 - j25 + j50 - 25(1 + j) = 0.$$

Whether it is real or reactive, energy must be conserved, leading to zero net energy for any given circuit.

Example 127: Consider the following circuit in steady state.

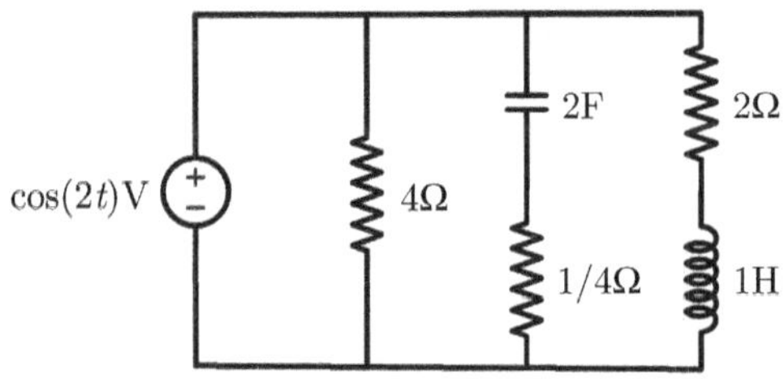

Find the complex powers of all elements.

Solution: In this case, we have $\omega = 2$ rad/s, leading to the following conversions:

- $\cos(2t) \longrightarrow \frac{1}{\sqrt{2}}\angle 0$ V
- $2\text{ F} \longrightarrow -j/4\ \Omega$,
- $1\text{ H} \longrightarrow j2\ \Omega$.

Hence, in the phasor domain, the circuit is as follows.

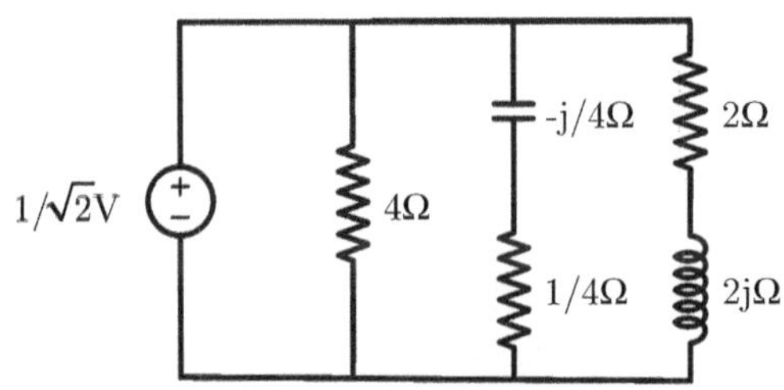

The complex power of the 4 Ω resistor can be found as

$$s_{4\Omega} = \underline{v}_{4\Omega} \times \underline{i}_{4\Omega}^* = \frac{|\underline{v}_{4\Omega}|^2}{Z_{4\Omega}^*} = 1/8\ \text{W}.$$

Next, the current across the capacitor and 1/4 Ω resistor can be obtained as

$$\underline{i}_C = \underline{i}_{1/4\Omega} = \frac{\frac{1}{\sqrt{2}}\angle 0}{\frac{1}{4} - \frac{j}{4}} = \frac{\frac{1}{\sqrt{2}}\angle 0}{\frac{\sqrt{2}}{4}\angle(-\pi/4)} = 2\angle \pi/4,$$

leading to

$$s_C = \underline{v}_C \times \underline{i}_C^* = Z_C|\underline{i}_C|^2 = (-j/4)\times 4 = -j \text{ V A},$$
$$s_{1/4\Omega} = \underline{v}_{1/4\Omega} \times \underline{i}_{1/4\Omega}^* = Z_{1/4\Omega}|\underline{i}_{1/4\Omega}|^2 = 1 \text{ W}.$$

Finally, the current across the inductor and 2 Ω resistor is obtained similarly as

$$\underline{i}_L = \underline{i}_{2\Omega} = \frac{\frac{1}{\sqrt{2}}\angle 0}{2+j2} = \frac{\frac{1}{\sqrt{2}}\angle 0}{2\sqrt{2}\angle \pi/4} = \frac{1}{4}\angle(-\pi/4).$$

Therefore, the corresponding complex powers can be calculated as

$$s_L = \underline{v}_L \times \underline{i}_L^* = Z_L|\underline{i}_L|^2 = j2 \times \frac{1}{16} = j/8 \text{ V A}$$
$$s_{2\Omega} = \underline{v}_{2\Omega} \times \underline{i}_{2\Omega}^* = Z_{2\Omega}|\underline{i}_{2\Omega}|^2 = 1/8 \text{ W}.$$

Using the conservation of energy, we can also obtain the power of the voltage source as

$$s_s = -s_{4\Omega} - s_C - s_{1/4\Omega} - s_L - s_{2\Omega}$$
$$= -1/8 + j - 1 - j/8 - 1/8 = -10/8 + j7/8 = (-10 + j7)/8 \text{ V A}.$$

Example 128: Consider the following circuit in steady state.

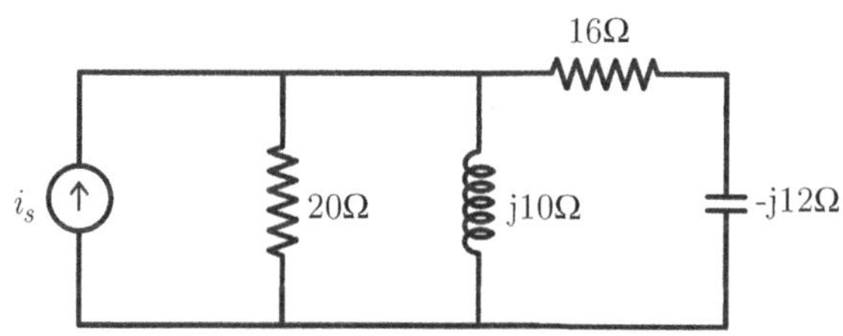

If the complex power of the current source is $-144 - j112$, find the complex power of each component.

Solution: First, we can find the impedance seen by the source as

$$Z_{\text{in}} = 20 \parallel j10 \parallel (16 - j12) = \frac{j20}{2+j} \parallel (16 - j12)$$
$$= (4 + j8) \parallel (16 - j12) = \frac{(4+j8)(16-j12)}{20 - j4}$$
$$= \frac{4(1+j2)(4-j3)}{5-j} = 20\frac{2+j}{5-j}\ \Omega.$$

Using this impedance and the given complex power, we can find the absolute value of the source voltage tro be

$$s_s = \underline{v}_s \underline{i}_s^* = \frac{|\underline{v}_s|^2}{-Z_{\text{in}}^*} = -144 - j112$$

$$\longrightarrow |\underline{v}_s|^2 = (-144 - j112) \times (-20) \frac{2-j}{5+j} = 1600,$$

leading to $|\underline{v}_s| = 40$ V. Then, using this voltage value, we obtain

- $s_{20\Omega} = 1600/20 = 80$ W,
- $s_{j10\Omega} = 1600/(-j10) = j160$ V A,
- $s_{16\Omega} = \text{Re}\{1600/(16 + 12j)\} = 64$ W,
- $s_{-j12\Omega} = \text{Im}\{1600/(16 + 12j)\} = -j48$ V A.

We also note that

$$s_{20\Omega} + s_{j10\Omega} + s_{16\Omega} + s_{-j12\Omega} = 144 + j112 = -s_s,$$

as expected from the conservation of energy.

Example 129: Consider the following circuit in steady state.

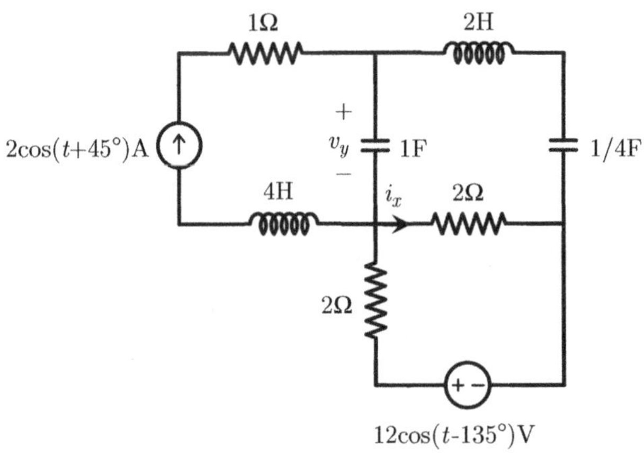

Find the time-average power of the voltage source.

Solution: In the phasor domain, the circuit is as follows.

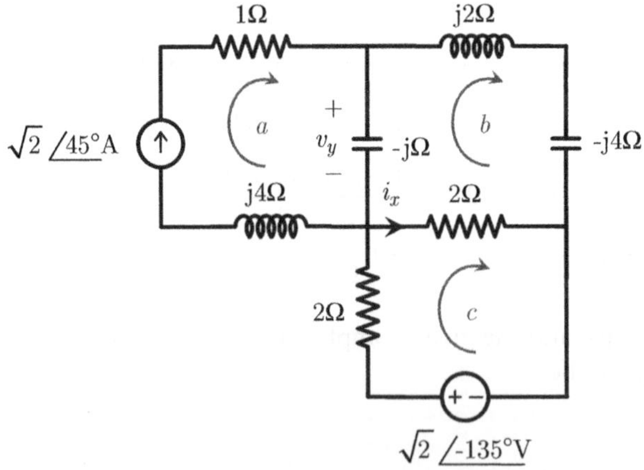

Applying mesh analysis, we first have

$$\underline{i}_a = \sqrt{2}\angle\pi/4 = (1+j)\text{ A}.$$

Then, applying KVL in meshes b and c, we obtain

- KVL(b): $-j(\underline{i}_b - \underline{i}_a) - j2\underline{i}_b + 2(\underline{i}_b - \underline{i}_c) = 0 \longrightarrow (2-j3)\underline{i}_b - 2\underline{i}_c = 1-j,$
- KVL(c): $2\underline{i}_c + 2(\underline{i}_c - \underline{i}_b) + (6+j6) = 0 \longrightarrow 2\underline{i}_c - \underline{i}_b = -3-j3.$

Combining the equations, we get

$$(1-j3)\underline{i}_b = -2-j4 \longrightarrow \underline{i}_b = \frac{-2-j4}{1-j3} = (1-j)\text{ A}$$

and

$$\underline{i}_c = (-1-j2)\text{ A}.$$

Therefore, the complex power of the voltage source can be found to be

$$s_s = (-6-j6)\times(-\underline{i}_c)^* = (-6-j6)\times(1-j2)$$
$$= -6+j12-j6-12 = (-18+j6)\text{ V A}.$$

Finally, the time-average power can be obtained as

$$p_{\text{avg},s} = \text{Re}\{s_s\} = -18\text{ W}.$$

Exercise 114: Consider the following circuit in steady state.

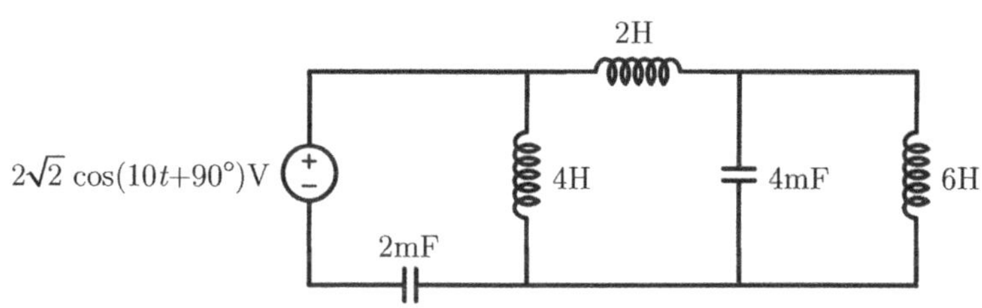

Find the complex power of the voltage source.

Exercise 115: Consider the following circuit in steady state.

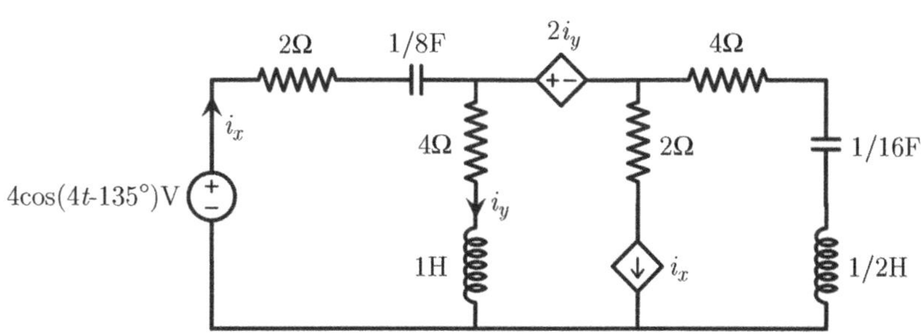

- Find the power of the current-dependent voltage source.
- Find the time-average power of the current-dependent current source.

7.6.3 Impedance Matching

Consider a load impedance $Z_l = R_l + jX_l$ connected in series to a voltage source $\underline{v}_s$ and a resistor $Z_s = R_s + jX_s$.

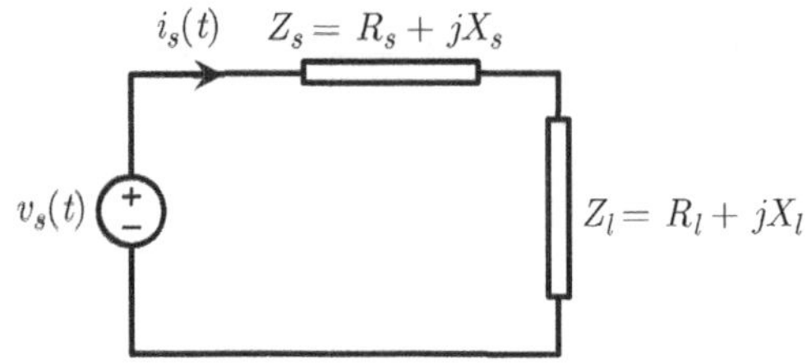

Our aim is to maximize the average power delivered to Z_l. We have

$$\underline{i}_s = \frac{\underline{v}_s}{(R_s + R_l) + j(X_s + X_l)},$$

leading to

$$s_l = Z_l |\underline{i}_l|^2 = (R_l + jX_l)\frac{|\underline{v}_s|^2}{(R_s + R_l)^2 + (X_s + X_l)^2}$$

and

$$p_{\text{avg},l} = \frac{R_l |\underline{v}_s|^2}{(R_s + R_l)^2 + (X_s + X_l)^2}.$$

In order to maximize the expression above, one must select $X_l = -X_s$, leading to

$$p_{\text{avg},l} = \frac{R_l |\underline{v}_s|^2}{(R_s + R_l)^2}.$$

This final form is the same as the maximization of the power delivered to a resistive load. Hence, we further have $R_l = R_s$ so that

$$p_{\text{avg},l}^{\text{max}} = \frac{|\underline{v}_s|^2}{4R_s^2}.$$

Consequently, for the maximum average power transferred to the load, its impedance should be

$$Z_l = Z_s^*.$$

This expression reduces to $R_l = R_s$, as expected, for purely resistive (real) impedances (see Section 5.2).

Example 130: Consider the following circuit in steady state.

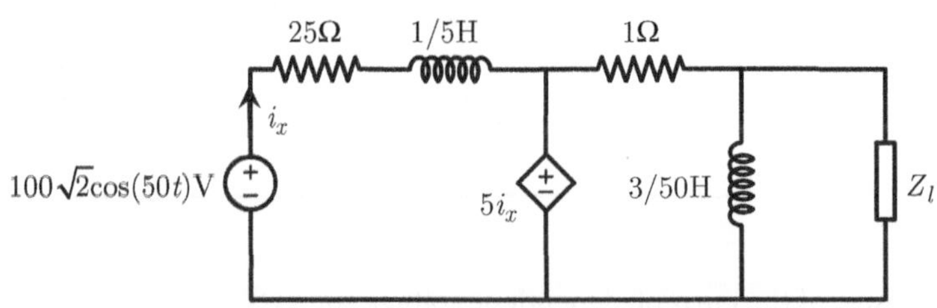

Find the impedance of the load for the maximum power transfer.

Solution: For the analysis of the circuit in the phasor domain, we note that $\omega = 50$ rad/s. Then the components can be converted as follows.

- $100\sqrt{2}\cos(50t)$ V $\longrightarrow 100\underline{/0}$ V
- $1/5$ H $\longrightarrow j10\ \Omega$
- $3/50$ H $\longrightarrow j3\ \Omega$

Hence, the circuit can be represented in the phasor domain as follows.

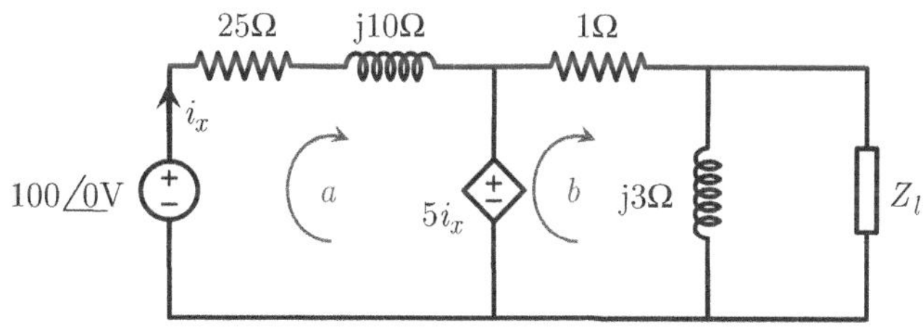

Next, the problem should be solved twice, for open circuit and for short circuit. Applying mesh analysis for the open-circuit case, we define two mesh currents and use KVL to derive

- KVL(a): $-100\underline{/0} + (25 + j10)\underline{i}_a + 5\underline{i}_a = 0 \longrightarrow \underline{i}_a = \dfrac{10}{3+j}$ A,
- KVL(b): $-5\underline{i}_a + \underline{i}_b + j3\underline{i}_b = 0 \longrightarrow \underline{i}_b = \dfrac{5\underline{i}_a}{1+j3}$,

leading to

$$\underline{i}_b = \frac{5\underline{i}_a}{1+j3} = \frac{50}{(3+j)(1+j3)} = \frac{50}{3+j9+j-3} = -j5 \text{ A}.$$

This way, we obtain the open-circuit voltage as

$$\underline{v}_{\text{oc}} = j3 \times (-j5) = 15 \text{ V}.$$

For the short-circuit case, the mesh current in mesh a remains the same,

$$\underline{i}_a = \frac{10}{3+j} \text{ A}.$$

However, we have

$$\underline{i}_{\text{sc}} = \underline{i}_b = \frac{5\underline{i}_a}{1} = \frac{50}{3+j} = (15 - j5) \text{ A}.$$

Therefore, the Thévenin equivalent seen by the load can be constructed by using

$$Z_{\text{th}} = \frac{\underline{v}_{\text{oc}}}{\underline{i}_{\text{sc}}} = \frac{15}{15-j5} = \frac{3}{3-j} = \frac{3}{10}(3+j)\ \Omega.$$

Then considering the Thévenin circuit, the load should be selected as

$$Z_l = Z_{\text{th}}^* = \frac{3}{10}(3-j)\ \Omega.$$

Such an impedance can be obtained by connecting a $9/10\ \Omega$ resistor to a capacitor C, where

$$\frac{1}{j\omega C} = -j\frac{3}{10} \longrightarrow C = \frac{10}{3\omega} = 1/15 \text{ F}.$$

Example 131: Consider the following circuit in steady state.

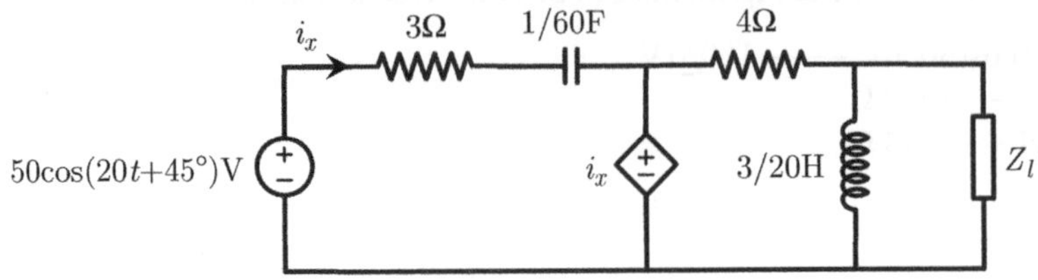

Find Z_l so as to maximize power transfer.

Solution: In the phasor domain, the open-circuit case is as follows.

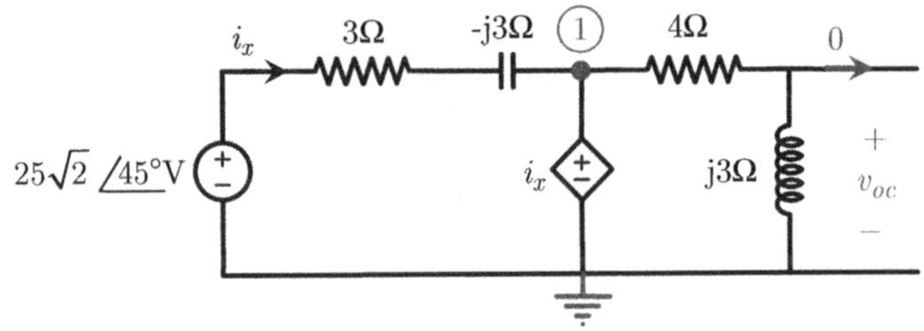

Using nodal analysis, we have $\underline{v}_1 = \underline{i}_x$ and

$$25(1+j) - \underline{v}_1 = (3 - j3)\underline{i}_x,$$

leading to $\underline{i}_x = (1 + j7)$ A. Therefore, the open-circuit voltage can be found to be

$$\underline{v}_{oc} = \underline{i}_x \times \frac{j3}{4+j3} = (-3 + j3) \text{ V}.$$

Next, we consider the short-circuit case as follows.

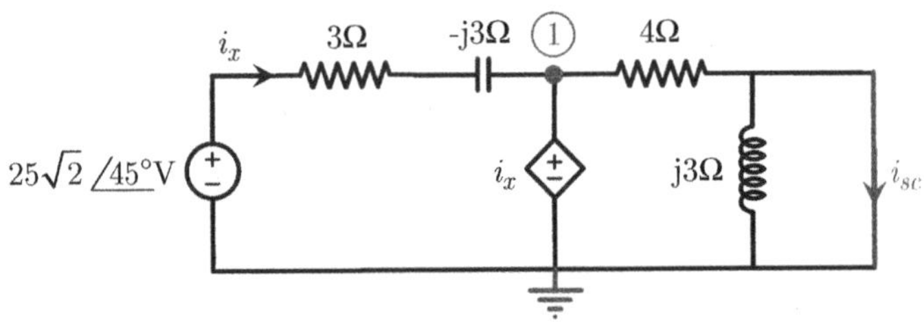

In this case, we still have $\underline{i}_x = (1 + j7)$ A; therefore,

$$\underline{i}_{sc} = \underline{i}_x/4 = (1 + j7)/4 \text{ A}.$$

Consequently, the Thévenin impedance is

$$Z_{th} = \frac{-3 + j3}{(1 + j7)/4} = 12\frac{(-1 + j)(1 - j7)}{50} = \frac{12}{25}(3 + j4)\ \Omega.$$

Then, to maximize the transferred power, we must select

$$Z_l = Z_{th}^* = \frac{12}{25}(3 - j4)\ \Omega.$$

This impedance can be obtained by using a 36/25 Ω resistor in series with a capacitor C, where

$$\frac{1}{j\omega C} = -\frac{48j}{25} \longrightarrow C = \frac{25}{48 \times 20} = \frac{5}{192} \text{ F.}$$

Example 132: Consider the following circuit in steady state.

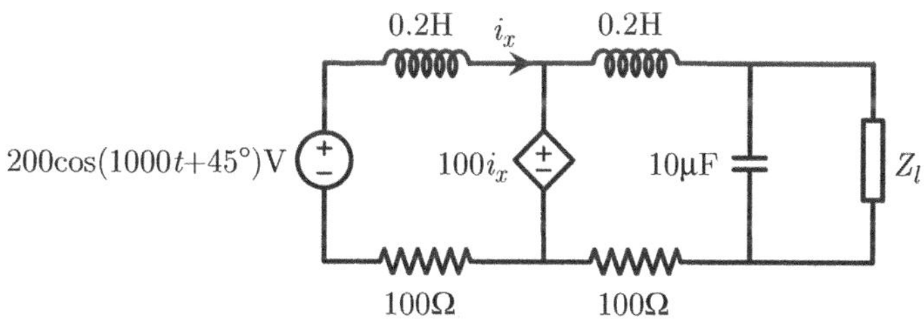

Design Z_l using a resistor and capacitor/inductor to maximize the power to this load.

Solution: In the phasor domain, the circuit is as follows.

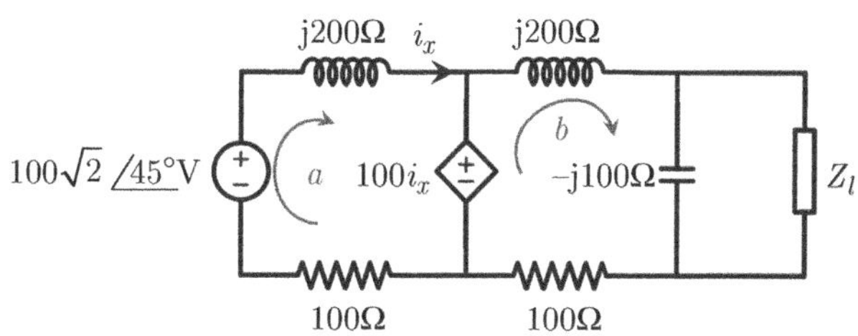

Using mesh analysis and noting that $\underline{i}_x = \underline{i}_a$, KVL in mesh a can be written as

- KVL(a): $-100 - j100 + j200\underline{i}_a + 100\underline{i}_x + 100\underline{i}_a = 0 \longrightarrow \underline{i}_a = 1/2\,\text{A}$.

We also note that $\underline{i}_a$ does not depend on whether an open or short circuit is used for Z_l. For the open-circuit case, KVL in mesh b can be derived as

- KVL(b): $-100\underline{i}_x + j200\underline{i}_b - j100\underline{i}_b + 100\underline{i}_b = 0$,

leading to

$$\underline{i}_b = \frac{\underline{i}_a}{1+j} = \frac{1-j}{4} \text{ A.}$$

Therefore, the open-circuit voltage can be found to be

$$\underline{v}_{\text{oc}} = -j100 \times \underline{i}_b = (-25 - j25) \text{ V.}$$

For the short-circuit case, we have

- KVL(b): $-100\underline{i}_x + j200\underline{i}_b + 100\underline{i}_b = 0$,

where the capacitor impedance is omitted due to the short circuit. Then the value of $\underline{i}_b$ can be found to be

$$\underline{i}_b = \frac{\underline{i}_a}{1+j2} = \frac{1-j2}{10} \text{ A.}$$

This value corresponds to the short-circuit current, that is,

$$\underline{i}_{sc} = \underline{i}_b = \frac{1-j2}{10} \text{ A.}$$

Therefore, the Thévenin impedance is

$$Z_{th} = \frac{-25-j25}{(1-j2)/10} = -250\frac{1+j}{1-j2} = -50(1+j)(1+j2) = 50(1-j3)\ \Omega.$$

Hence, in order to maximize the transferred power, one needs to select

$$Z_l = Z_{th}^* = 50(1+j3)\ \Omega.$$

Such a load can be designed by connecting a 50 Ω resistor in series to a $150/\omega = 0.15$ H inductor.

Exercise 116: Consider the following circuit in steady state.

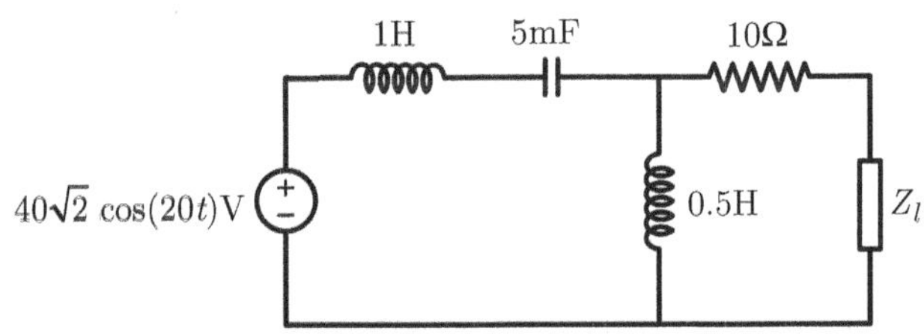

Find Z_l in the phasor domain to maximize the power transferred to this load.

Exercise 117: Consider the following circuit in steady state.

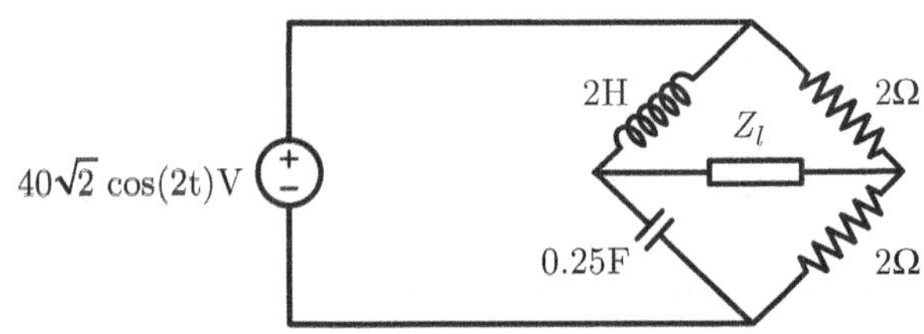

Find Z_l in the phasor domain to maximize the power transferred to this load. Also design Z_l using a resistor and capacitor/inductor.

Exercise 118: Consider the following circuit in steady state.

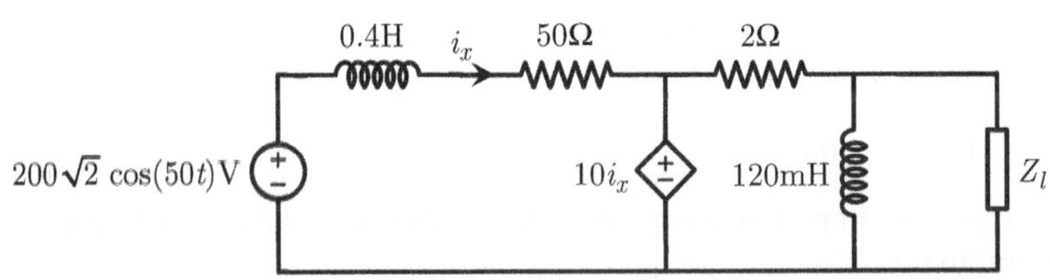

Find Z_l in the phasor domain to maximize the power transferred to this load. Also design Z_l using a resistor and capacitor/inductor.

Exercise 119: Consider the following circuit in steady state.

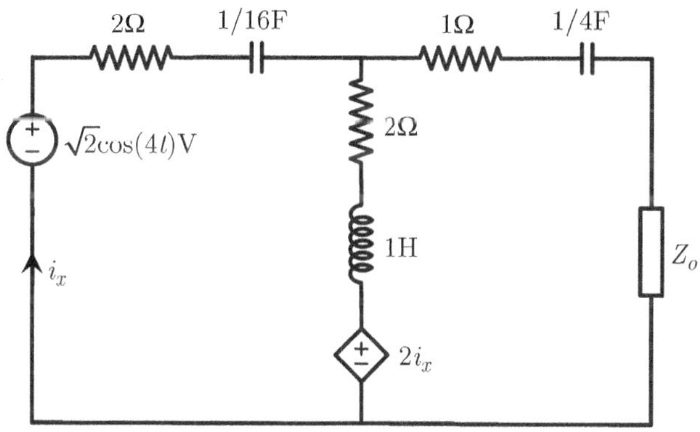

Find Z_o in the phasor domain to maximize the power transferred to this load. Also design Z_o using a resistor and capacitor/inductor.

7.7 When Things Go Wrong in Steady-State Analysis

As we already know, ideal components sometimes lead to impossible scenarios that involve conflicts or infinite values. In AC circuits, such impossible cases may occur depending on the frequency. Consider the following circuit involving a simple series connection of a capacitor and inductor.

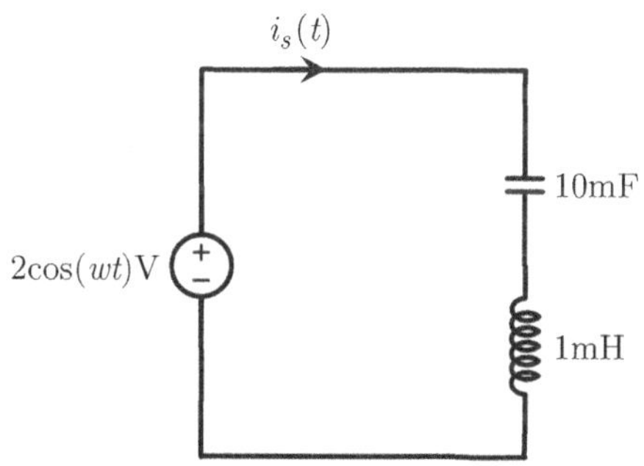

The current through the circuit in the phasor domain can be found to be

$$\underline{i}_s = \frac{\sqrt{2}}{j10^{-3}\omega + 1/(j10^{-2}\omega)} = \frac{j10^{-2}\sqrt{2}\omega}{1 - 10^{-5}\omega^2}.$$

The expression above is infinite when $\omega^2 = 10^5$ or $\omega = \sqrt{10^5}$ rad/s. The corresponding special frequency can be found to be

$$f = \frac{\sqrt{10^5}}{2\pi} \approx 50.33 \text{ Hz}.$$

Therefore, at this particular frequency, infinite current amplitude is established through the circuit. Such a case when a capacitor and an inductor perfectly match and cancel the effects of each other is called resonance.

It is no coincidence that the resonance frequency above corresponds to the natural frequency of the LC structure discussed in Section 6.6. At this frequency in steady state, the impedances of the capacitor and inductor perfectly match each other ($Z_C = -Z_L$) so that the overall impedance ($Z_C + Z_L$) becomes zero. Under this condition, the complex power of the capacitor is exactly the negative of the complex power of the inductor. Hence, with respect to time, these components store/release energy, which is transferred back and forth between them. Then the energy provided by the voltage source has nowhere to go. In real life, this energy is consumed by the internal resistances of the components, while the measured current can still be very large since these resistances are typically small.

Another source of confusion occurs when an AC circuit involves sources operating at different frequencies. Especially, in modern circuits, AC and DC signals are used together, where DC sources can be considered as AC sources with zero frequency. But then one cannot simply convert the circuit to the phasor domain, because it is not obvious how to convert the sources and components into phasor forms. If the circuit is linear, however, a superposition can be employed to find the overall response of the circuit. As an example, consider the following circuit involving a DC source and an AC source, both of which are voltage sources.

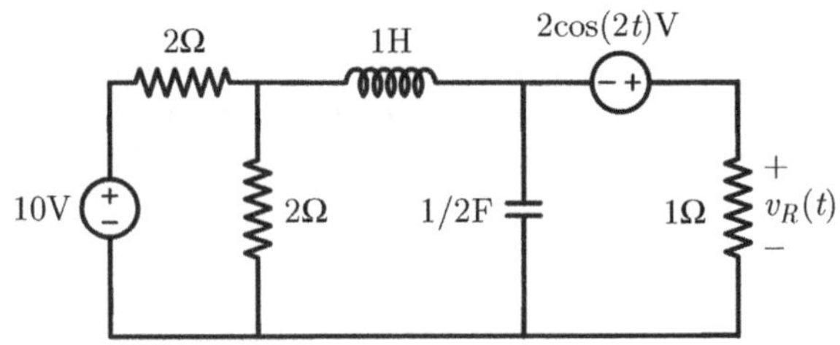

In order to find the voltage across the 1 Ω resistor in steady state, one can use superposition by solving the circuit twice and combining the results. In these solutions, the sources are considered to be active one by one. When one of the sources is active, the other sources must be closed down by replacing voltage sources with short circuits and current sources with open circuits. Considering also the capacitor and inductor at zero frequency, the DC version of the circuit can be shown as follows.

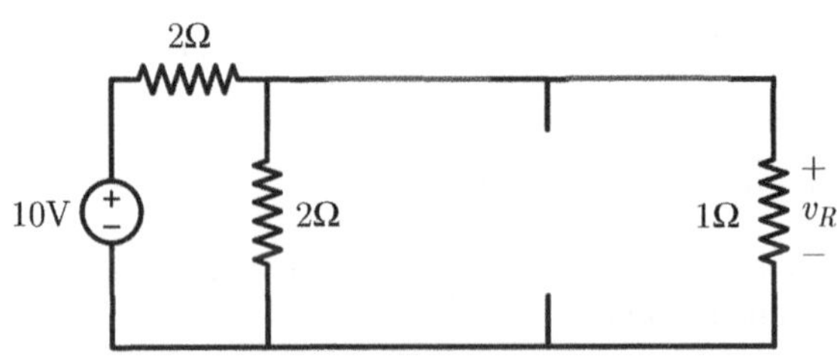

For this circuit, the voltage of the resistor is given by

$$v_{R,\mathrm{DC}} = 10 \times \frac{2 \parallel 1}{(2 \parallel 1) + 2} = 10 \times \frac{2/3}{2/3 + 2} = 10/4 = 5/2\ \mathrm{V}.$$

Next, the AC version of the circuit can be considered as follows.

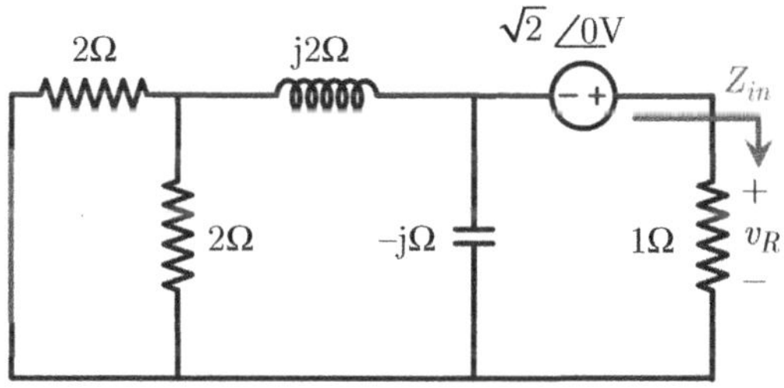

In this case, we have

$$Z_{\mathrm{in}} = 1 + (1 + j2) \parallel (-j) = \frac{3}{2}(1 - j)\ \Omega$$

and

$$\underline{v}_{R,\mathrm{AC}} = \frac{2\sqrt{2}}{3(1-j)} = \frac{\sqrt{2}(1+j)}{3} = \frac{2}{3}\angle 45^\circ \mathrm{V},$$

leading to

$$v_{R,\mathrm{AC}}(t) = \frac{2\sqrt{2}}{3}\cos(2t + 45^\circ)\ \mathrm{V}.$$

Consequently, the overall voltage across the resistor is

$$v_R(t) = \frac{5}{2} + \frac{2\sqrt{2}}{3}\cos(2t + 45^\circ)\ \mathrm{V}.$$

An common mistake occurs when dealing with series and parallel connections of capacitors when they are represented in the phasor domain. Consider a parallel connection of 1/2 F and 1/10 F capacitors when the frequency is $\omega = 2$ rad/s.

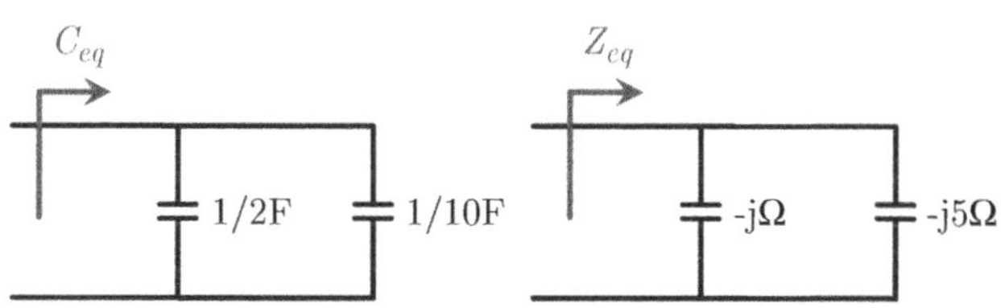

Once converted into impedances, series and parallel connections should be treated similarly to resistors, even when the origin of the impedance is a capacitor. For the parallel connection above, we have

$$Z_{\mathrm{eq}} = (-j) \parallel (-j5) = \frac{j^2 5}{-j6} = \frac{-5}{-j6} = -\frac{j5}{6}\ \Omega.$$

Alternatively, we can combine the capacitors in the original circuit as

$$C_{eq} = \frac{1}{2} + \frac{1}{10} = \frac{3}{5}\ \text{F},$$

leading to the same value as before,

$$Z_{eq} = \frac{1}{j2 \times 3/5} = -\frac{j5}{6}\ \Omega.$$

7.8 What You Need to Know before You Continue

As shown in this chapter, transformation to the phasor domain is essential to analyze AC circuits in steady state. A few important points are as follows.

- In time-harmonic circuits, once converted into the phasor domain, capacitors and inductors can be investigated as resistors with imaginary resistances. This leads to a generalized definition of resistance as impedance, which can have complex values.
- Unlike their behaviors for DC signals, capacitors and inductors behave like short and open circuits, respectively, for sufficiently large frequencies.
- RC, RL, and RLC circuits with time-harmonic sources behave like filters, allowing the transmission of signals in different ranges of frequencies while blocking others.
- In the phasor domain, a complex power can be defined to represent both dissipated (real) and stored/released (imaginary) powers.
- To maximize the real power transferred to a load, its impedance should be selected as the complex conjugate of the input impedance of the circuit.
- Phasor-domain analysis is only valid for single-frequency circuits, where all sources have the same frequency. For linear circuits, however, superposition can be used to analyze circuits involving sources with different frequencies (including DC as zero frequency).

With the steady-state analysis of time-harmonic circuits, we have completed all the basic topics of circuit analysis, which form the major focus of this book. In the next three extension chapters, we briefly consider some important components and practical technologies of modern circuits, as well as some final suggestions that might be useful when exploring electrical and electronic circuits beyond this book.

8

Selected Components of Modern Circuits

Previous chapters have presented fundamental laws (e.g., KVL and KCL) and techniques (e.g., nodal analysis, mesh analysis, and black-box equivalence) for the analysis of circuits involving independent and dependent sources, resistors, capacitors, and inductors. While they are used to solve basic circuits, the laws and techniques presented are applicable to all types of circuits and components. We have been particularly concerned with resistors, capacitors, and inductors, since they are not just components but also represent basic electrical properties of objects; any structure has a resistance, capacitance, and inductance, which characterize its electrical response. From this perspective, the topics covered in the previous chapters can be seen as universal. However, electrical and electronic circuits evolve continuously thanks to rapid developments in science and technology. As the first extension, this chapter presents some selected components in modern circuits. We emphasize that "modern" is a subjective term. For example, transistors are essential parts of modern circuits, whereas they have been conceptually known for decades. These modern components also evolve and are physically transformed (e.g., transistors become smaller and smaller), and perhaps some of them will completely diminish in the future, with the term "modern" moving on to cover other types of components.

In the following, we focus on five types of components, each of which demonstrates different electrical characteristics, making them essential parts of modern circuits in state-of-the-art technology. We also discuss the analysis of basic circuits involving these components, in order to provide hints on their analysis in more complex scenarios.

8.1 When Connections Are via Magnetic Fields: Transformers

In electrical circuits and components, electrical connectivity is mainly constructed by electrical conduction, or more generally, an electric field. This includes the terminals of the capacitor and the space between them filled with the electric field formed by accumulated charge. As seen in inductors, magnetic fields also play a major role in electrical circuits. On the other hand, an exact counterpart of a capacitor (where connectivity is established by an electric field) is not an inductor, but a proper combination of two inductors, namely, a transformer.

As shown below (see also Figure 11.4), a transformer is a two-sided component, leading to the definition of two pairs of voltage and current values.

Introduction to Electrical Circuit Analysis, First Edition. Özgür Ergül.

Companion Website: www.wiley.com/go/ergul4412

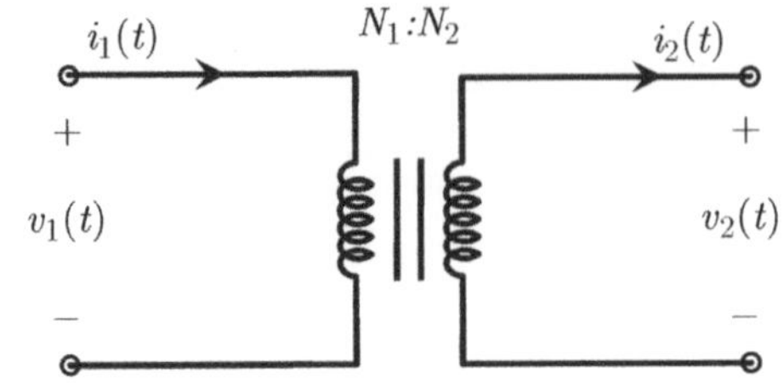

If the directions are as given above, we have

$$\frac{v_1(t)}{v_2(t)} = \frac{N_1}{N_2},$$

$$\frac{i_1(t)}{i_2(t)} = \frac{N_2}{N_1},$$

where N_1 and N_2 are the numbers of turns on the first (primary) and secondary sides, respectively. These expressions are valid only for AC voltage and current signals. This is because the two sides of a transformer are connected via a magnetic field. This field, which is created by an input current in one of the sides (e.g., primary side) and passes through the other side (e.g., secondary side), induces an electromotive force (voltage and current) only if it changes with respect to time. When a DC signal appears on one of the sides in steady state, it creates a static magnetic field that does not lead to any voltage/current on the other side. Consequently, DC signals are filtered out through transformers, which is one application of these components as isolators. Obviously, in a transient state, variations in DC signals may pass through the transformer, mostly as unwanted signals.

There are a couple of important results that can be derived from the expressions above. First, since the ratio $N_2/N_1 = N$ is a real number, phasor-domain quantities are also related by

$$\frac{\underline{v}_1}{\underline{v}_2} = \frac{N_1}{N_2} = \frac{1}{N},$$

$$\frac{\underline{i}_1}{\underline{i}_2} = \frac{N_2}{N_1} = N.$$

To be specific, an ideal transformer does not add any phase to the voltage and current values. When $N > 1$, indicating that the voltage on the secondary side is larger than that on the primary side, it is called a step-up transformer. Obviously, in this case, the current on the secondary side is smaller than the current on the primary side. In contrast, in a step-down transformer with $N < 1$, the voltage value drops while the current value increases. In any case, we have

$$s_1 = \underline{v}_1 \times \underline{i}_1^* = \frac{\underline{v}_2}{N} \times (N\underline{i}_2)^* = \underline{v}_2 \times \underline{i}_2^* = s_2,$$

indicating that an ideal transformer neither consumes nor delivers energy.

Using transformers, AC voltage and current values can be adjusted as needed. This controllability is indeed an important advantage of AC signals over DC signals. As high-voltage signals can be transmitted over longer distances with lower energy loss, high voltage is preferred in power lines across cities. Then, once transmitted, the electricity is transformed into lower voltage values without loss of power (or in practice with small power loss), for safe indoor usage.

Example 133: Consider the following circuit in steady state.

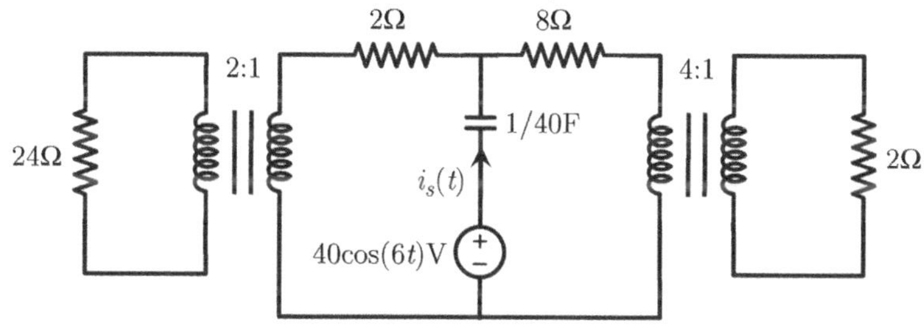

Find $i_s(t)$ in steady state.

Solution: This circuit can be analyzed in the phasor domain. At first glance, it may not be obvious how to simplify the transformers. For this purpose, we consider the following case, where an impedance Z_s is connected on the secondary side.

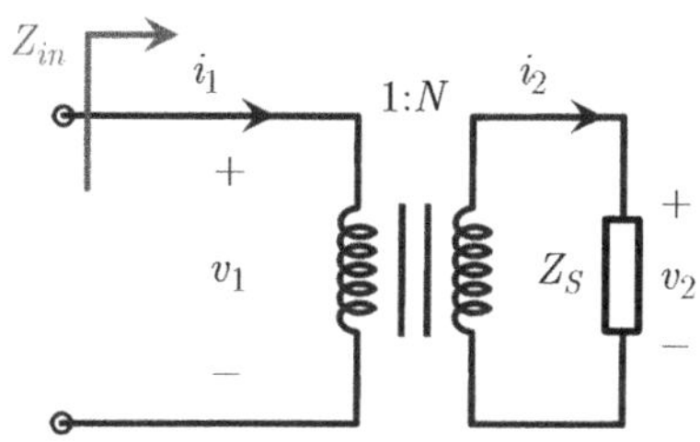

Using the definitions, we have $\underline{v}_1 = \underline{v}_2/N$ and $\underline{i}_1 = N\underline{i}_2$, leading to

$$Z_{\text{in}} = \frac{\underline{v}_1}{\underline{i}_1} = \frac{\underline{v}_2}{N^2\underline{i}_2} = \frac{Z_s}{N^2}.$$

Hence, using a transformer with turn ratio N, an impedance is transformed by a scaling with N^2 (e.g., reduced to a smaller value if $N > 1$).

Now, considering the impedance transformation described above, the circuit can be represented in the phasor domain as follows.

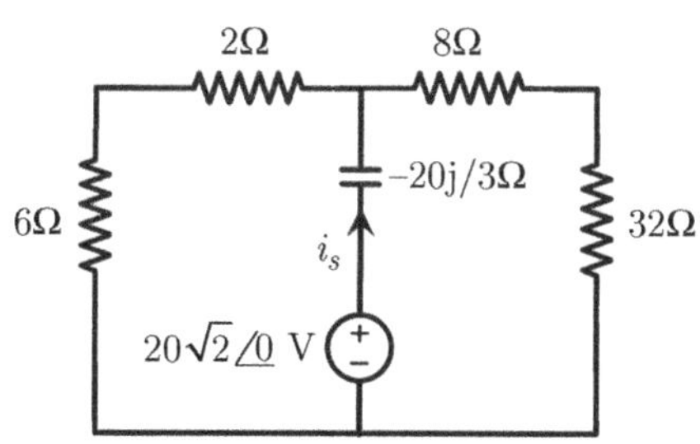

In the above, we use the following conversions:

- $40\cos(6t) \longrightarrow 20\sqrt{2}\angle 0$ V,
- $1/40\text{F} \longrightarrow -j20/3\,\Omega$,
- $24\,\Omega$ & Transformer$(1:2) \longrightarrow 24/4 = 6\,\Omega$,
- $2\,\Omega$ & Transformer$(4:1) \longrightarrow 2/(1/16) = 32\,\Omega$.

Then the current through the voltage source can easily be found to be

$$\underline{i}_s = \frac{20\sqrt{2}}{-j20/3 + 8 \parallel 40} = \frac{20\sqrt{2}}{-j20/3 + 20/3} = \frac{3\sqrt{2}}{1-j} = 3(1+j),$$

leading to

$$i_s(t) = 3\sqrt{2}\cos(6t + 45^\circ)\text{ A}$$

in the time domain.

Exercise 120: In the following circuit, find $i_s(t)$ in steady state, if $v_s(t) = 8\cos(4t + 135^\circ)$ V.

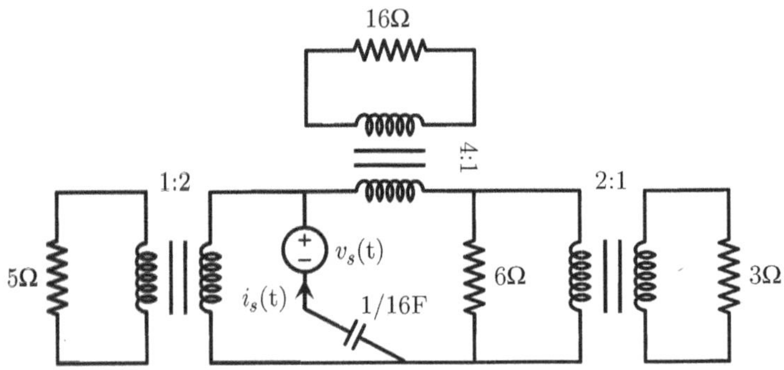

Exercise 121: In the following circuit, find $i_x(t)$ in steady state, if $v_x(t) = 4\sqrt{2}\cos(2t)$ V.

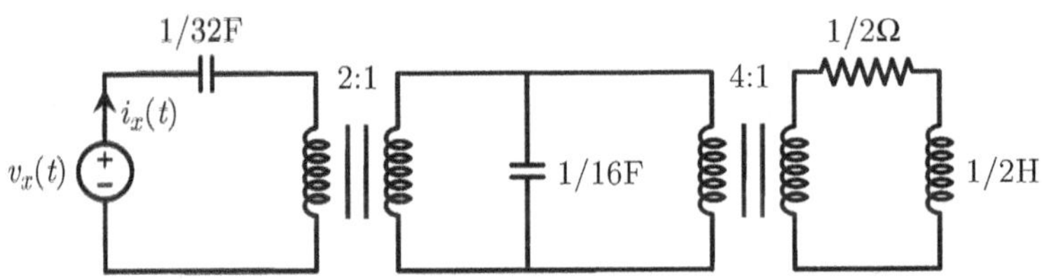

8.2 When Components Behave Differently from Two Sides: Diodes

When discussing resistors, capacitors, and inductors, we have not considered a particular orientation for them. For example, it does not matter how a standard resistor is connected to a voltage source; any of its terminals can be connected to the positive side of the voltage source and the same fixed resistance is observed. This symmetric property is common to most resistors, capacitors, and inductors, but there are also asymmetric (i.e., polarized) capacitors.

Diodes are two-terminal devices that are completely asymmetric (see Figure 11.5). In fact, this property of diodes makes them useful in many applications, where the electricity must flow only in one direction. In modern diodes, the one-directional property is obtained by using semiconductors. An ideal diode is defined as

- short circuit in the positive direction (on mode),
- open circuit in the negative direction (off mode).

But the question is how to decide the working mode when the diode is a part of a circuit. Let v_d and i_d represent the voltage and current though a diode when the sign convention is used by considering its positive and negative terminals. This is depicted in the following figure, with a common representation of diodes.

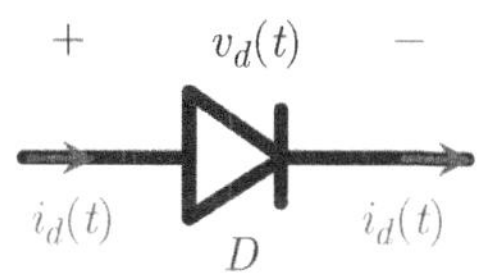

We need the following checks.

- On mode: If $i_d > 0$, then the diode is in on mode, leading to $v_d = 0$.
- Off mode: If $v_d < 0$, then then diode is in off mode, leading to $i_d = 0$.

We emphasize that checking $i_d < 0$ and $v_d > 0$ is not meaningful, as these conditions are never satisfied.

The modes described above represent the behavior of an ideal diode, which may not be sufficient in some analyses. As shown below, a typical diode has a rapidly increasing current with respect to voltage, which may not be perfectly zero in the on mode. There are also more accurate representations, including a reverse current, as well as a breakdown when the diode is exposed to high negative voltage. Engineers also use other intermediate models, where the voltage in the on mode is set to a positive value (e.g., 0.6–0.7 V) instead of zero, independent of the current.

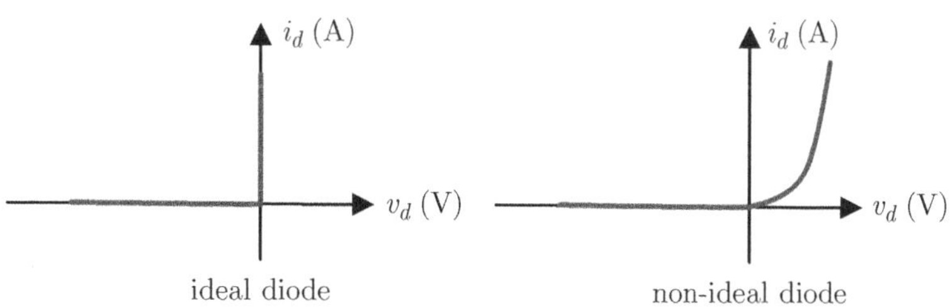

In the following, we consider only (very) ideal diodes with zero on voltages. We note that the power of such an ideal diode is zero, that is, it never consumes or delivers energy, since either its voltage or its current is zero.

Example 134: Consider the following simple circuit in steady state.

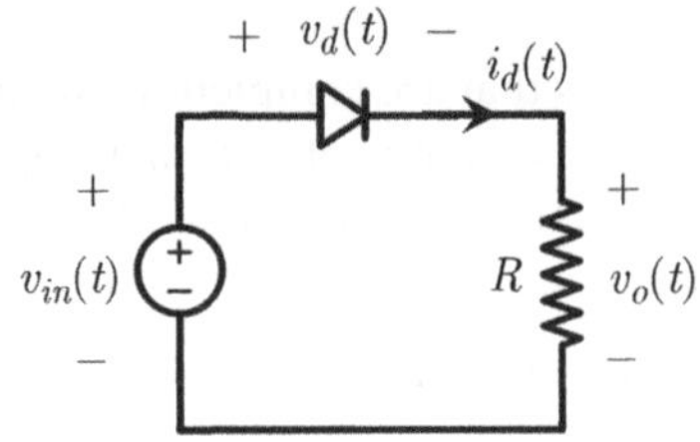

Find $v_o(t)$ as a function of time if $v_{\text{in}}(t) = v_m \sin(\omega t)$. The diode is ideal with zero on voltage.

Solution: First, we note that the input voltage can be plotted as follows.

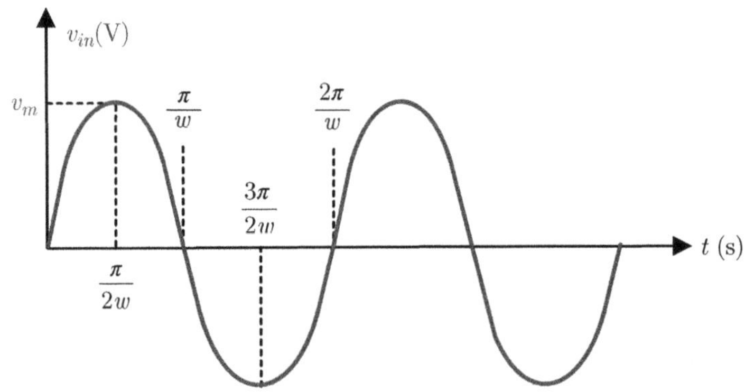

Since the input voltage is periodic and repeats itself, we need to consider just a period of it to understand how the circuit works. First, we focus on the time interval $0 \leq t \leq \pi/\omega$, where $v_{\text{in}} > 0$. We have two options: the diode can be on ($i_d > 0$) or off ($v_d < 0$). Both of these cases can be checked separately. For example, assuming that the diode is on, we have $i_d > 0$ (the condition that must be satisfied) and $v_d = 0$ (the result of being on). Solving the circuit with this assumption leads to the value of i_d that can be checked. If the solution consistently gives $i_d > 0$, the assumption is understood to be correct. Otherwise, the circuit must be solved again with the other (off) assumption.

Now, assuming that the diode is on for $0 \leq t \leq \pi/\omega$, we set $v_d = 0$, leading to

$$v_o = v_{\text{in}} - v_d = v_{\text{in}}.$$

Then the current through the circuit is given by

$$i_d = v_o/R = v_{\text{in}}/R,$$

which is positive since v_{in} is positive. Therefore, the initial assumption ($i_d > 0$) is correct and the diode really is on in this time interval.

Next, we consider the interval $\pi/\omega \leq t \leq 2\pi/\omega$, where $v_{\text{in}} < 0$. Assuming that the diode is on also in this case leads to $v_d = 0$, $v_o = v_{\text{in}}$, and $i_d = v_{\text{in}}/R$. However, this indicates a negative value for i_d (since v_{in} is negative), which contradicts the assumption that the diode is on. When a contradiction occurs, we can select the other assumption and check the circuit again under the new condition. Assuming that the diode is off when $v_{\text{in}} < 0$, we have $i_d = 0$, leading to $v_o = 0$. Then the voltage of the source appears across the diode, that is,

$$v_d = v_{\text{in}} - v_o = v_{\text{in}}.$$

Since $v_{\text{in}} < 0$ in this time interval, v_d is also less than zero, verifying that the diode is indeed off.

To sum up, we obtain the value of v_o as a function time as

$$v_o(t) = \begin{cases} v_{\text{in}}(t), & 0 \leq t \leq \pi/\omega \\ 0, & \pi/\omega \leq t \leq 2\pi/\omega, \end{cases}$$

which also repeats itself with period $2\pi/\omega$. This interesting voltage signal can be plotted as follows.

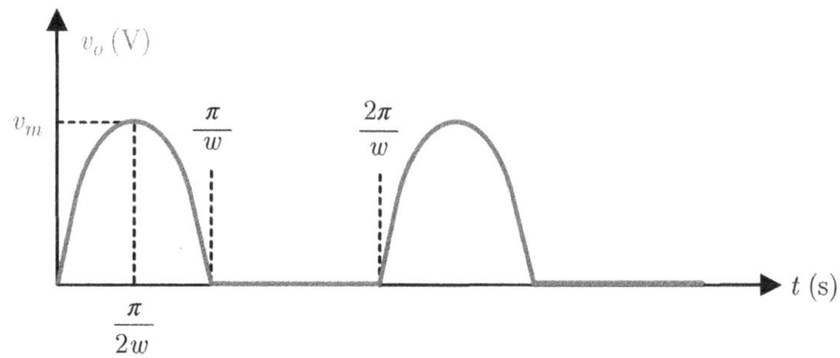

The circuit above is called a half-wave rectifier. As mentioned in Section 1.1.6, rectifiers convert AC signals into DC signals; and the most basic circuit for this operation is analyzed above. The output voltage in this case is not purely a DC signal; in fact, it is still an AC signal. However, it contains a DC part, considering that

$$v_{o,\,\text{avg}} = \frac{1}{T}\int_0^T v_o(t)dt = \frac{v_m}{\pi},$$

whereas $v_{\text{in, avg}} = 0$. This DC part can be improved by adding a parallel capacitor so that the output voltage does not quite reach zero in the off periods of the diode.

Example 135: Consider the following circuit, a full-wave rectifier, in steady state.

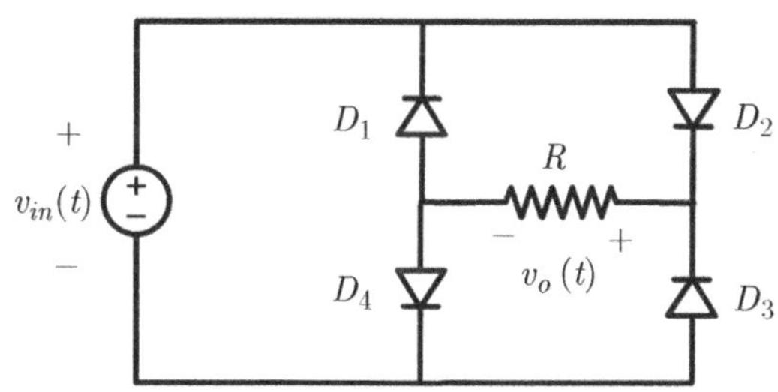

Find $v_o(t)$ as a function of time if $v_{\text{in}}(t) = v_m \sin(\omega t)$. All diodes are ideal with zero on voltages.

Solution: Once again, we need to consider separate time intervals.

- For $0 \leq t \leq \pi/\omega$, when $v_{\text{in}}(t) > 0$, we assume that D_2 and D_4 are on, while D_1 and D_3 are off. Therefore, we have $v_{D2} = 0$ (short circuit), $v_{D4} = 0$ (short circuit), $i_{D1} = 0$ (open circuit), and $i_{D3} = 0$ (open circuit). Then the circuit in this case can be redrawn as follows (note the voltage and current orientations according to the sign convention and diode definitions).

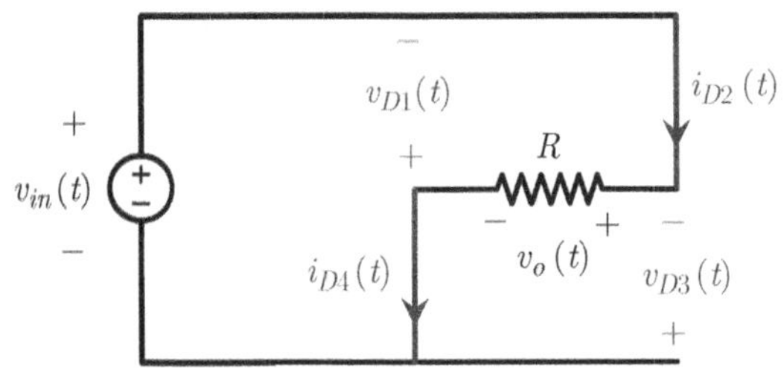

We can check the necessary voltage and current values: we have

$$v_{D1}(t) = -v_{\text{in}}(t) < 0,$$
$$i_{D2}(t) = v_{\text{in}}(t)/R > 0,$$
$$v_{D3}(t) = -v_{\text{in}}(t) < 0,$$
$$i_{D4}(t) = v_{\text{in}}(t)/R > 0,$$

verifying that all assumptions are correct. Hence, in this mode of operation, we have

$$v_o(t) = v_{\text{in}}(t) = v_m \sin(\omega t) \quad (0 \leq t \leq \pi/\omega).$$

- For $\pi/\omega \leq t \leq 2\pi/\omega$, when $v_{\text{in}}(t) < 0$, we assume that D_1 and D_3 are on, while D_2 and D_4 are off. Therefore, we have $v_{D1} = 0$ (short circuit), $v_{D3} = 0$ (short circuit), $i_{D2} = 0$ (open circuit), and $i_{D4} = 0$ (open circuit). The circuit is as follows.

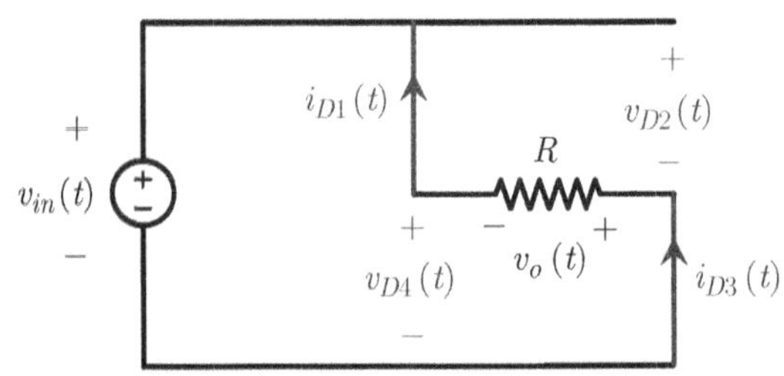

With $v_{\text{in}}(t) < 0$, we have

$$i_{D1}(t) = -v_{\text{in}}(t)/R > 0,$$
$$v_{D2}(t) = v_{\text{in}}(t) < 0,$$
$$i_{D3}(t) = -v_{\text{in}}(t)/R > 0,$$
$$v_{D4}(t) = v_{\text{in}}(t) < 0,$$

all of which satisfy the required conditions for the assumptions. Then

$$v_o(t) = -v_{\text{in}}(t) = -v_m \sin(\omega t) \quad (\pi/\omega \leq t \leq 2\pi/\omega).$$

It can be observed that the output voltage across the resistor is the same as the input voltage when the input voltage is positive, while it reversed when the input voltage is negative. The output voltage, which is periodic with period π/ω, can now be plotted as follows.

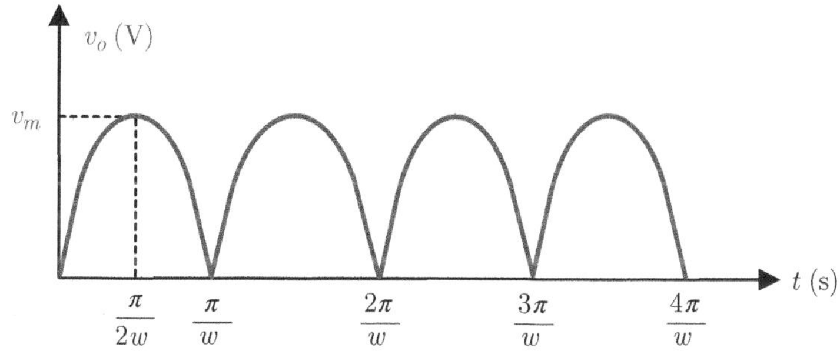

Unlike the output of the half-wave rectifier, the output voltage of the full-wave rectifier is nonzero in all time intervals, indicating a more effective conversion of the AC signal into a DC signal. It can be shown that

$$v_{o,\,\text{avg}} = \frac{1}{T}\int_0^T v_o(t)dt = \frac{2v_m}{\pi},$$

using $T = 2\pi/\omega$ as the period of the input voltage. Hence, using a full-wave rectifier, the DC content of the output signal is twice the DC content of output of the half-wave rectifier.

Exercise 122: In the following circuit, where the diode combination is used as a limiter (clipper), find $v_o(t)$ in steady state, if $v_{\text{in}}(t) = v_m \sin(\omega t)$ and $0 < v_{\text{cl}} < v_m$ is a DC source. Assume an ideal diode with zero on voltage.

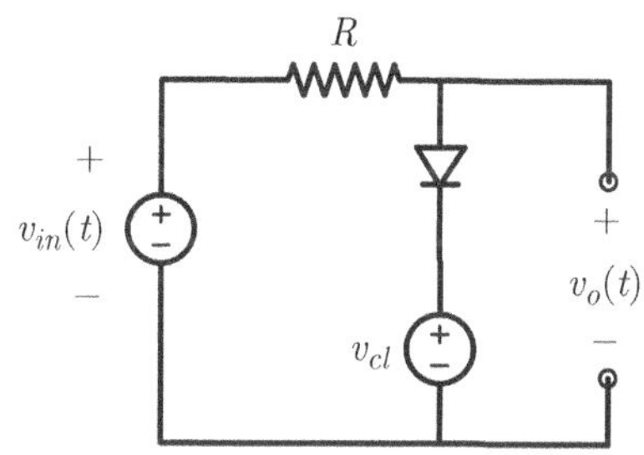

Exercise 123: In the following circuit, find $v_o(t)$ in steady state, if $v_{\text{in}}(t) = v_m \sin(\omega t)$, and $0 < v_{c1} < v_m$ and $0 < v_{c2} < v_m$ are DC sources. Assume ideal diodes with zero on voltages.

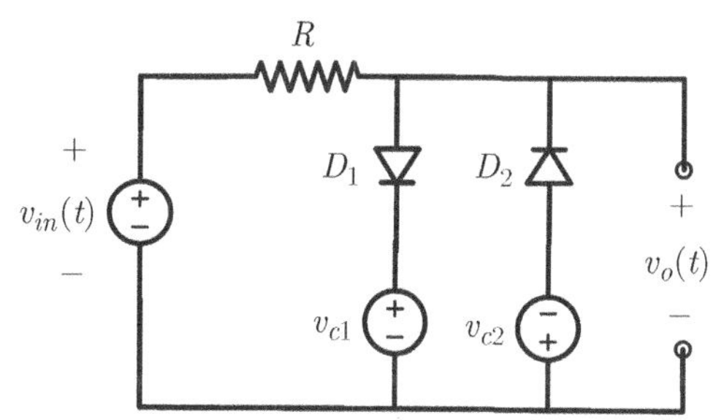

8.3 When Components Involve Many Connections: OP-AMPs

Until now, we have considered mostly two-terminal components (e.g., independent and dependent sources, resistors, capacitors, inductors, and diodes). Transformers are two-port devices, where a port can be defined as a pair of terminals with positive and negative values. In higher-level circuits, it is common to combine basic elements into more complex packages and devices. To take an important example, integrated circuits involve microchips that contain multiple elements (e.g., transistors). As the complexity increases, these packages, which can also be considered as components themselves, involve more and more terminals. Operational amplifiers (or OP-AMPs) are popular components of this kind with multiple connections (see Figure 11.6).

OP-AMPs usually consist of many transistors and resistors, but are often represented as multi-terminal components (without reference to what is inside them) in circuit analysis. As their name suggests, OP-AMPs are amplifiers, that is, they amplify given signals with smaller amplitudes into other forms with larger amplitudes. The input of an OP-AMP is provided through two terminals, namely inverting and noninverting inputs, while their difference is amplified and provided as the output. The amplification provides enables engineers to control signals and design electrical devices for various purposes.

As shown below, five terminals of an OP-AMP are usually shown, among which the DC terminals v_{Sp} (positive) and v_{Sm} (negative) that are used to power up the OP-AMP are usually neglected in their analysis.

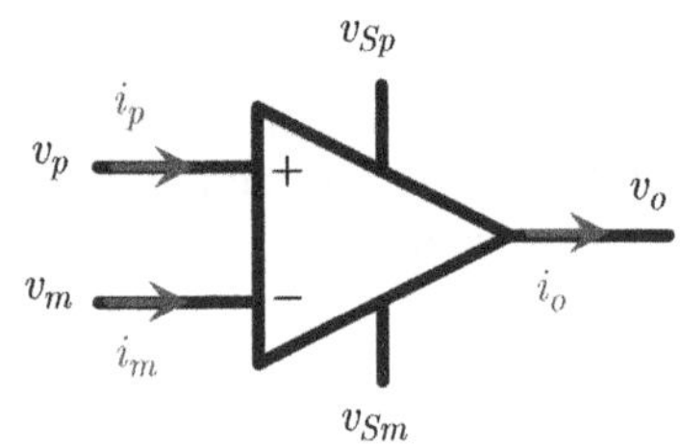

Similarly to diodes, there are different levels of approximations (idealizations) when analyzing OP-AMPs. In an ideal model, we have

$$v_o = A_g(v_p - v_m),$$

where A_g is the gain, typically very large (e.g., 100 000). Hence, the voltage of the output (defined with respect to a ground) is A_g times the difference between the voltages of the noninverting (v_p) and inverting (v_m) terminals. Obviously, some limitations are needed, in the form of saturation voltages, the highest and lowest values that can be obtained at the output terminal. These saturation voltages are often the same as v_{Sp} and v_{Sm}. An equivalent representation for this OP-AMP model can be depicted as follows.

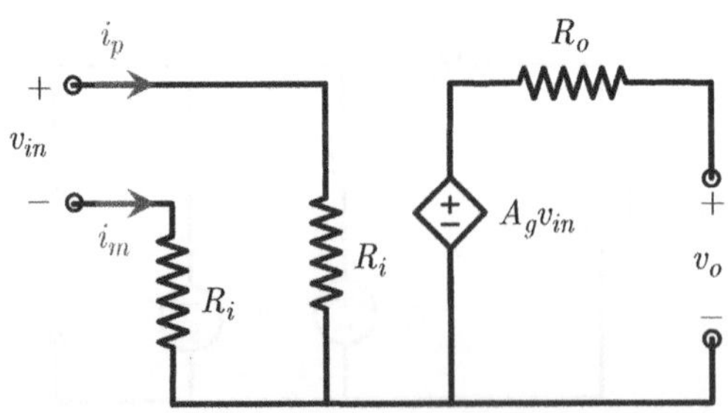

In this representation, the input resistances illustrated by R_i are typically very large, whereas the output resistance R_o is small.

While the representation above is already simple, there is even a more idealized OP-AMP model that is frequently used. In this more ideal representation, we further assume $R_i \to \infty$ so that the input currents i_p and i_m are zero. In addition, $R_o = 0$ and $A_g \to \infty$, that is, the gain of the OP-AMP is infinite. But, in order to keep the output voltage at finite values, we need equal voltage values at the input terminals, that is, the equality $v_p - v_m$ is enforced. While this ideal representation may seem to be an oversimplification, it is indeed useful and enough to understand the most basic circuits involving OP-AMPs.

Example 136: Consider the following circuit involving two resistors R_1 and R_2 connected to an OP-AMP.

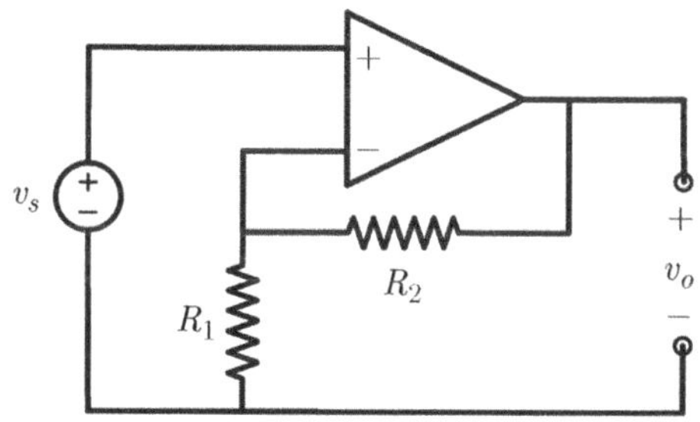

Find the output voltage v_o in terms of v_s and resistor values.

Solution: We can use nodal analysis for the solution of this circuit, as follows.

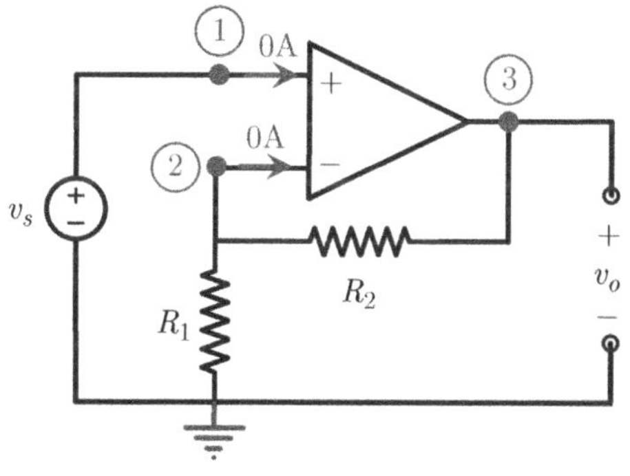

Considering an ideal OP-AMP, we have zero input currents, and

$$v_1 = v_2 = v_s.$$

Applying KCL at node 2, we further derive

$$(v_3 - v_2)/R_2 - v_2/R_1 = 0,$$

leading to

$$v_3 = \left(1 + \frac{R_2}{R_1}\right) v_2.$$

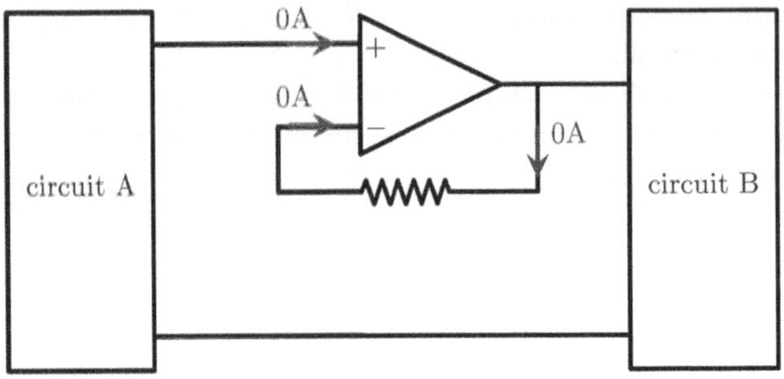

Figure 8.1 A voltage follower between two circuits, *A* and *B*.

Therefore, we obtain

$$v_o = \left(1 + \frac{R_2}{R_1}\right) v_s.$$

In this final expression, $1 + R_2/R_1$ can be interpreted as the finite gain of the circuit. This gain can easily be controlled by selecting the values of R_1 and R_2, in contrast to the huge gain of the OP-AMP itself. Therefore, the circuit above, which is called a noninverting amplifier in reference to its positive gain, is commonly used in electrical circuits (instead of a direct usage of a single OP-AMP). We note that selecting $R_2 = 0$ and/or $R_1 \to \infty$ leads to a gain of unity, leading to $v_o = v_s$. Such a circuit, where the output voltage is exactly the same as the input voltage, can be used as a buffer to isolate two circuits from each other. As shown above, the voltage is directly transferred in this case, whereas the circuits *A* and *B* do not share any current flow. This kind of buffer is called a voltage follower.

Example 137: Consider the following circuit, which is called a differential amplifier.

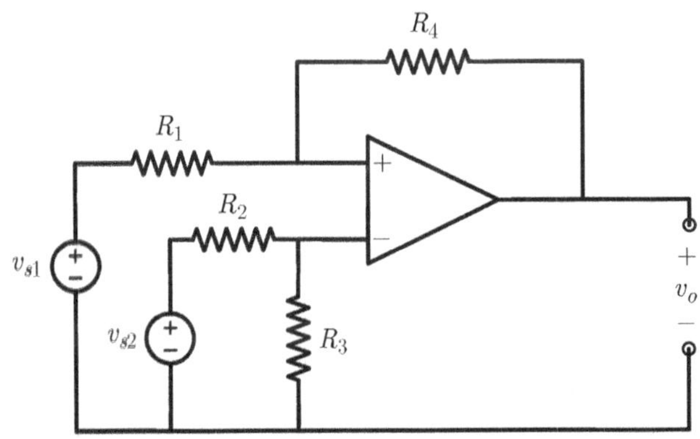

Find v_o in terms of v_{s1} and v_{s2}.

Solution: Using nodal analysis, we label the nodes as follows.

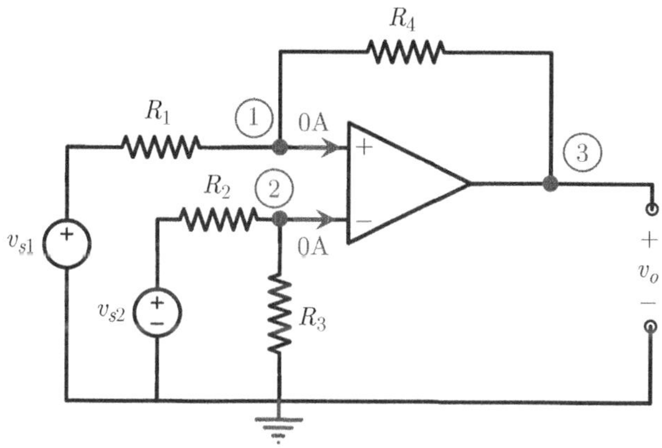

Considering an ideal OP-AMP, we have $v_1 = v_2$. Applying KCL at node 2, we further derive

- KCL(2): $(v_{s2} - v_2)/R_2 - v_2/R_3 = 0$,

leading to

$$v_2 = \frac{v_{s2}}{1 + R_2/R_3}.$$

In addition, applying KCL at node 1, we have

- KCL(1): $(v_{s1} - v_1)/R_1 - (v_1 - v_3)/R_4 = 0$,

which can be rewritten as

$$v_3 = \left(1 + \frac{R_4}{R_1}\right) v_1 - \frac{R_4}{R_1} v_{s1}.$$

Combining the two equations, we arrive at

$$v_3 = v_{s2} \frac{R_3}{R_2 + R_3} \left(1 + \frac{R_4}{R_1}\right) - \frac{R_4}{R_1} v_{s1}.$$

Therefore, the output voltage can be written in terms of input voltages as

$$v_o = -\frac{R_4}{R_1} v_{s1} + \left[\frac{R_3}{R_2 + R_3}\left(1 + \frac{R_4}{R_1}\right)\right] v_{s2}.$$

It can be observed that a weighted difference of v_{s1} and v_{s2} appears at the output. As an interesting set of choices, $R_1 = R_2$ and $R_3 = R_4$ lead to

$$v_o = -\frac{R_3}{R_1} v_{s1} + \frac{R_3}{R_1} v_{s2} = \frac{R_3}{R_1}(v_{s2} - v_{s1}).$$

Hence, by selecting R_3 larger than R_1, one can amplify the difference of the input voltages (as the name differential amplifier suggests). Finally, we note that, when using nodal analysis, KCL at the output of an ideal OP-AMP may not be useful (indeed not required).

In most circuits involving an OP-AMP, it is common to connect the output (e.g., by using a resistor) back to the negative input terminal (see all circuits above). This feedback loop is particularly important for obtaining stable behavior, which may not be achieved otherwise due to the very high (ideally infinite) gain of OP-AMPs.

Exercise 124: Consider the following circuit, which is called an inverting amplifier.

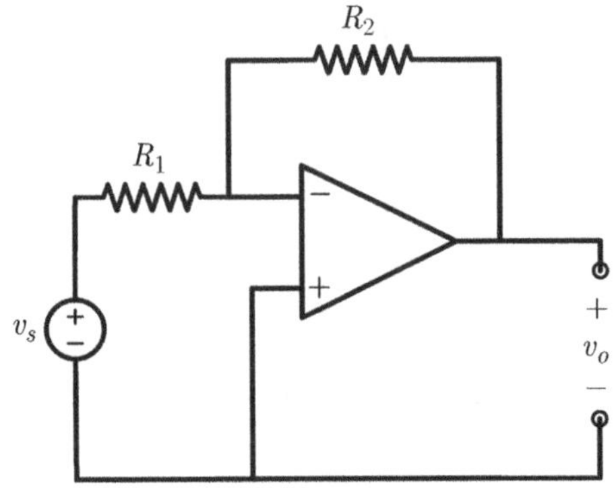

Find v_o in terms of v_s.

Exercise 125: In the following circuit, find v_o in terms of v_{s1} and v_{s2}.

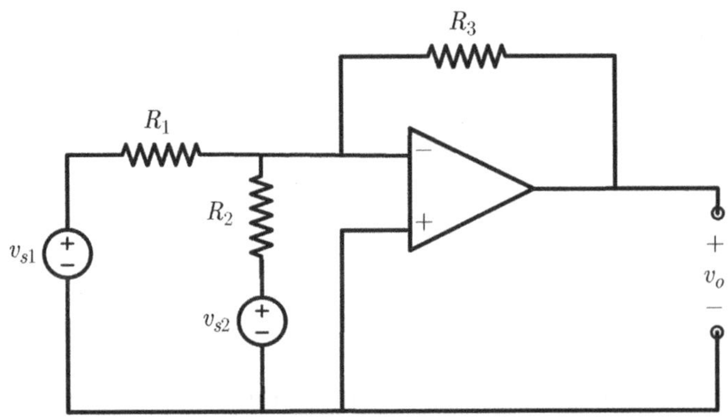

8.4 When Circuits Become Modern: Transistors

Transistors are essential components of modern circuits. Following their first practical implementation in 1947, they became key components of many electronic devices due to their favorable properties, for example, being inexpensive and suitable for mass production (as integrated circuits). Transistors are made of semiconductors, exploiting the flexibility and controllability of the conductivity of these special materials. Basically, a transistor is a kind of amplifier, which transforms a signal into another one with higher amplitude, depending on its operating mode that is determined by a DC bias. The fact that the response of a transistor depends on its operating mode makes it suitable especially for switching operations. A collection of transistors working together lead to electronic devices that are designed to perform given tasks rather than simply modifying electric signals. This is perhaps a major distinction between electronic devices (task performers) and electrical devices (electricity modifiers).

Due to their important roles in modern circuits, there are different types of transistors, such as bipolar junction transistors (BJTs; see Figure 11.7) and field-effect transistors (FETs), as well as their subcategories. In the following, we consider npn-type BJTs, which are based on a p-doped semiconductor sandwiched between two n-doped semiconductors. Indeed, a common semiconductor diode involves a combination of p-doped and n-doped materials, creating a junction allowing controllable and asymmetric conductivity. From this perspective, the existence of two junctions in a BJT provides more capability for controlling signals.

A BJT is a three-terminal device as follows.

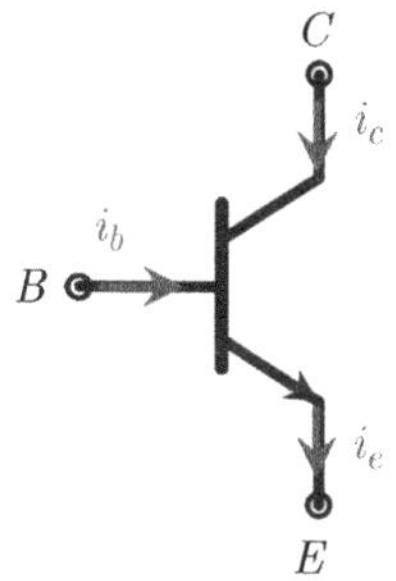

The terminals are called base (B), emitter (E), and collector (C). Typically, both AC and DC signals are connected to a BJT. For example, in a simple amplifier involving a single npn-type BJT (see below), a DC source is connected to the collector to switch on the transistor, whereas the signal to be amplified (usually AC) is connected between the base and emitter. Then the output of the circuit appears between the collector and emitter.

Even when it is connected to a couple of other components, an exact analysis of a BJT may not be straightforward since its response is nonlinear. On the other hand, in most cases, the response of a BJT can be approximated as linear. This enables a separate analysis of AC and DC sources, leading to the overall solution as their superposition. In a DC analysis, as summarized below, the operating mode of the BJT is determined. Three possible modes are cutoff, active, and saturation, in addition to the less common reverse active. When the transistor operates in the active region, AC analysis is carried out to derive the exact response of the circuit. In such an analysis, the AC signal is assumed to be small enough to avoid disturbing the linearity assumption. This is the reason why it is called small-signal analysis.

In the following, we only consider examples of the DC analysis of simple BJT circuits. We define the voltages as $v_{be} = v_b - v_e$, $v_{cb} = v_c - v_b$, and $v_{bc} = -v_{cb} = v_b - v_c$, as usual. The three main operating modes are described as follows.

- Cutoff: In this mode, the transistor is off, and we have $i_b = 0$, $i_c = 0$, and $i_e = 0$. This mode requires $v_{be} < 0$ and $v_{cb} > 0$, similar to diodes being off.
- Active: In this mode, we have a positive on voltage v_{be}, which depends on the properties of the BJT. A commonly used value is $v_{be} = 0.7$ V. This requires a consistency check of the base current, that is, we require $i_b > 0$. In addition, in the active mode, we

must have $v_{cb} > 0$. But, if i_b is nonzero, the analysis requires a relationship between the currents of the transistor. The required equality is

$$i_c = \beta i_b,$$

where β is called the common-emitter current gain. Obviously, using KCL, we further have

$$i_e = i_b + i_c = i_b + \beta i_b = (1 + \beta)i_b.$$

The value of β is typically large, of the order of hundreds.

- Saturation: In saturation mode, we still have $v_{be} = 0.7\,\text{V}$, which requires $i_b > 0$. In addition, due to the forward bias of the BC junction, we should have a nonzero v_{bc}, typically, $v_{bc} = 0.2\,\text{V}$. In addition to nonzero i_b, we must check if $i_c < \beta i_b$ in order to verify that the BJT is indeed in saturation mode.

Similarly to the analysis of diodes, we start with an assumption for the mode of a BJT and check whether the required conditions are satisfied. In case of conflicting values, the assumption is revised accordingly.

Example 138: Consider the following circuit, involving a BJT with 0.7 V on voltage for v_{be} and $\beta = 100$.

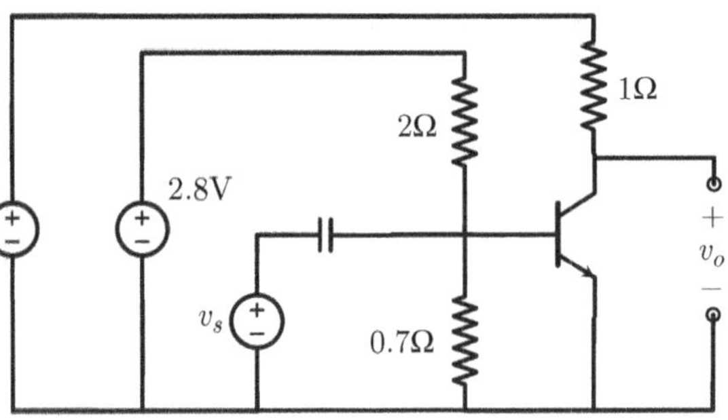

Find the DC signal at the output.

Solution: First, we draw the DC version of the circuit as follows.

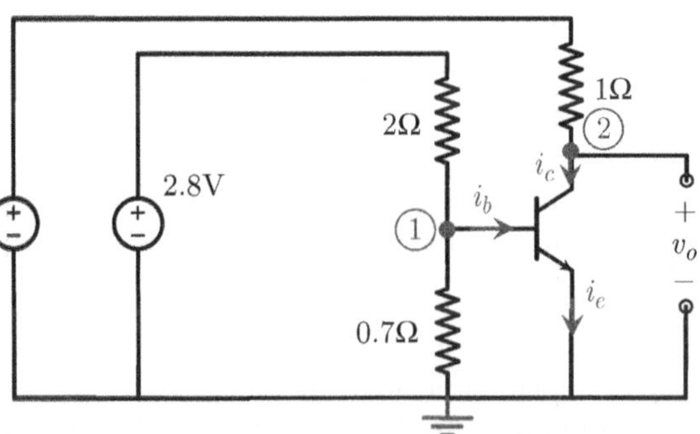

It is remarkable that the voltage source v_s is removed due to the open circuit created by the capacitor. This configuration, an input source and capacitor connected to the base, is very common in BJT circuits. The capacitor filters out any DC component of v_s, while the AC part is coupled to the base (by a careful selection of the capacitor value such that negligible phase is added).

As shown above, nodal analysis is usually suitable for the analysis of BJT circuits. Selecting a proper ground, we already have voltage values at some of the nodes so that they are not labeled. At this stage, we must select a mode for the transistor. Whatever we choose, we have to be careful about assumptions and checks. Assuming that the transistor is active, we immediately have $v_{be} = 0.7$ V (assumption), leading to

$$v_1 = 0.7 \text{ V}$$

since the emitter is grounded. Hence, the current through the 2 Ω and 0.7 Ω resistors can be obtained as

$$i_{2\Omega} = (2.8 - 0.7)/2 = 1.05 \text{ A},$$
$$i_{0.7\Omega} = 0.7/0.7 = 1 \text{ A}.$$

Therefore, the base current can be derived by applying KCL at node 1,

$$i_{2\Omega} - i_{0.7\Omega} - i_b = 0 \longrightarrow i_b = 0.05 \text{ A}.$$

This concludes our first check as $i_b > 0$, which is required if the transistor is active.

Next, using the gain of the transistor (still assuming that the transistor is in the active mode), we can find the collector current to be

$$i_c = \beta i_b = 100 \times 0.05 = 5 \text{ A}.$$

This allows us to find the voltage at the collector node (with respect to the ground),

$$v_2 = 7 - 1 \times i_c = 2 \text{ V}.$$

Finally, we perform the second check on the value of v_{cb},

$$v_{cb} = v_2 - v_1 = 2 - 0.7 = 1.3 \text{ V} > 0.$$

Hence, we verify that the active transistor assumption is perfectly correct. Then we have

$$v_o = v_2 = 2 \text{ V}$$

as the DC signal at the output.

Example 139: Consider the following circuit, involving a BJT with 0.7 V on voltage for v_{be}, 0.2 V on voltage for v_{bc}, and $\beta = 100$.

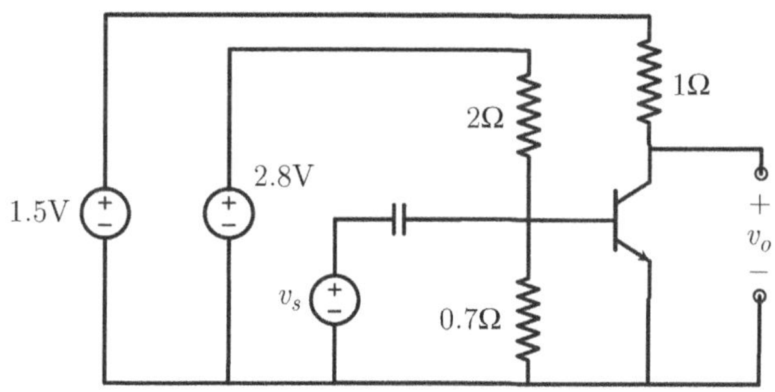

Find the DC signal at the output.

Solution: We note that this circuit is very similar to the previous one, except for the value of the leftmost voltage source, which is now 1.5 V.

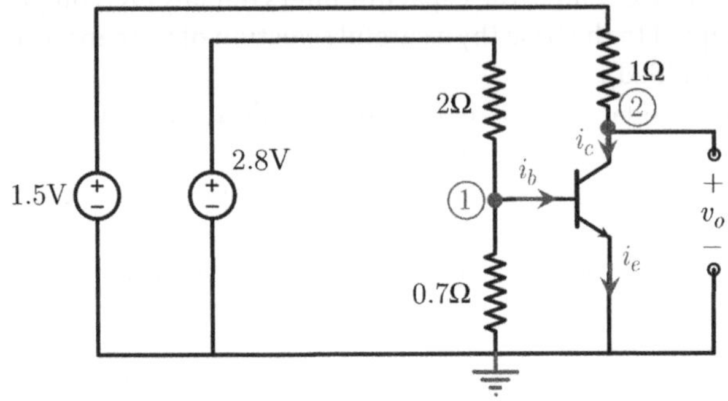

Assuming that the transistor is active, we have $v_{be} = 0.7$ V, leading to

$$v_1 = 0.7 \text{ V},$$
$$i_{2\,\Omega} = (2.8 - 0.7)/2 = 1.05 \text{ A},$$
$$i_{0.7\,\Omega} = 0.7/0.7 = 1 \text{ A},$$
$$i_b = 0.05 \text{ A},$$

exactly as before. Then, using the gain, we should also have

$$i_c = \beta i_b = 100 \times 0.05 = 5 \text{ A}.$$

However, this leads to a collector voltage of

$$v_2 = 2.8 - 1 \times i_c = -2.2 \text{ V},$$

leading to $v_{cb} = -2.2 - 0.7 = -2.9$ V which is less than zero. Therefore, our active transistor assumption is incorrect and we must revise the analysis.

We can restart the analysis by assuming that the transistor is in saturation mode. In this case, we still have $v_{be} = 0.7$ V, and the following steps are still valid:

$$v_1 = 0.7 \text{ V},$$
$$i_{2\,\Omega} = (2.8 - 0.7)/2 = 1.05 \text{ A},$$
$$i_{0.7\,\Omega} = 0.7/0.7 = 1 \text{ A},$$
$$i_b = 0.05 \text{ A}.$$

On the other hand, we cannot use $i_c = \beta i_b$. Instead, supposing that the transistor is saturated, we assume that $v_{bc} = 0.2$ V, leading to

$$v_2 = v_1 - 0.2 = 0.5 \text{ V}.$$

Consequently, the collector current can be obtained as

$$i_c = (1.5 - 0.5)/1 = 1 \text{ A}.$$

As a final check, we note that

$$i_c = 1 < \beta i_b = 5,$$

showing the correctness of the assumption. The DC signal at the output voltage can be written as

$$v_o = v_2 = 0.5\ \text{V}.$$

Exercise 126: Consider the following circuit, involving a BJT with 0.7 V on voltage for v_{be}, 0.2 V on voltage for v_{bc}, and $\beta = 199$.

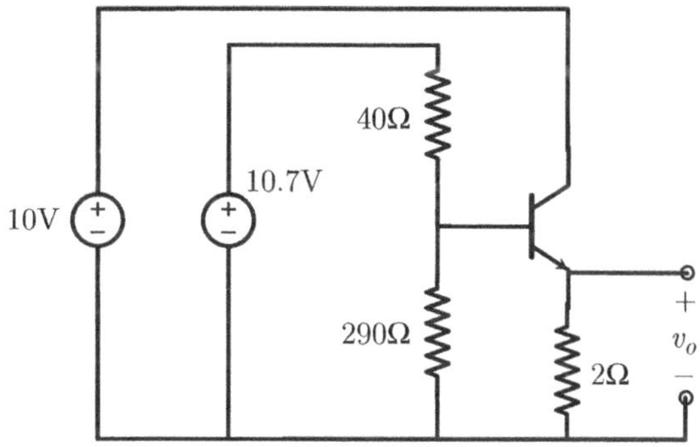

Find v_o.

8.5 When Components Generate Light: LEDs

Electrical energy can be converted into different forms, especially heat, in electrical circuits. In fact, resistors are direct converters of electrical energy into heat, whereas the internal resistances of many components lead to undesired heating and rise of temperature (as is happening in my laptop as I write this). However, across the range of components, heat is not the only product. For example, electric motors are machines that convert electrical energy into useful mechanical energy. Here, we briefly consider another interesting type of component, namely, light-emitting diodes (LEDs; see Figure 11.8), which convert electrical energy into light.

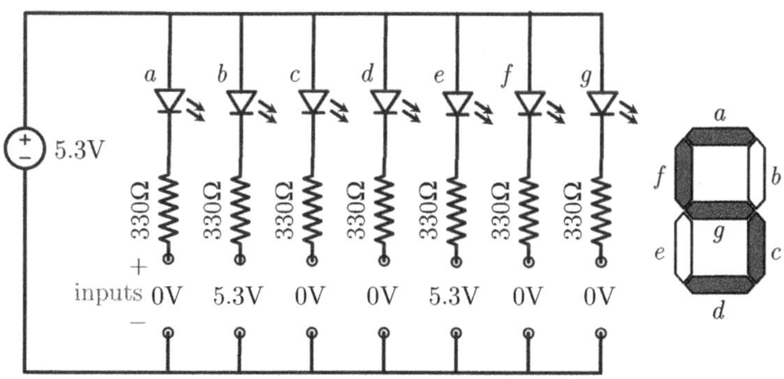

Figure 8.2 A representation of a seven-segment display consisting of seven LEDs.

A typical LED is a pn-junction diode with a relatively large on-voltage value such that it effectively consumes energy to be released as light. Usually, the on-voltage value and the desired current through the LED, hence the power of the LED, are known. Therefore, given a source voltage, the current through the LED is controlled by a resistor. There are also reverse components called photodiodes and photoresistors, whose electrical response depend on the light illuminating them. LEDs, photodiodes, and photoresistors can be used together to build electronic devices, for example isolators, where the components interact via light. This is similar to transformers using the magnetic field instead of electric current for operation.

An iconic and probably one of the most common applications of LEDs is the seven-segment display in the form of the number 8. A total of seven LEDs are used to represent numeric values from 0 to 9, depending on the LEDs that are on and off. As shown below, this can be achieved by controlling the LED voltages, according the number to be displayed.

In the circuit above, each LED has an on voltage of 2 V. The inputs, which are controlled by a logic gate, are either 5.3 V or zero, making the LEDs off and on, respectively. Specifically, when a LED is on with 2 V across its terminal, the current through it is

$$i_{on} = \frac{5.3 - 2}{330} = 10 \text{ mA}.$$

Then the power of the LED can be obtained as $p_{on} = 2 \times i_{on} = 20$ mW, which represents the amount of energy released mostly as light per second.

8.6 Conclusion

Five different types of components of modern circuits are briefly considered in this chapter. All these components have special properties, making them useful as components of electrical and electronic devices. Transformers use magnetic induction between their ports and are employed for controlling the amplitudes of voltage and currents, as well as for isolation. Diodes have unidirectional characteristics, providing engineers a control ability that can be used in constructing rectifiers, clippers, and logic gates. LEDs are also diodes with distinct light-emitting properties that can be used in display technologies, lighting, and optical controllers. OP-AMPs are particularly useful for amplifiers, filters, and oscillators. Finally, transistors are fundamental components of modern circuits, enabling engineers to build smaller and cheaper electronic circuits with more capabilities. A typical smartphone, at the time of writing, involves a chip with 2 billion transistors, each of which has dimensions of the order of tens of nanometers. While we focus on representative types, there are also different subcategories of all these components with diverse properties.

Our exploration of modern circuits continues in the next chapter, where we consider several practical technologies in modern circuits.

9

Practical Technologies in Modern Circuits

In this second extension chapter, we consider some practical technologies in modern electrical and electronic circuits. These selected technologies, which have very diverse properties and application areas, making them difficult to confine to a single book (not to mention a single chapter), are considered very briefly at the circuit analysis level. Similarly to the previous chapter, we focus on only basic concepts of these technologies as essential parts of modern circuits, along with simple examples for their analysis.

9.1 Measurement Instruments

So far, we have considered many circuits involving diverse components with given voltage and current values across them. However, we have not discussed how these voltage and current values are actually measured in real-life circuits. In this section, we briefly discuss the most popular measurement instruments, namely, the voltmeter and ammeter, which are used to measure voltage and current, respectively, in circuits. We start with the component representation of these instruments as follows.

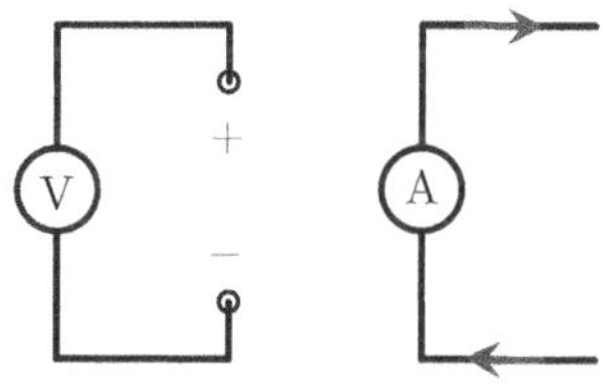

A voltmeter (shown on the left) measures the voltage difference between its terminals. If a voltage across a component needs to be measured, the voltmeter must be connected in parallel to it. Ideally, a voltmeter has infinite resistance so that no current can flow through it. Hence, an ideal voltmeter does not affect the voltage across the component to which it is connected. In practice, however, a voltmeter has a large (but not infinite) internal resistance, leading to a small current flow across it and affecting the measured voltage value. This internal resistance can be represented as a large resistor connected in parallel to an ideal voltmeter.

As an example, consider a nonideal voltmeter with 1 MΩ internal resistance that is used to measure the voltage across a 5 kΩ resistor as follows.

Introduction to Electrical Circuit Analysis, First Edition. Özgür Ergül.

Companion Website: www.wiley.com/go/ergul4412

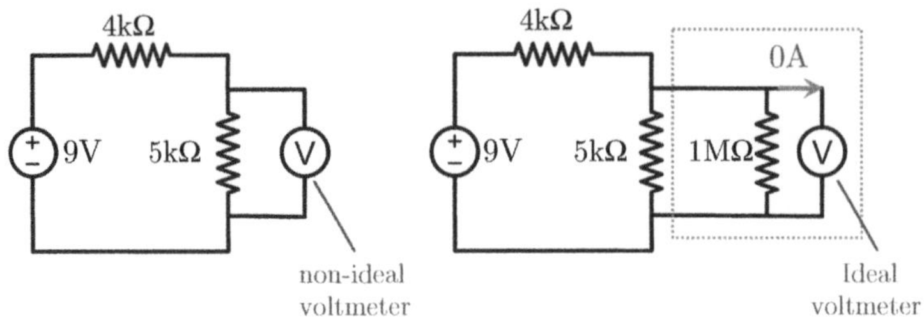

Due to the nonideal voltmeter, a 1 MΩ resistor is effectively connected in parallel to the 5 kΩ resistor. The overall resistance seen by the voltage source can be found to be

$$R_{eq} = 4 + 5 \parallel 10^3 \approx 8.975 \text{ k}\Omega.$$

Then the current through the voltage source can be obtained as

$$i_{9V} \approx \frac{9}{8.975} \approx 1.0028 \text{ mA},$$

instead of the 1 mA that would flow if the voltmeter was ideal. This current is divided between the 5 kΩ resistor and the internal resistance of the voltmeter, leading to

$$i_{5k\Omega} = i_{9V} \times \frac{1000}{1000 + 5} \approx 0.9978 \text{ mA}.$$

Then the measured voltage across the resistor can be found to be

$$v_{5k\Omega} = 5 \times i_{5k\Omega} \approx 4.989 \text{ V}.$$

Consequently, the deviation from the correct voltage value (when the voltmeter is not connected) of 5 V is only 0.22%. Obviously, this deviation would be larger if the value of the resistor, to which the voltmeter is connected, were larger.

An ammeter measures the current flowing through itself; hence, it must be connected in series to components whose current values are to be measured. An ideal ammeter has a zero internal resistance such that no voltage appears across it. However, in real life, a nonzero resistance exists across the terminals of the ammeter, leading to a voltage drop. Therefore, a nonideal ammeter can be modeled as an ideal one connected in series to a resistor representing the small internal resistance. Due to the internal resistance, a nonideal ammeter may lead to a small deviation of the measured current from the actual current that flows when the ammeter is not connected.

As an example, consider the following circuit where the current flowing through a series connection of a voltage source and a 10 Ω resistor must be measured via an ammeter with an internal resistance of 1 mΩ.

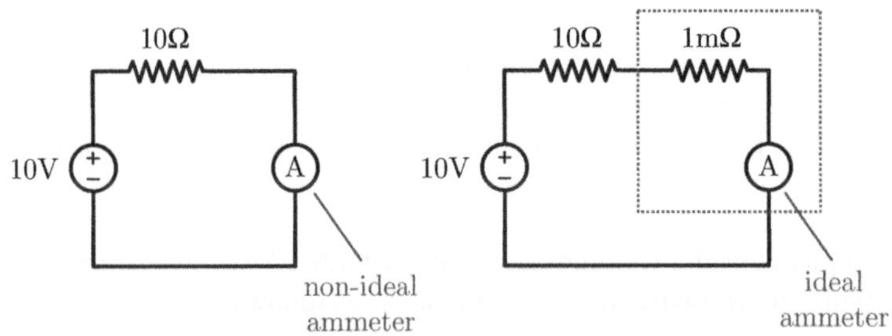

Instead of 1 A, which is the actual current flowing through the original circuit, the ammeter measures

$$i_{10\Omega} = \frac{10}{10 + 0.001} \approx 0.9999 \text{ A},$$

which is a slight deviation. Once again, the deviation of the measured value from the actual current depends on the components. In the circuit above, resistors smaller than 10 Ω would lead to larger deviations. However, for both voltmeter and ammeter, knowing the internal resistance can be useful to deduce the actual values of the quantities that are required to be measured.

Most of the early examples of voltmeters and ammeters used a galvanometer as the core mechanism to measure the voltage and current values. These instruments are analog devices, which are based on continuous signals. In many modern applications, analog voltmeters and ammeters are replaced by digital versions with numeric displays. These instruments are based on the digital processing of signals, making the measurement usually more reliable. Digital components also make it possible to combine a voltmeter and ammeter into a single device, called a multimeter (see Figure 11.9).

9.2 Three-Phase Power Delivery

Modern power transmissions are carried out by using three-phase systems, which have significant advantages over single-phase systems. In order to understand the concept of phase in power delivery, we first consider a load connected to a voltage source.

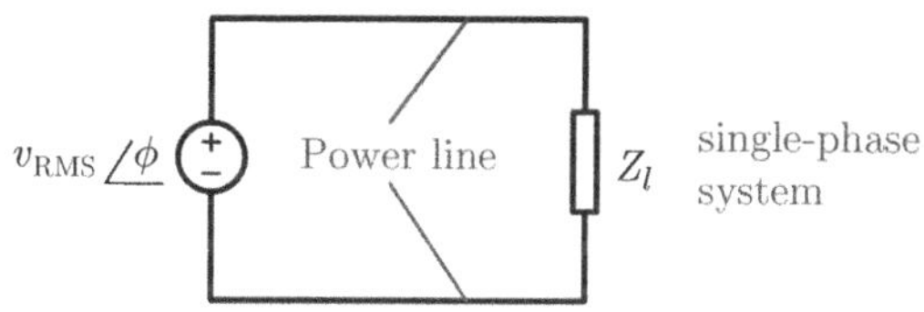

In the above, Z_l can be a device involving a complex circuit itself. In addition, the voltage source can be a complex generator involving multiple electrical components. However, in a power delivery scenario, we consider the source and load as black boxes with a connection (power line) between them. In this context, the circuit above is a single-phase system because there is a single voltage waveform created by the source and carried by the power line to the load. If there were no concerns about other factors (see below), the scheme above could be the ultimate way to transfer power, as it is simple and easy to realize.

In real-life power delivery systems, however, one must consider two important factors, namely, safety and cost, which are beyond the standard circuit design. When analyzing circuits, we neglect the wiring between components, but they become important components to be studied in power delivery systems. In the single-phase system above, let $v_{\text{RMS}} = 120$ V and $Z_l = 2\ \Omega$ (purely resistive). Then the real power delivered is

$$p_l = \frac{120 \times 120}{2} = 7.2 \text{ kW}.$$

While delivering this power to the load, the RMS of the current flowing through the circuit is $i_{\text{RMS}} = 120/2 = 60$ A. In order to carry this amount of current, the wire needed may be heavy and expensive, especially if the power line is long. Indeed, in order to reduce ohmic losses, much higher voltage values are used when transporting electrical power across long distances; however, the voltage is eventually reduced to lower values (e.g., 120 V) for safety in household appliances. Now, in order to decrease the current while keeping the power at the same level, we can increase the voltage to 240 V and load resistance to 8 Ω, leading to $i_{\text{RMS}} = 240/8 = 30$ A. Reducing the current by half can make significant improvements in weight and cost of the wiring. However, increasing the voltage in a power line reduces the safety. This tradeoff between the safety and weight/cost is the main motivation for engineers seek for alternative solutions for power delivery, not only for home wiring but also in high-voltage power transmissions.

Next, we consider the following modified power delivery method, namely a split-phase system.

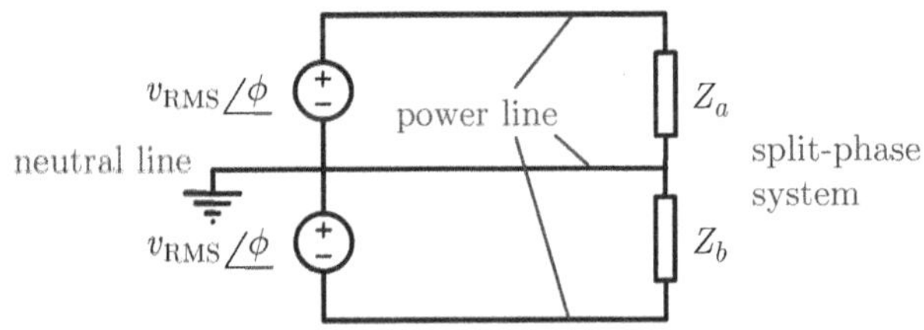

In the above, the overall load is represented by the combination of Z_a and Z_b. For comparison with the previous system, we select $v_{\text{RMS}} = 120$ V and $Z_a = Z_b = 4$ Ω. Therefore, the current through Z_a and Z_b is $i_{\text{RMS}} = 120/4 = 30$ A, which is good for the weight and cost of the wiring. At the same time, the total amount of power consumed by Z_a and Z_b can be found to be

$$p_a = p_b = \frac{120 \times 120}{4} = 3.6 \text{ kW},$$

leading to

$$p_l = p_a + p_b = 7.2 \text{ kW}$$

as the desired power value. Hence, the split-phase system enables the delivery of the same power with reduced current, while the voltage across each sub-load (Z_a or Z_b) is kept at 120 V (which is safer than 240 V).

In the split-phase system above, the neutral line, which is grounded not only in terms of circuit analysis but also physically to earth, does not carry any current, thanks to the identical, symmetrically located, and in-phase sources connected to equal sub-loads. Therefore, this power delivery scheme is also called a balanced system. We note that, to achieve a balance, the sources must be in phase. Otherwise, even when the RMS values of the sources are the same (and still using equal loads), the neutral line carries nonzero current, whereas the RMS of the total voltage across the loads becomes less than 240 V.

The three-phase system is a higher-level power delivery method, which can be illustrated as follows.

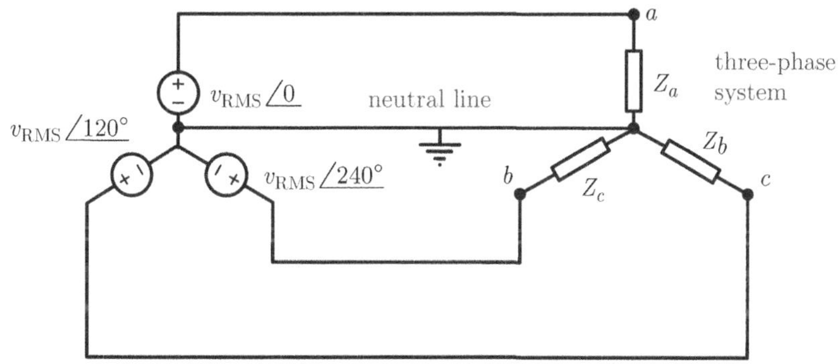

The above system is particularly called the wye (Y) configuration, while another popular one is the delta configuration involving a connection of sources in a Δ shape. For a balanced system, the loads are equal, as well as the RMS values of the voltage sources, while there are 120° phase differences between the sources (so between the 'hot' wires). Under these conditions, the neutral line does not carry any current. The voltage differences between the hot lines are

$$\underline{v}_a - \underline{v}_b = \underline{v}_{\mathrm{RMS}}\angle 0 - \underline{v}_{\mathrm{RMS}}\angle 24° = \sqrt{3}\underline{v}_{\mathrm{RMS}}\angle 30°,$$
$$\underline{v}_b - \underline{v}_c = \underline{v}_{\mathrm{RMS}}\angle 240° - \underline{v}_{\mathrm{RMS}}\angle 120° = \sqrt{3}\underline{v}_{\mathrm{RMS}}\angle -90°,$$
$$\underline{v}_c - \underline{v}_a = \underline{v}_{\mathrm{RMS}}\angle 120° - \underline{v}_{\mathrm{RMS}}\angle 0 = \sqrt{3}\underline{v}_{\mathrm{RMS}}\angle 150°.$$

We note that the amplitude of any line-to-line voltage is $\sqrt{3}$ times the source voltage, while we have new phase values 30°, 150°, and −90° = 270°.

Compared to a single-phase system, a three-phase system has lower cost. Considering the numerical examples above, let 7.2 kW power be delivered to the overall load (now composed of Z_a, Z_b, and Z_c) using sources with $v_{\mathrm{RMS}} = 120$ V. Dividing the power equally among the sub-loads, the current through a hot line is only 20 A, which is simply one third of the current in a single-phase system. Hence, with a lower current, the wiring required for a three-phase system is cheaper and lighter. This is the main reason why three-phase systems are a crucial part of power generation, transmission, and distribution in modern electrical networks.

In general, in order to draw the power in a three-phase system, a single-phase load Z_l can be connected between any two hot wires or between a hot wire and the neutral line. On the other hand, dividing the overall load equally into three parts as Z_a, Z_b, and Z_c, as described above, leads to a balanced circuit that is more efficient. Of course, even in a balanced system, if one of these loads is switched off, the circuit becomes unbalanced with a nonzero current in the neutral line.

Another advantage of three-phase systems appears when three-phase loads are used, that is, when Z_a, Z_b, and Z_c are combined in a single device. As a popular example, large AC electric motors are three-phase devices that exploit the fact that three-phase systems can provide constant-amplitude magnetic flux with rotating direction. Three-phase motors vibrate less than single-phase motors, making them suitable for many industrial applications.

9.3 AD and DA Converters

When we discussed LEDs in Section 8.5, we considered two states for these components: they were either on or off. The related circuits, for example, the separate lines in a seven-segment display, are designed such that only these two states may occur. Indeed, in such a display, an LED with fading or flickering light due to a nonstandard voltage value is considered faulty. In general, components and circuits that are based on a limited number of signal states are called digital. Hence, a digital signal is a sequence of discrete values used in digital circuits. When there are only two states in the digital signal, for example, on (1) and off (0) as in LEDs, it is also called a logic signal. From this perspective, all other circuits and signals, involving voltage and current values that are not limited to a few discrete states, are called analog. Watches are good examples, where both analog and digital technologies exist together due to the different advantages of both types.

Today's electronic devices, including computers, smartphones, and televisions, are mainly based on digital signals and circuits. This dominance is particularly due to the advantages of digital operations that are very tolerant to errors due to fabrication faults, signal deteriorations resulting from outer conditions, and especially noise. In the analog world, voltage and current values must match the proposed levels more precisely; a voltage drop due to a switching operation may cause undesired consequences and failures. In the digital world, however, signals are categorized into a few possible states (e.g., 0 and 1), and these states are well defined and separated from each other so that a signal is difficult to contaminate and misinterpret. Digital circuits and components also enable the user (not only the producer but also the customer) to redesign the operation of a device at the software level without rewiring the hardware.

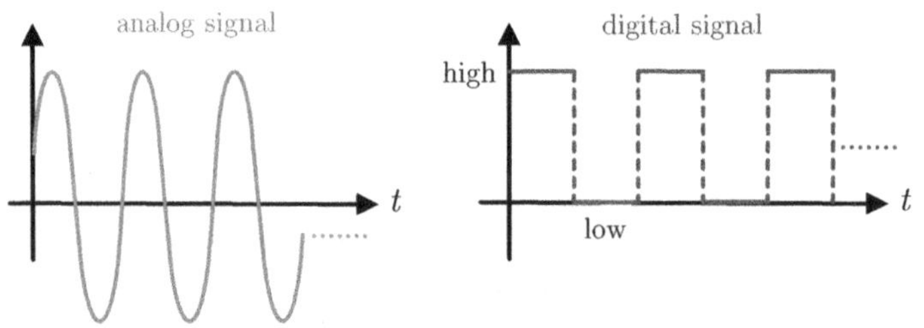

Figure 9.1 Analog and digital signals with respect to time. The digital signal shown has only two permitted values (logic signal), high (1) and low (0).

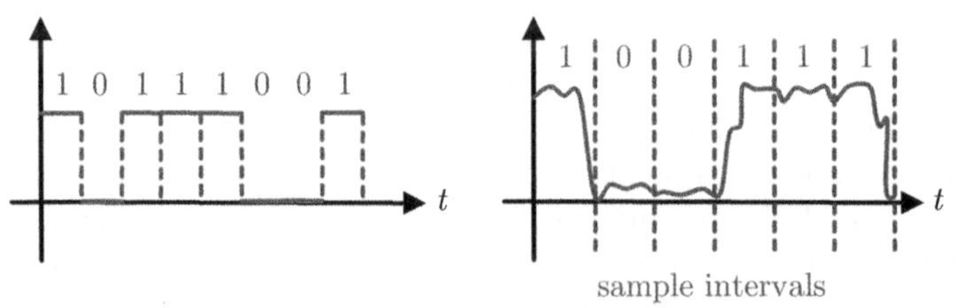

Figure 9.2 An ideal digital signal has only discrete values that can be encoded as numbers (e.g., binaries). In real life, the signal may be distorted due to noise and other effects, but still be encoded correctly.

Since there are only two or several states in digital signals, analysis of digital circuits is often reduced to studying inputs and outputs of higher-level components (which contain more basic elements, such as transistors). For example, in logic circuits, inputs and outputs of components are only a set of 0s and 1s, leading to a finite number of possible scenarios. This simplification allows engineers to combine many digital components together to build complex electronic devices that perform given tasks that are beyond the capabilities of analog circuits. Nevertheless, analog circuits maintain their position, at least as alternatives to digital counterparts, in various areas. While digital circuits are tolerant of operational errors, such as noise, their failure may have larger impacts. For example, a single faulty component in a digital circuit may turn off a huge electronic device, system, or network. Replacing such a faulty component can be expensive and difficult; in many cases, replacing the overall circuit and even the device is preferred. In addition, when a real-life analog signal (e.g., voice) is converted into a digital signal, there is always some loss of information that may not be recovered.

In the next two sections of this chapter, we consider some popular components of digital circuits. In the following, we study the conversion of analog and digital signals into each other, using analog-to-digital (AD) and digital-to-analog (DA) converters. Since we are living in an analog world, practically speaking, these conversions are essential to interact with digital circuits and devices. We consider AD and DA converters, once again, from the perspective of circuit analysis.

Conventional AD converters are packed as small chips that contain many elements working together to convert the given analog signals into digital forms. Among alternatives, integrating converters are relatively slow but quite precise in digitization. A complete integrating converter usually consists of an integrating unit, comparator, clock, and controller. Here, we focus on the integrating unit as the heart of the converter. In order to understand why we need integration, the following figure depicts how a continuous (analog) signal can be discretized.

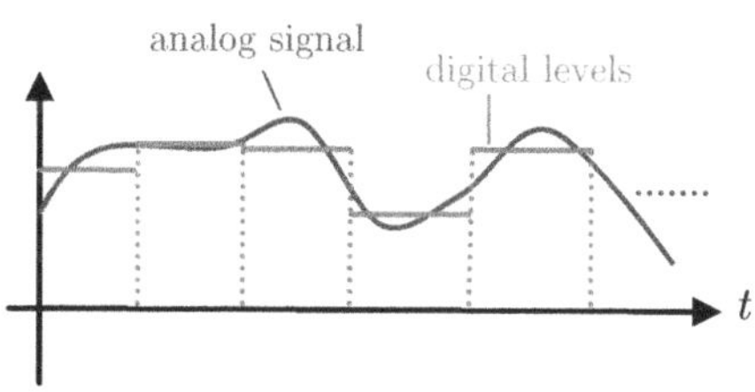

In the above, the smaller the time intervals, the better the discretized waveform approximates the analog signal. The digital levels, which are determined via integration, may not be the final digital signal. However, these levels can easily be categorized (using a comparator and clock) according to the number of configurations to be used in the digital output. For example, signal levels above a given threshold can be set to 1 while others are set to 0, leading to a 1-bit device. Conventional AD converters usually have larger numbers of bits (e.g., 8–24) that allow for many digital levels in the discretization of analog signals.

Obviously, in order to implement an AD converter, we need to find the total amount of signal (integral of the analog function) in given time intervals. To perform this operation, the following integrator based on an OP-AMP can be used.

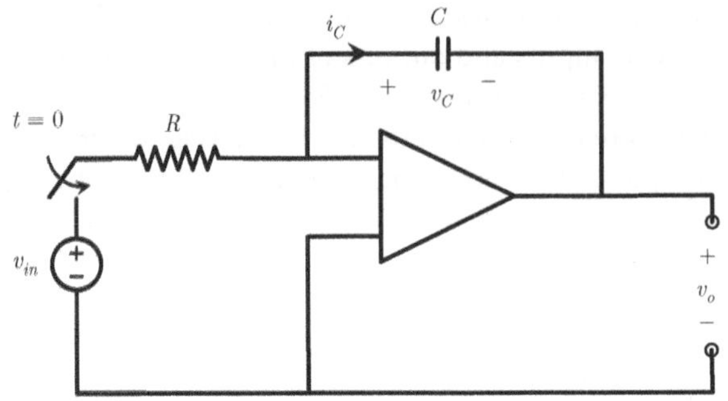

When the switch is closed and the input voltage (analog signal) is connected to the OP-AMP, electric current flows through the resistor and charges the capacitor. For an ideal OP-AMP, the output is simply the negative of the capacitor voltage. Hence, we have

$$\frac{dv_o(t)}{dt} = -\frac{dv_C(t)}{dt} = -\frac{1}{C}i_C(t) = -\frac{v_{\text{in}}(t)}{RC},$$

leading to

$$v_o(t) = -\frac{1}{RC}\int_0^t v_{\text{in}}(t')dt',$$

assuming that $v_o(t = 0) = 0$. Therefore, the output voltage is effectively the integral of the input voltage over time. In an AD converter, the switching operation is performed repetitively, allowing the input voltage to charge the capacitor in fixed time periods and another reference (negative) voltage to discharge it. This way, by integrating the input voltage in fixed time intervals, the digital levels are determined to be converted into the output digital signal.

Similarly to AD converters, there are different versions of DA converters, which are used to convert digital signals generated by digital circuits and devices into analog signals. Here, we consider one of the most basic converters using an OP-AMP and a sequence of resistors, often called a resistor ladder. Such a DA converter with a 4-bit digital input can be depicted as follows.

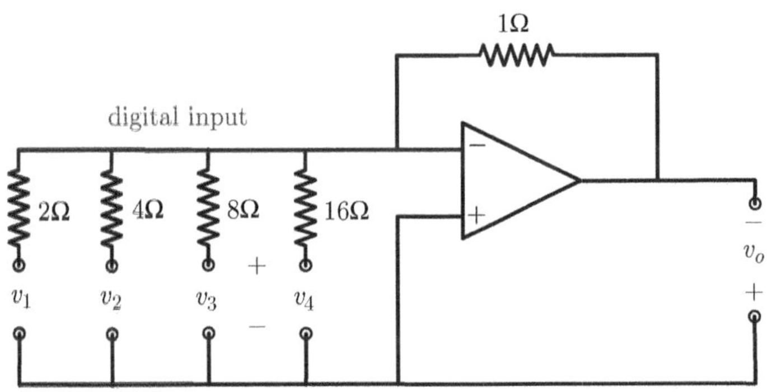

In the circuit above, v_1, v_2, v_3, and v_4 are voltages representing the overall digital input. For example, in the signal, 0 and 1 states may correspond to 0 and 5 V, respectively.

The purpose of the DA converter is to generate a unique output voltage corresponding to a given digital signal. The ladder structure above enables such an ability to distinguish different digital signals represented by four digits. Assuming an ideal OP-AMP, the output voltage can be found to be

$$v_o = \frac{v_1}{2} + \frac{v_2}{4} + \frac{v_3}{8} + \frac{v_4}{16}.$$

Then, if the digital signal is $[d_1, d_2, d_3, d_4] = [1, 0, 0, 1]$, we have $v_1 = v_4 = 5\,\text{V}$ and $v_2 = v_3 = 0$, leading to

$$v_o = 5/2 + 0 + 0 + 5/16 = 45/16 = 2.8125\ \text{V}.$$

We note that this voltage value corresponds to $[1, 0, 0, 1]$ and no other combinations of four bits. As further examples, $[0, 1, 1, 0]$ leads to $v_o = 1.875\,\text{V}$, whereas $[1, 1, 1, 1]$ and $[0, 0, 0, 0]$ lead to $v_o = 4.6875\,\text{V}$ and $v_o = 0$ as the maximum and minimum values, respectively.

We now know ways to convert analog signals into digital and vice versa. But how can digital circuits be analyzed? In the next sections, we consider logic gates and memory units as popular components of digital circuits.

9.4 Logic Gates

Logic gates are building blocks of most digital circuits. They are usually made of diodes and transistors, which are packed into single components that can carry out the required logic operations. Since the signal levels are reduced to a set of values (i.e., 0 and 1) in logic circuits, the input/output functions of logic gates are generally defined in terms of these discrete values, without considering the actual strengths of voltages and currents. This simplification allows engineers to combine and cascade many logic gates to build electronic circuits and devices that can perform complex tasks.

Among various types, the most basic logic gates with one or two input and single output terminals can be described as follows.

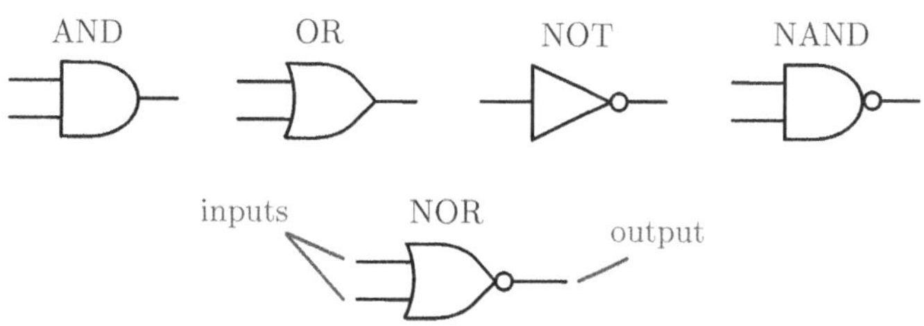

For all these components, inputs and outputs are defined as 0 or 1. Considering logic operations, these low and high states are also called FALSE and TRUE, respectively. The behavior of each component is well defined and tabulated as a truth table. For example, the output of an AND gate for different input combinations can be shown as follows.

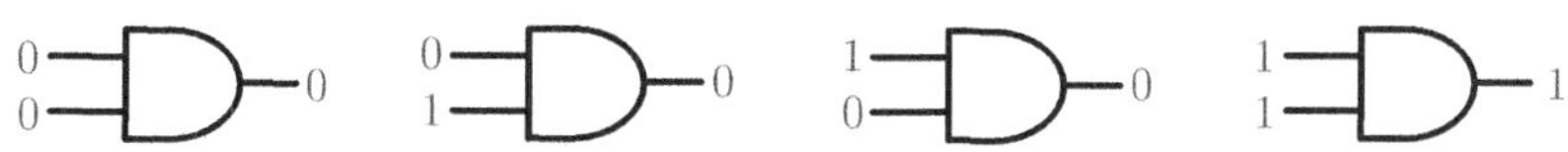

It can be observed that, in order to have a TRUE output, both inputs of an AND gate must be TRUE, hence its name. On the other hand, an OR gate gives a TRUE output whenever one of its inputs is TRUE.

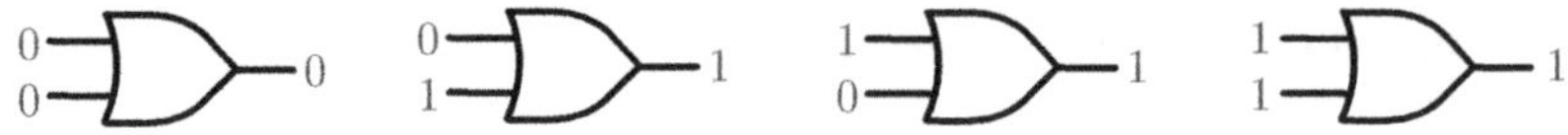

As its name suggests, a NOT gate inverts the signal, that is, a TRUE input gives a FALSE output while a FALSE input gives a TRUE output. A NAND gate (see Figure 11.10) can be seen as a combination of an AND gate and a NOT gate; its output is FALSE if and only if both inputs are TRUE. Similarly, a NOR gate is a combination of an OR gate and a NOT gate, leading to a TRUE output if and only if both inputs are FALSE.

In a logic circuit, many gates are cascaded to produce the desired output for given input combinations. As an example, we consider the following circuit involving four logic gates.

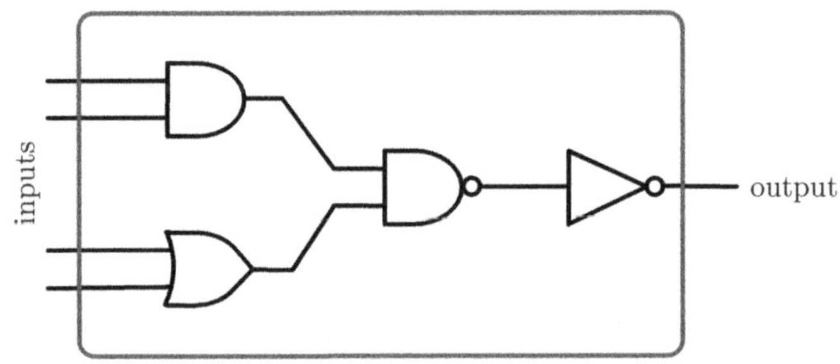

The overall structure has four inputs and a single output. When analyzing logic circuits, engineers use boolean expressions and arithmetic in order to derive a single formula, if possible, that represents the output for any combination of input values. The gate operations are well defined in boolean arithmetic, enabling the analysis of complex combinations without dealing with input combinations one by one. Since these calculations are outside the scope of this book, and the circuit above is relatively simple to analyze, we investigate it in detail using a truth table.

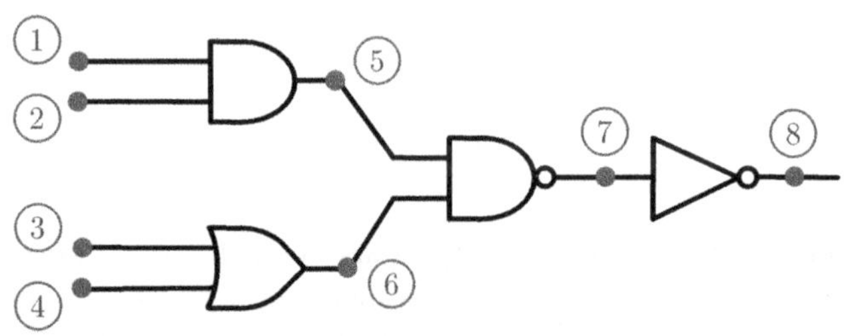

First, we start with an incomplete truth table, initially filled in with possible input combinations. In addition, we add the values at nodes 5 and 6 by using the properties of the AND and OR gates, as follows.

n_1	n_2	n_3	n_4	n_5	n_6	n_7	n_8
0	0	0	0	0	0		
0	0	0	1	0	1		
0	0	1	0	0	1		
0	0	1	1	0	1		
0	1	0	0	0	0		
0	1	0	1	0	1		
0	1	1	0	0	1		
0	1	1	1	0	1		
1	0	0	0	0	0		
1	0	0	1	0	1		
1	0	1	0	0	1		
1	0	1	1	0	1		
1	1	0	0	1	0		
1	1	0	1	1	1		
1	1	1	0	1	1		
1	1	1	1	1	1		

We note that n_5 is TRUE if and only if both n_1 and n_2 are TRUE. On the other hand, n_6 is FALSE if and only if both n_3 and n_4 are FALSE. Next, for given values of n_5 and n_6, as well as the properties of the NAND gate, we find the value at node 7 for different cases. The output of the NAND gate is FALSE if and only if both inputs are TRUE (only the last three cases). The output at node 8 is simply the reverse of the value at node 7; hence, we obtain the following complete table.

n_1	n_2	n_3	n_4	n_5	n_6	n_7	n_8
0	0	0	0	0	0	1	0
0	0	0	1	0	1	1	0
0	0	1	0	0	1	1	0
0	0	1	1	0	1	1	0
0	1	0	0	0	0	1	0
0	1	0	1	0	1	1	0
0	1	1	0	0	1	1	0
0	1	1	1	0	1	1	0
1	0	0	0	0	0	1	0
1	0	0	1	0	1	1	0
1	0	1	0	0	1	1	0
1	0	1	1	0	1	1	0
1	1	0	0	1	0	1	0
1	1	0	1	1	1	0	1
1	1	1	0	1	1	0	1
1	1	1	1	1	1	0	1

It can be observed that the output of the circuit is TRUE only for three (out of 16) combinations of the inputs. Specifically, these are $[n_1, n_2, n_3, n_4] = [1, 1, 0, 1]$, $[n_1, n_2, n_3, n_4] = [1, 1, 1, 0]$, and $[n_1, n_2, n_3, n_4] = [1, 1, 1, 1]$. The output is FALSE for all other cases. This response of the circuit can be represented by using boolean operations on the inputs. Then, with a well-defined input–output relationship, the circuit can be combined with other circuits to implement larger and more technological circuits and electronic devices.

As briefly mentioned above, logic gates are mainly made of diodes and transistors, as well as resistors for coupling and adjusting current values. Here we consider one of the most basic gates, namely, the resistor-transistor-logic (RTL) NOR gate, which involves a BJT transistor combined with a few resistors. There are also RTL gates with multiple BJTs, but we investigate the single-transistor version for simplicity while understanding the working principles of transistors as logic elements.

Consider the following RTL NOR gate with two inputs and an output.

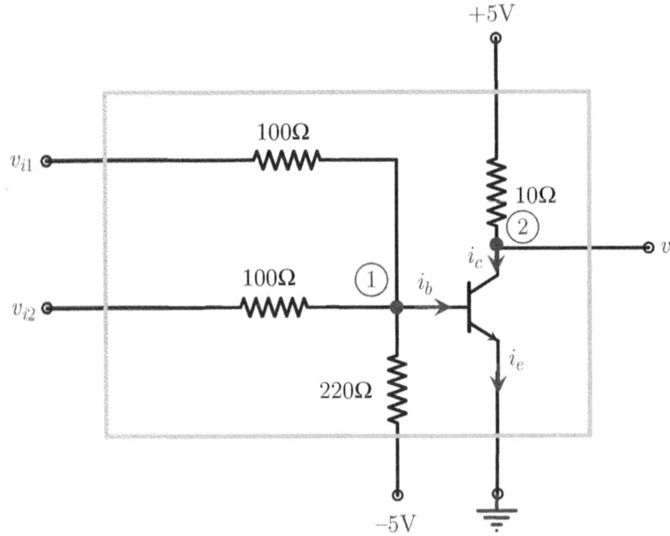

We assume that the transistor has $\beta = 100$, 0.7 V on voltage for v_{be}, and 0.2 V on voltage for v_{bc}. In addition, all node voltages are defined with respect to a common ground, which is also depicted in the figure. First, we consider the asymmetric case where one of the inputs is TRUE and the other is FALSE. Using $v_{i1} = 5\,\text{V}$ (for TRUE) and $v_{i2} = 0\,\text{V}$ (for FALSE), the transistor is in saturation mode, leading to $v_1 = v_{be} = 0.7\,\text{V}$ and $v_2 = v_1 - v_{bc} = 0.5\,\text{V}$. Applying KCL at node 1, we have

- KCL(1): $(5 - 0.7)/100 + (0 - 0.7)/100 - (0.7 + 5)/220 - i_b = 0$,

leading to $i_b \approx 10.1\,\text{mA}$. We can also find the collector current, $i_c = (5 - 0.5)/10 = 0.45\,\text{A}$. Therefore, we verify that $i_c < \beta i_b = 1.01\,\text{A}$ as a requirement of the saturation mode. In this case, the output voltage (i.e., the collector voltage) is simply $v_o = 0.5\,\text{V}$. While this value is not a perfect zero, it is a low value that can be interpreted as a FALSE output. The solution of the circuit is very similar when $v_{i1} = 0\,\text{V}$ and $v_{i2} = 5\,\text{V}$, leading again to a FALSE output.

Next, we consider what happens when both inputs are TRUE. Using $v_{i1} = 5$ V and $v_{i2} = 5$ V, the transistor is again in saturation mode. KCL at node 1 can now be written as

- KCL(1): $(5 - 0.7)/100 + (5 - 0.7)/100 - (0.7 + 5)/220 - i_b = 0$,

leading to $i_b \approx 60$ mA. Once again, $i_c = 0.45 < \beta i_b = 6$ A, verifying the saturation mode. Hence, the output for this case is also FALSE.

Finally, we consider only FALSE inputs; $v_{i1} = 0$ and $v_{i2} = 0$. In this case, the transistor is in cutoff mode and $i_b = i_c = i_e = 0$ A. Then KCL at node 1 is written as

- KCL(1): $(0 - v_1)/100 + (0 - v_1)/100 - (v_1 + 5)/220 = 0$,

leading to the node voltage $v_1 \approx -0.926$ V, hence $v_{be} = -0.926$ V. In addition, since there is no collector current, $v_2 = 5$ V. Therefore, we obtain $v_{bc} \approx -5.926$ V, which is also negative, verifying the cutoff mode. In this case, the output voltage is 5 V, that is, TRUE. As a result, the circuit above provides TRUE output only when both inputs are FALSE, and it operates as a NOR gate.

In general, NOR and NAND gates are universal gates, which can be combined to generate any logic operation (not only complex ones, but also basic AND, OR, and similar operations). In the following section, we discuss how NOR gates can be used to construct a memory unit.

9.5 Memory Units

Memory units are essential parts of computers and smart devices that need to store information. The term "memory" is generally used to refer primary types, which are directly accessible by the processing unit (e.g., CPU). Internal or external hard disks are categorized as secondary types, and they are mainly used to store data to be accessed later. Among primary memory units, one can mention volatile and nonvolatile types, depending on the dependence of the storage on the power. Volatile units (e.g., random access memory (RAM)) keep the data as long as the power is supplied, whereas nonvolatile units (e.g., read-only memory (ROM)) have the ability to store the data without external power. In the following, we briefly discuss some building blocks of volatile primary memory units and their working principles.

When a computer or smart device is on, it needs storage for temporary data. For example, a software program in use is temporarily copied to RAM for quick response to the user's requests. RAMs are fast, but they are volatile; once the power is down, all data stored in them are lost. This is the reason why the data must be saved to storage disks, which use magnetism to keep data for long periods without requiring external power. RAMs are also categorized based on their building blocks. Dynamic RAMs (DRAMs) include capacitors to store data, which are organized by transistor switches. Depending on the state of a capacitor as full (charged) or empty, it can simply indicate 1 and 0 as a single bit. On the other hand, static RAMs (SRAMs) are generally based on flip-flop circuits, as discussed below.

A flip-flop is a circuit that has two stable states (hence, they are called bi-stable) such that they can be used to store a single bit of information. They are usually made of transistors, while in circuit analysis they can easily be represented in terms of logic gates.

A flip-flop usually contains a clock for synchronization; but there are also asynchronous flip-flops, namely, latches. The following figure depicts a set–reset NOR latch.

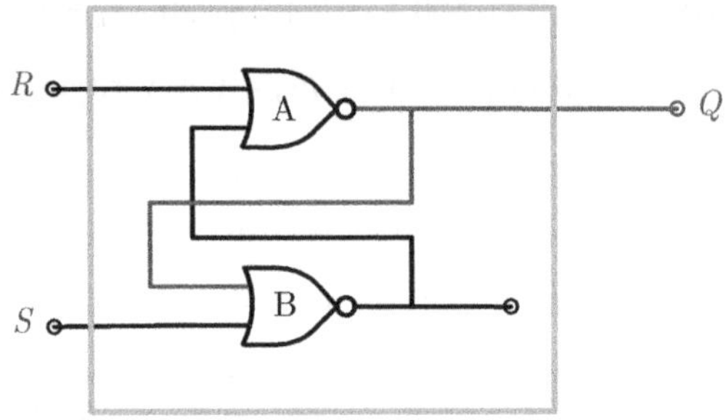

This latch has two inputs, namely, SET (S) and RESET (R), and we focus on a single output Q. First, we consider the case where both S and R are FALSE. There are two options.

- If the output Q is FALSE, both inputs of the B gate are FALSE, leading to a TRUE output signal. Then the A gate has FALSE and TRUE inputs, leading to FALSE as the overall output. Consequently, everything is consistent, and the output is stable as FALSE.
- If the output Q is TRUE, the B gate gives a FALSE output signal. Then the A gate has two FALSE inputs, leading to a TRUE overall output. This option is also stable and self-consistent.

Therefore, we can conclude that, if both S and R are FALSE, the output of the latch remains as it is, either as FALSE and TRUE. This can be considered as the storage of a single bit of data.

But how can we insert data into the latch above? Assume that we would like to set the status of the output as TRUE. Then all we need to do is to make S = TRUE and R = FALSE. Let us see what happens in this case.

- If the output Q was FALSE before the set operation, S = TRUE forces the output of the B gate to FALSE, which in turns make Q = TRUE. Therefore, the output of the latch becomes TRUE, even if it was FALSE before.
- If the output Q was already TRUE, S = TRUE still leads to a FALSE signal as output of the B gate. Then the A gate again has two FALSE inputs, leading to a TRUE overall output. Hence, the TRUE output is preserved as it is.

Similarly, it can be shown that S = FALSE and R = TRUE leads to Q = FALSE, that is, it resets the latch whatever its initial value is.

An interesting case occurs if one sets both inputs to TRUE. Whatever the previous outputs of the NOR gates, they are forced to provide FALSE outputs due to the TRUE inputs. Using a single output, this does not seem inappropriate, with a global FALSE output. However, latches are usually used with two outputs, where the second output (the output of the B gate) must always be the reverse of the first output (the output of the A gate). When both outputs are FALSE, this contradicts with the definition of the latch. Therefore, setting both inputs as TRUE is not common practice for this type of memory unit.

There are many combinations of NOR and NAND gates, or more precisely transistors, to construct latches as the building blocks of memory devices, particularly SRAMs. We emphasize that memory is merely one application of logic gates, which are basic units of all electronic circuits in modern technology.

9.6 Conclusion

Some practical technologies are considered in this extension chapter. Measurement tools allow engineers to test voltage and current values in circuits and to detect faulty components easily. Three-phase systems are used for distributing electric power safely and less expensively, across long distances in the AC form. The last three sections tell us how analog circuits evolve into digital circuits, involving logic gates as building blocks. This is also how electrical circuits become electronic circuits as decision-making and programmable systems via logic operations.

We are now very close to the end of this book. In the final extension chapter, we provide some advice on how to approach complex electrical and electronic circuits and devices in real life.

10

Next Steps

We conclude this book with some final notes. In this short farewell chapter, we discuss a couple of important points when we deal with complex circuits and devices. As the purpose of this book is to cover the fundamentals of analyzing circuits, but not the circuits themselves, we clearly emphasize the main ideas before the reader enters into the complex world of electrical and electronic circuits in specialized areas. Whatever the area is, we keep in mind that

- energy is always conserved,
- complex circuits can be analyzed by considering them as combinations of subcircuits, whose functions are well known,
- state-of-the-art components are usually squeezed into integrated microchips, allowing us to build more complex circuits,
- the human body itself can be considered as a circuit element,
- and electricity is a natural phenomenon that must be handled with care.

Good luck!

10.1 Energy Is Conserved, Always!

When we analyze a circuit, we keep in mind that the total power of all elements must be zero, as a result of the conservation of energy. Specifically, some components may have positive power values (consuming energy) while others have negative (delivering energy), but their algebraic sum must be zero. A component that consumes energy at a particular time may store it (capacitors and inductors) or convert it into other forms (e.g., heat in resistors, light in LEDs) that may not be recovered. Similarly, a component that delivers energy may actually produce it (voltage and current sources) from other forms (e.g., chemical energy in batteries) or use energy that is stored earlier (again, capacitors and inductors). Energy is always conserved; there is no way to produce it from nothing and consume it without converting it into another form. This is not because of a special balance mechanism; it is due to how we define energy.

The conservation of energy must hold when electrical circuits are combined with other mechanisms and devices. In some cases, however, the balance of energy among the elements may not be trivial. Consider the following circuit involving a DC motor.

Introduction to Electrical Circuit Analysis, First Edition. Özgür Ergül.

Companion Website: www.wiley.com/go/ergul4412

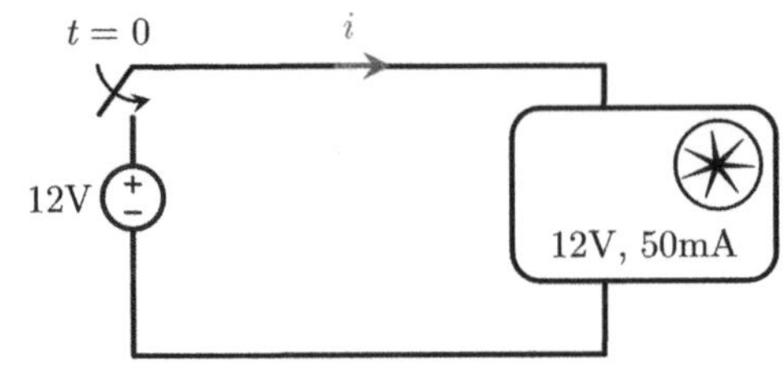

DC motors convert electrical energy into mechanical energy, using DC voltage and current, unlike AC motors. Each DC motor is described with voltage/current/power values that are specified when the motor is fully operational. In the circuit above, a motor with 12 V and 50mA is used, leading to $p_m = 12 \times 0.05 = 0.6$ W. Hence, when it is fully operational, this motor converts 0.6 J electrical energy into mechanical energy (torque) per second. The energy required is provided by the voltage source (e.g., battery).

But now let us consider the instant at which the switch is closed and the motor (which is at rest) just becomes connected to the voltage source. At $t = 0^+$, the motor is still not working (e.g., its rotor does not rotate). Hence, it should not consume any energy at this particular time. Then, due to the conservation of energy, the voltage source should not deliver any energy, leading to $i(0^+) = 0$ current through the circuit. But here is the dilemma: If the current is zero, how the motor can start gaining energy such that its rotor accelerates? Should not the motor be at rest infinitely?

Indeed, when the switch is closed in the circuit above, there is a large current flowing through the circuit. This inrush current, which decays to the steady-state value of 50 mA as time passes, is responsible for starting up the motor and accelerating the rotor until it gains full angular velocity. In order to understand the circuit at $t = 0^+$, we need to consider the small internal resistance of the motor.

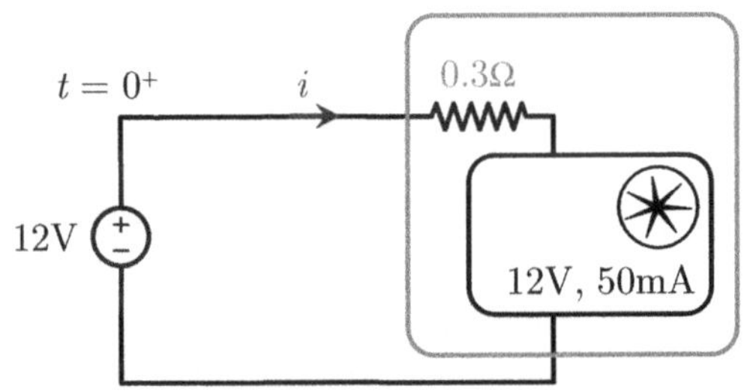

At $t = 0^+$, the rotor does not rotate, that is, no electrical energy is converted into mechanical energy, but due to the small internal resistance of the motor, a huge amount of current flows: $i = 12/0.3 = 40$ A. As the current flows through the wires on the rotor, a magnetic force is generated between the rotor and stationary magnets, leading to acceleration. Hence, the mechanical power increases with time. In the limit case, when $i = 50$ mA, most of the energy provided by the voltage source is converted into mechanical energy. In this case, the power consumed by the internal resistance of the motor is $0.3 \times 0.05 = 0.015$ W, which is a small percentage of the power consumed by the motor for mechanical operation.

In the circuit above, one can also consider the internal resistance of the voltage source, as well as the wires, for more a realistic analysis. In addition, the heat generated due to the friction between the rotor and stationary part can be included as another source of power consumption. In any case, the conservation of energy is a universal law; it should be always satisfied, independent of the complexity of the circuit analysis.

10.2 Divide and Conquer Complex Circuits

Most circuits used in real life contain many components and connections between them. While they may seem complex, they can be analyzed by following the techniques discussed in this book. Specifically, mesh analysis and nodal analysis, together with the black-box concept, can be employed to analyze any circuit with an arbitrary number of components. In some cases, however, it can be beneficial to understand how the circuit works, before embarking on a full analysis. For this purpose, most circuits can be decomposed into sub-circuits, whose functions may be well known from earlier analysis.

As an example, consider the following circuit, which represents a single-transistor AM radio receiver. The values of the components are omitted.

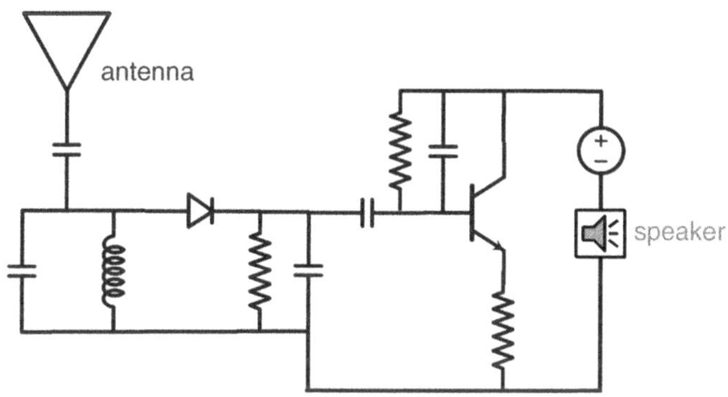

This circuit is much simpler than other AM receivers that contain multiple transistors. Nevertheless, it does not seem trivial to analyze. The circuit contains an antenna, which converts electromagnetic signals propagating in the air into electrical signals. Hence, it can be considered as a source that generates an AC signal (e.g., the signal to be captured). On the right-hand side, there is a speaker that converts the electrical signal into sound, which can be considered as a power-consuming resistor. But, between the antenna and speaker, there are many components, including a diode and transistor, in addition to resistors, capacitors, and an inductor. For a rigorous analysis in steady state, the circuit should be converted into the phasor form. Then parallel connections of resistors and capacitors/inductor can be simplified into impedances. On the other hand, AC analysis of the transistor would require a small-signal assumption. Even in the phasor form, we may find ourselves in trouble.

But how does the circuit above work? Investigating it in detail, we can identify some of the parts as follows.

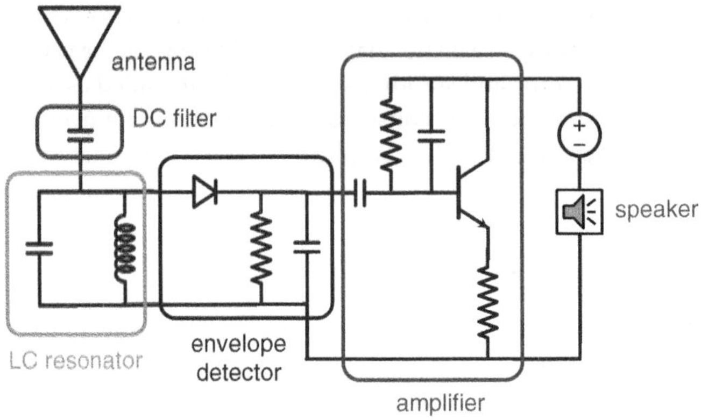

First, we note that the antenna is coupled to the circuit via a capacitor. This capacitor helps us to get rid of noise with low-frequency content. This is because the capacitor acts like an open circuit as the frequency goes to zero. Then we have an LC resonator, which behaves as a band-pass filter, in order to select the frequency of the signal to be listened to. In practice, the capacitor and inductor in the LC resonator are adjustable so that their resonance frequency can be tuned to the desired frequency. The output of the LC resonator goes through an envelope detector, which is a combination of a diode, resistor, and capacitor. This is needed because AM radio signals carry the information as an envelope of the carrier frequency. Hence, this envelope must be detected and converted into meaningful sound. The original AM signals propagating in the environment are naturally small, of the order of microwatts. Therefore, an amplifier based on a transistor is used to amplify the signal before the speaker. In most applications, a single amplifier may not be sufficient so that a cascade of transistors are used to amplify the signal to reasonable levels. Then the amplified signal can be converted into sound via the speaker. In the circuit above, the DC voltage source is used to power up the overall circuit (e.g., to keep the transistor in on mode).

As briefly discussed, a complex circuit can be explained by identifying its sub-circuits. This is so common in electrical and electronic engineering that most engineers immediately recognize a resonator, an envelope detector, or an amplifier, in order to understand the function of a complete circuit. A detailed analysis, for example to determine the component values, can be achieved once sub-circuits are clearly identified. In such an analysis, it may also be possible to study the sub-circuits separately by considering them as black boxes with well-defined inputs and outputs.

10.3 Appreciate the Package

Integrated-circuit technology enables the confinement of many components, particularly transistors, into a single small structure, namely, a microchip. This leads to the construction of complex circuits using microchips as black boxes. As the technology develops, more and more transistors are being fitted onto microchips, making them more capable of performing complicated tasks in physically small regions. Many digital circuits involve complex combinations of numerous chips (without any visible transistor), where the interactions are reduced to binary operations via logic gates.

As an example, we consider the famous 7400 chip, which consists of four NAND gates as follows.

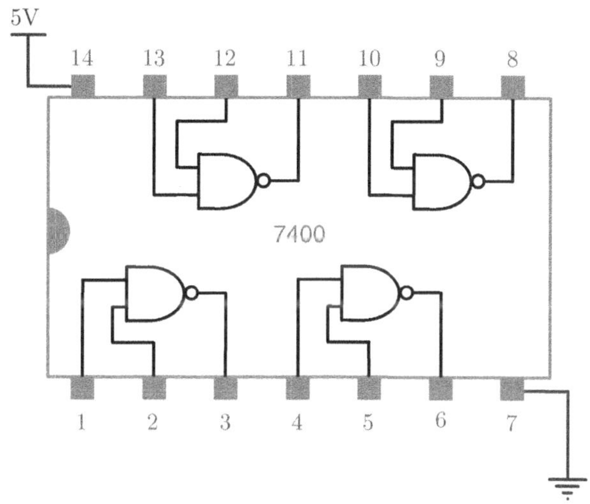

In this schematic, the NAND gates are only for illustration purposes; these gates are actually combinations of transistors, which are located together on the same structure. Therefore, a 7400 chip actually contains several (e.g., 16) transistors that are fabricated together on a single semiconductor (e.g., silicon) material. However, an engineer using this chip cannot see even these transistors; s/he uses them via the visible pins that are numbered from 1 to 14. Pins 7 and 14 are always connected to the ground and high voltage (+5 V) in order to power up the chip (i.e., for biasing the transistors).

Now suppose that we would like to construct an AND gate using a single 7400 chip. Using logic operations, we can cascade two NAND gates as follows.

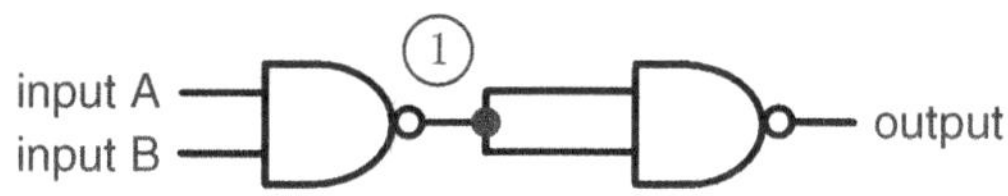

One can verify that the output of the above combination is TRUE only if both inputs A and B are TRUE; so it is an AND gate. In order to implement it, we can use the following connections on the 7400 chip.

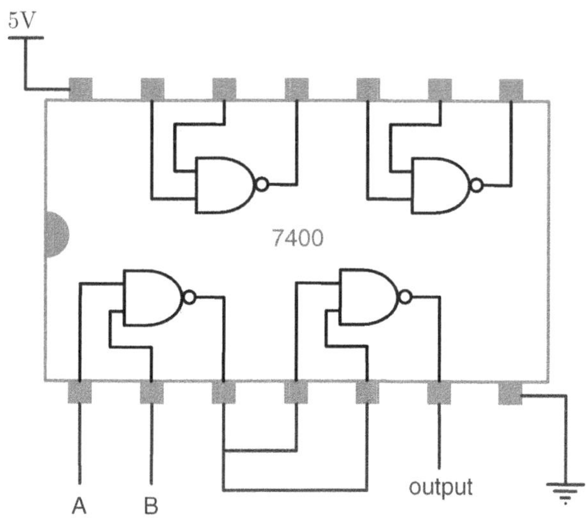

In this circuit, two NAND gates of the 7400 chip are not used. In many circuits, the gates in the microchips are used more economically, minimizing the number of microchips required to perform all required operations. Obviously, due to such economic usage, any electrical problem (e.g., a permanent short circuit) in a microchip requires the replacement of the overall chip, since the gates in the same package are inseparable parts of the same structure. Nevertheless, one major advantage of integrated circuits is their low cost, enabling their mass production. Nowadays, it is not uncommon to replace a whole faulty electronic card involving multiple microchips, and even a device, rather than repairing or replacing a single microchip.

10.4 Consider Yourself as a Circuit Element

The human body can be quite conductive under certain conditions. While it depends on the frequency and the voltage itself (due to the breakdown of the skin for large voltage values), hand-to-hand resistance can be of the order of several kilohms for large contact areas. This value can be even smaller for wet skin and voltage values larger than the typical electricity used in homes. In the next section, we briefly discuss the safety issues around electric shock. In this section, we consider the conductive property of the human body (its capacitive nature) for a useful application, that of touchscreens.

As most of us frequently use them in our smartphones, touchscreens are interfaces between us and smart devices, including tablets, consoles, and some personal computers. As they have become increasingly popular, several types of touchscreens have dominated the electrical and electronics industry so far. These are specifically resistive and capacitive touchscreens, which are based on resistor and capacitor arrays, respectively, as their names suggest. Most resistive touchscreens do not use the electrical properties of the body. They are merely based on the pressure of the finger that physically deforms and changes the electrical properties (resistance distribution) of the screen. Therefore, resistive touchscreens can be used with gloves on fingers or by using anything that can apply the required pressure. On the other hand, capacitive touchscreens directly use the electrical characteristics of the human body, as described below.

Basically, the working principles of a capacitive touchscreen can be summarized as follows.

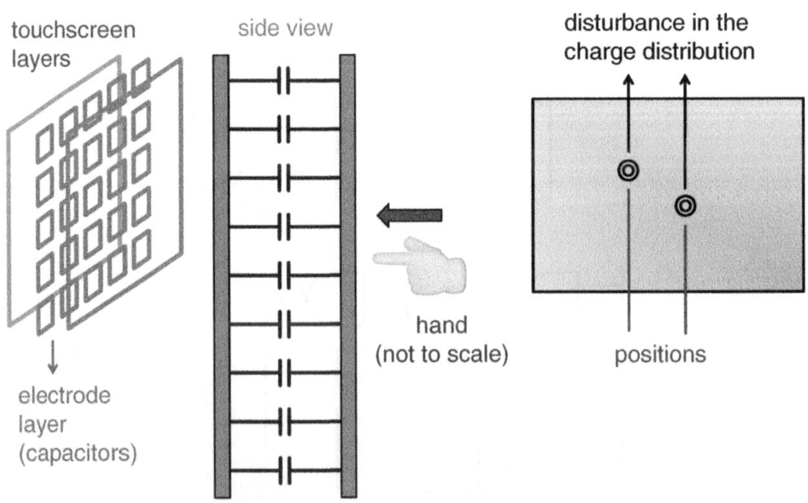

In a capacitive touchscreen, there is an electrode layer, which can be represented as a two-dimensional array of capacitors. In the inactive mode, the charges are distributed uniformly among these capacitors. Then a touch on the screen leads to a disturbance of the charge distribution, since human tissue is conductive. As an electrical circuit model, this can be interpreted as an additional capacitance between the human finger and the electrode layer being inserted into the capacitor array from outside, creating more charge accumulation close to the finger. Such an imbalance in the charge distribution can be detected electrically. We note that touching at multiple positions can also be identified when the electrode layer consists of multiple conductive parts. In some touchscreens, the electrode layer is a single body, and the required voltage is provided from the conductors located at the corners. In this case, the touch position is calculated by considering the distances from the corners to the disturbance point. This kind of touchscreen may not identify multiple touches.

The capacitance of the body is also used in some other applications, such as touch sensors, touch-sensitive lamps, and, interestingly, in a musical instrument, the theremin. In electrostatic calculations, the whole human body is often modeled as a 100 pF capacitor in series to a resistor. In certain conditions, a human body may be charged up to several tens of kilovolts with respect to its surroundings (e.g., earth). This static charge accumulation causes small-scale arcs when the person touches a grounded conductive (e.g., a doorknob). While they are harmless to the human body, these static shocks must be prevented in many industrial processes involving flammable materials. Integrated circuits are very sensitive to electrostatic discharges such that special antistatic procedures and tools must be employed to avoid damaging them while constructing electronic circuits.

10.5 Safety First

Electricity can be dangerous. Electric shock, which happens when a large amount of current passes through a part of the body, may cause ventricular fibrillation, neurological problems, and severe burns, depending on the strength of the electricity. Contraction of muscles (tetanus) due to the effects of electricity on the nervous system can make it difficult for the victim to detach from the source of the electricity. The resistance of the human body is extremely nonlinear and highly dependent on environmental conditions. Therefore, it is not straightforward to say whether electricity of a given strength is harmless or harmful. It also depends on the frequency of the electricity; for example, AC is generally more dangerous than DC. At 50–60 Hz, a 20 mA current through a part of the body can be very painful and dangerous. In extreme conditions with very dry and clean skin, leading to around 1000 kΩ hand-to-hand resistance, this would require 20 000 V, which seems quite high. However, as briefly mentioned in the previous section, the body's resistance can easily drop to several kilohms. When the value of the resistance is 10 kΩ, 200 V voltage can produce a dangerous 20 mA current, which can be deadly. Obviously, the lethality of the electric shock also depends on the duration; but we can easily conclude that domestic electricity can be fatal.

In the following, we consider some scenarios where electric shock may occur or not. These are very simplified analyses and they should not be used as a guide to distinguishing safe and unsafe cases in real life. We only consider the dynamics of the electric shock in terms of a circuit analysis. As a rule of thumb, electric shock occurs when a high amount of current passes through a part of the body. From a circuit-analysis perspective, this needs a sufficiently high voltage difference to occur between two separate points of the body. Hence, the current is formed on the pathway between these two points, passing through tissues and organs, some of which can be critical (e.g., the heart and lungs between two hands). This is important; a high voltage value at a single point may not lead to a current flow and electric shock. We need two points with different potentials so that current may flow.

Consider the following figure.

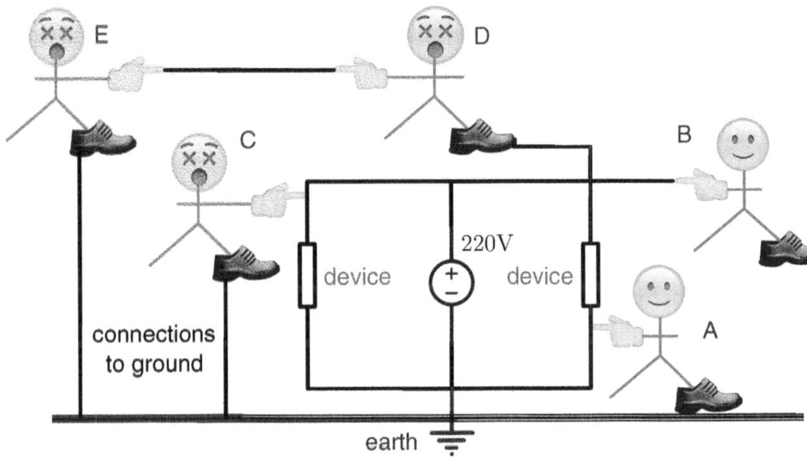

We have an electrical circuit involving two devices and a 220 V voltage source. The ground of the circuit is defined not only as a mathematical reference but also physically (via a connection to the earth), as commonly practiced in real life. The following cases can be considered.

- Case A: If a person touches the ground line of the circuit and if s/he is standing on the ground, no voltage difference occurs across the body and s/he does not receive a shock. Most electrical and electronic devices have insulator cases, which are also grounded. Hence, it is safe to touch them, unless there is an electrical fault.
- Case B: If a person touches the hot line of the circuit and s/he is isolated from the ground, no electrical shock occurs. This can be a confusing case because the voltage of the person with respect to the ground is around 220 V. However, since the person is isolated, there is no current flow across her/his body. It should be emphasized that the isolation must be very good to prevent any current flow. For example, most shoes are not designed for electrical insulation, and it is not safe to touch a hot line without rigorous isolation.
- Case C: This is a typical case for most electric shocks. If a person standing on the ground touches the hot line of the circuit, the voltage difference between her/his hand and foot can lead to a serious current flow through her/his body. This case may occur due to a faulty circuit, leading to positive voltage values at conductors that are supposed to be grounded.
- Case D and E: It does not matter how the connection occurs for an electric shock. If there is a conductive path, even involving two persons as depicted above, the electric current flows through them. In this case (one person is grounded and the other person touches the hot line), the overall voltage is divided between the persons, which may reduce the effect of the shock.

We can now consider two further cases.

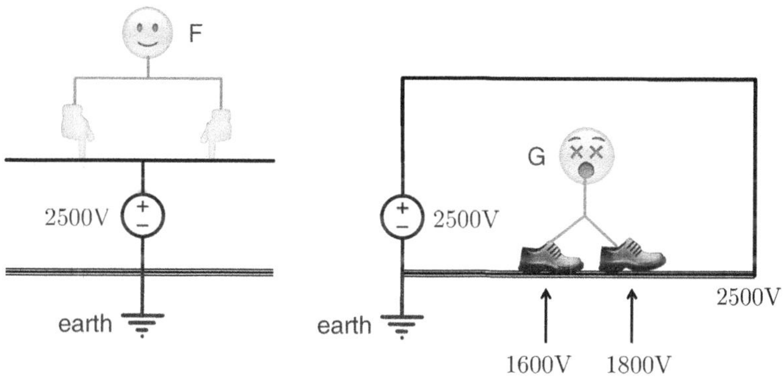

Here, we have a hot line of 2500 V with respect to earth.

- Case F: If a person touches the hot line with two hands, following the discussion above, we can easily claim that s/he has a potential of 2500 V with respect to the ground/earth; but s/he is not shocked as there is no current flow across her/his body. We again assume that s/he is perfectly isolated from the ground.
- Case G: On the other hand, a person who does not touch anywhere is not always safe. In the case of an electric fault (e.g., an accidental connection of the hot line to the ground), there can be a voltage difference across the ground itself. If the hot voltage is high enough, the voltage across the two feet of the person can be sufficient to generate a risky level of current flow.

In order to prevent electric shocks in daily life, as well as to avoid excessive faulty currents that may damage circuits and devices (even leading to fires), electrical engineers design and use various devices, such as circuit breakers, fault current interrupters, and switches. But eventually, it becomes consumers' responsibility to ensure the safe use of electricity.

Humankind did not invent electricity. We merely discovered how to use it, and it depends on us to make it beneficial.

11

Photographs of Some Circuit Elements

The following photographs were taken by Barışcan Karaosmanoğlu and Sadri Güler of the Middle East Technical University, Ankara, Turkey.

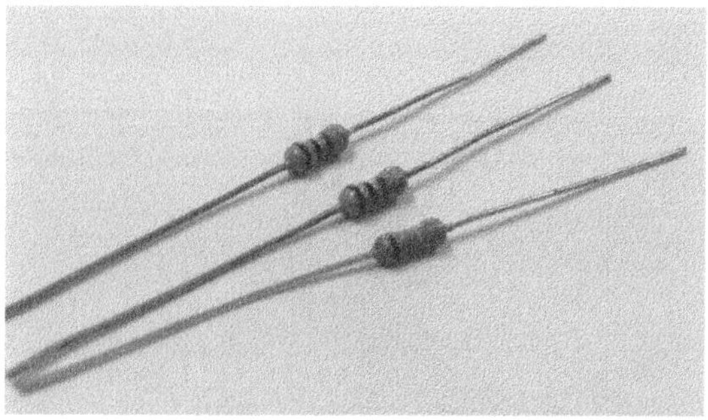

Figure 11.1 Resistors.

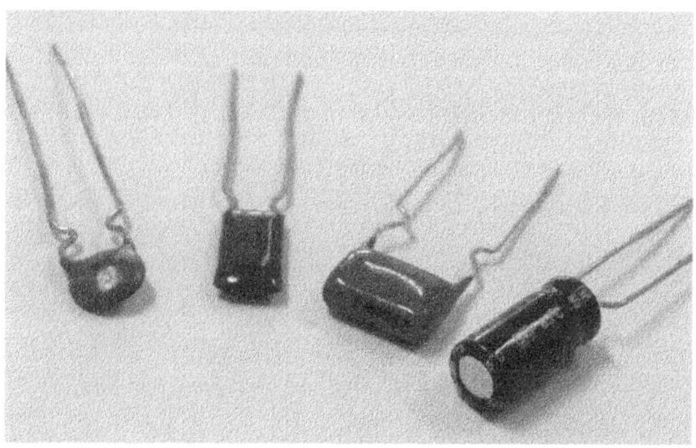

Figure 11.2 Capacitors.

Introduction to Electrical Circuit Analysis, First Edition. Özgür Ergül.

Companion Website: www.wiley.com/go/ergul4412

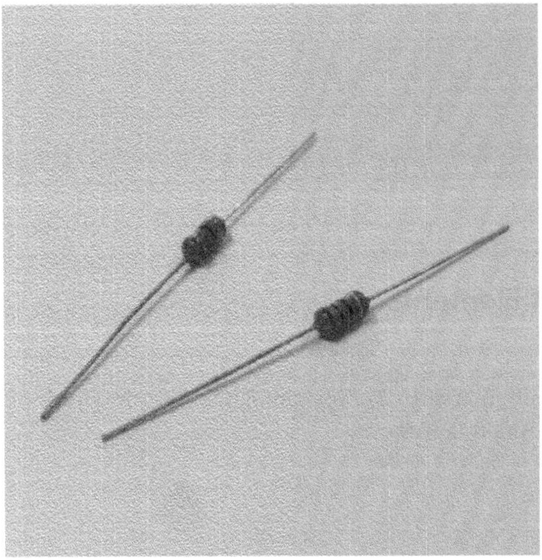

Figure 11.3 Inductors.

Figure 11.4 A transformer.

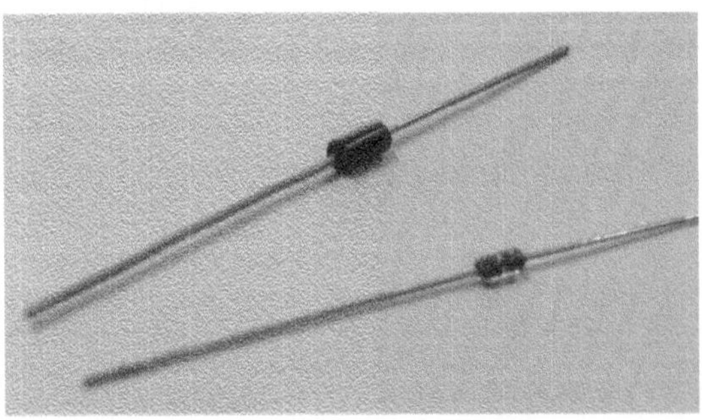

Figure 11.5 Diodes.

Figure 11.6 An OP-AMP.

Figure 11.7 A bipolar junction transistor.

Figure 11.8 A light-emitting diode.

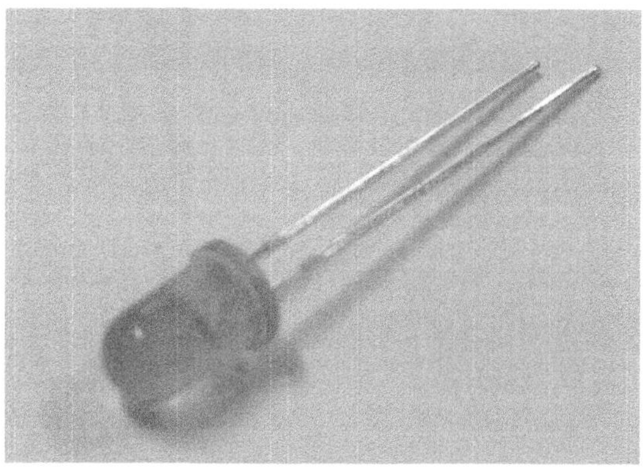

Figure 11.9 A digital multimeter.

Figure 11.10 A chip involving NAND gates.

A

Appendix

This appendix contains the brief mathematical background required to understand the topics covered by this book.

A.1 Basic Algebra Identities

- Difference of squares: $a^2 - b^2 = (a - b)(a + b)$
- Square of sum: $(a + b)^2 = a^2 + 2ab + b^2$
- Roots of quadratic equation: $ax^2 + bx + c = 0 \longrightarrow x = \dfrac{-b \pm \sqrt{b^2 - 4ac}}{2a}$
- Infinity: $\dfrac{a}{0} = \infty$ if $a \neq 0$ and $\dfrac{a}{\infty} = 0$ if $a \neq \infty$

A.2 Trigonometry

- Radian to degree conversion: π radians $= 180°$
- Sine to cosine conversion: $\sin x = \cos(\pi/2 - x) = \cos(x - \pi/2)$
- Cosine to sine conversion: $\cos x = \sin(\pi/2 - x) = -\sin(x - \pi/2)$
- Argument sum for sine: $\sin(x \pm y) = \sin x \cos y \pm \cos x \sin y$
- Argument sum for cosine: $\cos(x \pm y) = \cos x \cos y \mp \sin x \sin y$
- Half-argument formulas: $\sin(2x) = 2 \sin x \cos x$ and $\cos(2x) = 2\cos^2 x - 1$
- Product of cosines: $2 \cos x \cos y = \cos(x - y) + \cos(x + y)$
- Product of sines: $2 \sin x \sin y = \cos(x - y) - \cos(x + y)$
- Sum to product for sines: $\sin x + \sin y = 2 \sin\left(\dfrac{x + y}{2}\right) \cos\left(\dfrac{x - y}{2}\right)$
- Sum to product for cosines: $\cos x + \cos y = 2 \cos\left(\dfrac{x + y}{2}\right) \cos\left(\dfrac{x - y}{2}\right)$

A.3 Complex Numbers

- Standard form: $z = a + jb$, where $a = \text{Re}\{z\}$, $b = \text{Im}\{z\}$, and $j = \sqrt{-1}$
- Polar form: $z = |z|e^{j\phi} = |z| \exp(j\phi) = |z|\underline{/\phi}$, where $|z| = \sqrt{a^2 + b^2}$ and $\phi = \tan^{-1}(b/a)$

Introduction to Electrical Circuit Analysis, First Edition. Özgür Ergül.

Companion Website: www.wiley.com/go/ergul4412

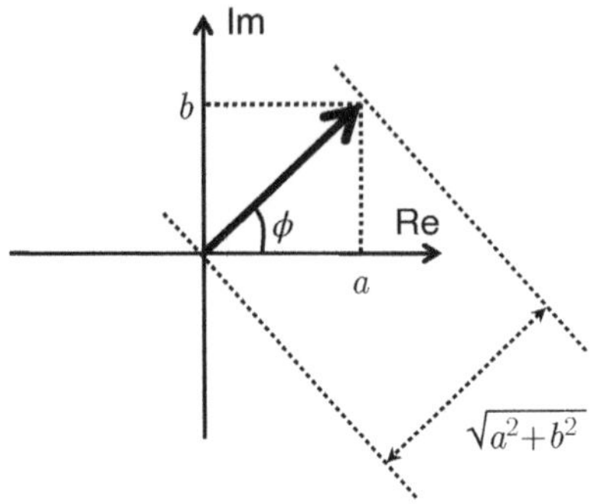

- Complex conjugate: $z^* = a - jb$ and $(z^*)^* = a + jb = z$
- Sum: $z_1 + z_2 = (a_1 + jb_1) + (a_2 + jb_2) = (a_1 + a_2) + j(b_1 + b_2)$
- Multiplication: $z_1 z_2 = (a_1 + jb_1)(a_2 + jb_2) = (a_1 a_2 - b_1 b_2) + j(a_1 b_2 + b_1 a_2)$
- Multiplication (polar form): $z_1 z_2 = |z_1 z_2| \exp[j(\phi_1 + \phi_2)] = |z_1 z_2| \underline{/(\phi_1 + \phi_2)}$
- Identity for amplitude: $|z|^2 = a^2 + b^2 = (a + jb)(a - jb) = zz^*$
- Powers: $z^n = |z|^n \underline{/n\phi}$ and $j^n = |j|^n \underline{/n\pi/2} = \cos(n\pi/2) + j\sin(n\pi/2)$
- Euler's identity: $e^{j\phi} = \cos\phi + j\sin\phi$ and $e^{-j\phi} = \cos\phi - j\sin\phi$
- Real part: $\mathrm{Re}\{e^{j\phi}\} = \cos\phi = (e^{j\phi} + e^{-j\phi})/2$
- Imaginary part: $\mathrm{Im}\{e^{j\phi}\} = \sin\phi = (e^{j\phi} - e^{-j\phi})/(2j)$
- Negative phases: $\exp(-j\phi) = \cos(-\phi) + j\sin(-\phi) = \cos\phi - j\sin\phi$
- Identities for π phases: $\exp(j\pi) = e^{j\pi} = -1$ and $\exp(-j\pi) = e^{-j\pi} = -1$
- $\pi/2$ phases: $\exp(j\pi/2) = e^{j\pi/2} = j$ and $\exp(-j\pi/2) = e^{-j\pi/2} = -j$

B

Solutions to Exercises

This chapter presents the solutions of exercises considered in this book. In most cases, the solution steps are provided as recipes.

Chapter 1

Solution to Exercise 1:

10 hours

Chapter 2

Solution to Exercise 2:

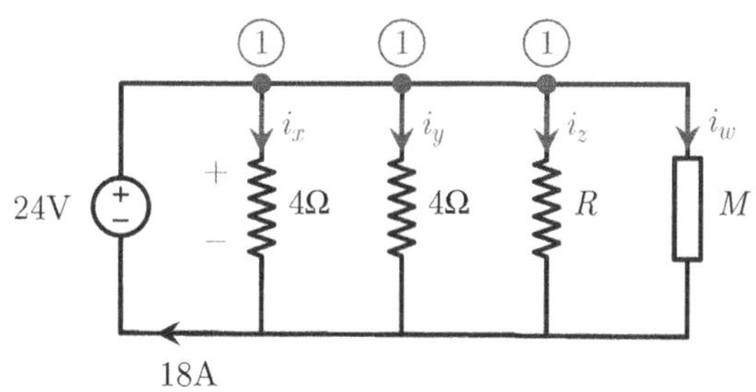

- KCL(1): $18 - i_x - i_y - i_z - i_w = 0 \longrightarrow 6 - 24/R - i_w = 0 \longrightarrow i_w = 6 - 24/R$
- Then $4 \leq 6 - 24/R \leq 6 \longrightarrow 24/R \leq 2 \longrightarrow R \geq 12$

Introduction to Electrical Circuit Analysis, First Edition. Özgür Ergül.

Companion Website: www.wiley.com/go/ergul4412

Solution to Exercise 3:

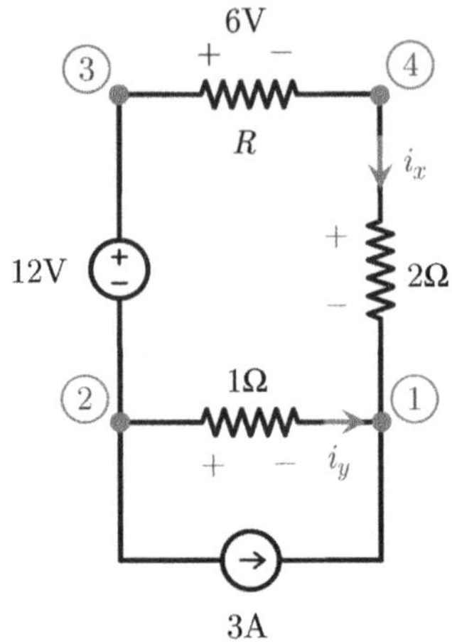

- KVL($1 \rightarrow 2 \rightarrow 3 \rightarrow 4 \rightarrow 1$): $-1i_y - 12 + 6 + 2i_x = 0 \longrightarrow 2i_x - i_y = 6$
- KCL(a): $i_x + i_y + 3 = 0 \longrightarrow i_x + i_y = -3$
- Then $3i_x = 3 \longrightarrow i_x = 1\,\text{A} \longrightarrow R = 6/1 = 6\,\Omega$

Solution to Exercise 4:

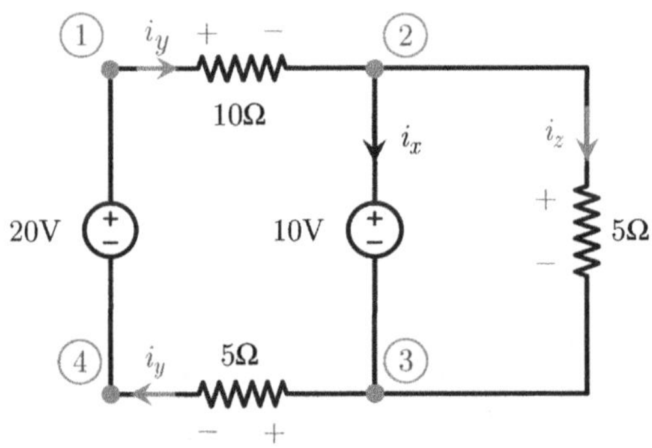

- KVL($1 \rightarrow 2 \rightarrow 3 \rightarrow 4 \rightarrow 1$): $10i_y + 10 + 5i_y - 20 = 0 \longrightarrow i_y = 10/15 = 2/3\,\text{A}$
- KVL($2 \rightarrow 3 \rightarrow 2$): $5i_z - 10 = 0 \longrightarrow i_z = 2\,\text{A}$
- KCL(2): $i_y - i_x - i_z = 0 \longrightarrow i_x = i_y - i_z = -4/3\,\text{A}$

Solution to Exercise 5:

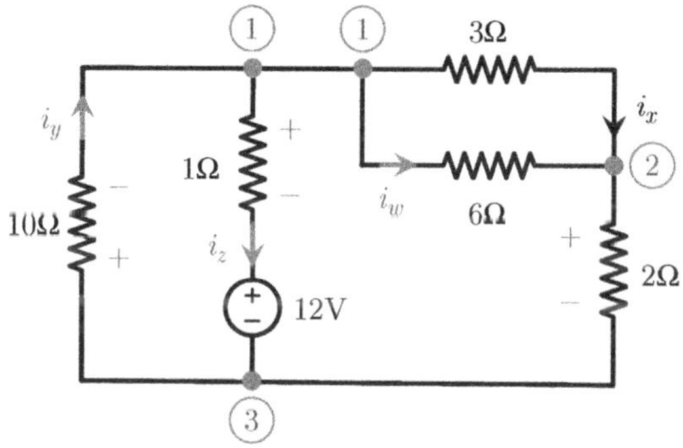

- KVL(1 → 2 → 1): $3i_x - 6i_w = 0 \longrightarrow i_x = 2i_w$
- KVL(1 → 3 → 1): $1i_z + 12 + 10i_y = 0 \longrightarrow i_z + 10i_y = -12$
- KVL(1 → 2 → 3 → 1): $6i_w + 2(i_x + i_w) - 12 - 1i_z = 0 \longrightarrow -i_z + 6i_x = 12$
- KCL (1): $i_y - i_z - (i_w + i_x) = 0 \longrightarrow i_y - i_z - (3/2)i_x = 0$
- Then $i_x = 40/27$ A

Solution to Exercise 6:

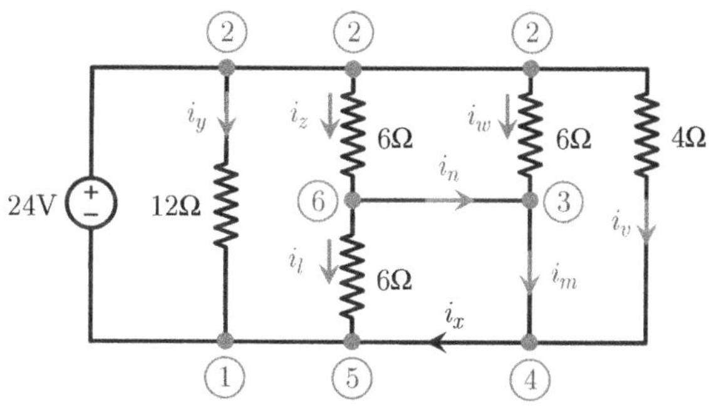

- KVLs(2 → ⋯ → 2): $i_y = 2$ A, $i_z = i_w = 4$ A, $i_v = 6$ A
- KVL(6 → 3 → 4 → 5 → 6): $i_l = 0$
- KCL(6): $i_z - i_n - i_l = 0 \longrightarrow i_n = 4$ A
- KCL(3): $i_w + i_n - i_m = 0 \longrightarrow i_m = 8$ A
- KCL(4): $i_v + i_m - i_x = 0 \longrightarrow i_x = 14$ A

Solution to Exercise 7:

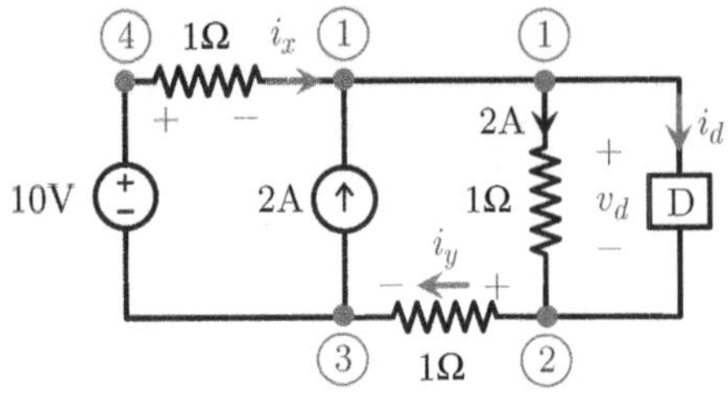

- $v_d = 2\,\text{V}$
- KVL($4 \to 1 \to 2 \to 3 \to 4$): $i_x + 2 + i_y - 10 = 0 \longrightarrow i_x + i_y = 8$
- KCL(2): $2 + i_d - i_y = 0 \longrightarrow i_d = i_y - 2$
- KCL(1): $i_x + 2 - 2 - i_d = 0 \longrightarrow i_x = i_d = i_y - 2$
- Then $i_y = 5\,\text{A}$, $i_x = 3\,\text{A}$, and $i_d = 3\,\text{A}$
- And $p_d = 6\,\text{W}$

Solution to Exercise 8:

- $R_{\text{eq}} = 3 + 10 \parallel [4 + 6 \parallel (1+2) + 20 \parallel 5] = 3 + 10 \parallel [4 + 2 + 4] = 8\,\Omega$
- $i_x = 12/8 = 3/2\,\text{A}$

Solution to Exercise 9:

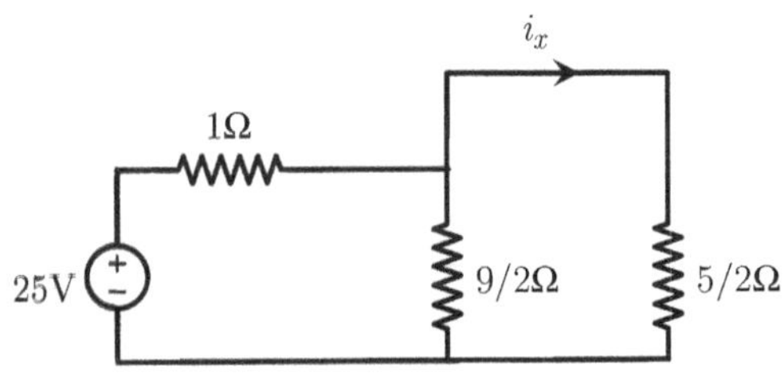

- $Z_{\text{in}} = 1 + (9/2) \parallel (5/2) = 1 + 45/28 = 73/28\,\Omega$
- $i_{\text{in}} = 25/Z_{\text{in}} = 25(28/73)\,\text{A}$
- Then $i_x = (9/2)/(9/2 + 5/2)i_{\text{in}} = (9/14)i_{\text{in}} = 450/73\,\text{A}$

Solution to Exercise 10:

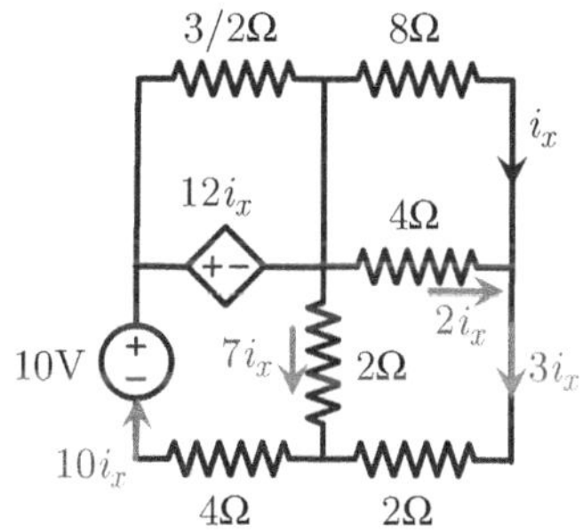

- KVL: $-10 + 12i_x + 8i_x + 6i_x + 40i_x = 0 \longrightarrow i_x = 5/33\,\text{A}$

Chapter 3

Solution to Exercise 11:

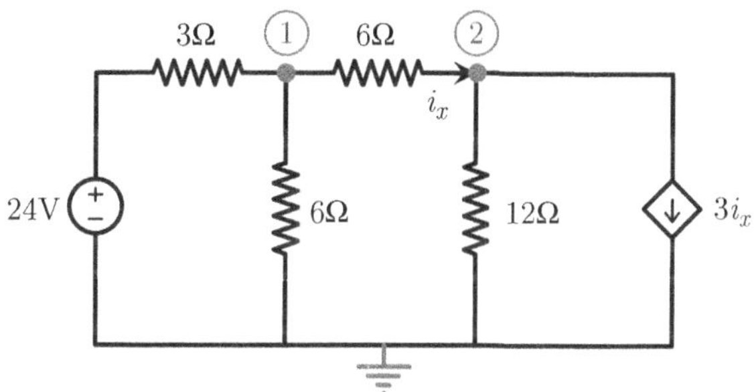

- $i_x = (v_1 - v_2)/6$
- KCL(1): $(24 - v_1)/3 - v_1/6 - (v_1 - v_2)/6 = 0 \longrightarrow 4v_1 - v_2 = 48$
- KCL(2): $(v_1 - v_2)/6 - v_2/12 - 3i_x = 0 \longrightarrow v_2 = 4v_1/3$
- Then $v_1 = 18\,\text{V}$ and $v_2 = 24\,\text{V} \longrightarrow i_x = -1\,\text{A}$

Solution to Exercise 12:

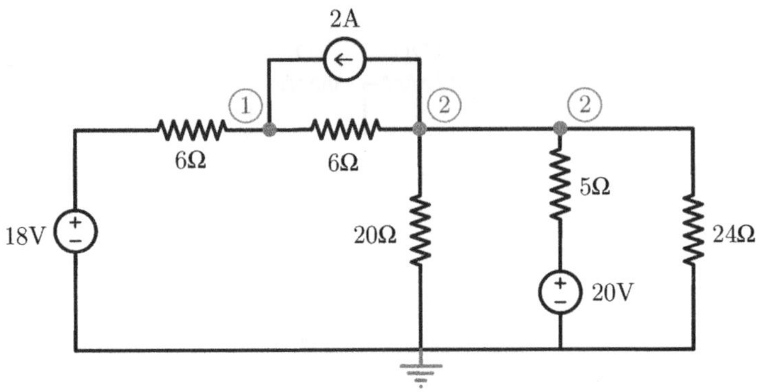

- KCL(1): $(18 - v_1)/6 + 2 + (v_2 - v_1)/6 = 0 \longrightarrow 2v_1 - v_2 = 30$
- KCL(2): $(v_1 - v_2)/6 - 2 - v_2/20 - (v_2 - 20)/5 - v_2/24 = 0 \longrightarrow 4v_1 - 11v_2 = -48$
- Then $v_1 = 21\,\text{V}$ and $v_2 = 12\,\text{V}$
- $p_{24\Omega} = 12^2/24 = 6\,\text{W}$

Solution to Exercise 13:

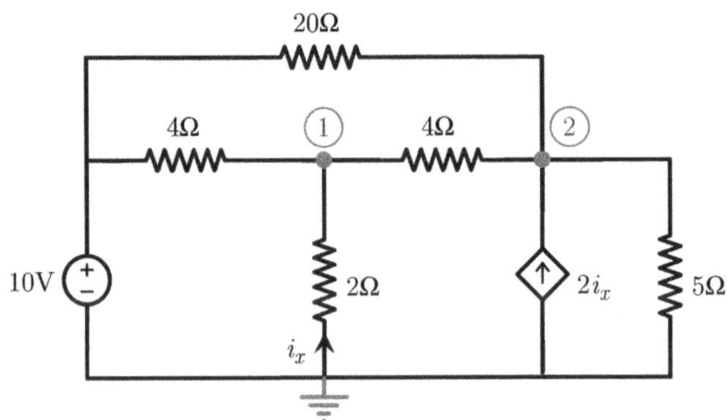

- $i_x = -v_1/2$
- KCL(1): $(10 - v_1)/4 - v_1/2 - (v_1 - v_2)/4 = 0 \longrightarrow 4v_1 - v_2 = 10$
- KCL(2): $(v_1 - v_2)/4 + (10 - v_2)/20 + 2i_x - v_2/5 = 0 \longrightarrow 3v_1 + 2v_2 = 2$
- Then $v_1 = 2\,\text{V}$ and $v_2 = -2\,\text{V}$
- $p_{5\Omega} = 4/5\,\text{W}$

Solution to Exercise 14:

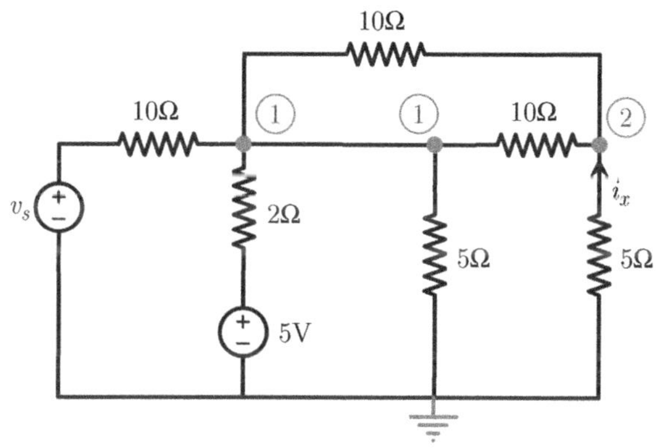

- $v_2 = -5i_x = -10\,\text{V}$
- KCL(2): $(v_1 - v_2)/10 + (v_1 - v_2)/10 + i_x = 0 \longrightarrow v_1/5 + 4 = 0 \longrightarrow v_1 = -20\,\text{V}$
- KCL(1): $(v_s - v_1)/10 + (5 - v_1)/2 - v_1/5 - 2(v_1 - v_2)/10 = 0 \longrightarrow v_s/10 = -41/2$
- Then $v_s = -205\,\text{V}$

Solution to Exercise 15:

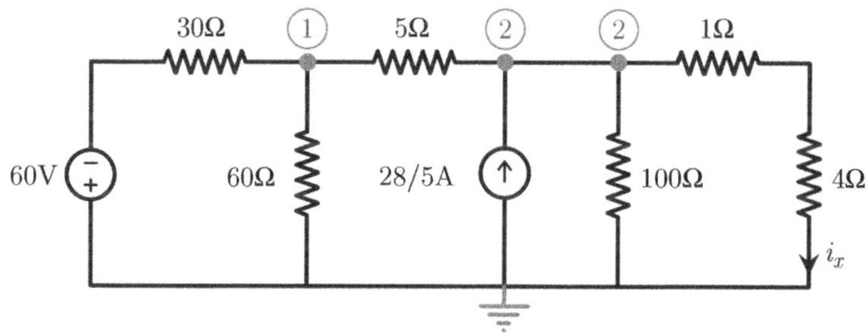

- KCL(1): $(-60 - v_1)/30 - v_1/60 - (v_1 - v_2)/5 = 0 \longrightarrow 5v_1 - 4v_2 = -40$
- KCL(2): $(v_1 - v_2)/5 + 28/5 - v_2/100 - v_2/5 = 0 \longrightarrow 20v_1 - 41v_2 = -560$
- Then $v_1 = 24/5\,\text{V}$, $v_2 = 16\,\text{V}$, and $i_x = 16/5\,\text{A}$

Solution to Exercise 16:

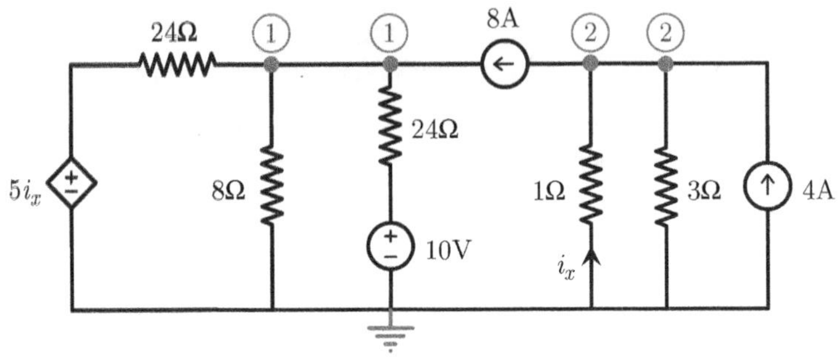

- $i_x = -v_2$
- KCL(1): $(5i_x - v_1)/24 - v_1/8 - (v_1 - 10)/24 + 8 = 0 \longrightarrow 5v_1 + 5v_2 = 202$
- KCL(2): $-8 - v_2/1 - v_2/3 + 4 = 0 \longrightarrow v_2 = -3\,\text{V}$
- Then $v_1 = 217/5\,\text{V}$
- And $p_{8A} = 8(-3 - 217/5) = -1856/5\,\text{W}$

Solution to Exercise 17:

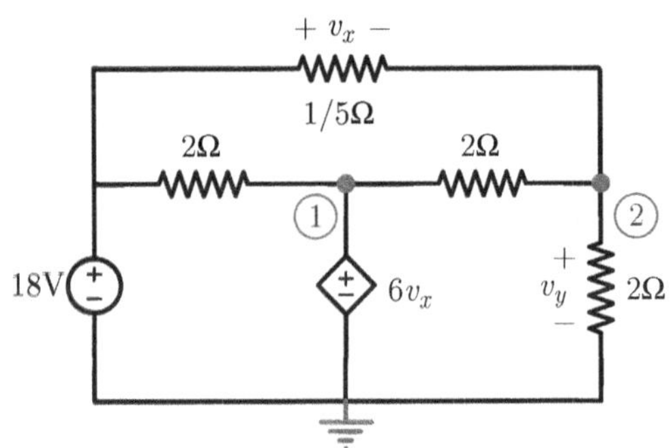

- $v_x = 18 - v_2$ and $v_1 = 108 - 6v_2$
- KCL(2): $(v_1 - v_2)/2 + (18 - v_2)/(1/5) - v_2/2 = 0 \longrightarrow v_2 = 16\,\text{V}$
- Then $v_y = v_2 = 16\,\text{V}$

Solution to Exercise 18:

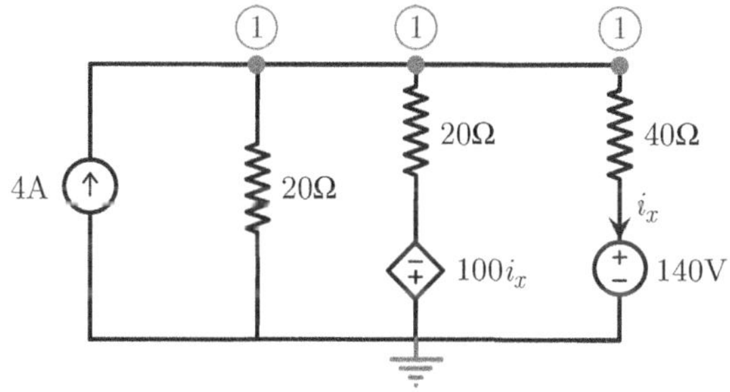

- $i_x = (v_1 - 140)/40$
- KCL(1): $4 - v_1/20 - (v_1 + 100i_x)/20 - (v_1 - 140)/40 = 0 \longrightarrow v_1 = 100\,\text{V}$
- Then $i_x = -1\,\text{A}$

Solution to Exercise 19:

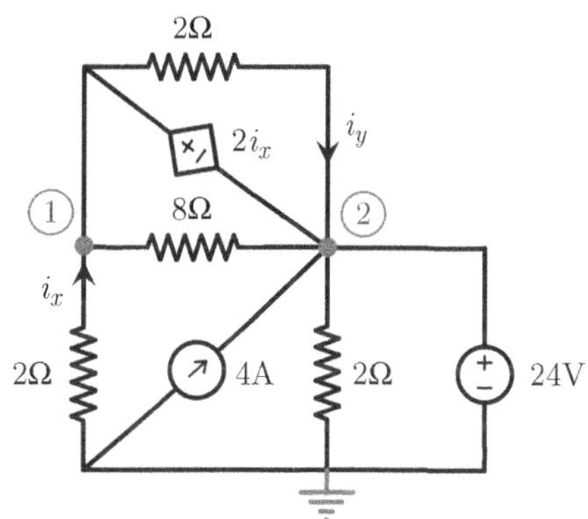

- $v_2 = 24\,\text{V}$
- $v_1 = v_2 + 2i_x = 24 + 2i_x$ and $v_1 = -2i_x$
- Then $-2i_x = 24 + 2i_x \longrightarrow i_x = -6\,\text{A}$

Solution to Exercise 20:

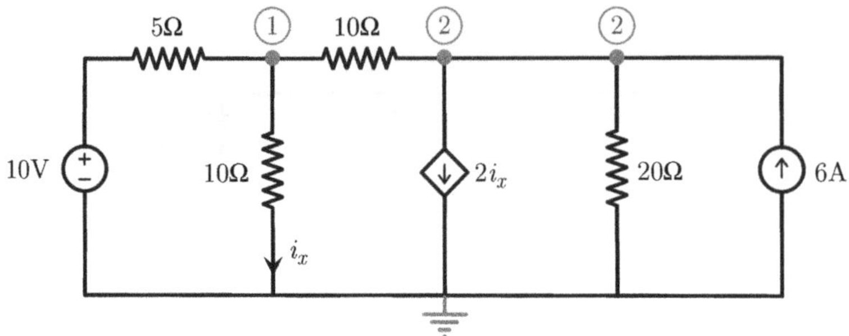

- $i_x = v_1/10$
- KCL(1): $(10 - v_1)/5 - v_1/10 - (v_1 - v_2)/10 = 0 \longrightarrow 4v_1 - v_2 = 20$
- KCL(2): $(v_1 - v_2)/10 - 2i_x - v_2/20 + 6 = 0 \longrightarrow 2v_1 + 3v_2 = 120$
- Then $v_1 = 90/7$ V, $v_2 = 220/7$ V, and $i_x = 9/7$ A

Solution to Exercise 21:

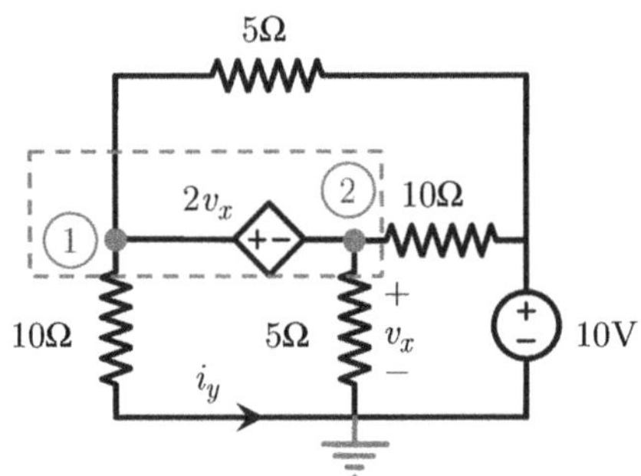

- $v_x = v_2$
- KCL(1&2): $-v_1/10 - v_2/5 + (10 - v_1)/5 + (10 - v_2)/10 = 0 \longrightarrow v_1 + v_2 = 10$
- Supernode: $v_1 - v_2 = 2v_x \longrightarrow v_1 = 3v_2 \longrightarrow v_1 = 15/2$ V
- Then $i_y = v_1/10 = 3/4$ A

Solution to Exercise 22:

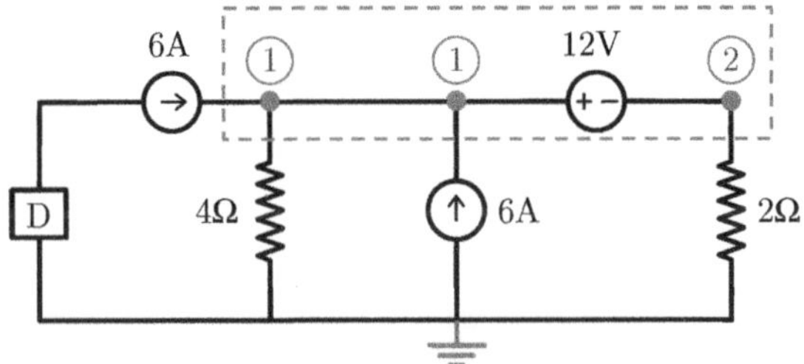

- Supernode: $v_1 = v_2 + 12$
- KCL(1&2): $6 - v_1/4 + 6 - v_2/2 = 0 \longrightarrow v_2 = 12\,\text{V}$
- Then $p_{2\Omega} = 144/2 = 72\,\text{W}$

Solution to Exercise 23:

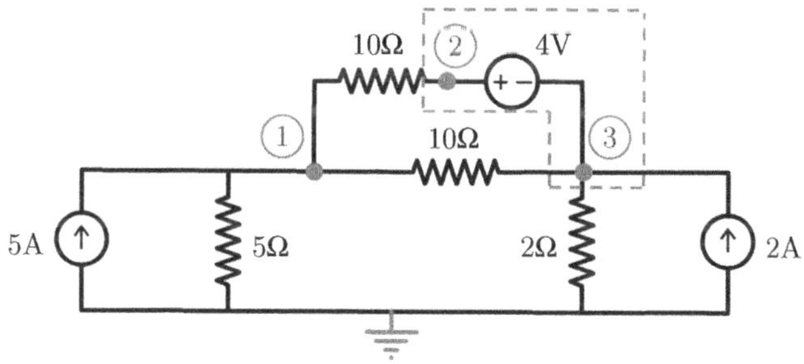

- KCL(1): $5 - v_1/5 - (v_1 - v_3)/10 - (v_1 - v_2)/10 = 0 \longrightarrow 4v_1 - v_2 - v_3 = 50$
- KCL(2&3): $(v_1 - v_2)/10 + (v_1 - v_3)/10 - v_3/2 + 2 = 0 \longrightarrow 2v_1 - v_2 - 6v_3 = -20$
- Supernode: $v_2 - v_3 = 4$
- Then $2v_1 - v_3 = 27$ and $2v_1 - 7v_3 = -16$
- And $v_3 = 43/6\,\text{V}$, $v_1 = 205/12\,\text{V}$, $v_2 = 67/6\,\text{V}$
- And $p_{10\Omega} = [(v_1 - v_2)^2]/10 = (71/12)^2/10\,\text{W}$

Solution to Exercise 24:

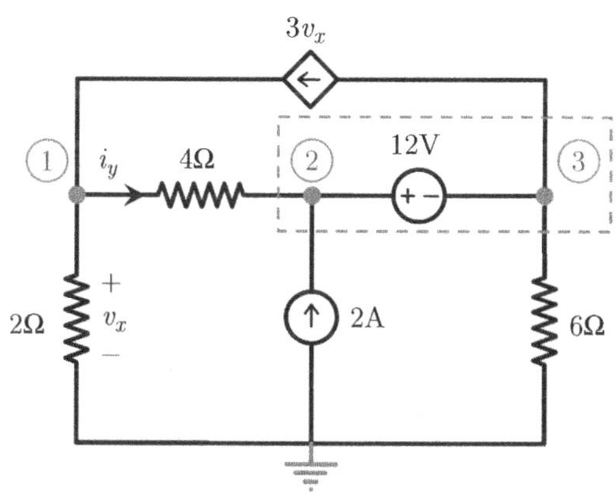

- $v_x = v_1$
- Supernode: $v_3 = v_2 - 12$
- KCL(1): $3v_x - v_1/2 - (v_1 - v_2)/4 = 0 \longrightarrow 9v_1 + v_2 = 0$
- KCL(2&3): $(v_1 - v_2)/4 + 2 - 3v_x - v_3/6 = 0 \longrightarrow 66v_1 + 10v_2 = 96$
- Then $v_1 = -4\,\text{V}$, $v_2 = 36\,\text{V}$, and $i_y = -10\,\text{A}$

Solution to Exercise 25:

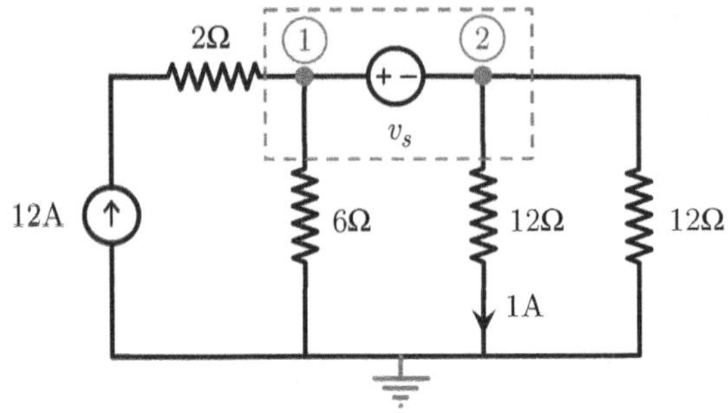

- KCL(1&2): $12 - v_1/6 - v_2/12 - v_2/12 = 0 \longrightarrow v_1 + v_2 = 72$
- Resistor: $v_2 = 12\,\text{V} \longrightarrow v_1 = 60\,\text{V}$
- Then $v_s = v_1 - v_2 = 48\,\text{V}$

Solution to Exercise 26:

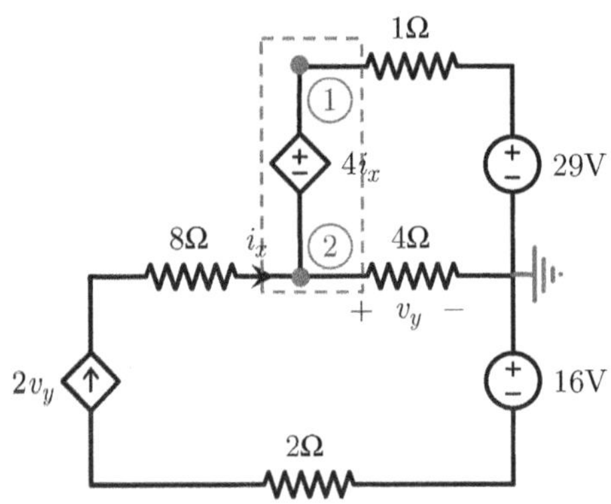

- $i_x = 2v_y = 2v_2$
- KCL(1&2): $i_x - v_2/4 - (v_1 - 29)/1 = 0 \longrightarrow 4v_1 - 7v_2 = 116$
- Supernode: $v_1 - v_2 = 4i_x = 8v_2 \longrightarrow v_1 = 9v_2$
- Then $v_2 = 4\,\text{V}$, $v_1 = 36\,\text{V}$, $i_x = 8\,\text{A}$
- And $p_s = 4i_x \times (29 - v_1)/1 = -224\,\text{W}$

Solution to Exercise 27:

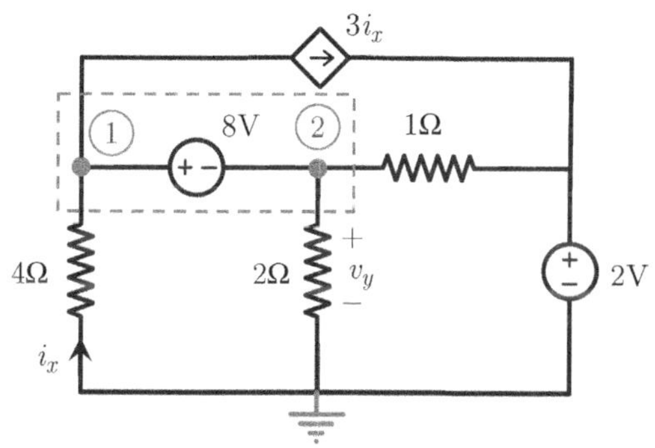

- $i_x = -v_1/4$
- Supernode: $v_1 - v_2 = 8$
- KCL(1&2): $-v_1/4 - v_2/2 - (v_2 - 2)/1 - 3i_x = 0 \longrightarrow v_1 - 3v_2 = -4$
- Then $v_1 = 14\,\text{V}$, $v_2 = 6\,\text{V}$, and $v_y = 6\,\text{V}$

Solution to Exercise 28:

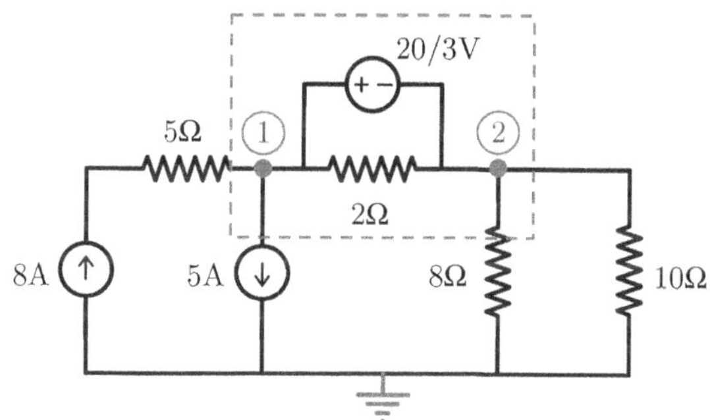

- Supernode: $v_1 - v_2 = 20/3$
- KCL(1&2): $8 - 5 - v_2/8 - v_2/10 = 0 \longrightarrow v_2 = 40/3\,\text{V}$
- Then $v_1 = 20\,\text{V}$ and $p_{5\text{A}} = 20 \times 5 = 100\,\text{W}$

Solution to Exercise 29:

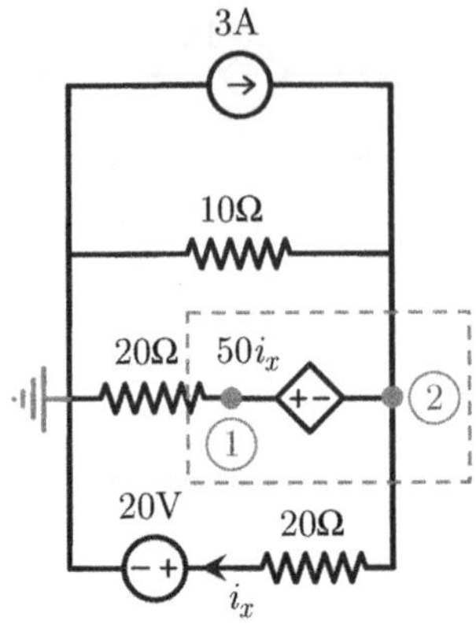

- KCL(1&2): $-v_1/20 + 3 - v_1/10 - (v_2 - 20)/20 = 0 \longrightarrow 3v_1 + v_2 = 80$
- Supernode: $v_1 - v_2 = 50i_x = 50(v_2 - 20) \longrightarrow 51v_2 - v_1 = 1000$
- Then $v_1 = 20$ V, $v_2 = 20$ V, and $i_x = 0$ A

Solution to Exercise 30:

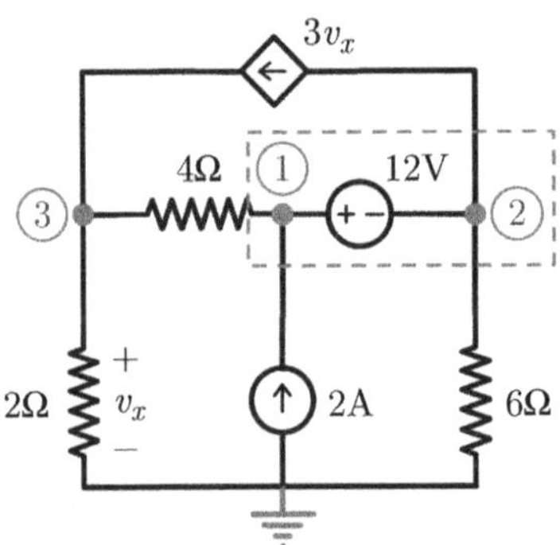

- $v_3 = v_x$
- KCL(3): $3v_x + (v_1 - v_3)/4 - v_3/2 = 0 \longrightarrow v_1 + 9v_3 = 0$
- KCL(1&2): $-(v_1 - v_3)/4 + 2 - v_2/6 - 3v_x = 0 \longrightarrow 3v_1 + 2v_2 + 33v_3 = 24$
- Supernode: $v_1 - v_2 = 12$
- Then $v_1 = 36$ V, $v_2 = 24$ V, and $v_3 = -4$ V
- And $v_x = v_3 = -4$ V

Solution to Exercise 31:

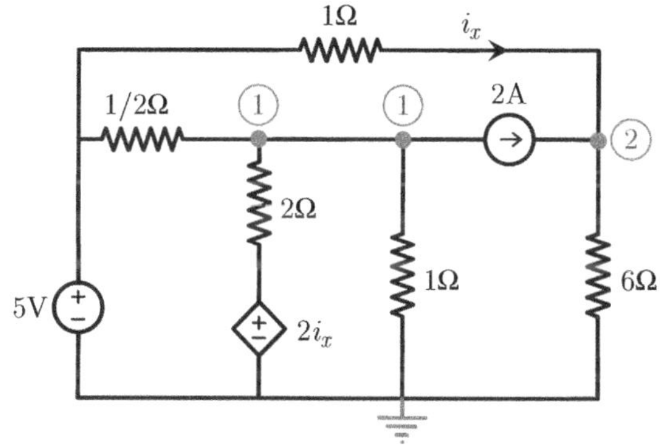

- KCL(1): $(5 - v_1)/(1/2) - (v_1 - 2i_x)/2 - v_1/1 - 2 = 0 \longrightarrow 7v_1 - 2i_x = 16$
- KCL(2): $(5 - v_2)/1 + 2 - v_2/6 = 0 \longrightarrow v_2 = 6\,\text{V}$ and $i_x = (5 - v_2)/1 = -1\,\text{A}$
- Then $v_1 = 2\,\text{V}$ and $i_{2\Omega} = (v_1 - 2i_x)/2 = 2\,\text{A}$
- And $p_s = 2i_x \times 2 = -4\,\text{W}$

Solution to Exercise 32:

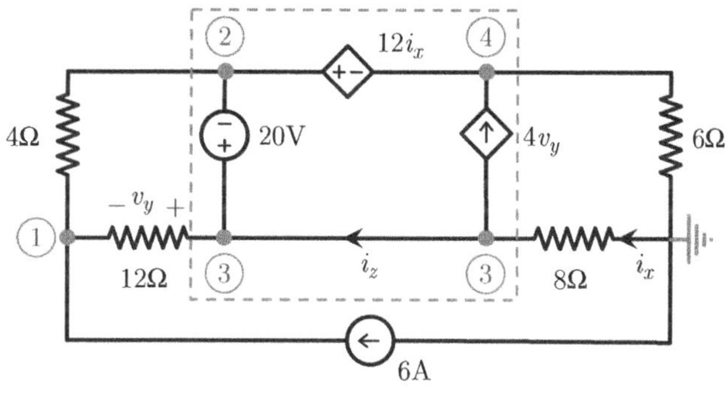

- $v_3 = v_2 + 20$
- KCL(1): $6 - (v_1 - v_2)/4 - (v_1 - v_3)/12 = 0 \longrightarrow v_1 = v_2 + 23$
- KCL(2&3&4): $(v_1 - v_2)/4 + (v_1 - v_3)/12 - v_4/6 - v_3/8 = 0 \longrightarrow 4v_4 + 3v_3 = 144$
- Supernode: $v_3 = v_4 + 20 + 12i_x = v_4 + 20 - 3v_3/2 \longrightarrow 5v_3 - 2v_4 = 40$
- Then $v_1 = 263/13\,\text{V}$, $v_2 = -36/13\,\text{V}$, $v_3 = 224/13\,\text{V}$, $v_4 = 300/13\,\text{V}$
- And $i_z = i_x - 4v_y = -28/13 + 12 = 128/13\,\text{A}$

Solution to Exercise 33:

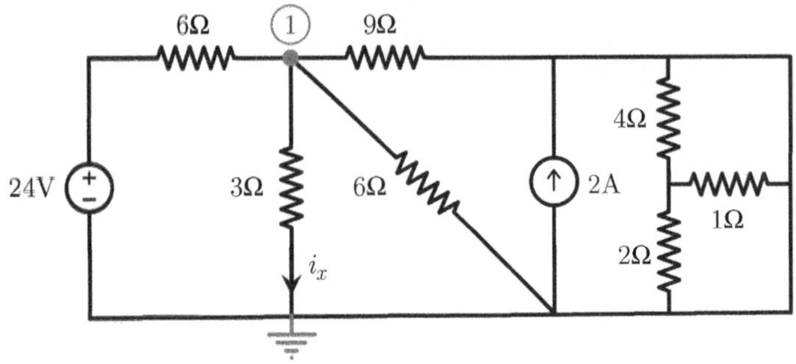

- $i_x = v_1/6$
- KCL(1): $(24 - v_1)/6 - v_1/3 - v_1/6 - v_1/9 = 0 \longrightarrow v_1 = 36/7\,\text{V}$
- Then $i_x = 12/7\,\text{A}$

Solution to Exercise 34:

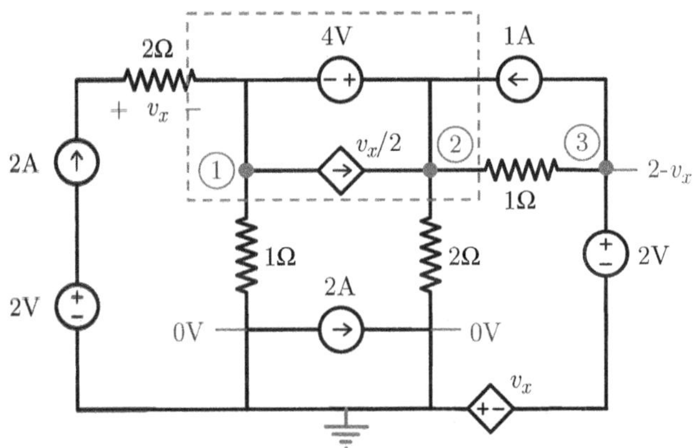

- $v_x = 4\,\text{V}$
- KCL(1&2): $2 - v_1/1 - v_2/2 + (2 - v_x - v_2)/1 + 1 = 0 \longrightarrow 2v_1 + 3v_2 = 2$
- Supernode: $v_2 - v_1 = 4$
- Then $v_1 = -2\,\text{V}$, $v_2 = 2\,\text{V}$, and $v_3 = -2\,\text{V}$
- KCL(3): $-1 - (v_3 - v_2)/1 + i_s = 0 \longrightarrow i_s = -3\,\text{A}$
- And $p_s = v_x \times i_s = 4 \times (-3) = -12\,\text{W}$

Solution to Exercise 35:

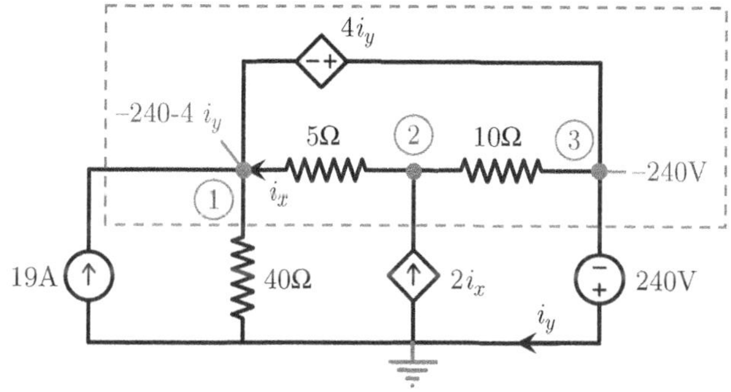

- $v_1 = -240 - 4i_y$ and $i_x = (v_2 - v_1)/5 = v_2/5 + 48 + 4/5 i_y$
- KCL(2): $-i_x + 2i_x + (-240 - v_2)/10 = 0 \longrightarrow 10i_x - v_2 = 240 \longrightarrow v_2 + 8i_y = -240$
- KCL(1&2&3): $19 - v_1/40 + 2i_x - i_y = 0 \longrightarrow 7i_y + 4v_2 = -1210$
- Then $i_y = 10$ A, $v_2 = -320$ V, $i_x = -8$ A, and $v_1 = -280$ V
- KCL(3): $-i_y + (v_2 - v_3)/10 - i_s = 0 \longrightarrow i_s = -18$ A
- And $p_s = 4i_y \times i_s = 4 \times (-3) = -720$ W

Solution to Exercise 36:

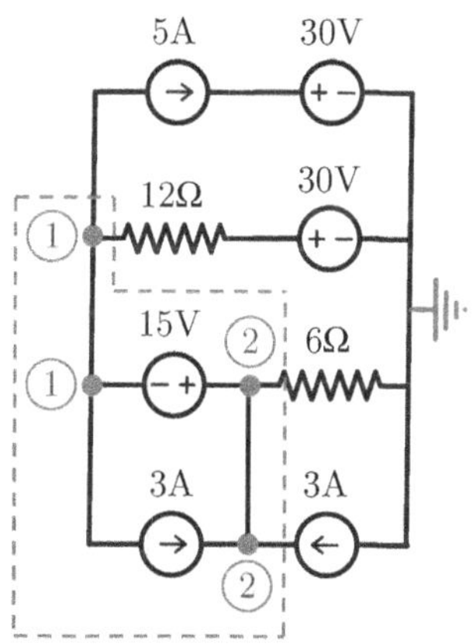

- KCL(1&2): $-5 - (v_1 - 30)/12 - v_2/6 + 3 = 0 \longrightarrow v_1 + 2v_2 = -6$
- Supernode: $v_2 - v_1 = 15$
- Then $v_1 = -8$ V and $v_2 = 7$ V
- And $p_{5A} = (v_1 - 30) \times 5 = -190$ W

Solution to Exercise 37:

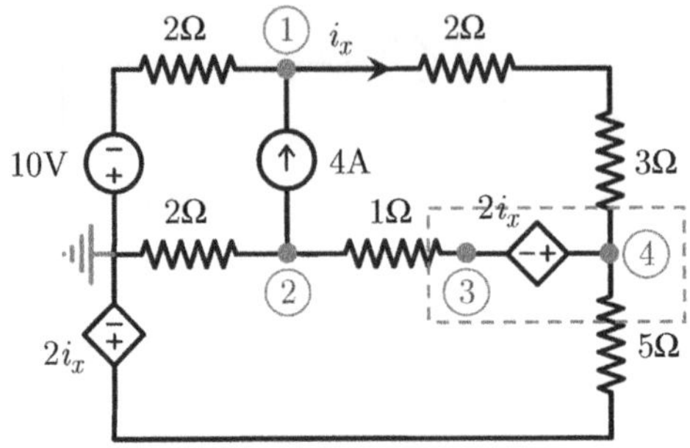

- $i_x = (v_1 - v_4)/5$
- Supernode: $v_4 - v_3 = 2i_x \longrightarrow 2v_1 - 7v_4 + 5v_3 = 0$
- KCL(1): $(-10 - v_1)/2 + 4 - (v_1 - v_4)/5 = 0 \longrightarrow 7v_1 - 2v_4 = -10$
- KCL(2): $-v_2/2 - 4 + (v_3 - v_2)/1 = 0 \longrightarrow 3v_2 - 2v_3 = -8$
- KCL(3&4): $(v_1 - v_4)/5 - (v_3 - v_2)/1 - (v_4 - 2i_x)/5 = 0 \longrightarrow 7v_1 + 25v_2 - 25v_3 - 12v_4 = 0$
- Then $v_1 = -74/29$ V, $v_2 = -164/29$ V, $v_3 = -130/29$ V
- And $v_4 = -114/29$ V, $i_x = (v_1 - v_4)/5 = 8/29$ A

Solution to Exercise 38:

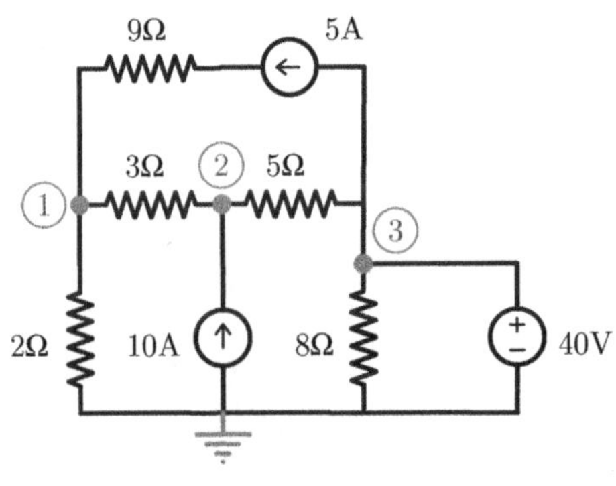

- KCL(1): $5 - v_1/2 + (v_2 - v_1)/3 = 0 \longrightarrow 5v_1 - 2v_2 = 30$
- KCL(2): $(v_1 - v_2)/3 + 10 + (40 - v_2)/5 = 0 \longrightarrow 5v_1 - 8v_2 = -270$
- Then $v_1 = 26$ V and $v_2 = 50$ V
- KCL(3): $-5 - (40 - v_2)/5 - 40/8 - i_{40V} = 0 \longrightarrow i_{40V} = -8$ A
- And $p_{40V} = 40 \times (-8) = -320$ W

Solution to Exercise 39:

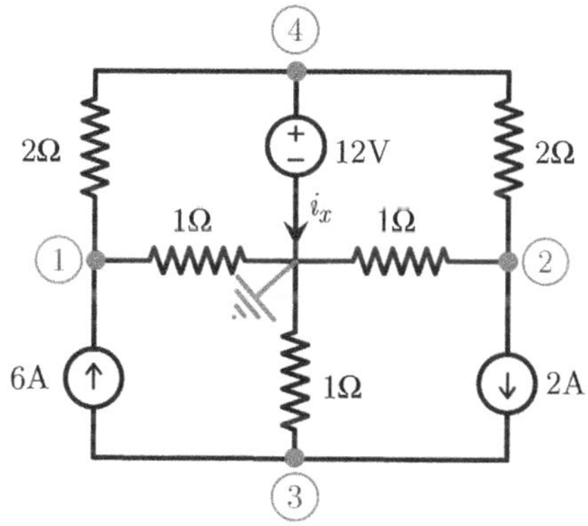

- $v_4 = 12\,\text{V}$
- KCL(1): $6 - (v_1 - v_4)/2 - v_1/1 = 0 \longrightarrow v_1 = 8\,\text{V}$
- KCL(2): $-2 + (v_4 - v_2)/2 - v_2/1 = 0 \longrightarrow v_2 = 8/3\,\text{V}$
- KCL(4): $(v_1 - v_4)/2 + (v_2 - v_4)/2 - i_x = 0 \longrightarrow i_x = -20/3\,\text{A}$
- Note that node 3 is not used!

Solution to Exercise 40:

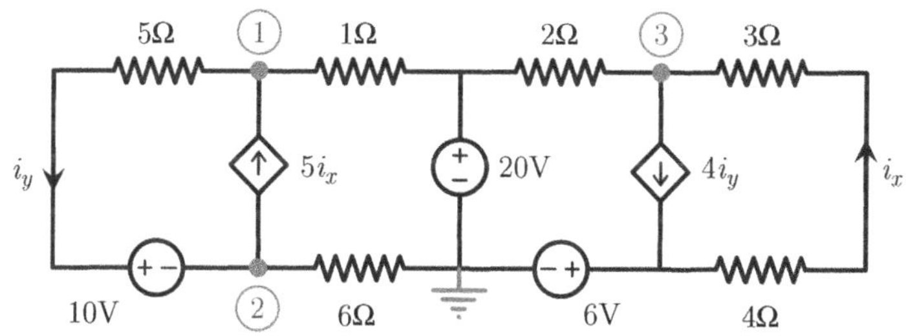

- $i_y = (v_1 - v_2 - 10)/5$
- $i_x = (6 - v_3)/7$
- KCL(1): $-(v_1 - v_2 - 10)/5 + 5i_x - (v_1 - 20) = 0 \longrightarrow 42v_1 - 7v_2 + 25v_3 = 920$
- KCL(2): $i_y - 5i_x - v_2/6 = 0 \longrightarrow 42v_1 - 77v_2 + 150v_3 = 1320$
- KCL(3): $-(v_3 - 20)/2 - 4i_y + (6 - v_3)/7 = 0 \longrightarrow 56v_1 - 56v_2 + 45v_3 = 1320$
- Then $v_1 = 870/43\,\text{V}$, $v_2 = -60/43\,\text{V}$, and $v_3 = 104/43\,\text{V}$
- And $i_x = 22/43\,\text{A}$

Chapter 4

Solution to Exercise 41:

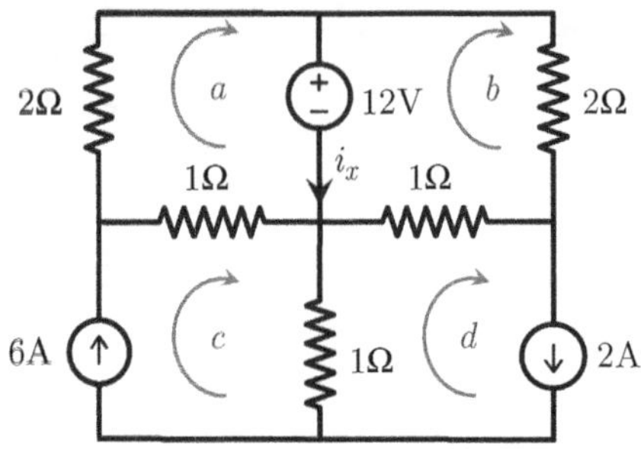

- $i_c = 6\,\text{A}$ and $i_d = 2\,\text{A}$
- KVL(a): $2i_a + 12 + 1(i_a - i_c) = 0 \longrightarrow i_a = -2\,\text{A}$
- KVL(b): $-12 + 2i_b + 1(i_b - i_d) = 0 \longrightarrow i_b = 14/3\,\text{A}$
- Then $i_x = i_a - i_b = -2 - 14/3 = -20/3\,\text{A}$

Solution to Exercise 42:

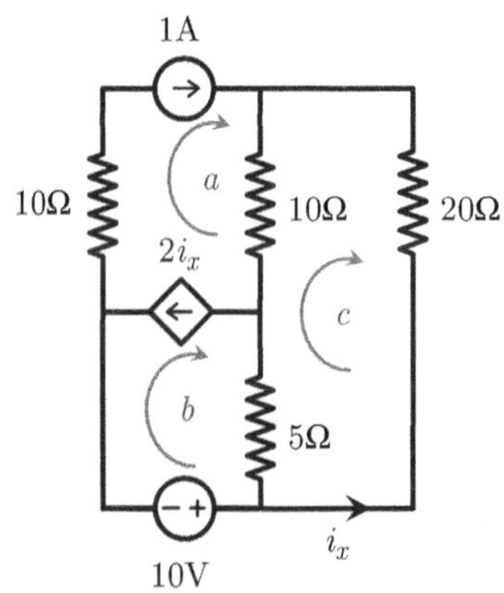

- $i_a = 1\,\text{A}$ and $i_a - i_b = 2i_x = -2i_c \longrightarrow i_b - 2i_c = 1$
- KVL(c): $5(i_c - i_b) + 10(i_c - i_a) + 20i_c = 0 \longrightarrow -i_b + 7i_c = 2$
- Then $i_c = 3/5\,\text{A}$ and $i_b = 11/5\,\text{A}$
- And $p_{10\text{V}} = 10 \times 11/5 = 22\,\text{W}$

Solution to Exercise 43:

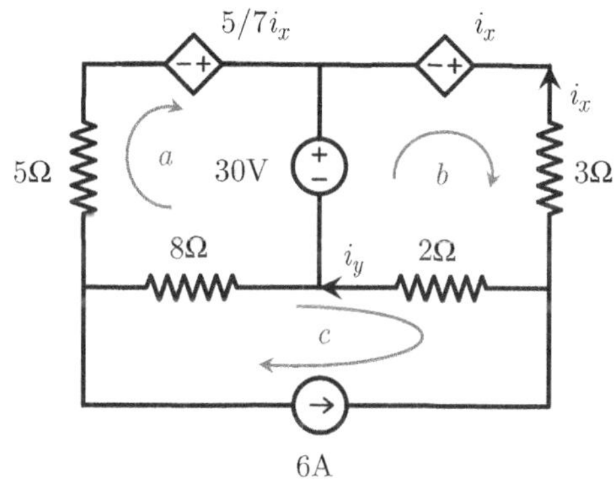

- $i_x = -i_b$
- Mesh(c): $i_c = -6$ A
- KVL(b): $-30 - i_x + 3i_b + 2(i_b - i_c) = 0 \longrightarrow 6i_b = 18 \longrightarrow i_b = 3$ A
- Then $i_y = i_b - i_c = 9$ A

Solution to Exercise 44:

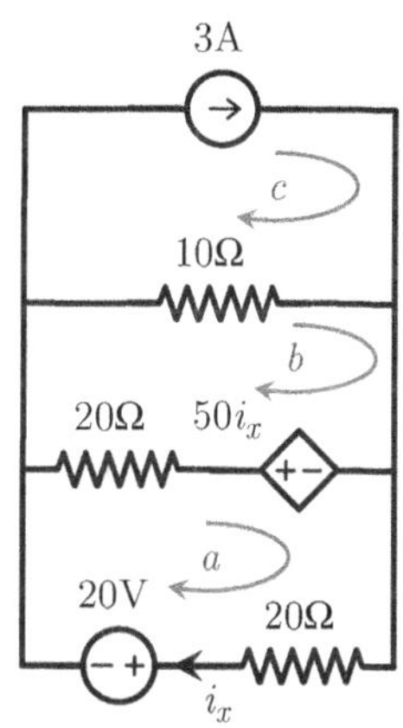

- $i_c = 3$ A and $i_x = i_a$
- KVL(a): $20(i_a - i_b) + 50i_x + 20i_a + 20 = 0 \longrightarrow 9i_a - 2i_b = -2$
- KVL(b): $10(i_b - i_c) - 50i_x - 20(i_a - i_b) = 0 \longrightarrow -7i_a + 3i_b = 3$
- Then $i_x = i_a = 0$ A

Solution to Exercise 45:

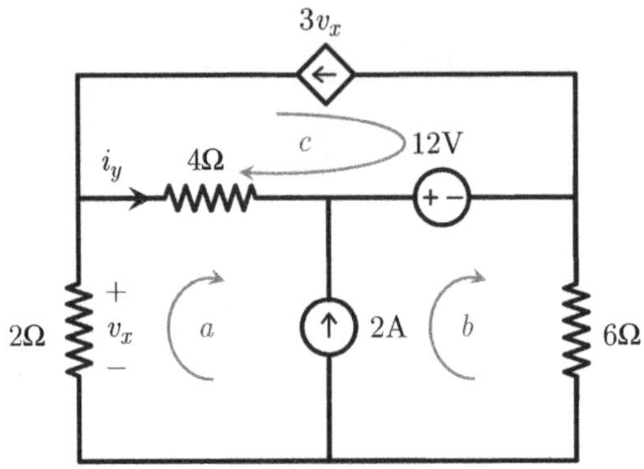

- $v_x = -2i_a$
- $i_b - i_a = 2\,\text{A}$
- KVL(a): $2i_a + 4(i_a - i_c) - v_{2A} = 0$
- KVL(b): $v_{2A} + 12 + 6i_b = 0$
- Then $2i_a + 4(i_a - i_c) + 12 + 6i_b = 0 \longrightarrow i_a = 2\,\text{A}$
- Then $i_b = 4\,\text{A}$, $i_c = 12\,\text{A}$, $i_y = -10\,\text{A}$

This question can be solved better using the concept of the supermesh.

Solution to Exercise 46:

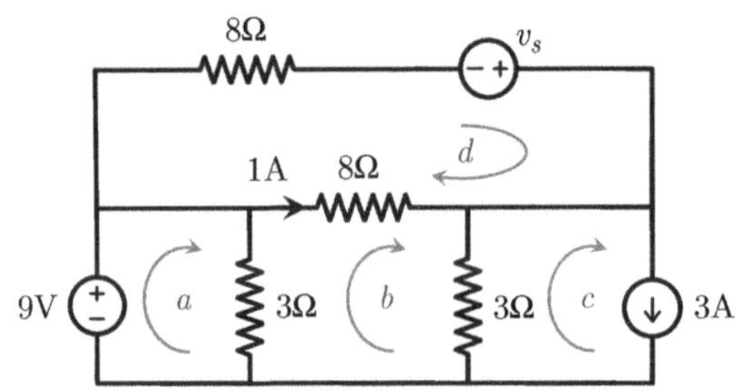

- $i_c = 3\,\text{A}$
- KVL(b): $3(i_b - i_a) + 8 + 3(i_b - i_c) = 0 \longrightarrow -3i_a + 6i_b = 1\,\text{A}$
- KVL(a): $-9 + 3(i_a - i_b) = 0 \longrightarrow i_a - i_b = 3\,\text{A}$
  $\longrightarrow i_a = 19/3\,\text{A}$ and $i_b = 10/3\,\text{A}$
- $i_b - i_d = 1\,\text{A} \longrightarrow i_d = 7/3\,\text{A}$
- KVL(d): $8i_d - v_s - 8 = 0 \longrightarrow v_s = 32/3\,\text{V}$

Solution to Exercise 47:

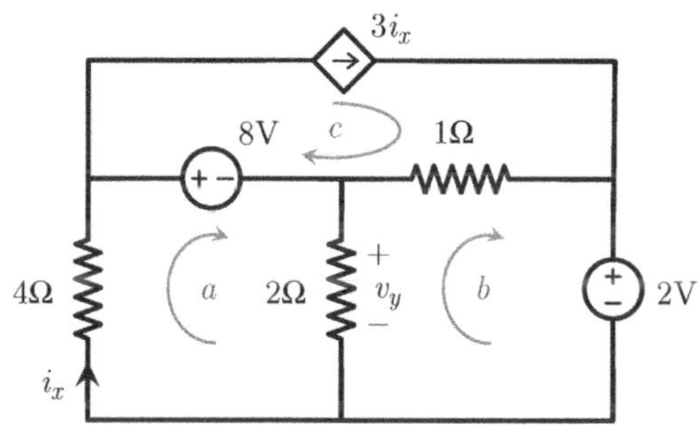

- $i_c = 3i_a$
- KVL(a): $4i_a + 8 + 2(i_a - i_b) = 0 \longrightarrow 3i_a - i_b = -4$
- KVL(b): $2(i_b - i_a) + 1(i_b - i_c) + 2 = 0 \longrightarrow -5i_a + 3i_b = -2$
- Then $i_a = -7/2$ A and $i_b = -13/2$ A
- And $v_y = 2(-7/2 + 13/2) = 6$ V

Solution to Exercise 48:

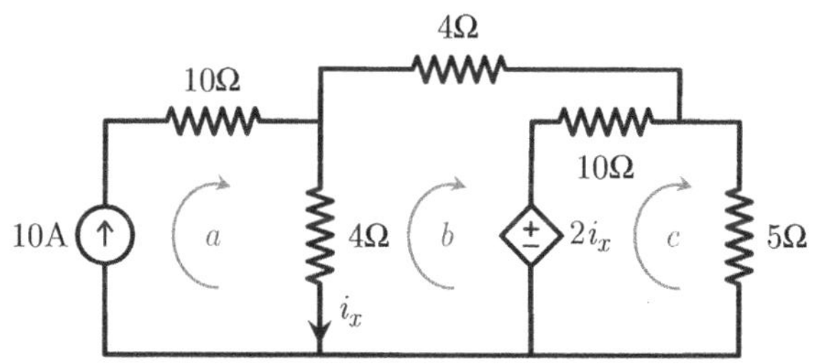

- $i_a = 10$ A and $i_x = i_a - i_b$
- KVL(b): $4(i_b - i_a) + 4i_b + 10(i_b - i_c) + 2i_x = 0 \longrightarrow 16i_b - 10i_c = 20$
  $\longrightarrow 8i_b - 5i_c = 10$
- KVL(c): $-2i_x + 10(i_c - i_b) + 5i_c = 0 \longrightarrow -8i_b + 15i_c = 20$
- Then $i_c = 3$ A and $p_{5\Omega} = 45$ W

Solution to Exercise 49:

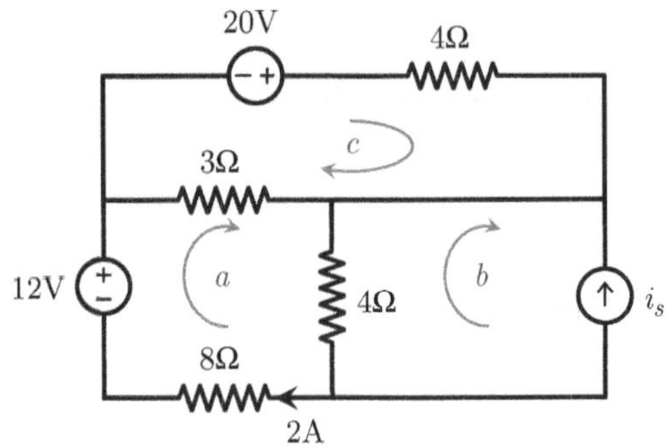

- $i_a = 2\,\text{A}$
- KVL(c): $-20 + 4i_c + 3(i_c - i_a) = 0 \longrightarrow i_c = 26/7\,\text{A}$
- KVL(a): $-12 + 3(i_a - i_c) + 4(i_a - i_b) + 16 = 0 \longrightarrow i_b = 12/7\,\text{A}$
- Then $i_s = -12/7\,\text{A}$

Solution to Exercise 50:

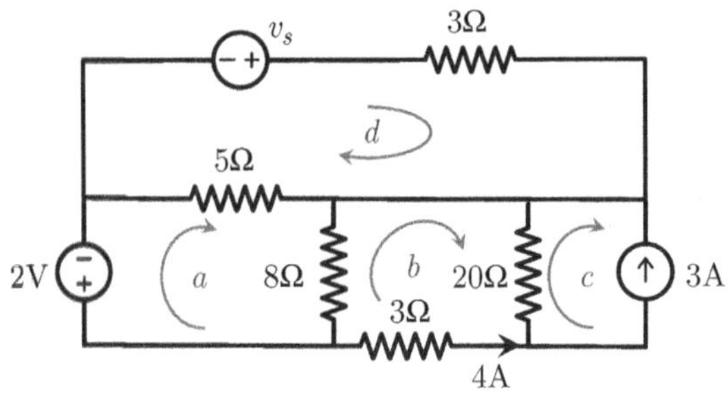

- $i_b = -4\,\text{A}$ and $i_c = -3\,\text{A}$
- KVL(b): $20(i_b - i_c) - 12 + 8(i_b - i_a) = 0 \longrightarrow i_a = -8\,\text{A}$
- KVL(a): $2 + 5(i_a - i_d) + 8(i_a - i_b) = 0 \longrightarrow i_d = -14\,\text{A}$
- KVL(d): $-v_s + 3i_d + 5(i_d - i_a) = 0 \longrightarrow v_s = -72\,\text{V}$

Solution to Exercise 51:

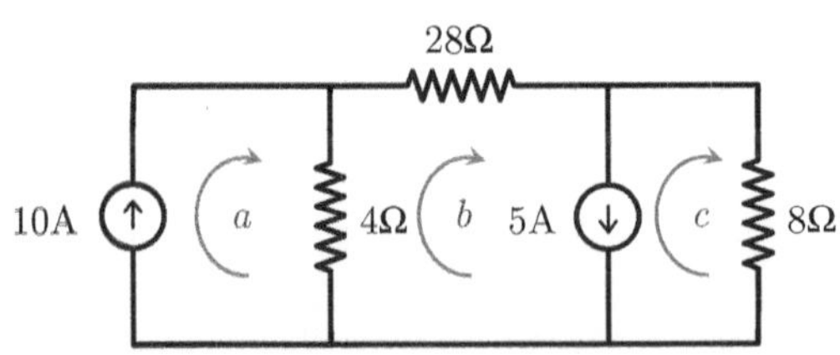

- $i_a = 10\,\text{A}$
- Supermesh: $i_b - i_c = 5$
- KVL(b&c): $4(i_b - i_a) + 28i_b + 8i_c = 0 \longrightarrow 4i_b + i_c = 5$
- Then $i_b = 2\,\text{A}$ and $i_c = -3\,\text{A}$
- And $p_{5\text{A}} = v_{5\text{A}} \times 5 = 8i_c \times 5 = -120\,\text{W}$

Solution to Exercise 52:

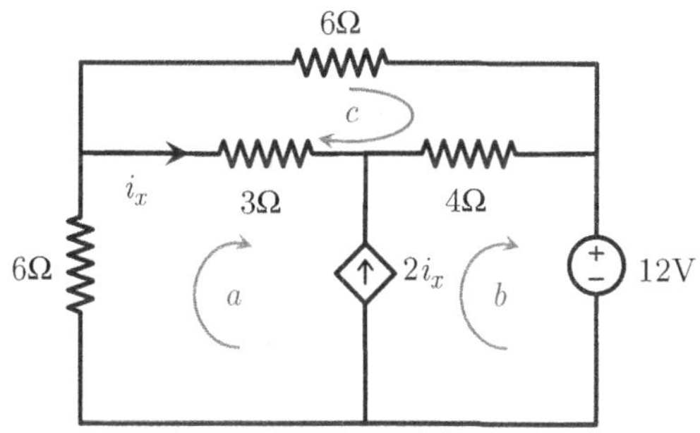

- $i_x = i_a - i_c$
- Supermesh: $2i_x = i_b - i_a \longrightarrow 3i_a - i_b - 2i_c = 0$
- KVL(a&b): $6i_a + 3(i_a - i_c) + 4(i_b - i_c) + 12 = 0 \longrightarrow 9i_a + 4i_b - 7i_c = -12$
- KVL(c): $3(i_c - i_a) + 6i_c + 4(i_c - i_b) = 0 \longrightarrow -3i_a - 4i_b + 13i_c = 0$
- Then $i_a = -7/6\,\text{A}$, $i_b = -11/5\,\text{A}$, and $i_c = -5/6\,\text{A}$
- And $i_x = i_a - i_c = -1/3\,\text{A}$

Solution to Exercise 53:

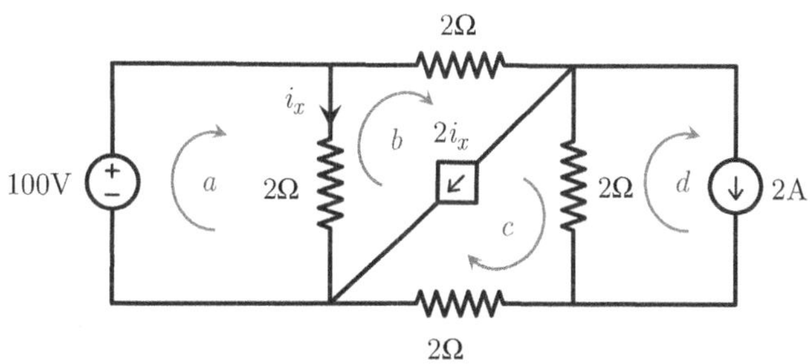

- $i_d = 2\,\text{A}$
- KVL(a): $-100 + 2(i_a - i_b) = 0 \longrightarrow i_a - i_b = 50$
- Supermesh: $i_b - i_c = 2i_x = 2(i_a - i_b) \longrightarrow 2i_a - 3i_b + i_c = 0$
- KVL(b&c): $2(i_b - i_a) + 2i_b + 2(i_c - i_d) + 2i_c = 0 \longrightarrow i_a - 2i_b - 2i_c = -2$
- Then $i_a = 134\,\text{A}$, $i_b = 84\,\text{A}$, $i_c = -16\,\text{A}$
- And $p_{100\text{V}} = -100 \times 134 = -13.4\,\text{kW}$

Solution to Exercise 54:

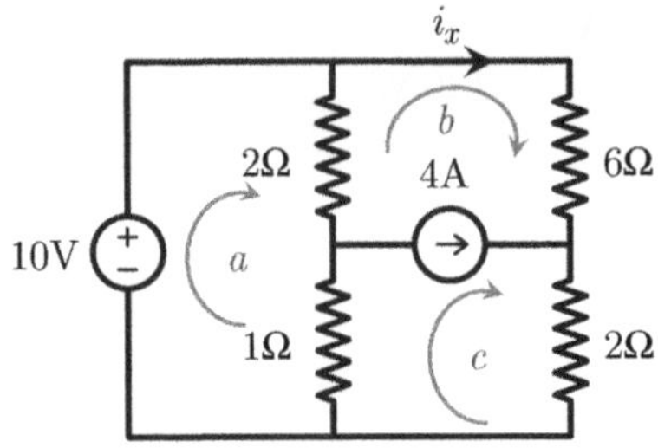

- Supermesh: $i_c - i_b = 4$
- KVL(a): $-10 + 2(i_a - i_b) + (i_a - i_c) = 0 \longrightarrow 3i_a - 2i_b - i_c = 10$
  $\longrightarrow 3i_a - 3i_b = 14$
- KVL(b&c): $6i_b + 2i_c + (i_c - i_a) + 2(i_b - i_a) = 0$
  $\longrightarrow -3i_a + 8i_b + 3i_c = 0 \longrightarrow -3i_a + 11i_b = -12$
- Then $i_b = 1/4$ A and $i_x = 1/4$ A

Solution to Exercise 55:

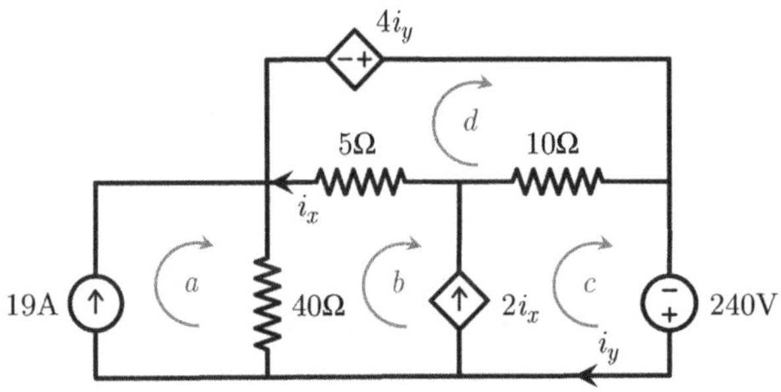

- $i_a = 19$ A, $i_x = i_d - i_b$, and $i_y = i_c$
- Supermesh: $i_c - i_b = 2i_x = 2i_d - 2i_b \longrightarrow i_b + i_c - 2i_d = 0$
- KVL(d): $-4i_y + 10(i_d - i_c) + 5(i_d - i_b) = 0 \longrightarrow -5i_b - 14i_c + 15i_d = 0$
- KVL(b&c): $40(i_b - i_a) + 5(i_b - i_d) + 10(i_c - i_d) - 240 = 0$
  $\longrightarrow 9i_b + 2i_c - 3i_d = 200$
- Then $-9i_c + 5i_d = 0 \longrightarrow 15i_d - 35/9i_d = 200 \longrightarrow i_d = 18$ A
- And $i_y = i_c = 10$ A and $p_{4i_y} = 40(-18) = -720$ W

Solution to Exercise 56:

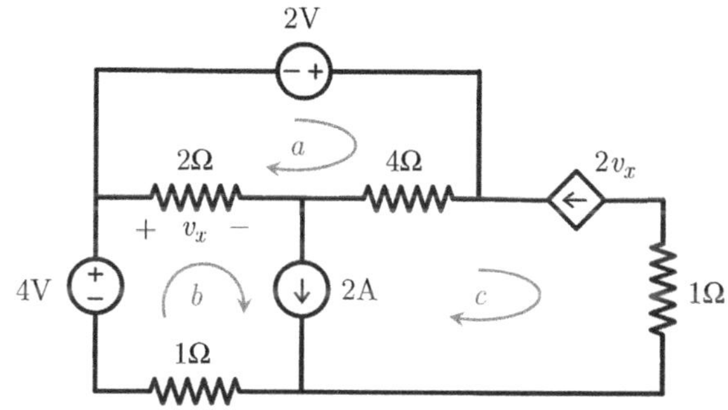

- $i_c = -2v_x = 4(i_a - i_b)$
- $-2 + 4(i_a - i_c) + 2(i_a - i_b) = 0 \longrightarrow 3i_a - i_b - 2i_c = 1 \longrightarrow -5i_a + 7i_b = 1$
- Supermesh: $i_b - i_c = 2 \longrightarrow -4i_a + 5i_b = 2$
- Then $i_a = -3$ A and $i_b = -2$ A
- KVL(b): $-4 + 2(i_b - i_a) + v_s + i_b = 0 \longrightarrow -2 + v_s - 2 = 0 \longrightarrow v_s = 4$ V
- And $p_s = 8$ W

Solution to Exercise 57:

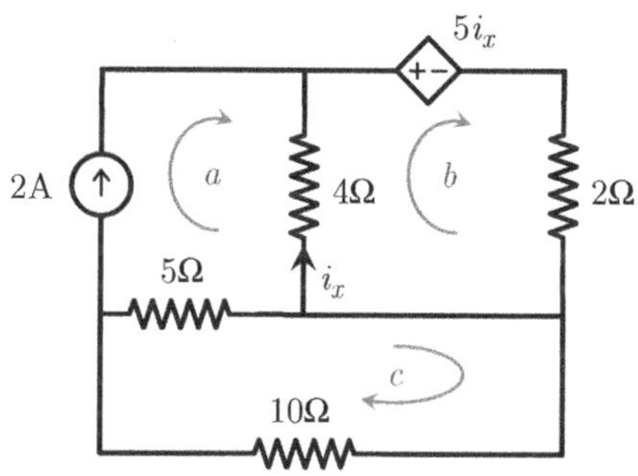

- $i_a = 2$ A and $i_x = i_b - i_a = i_b - 2$
- KVL(b): $4(i_b - i_a) + 5i_x + 2i_b = 0$
  $\longrightarrow i_b = 18/11$ A and $i_x = i_b - i_a = -4/11$ A
- KVL(c): $10i_c + 5(i_c - i_a) = 0 \longrightarrow i_c = 2/3$ A
- KVL($a\&b$): $v_{2A} + 5i_x + 2i_b + 5(i_a - i_c) = 0$
  $\longrightarrow v_{2A} = -16/11 - 20/3 = -268/33$ V
- Then $p_{2A} = -536/33$ W

Solution to Exercise 58:

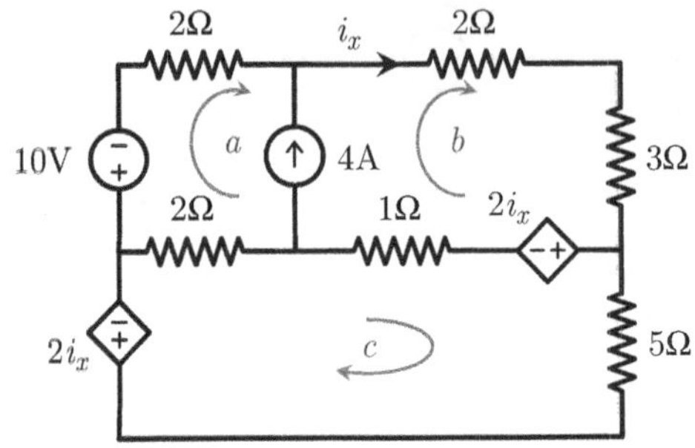

- $i_x = i_b$ and $i_b - i_a = 4$
- KVL(c): $2i_x + 2(i_c - i_a) + (i_c - i_b) - 2i_x + 5i_c = 0$
  $\longrightarrow -2i_a - i_b + 8i_c = 0$
- KVL(a&b): $10 + 2i_a + 5i_b + 2i_x + (i_b - i_c) + 2(i_a - i_c) = 0$
  $\longrightarrow 4i_a + 8i_b - 3i_c = -10$
- Then $-3i_a + 8i_c = 4$ and $4i_a - i_c = -14$
- And $i_a = -108/29$ A and $i_x = i_b = 8/29$ A

Solution to Exercise 59:

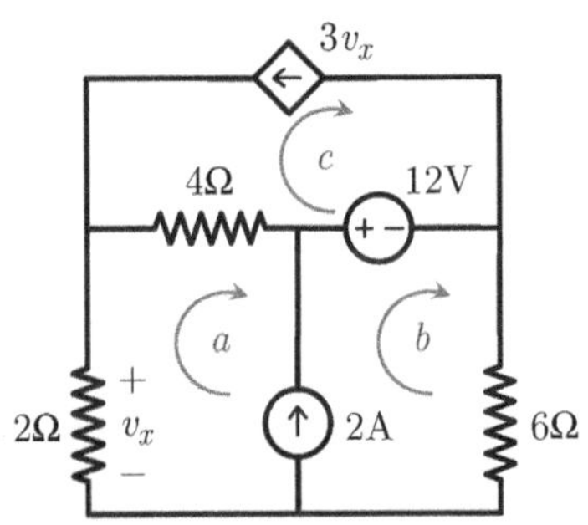

- $i_b - i_a = 2$ and $i_c = -3v_x = 6i_a$
- KVL(a&b): $2i_a + 4(i_a - i_c) + 12 + 6i_b = 0 \longrightarrow 3i_a - i_b = 2$
- Then $i_a = 2$ A and $v_x = -4$ V

Solution to Exercise 60:

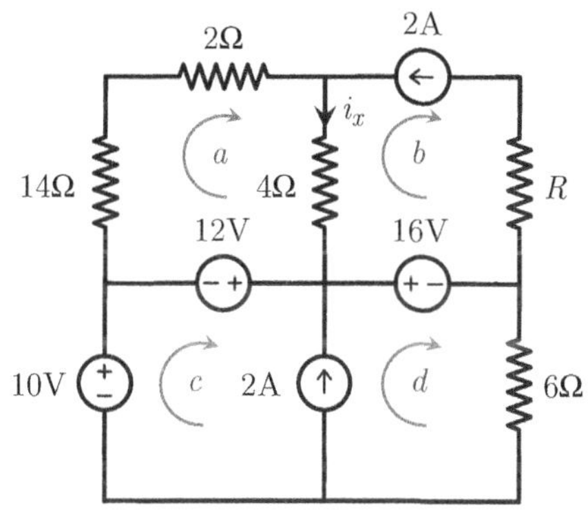

- $i_b = -2$ A
- Supermesh: $i_d - i_c = 2$
- KVL(a): $16i_a + 4(i_a - i_b) + 12 = 0 \longrightarrow 5i_a - i_b = -3 \longrightarrow i_a = -1$ A
- KVL($c\&d$): $-10 - 12 + 16 + 6i_d = 0 \longrightarrow i_d = 1$ A
- Then $i_x = i_a - i_b = 1$ A

Solution to Exercise 61:

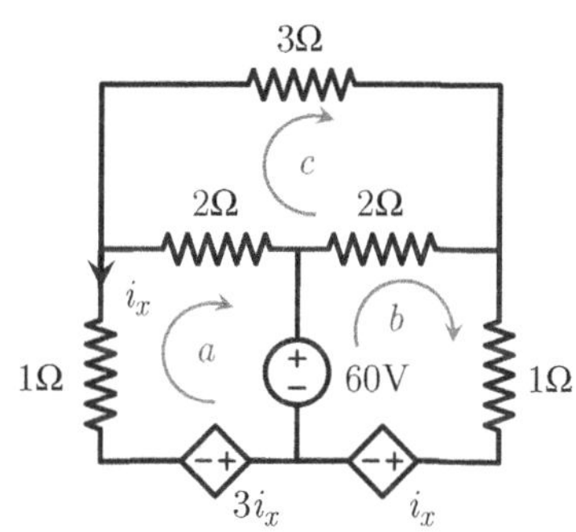

- $i_x = -i_a$
- KVL(a): $i_a + 2(i_a - i_c) + 60 + 3i_x = 0 \longrightarrow i_c = 30$ A
- KVL(b): $-60 + 2(i_b - i_c) + i_b + i_x = 0 \longrightarrow 3i_b - i_a = 120$
- KVL(c): $2(i_c - i_a) + 3i_c + 2(i_c - i_b) = 0 \longrightarrow i_a + i_b = 105$
- Then $i_b = 225/4$ A, $i_a = 105 - 225/4 = 195/4$ A, and $i_x = -195/4$ A

Solution to Exercise 62:

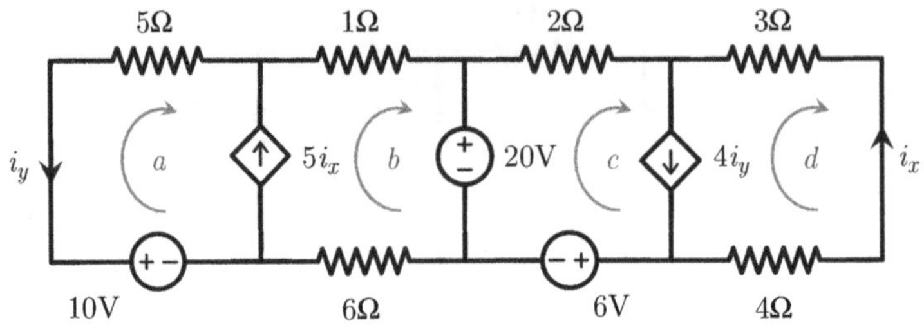

- $i_y = -i_a$ and $i_x = -i_d$
- $i_b - i_a = 5i_x = -5i_d \longrightarrow i_a - i_b - 5i_d = 0$
- $i_c - i_d = 4i_y = -4i_a \longrightarrow 4i_a + i_c - i_d = 0$
- KVL(a&b): $5i_a + i_b + 20 + 6i_b - 10 = 0 \longrightarrow 5i_a + 7i_b = -10$
- KVL(c&d): $-20 + 2i_c + 3i_d + 4i_d + 6 = 0 \longrightarrow 2i_c + 7i_d = 14$
- Then $4i_a + 7 - (7/2)i_d - i_d = 0 \longrightarrow 8i_a - 9i_d = -14$
- And $i_a - i_b - (40/9)i_a - 70/9 = 0 \longrightarrow -31i_a - 9i_b = 70$
- And $i_a = -100/43$ A, $i_b = 10/43$ A, and $i_x = 22/43$ A

Solution to Exercise 63:

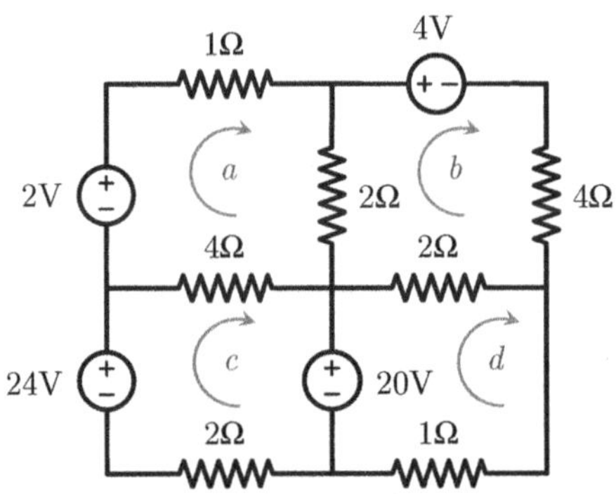

- KVL(a): $-2 + i_a + 2(i_a - i_b) + 4(i_a - i_c) = 0 \longrightarrow 7i_a - 2i_b - 4i_c = 2$
- KVL(b): $4 + 4i_b + 2(i_b - i_d) + 2(i_b - i_a) = 0 \longrightarrow -i_a + 4i_b - i_d = -2$
- KVL(c): $-24 + 4(i_c - i_a) + 20 + 2i_c = 0 \longrightarrow -2i_a + 3i_c = 2$
- KVL(d): $-20 + 2(i_d - i_b) + i_d = 0 \longrightarrow -2i_b + 3i_d = 20$
- Then $i_a = 2$, $i_b = 2$, $i_c = 2$, and $i_d = 8$
- And $p_{2V} = -2i_a = -4$ W, $p_{4V} = 4i_b = 8$ W, $p_{24V} = -24i_c = -48$ W
- And $p_{20V} = 20(i_c - i_d) = -120$ W
- And $p_R = -(-4 + 8 - 48 - 120) = 164$ W

Solution to Exercise 64:

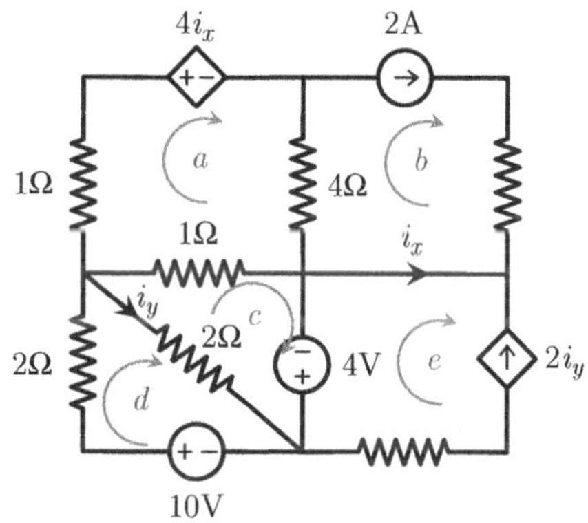

- $i_b = 2\,\text{A}, i_x = i_e - 2$
- $i_e = -2i_y = -2(i_d - i_c) \longrightarrow i_e = 2i_c - 2i_d$
- KVL(a): $i_a + 4i_x + 4(i_a - i_b) + i_a - i_c = 0 \longrightarrow 6i_a - i_c + 4i_e = 16$
- KVL(c): $(i_c - i_a) - 4 + 2(i_c - i_d) = 0 \longrightarrow -i_a + 3i_c - 2i_d = 4$
- KVL(d): $-10 + 2i_d + 2(i_d - i_c) = 0 \longrightarrow 2i_d - i_c = 5$
- Then $i_a = 3\,\text{A}, i_c = 6\,\text{A}, i_d = 11/2\,\text{A}, i_e = 1\,\text{A}$
- And $p_{10\text{V}} = 10(-i_d) = -55\,\text{W}$

Solution to Exercise 65:

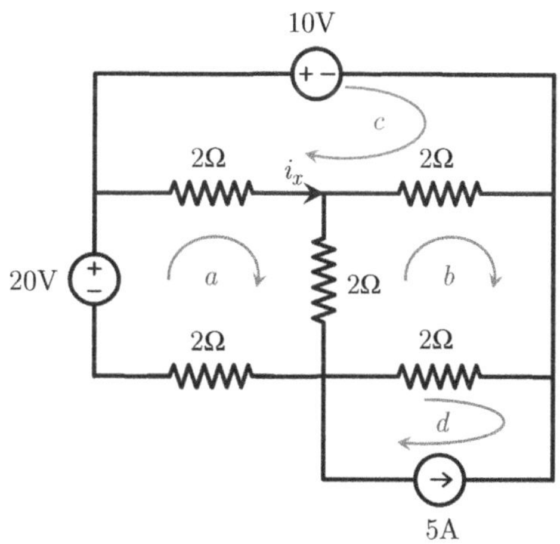

- Mesh(d): $i_d = -5\,\text{A}$
- KVL(a): $-20 + 2(i_a - i_c) + 2(i_a - i_b) + 2i_a = 0 \longrightarrow 3i_a - i_b - i_c = 10$
- KVL(b): $2(i_b - i_a) + 2(i_b - i_c) + 2(i_b - i_d) = 0 \longrightarrow -i_a + 3i_b - i_c = -5$

- KVL(c): $10 + 2(i_c - i_b) + 2(i_c - i_a) = 0 \longrightarrow -i_a - i_b + 2i_c = -5$
- Then $i_a = 15/8$ A, $i_b = -15/8$ A, $i_c = -20/8$ A
- And $i_x = i_a - i_c = 35/8$ A

Solution to Exercise 66:

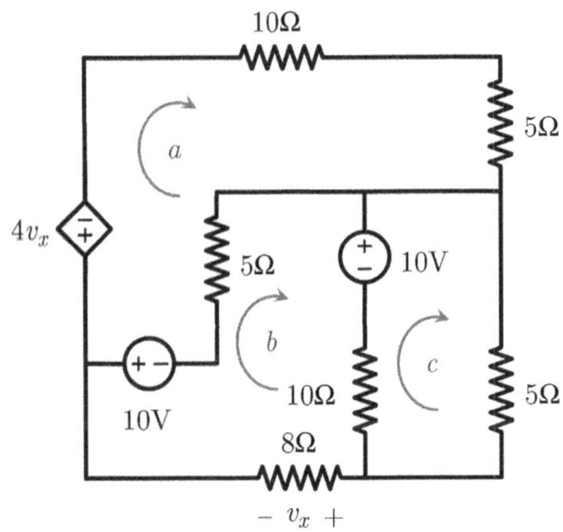

- $v_x = 8i_b$
- KVL(a): $4v_x + 15i_a + 5(i_a - i_b) - 10 = 0 \longrightarrow 20i_a - 27i_b = 10$
- KVL(b): $10 + 5(i_b - i_a) + 10 + 10(i_b - i_c) + 8i_b = 0 \longrightarrow -5i_a + 23i_b - 10i_c = -20$
- KVL(c): $10(i_c - i_b) - 10 + 5i_c = 0 \longrightarrow -2i_b + 3i_c = 2$
- Then $-27/4i_b + 23i_b - 20/3i_b = -65/6 \longrightarrow i_b = -26/23$ A
- And $v_x = -208/23$ V

Solution to Exercise 67:

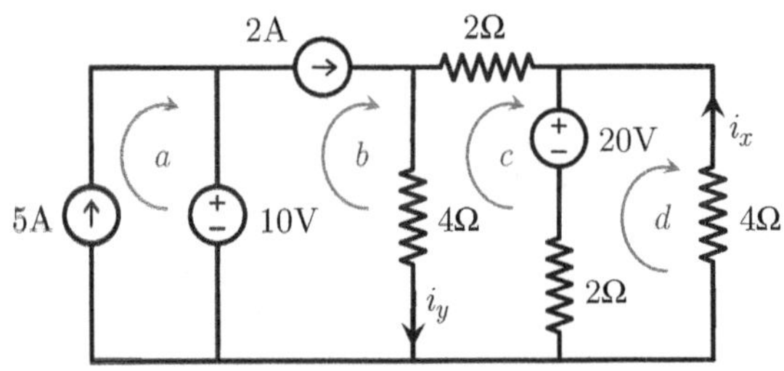

- $i_b = 2$ A
- KVL(d): $2(i_d - i_c) - 20 + 4i_d = 0 \longrightarrow -i_c + 3i_d = 10$
- KVL(c): $4(i_c - i_b) + 2i_c + 20 + 2(i_c - i_d) = 0 \longrightarrow 4i_c - i_d = -6$
- Then $i_d = 34/11$ A and $i_x = -34/11$ A

Solution to Exercise 68:

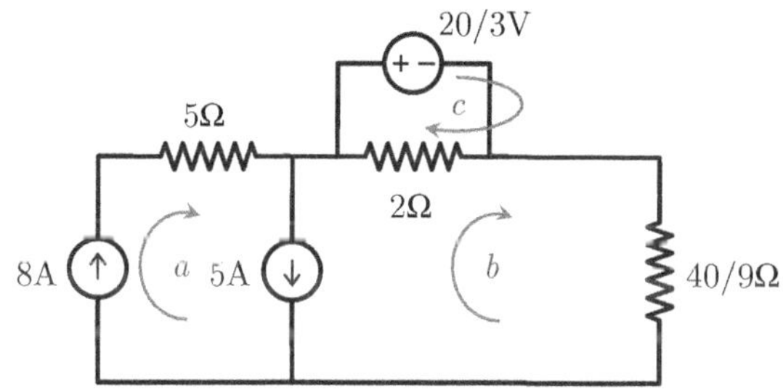

- $i_a = 8$ A and $i_a - i_b = 5 \longrightarrow i_b = 3$ A
- KVL(c): $20/3 + 2(i_c - i_b) = 0 \longrightarrow i_c = -1/3$ A
- Then $v_{5A} = (40/9)i_b + 2(i_b - i_c) = 40/3 + 20/3 = 20$ A
- And $p_{5A} = 100$ W

Solution to Exercise 69:

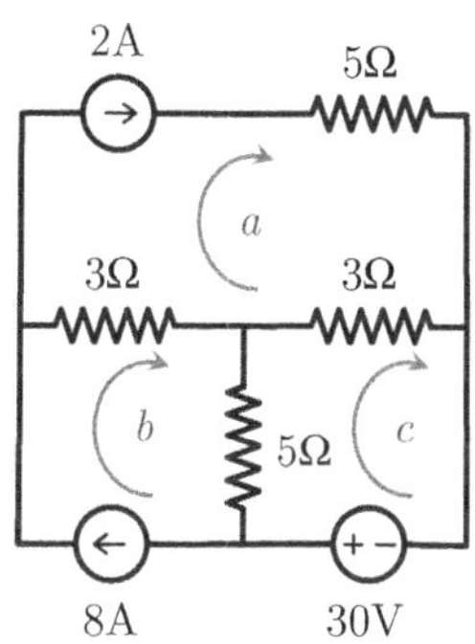

- $i_a = 2$ A and $i_b = 8$ A
- KVL(c): $5(i_c - i_b) + 3(i_c - i_a) - 30 = 0 \longrightarrow i_c = 19/2$ A
- Then $p_{30V} = 30(-i_c) = -285$ W

Solution to Exercise 70:

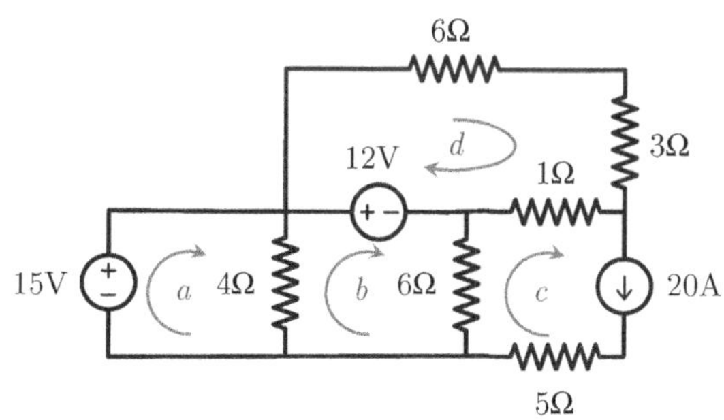

- $i_c = 20\,\text{A}$
- KVL(a): $-15 + 4(i_a - i_b) = 0 \longrightarrow 4i_a - 4i_b = 15$
- KVL(b): $4(i_b - i_a) + 12 + 6(i_b - i_c) = 0 \longrightarrow -2i_a + 5i_b = 54$
- Then $i_b = 41/2\,\text{A}$ and $i_a = 97/4\,\text{A}$
- And $p_{15\text{V}} = 15 \times (-i_a) = -1455/4\,\text{W}$

Chapter 5

Solution to Exercise 71:

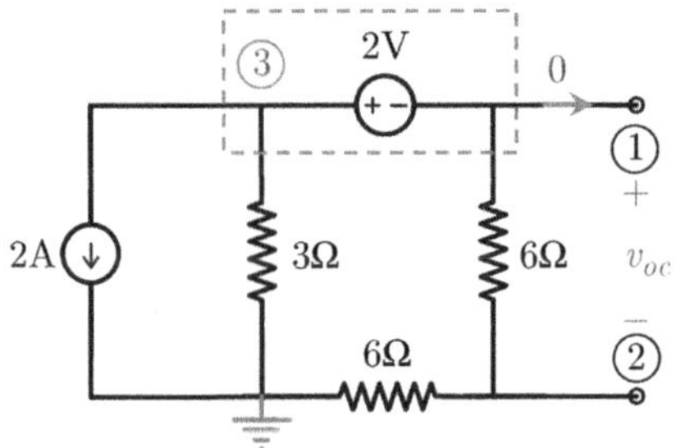

For v_{oc}:

- Supernode: $v_3 = v_1 + 2$
- KCL(1&3): $-2 - v_3/3 - v_1/12 = 0 \longrightarrow v_1 + 4v_3 = -24$
- Then $v_1 = -32/5\,\text{V}$, $v_{\text{oc}} = (-32/5)/2 = -16/5\,\text{V}$

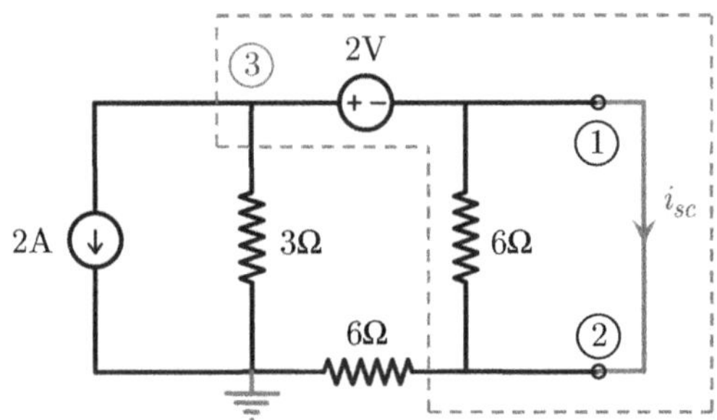

For i_{sc}:

- $v_1 = v_2$
- Supernode: $v_3 = v_1 + 2 = v_2 + 2$
- KCL(1&2&3): $-2 - v_3/3 - v_2/6 = 0 \longrightarrow v_2 + 2v_3 = -12$
- Then $v_1 = v_2 = -16/3\,\text{V}$, $v_3 = -10/3\,\text{V}$, $i_{\text{sc}} = v_2/6 = -8/9\,\text{A}$
- $R_{\text{th}} = 18/5\,\Omega$

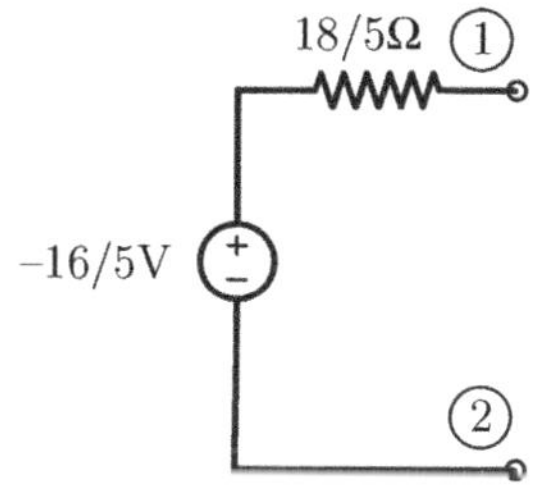

Solution to Exercise 72:

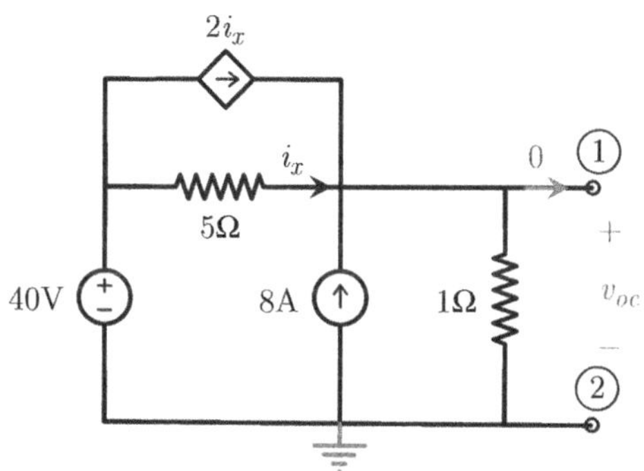

For v_{oc}:

- $i_x = (40 - v_1)/5 = 8 - v_1/5$
- KCL (1): $(40 - v_1)/5 + 2i_x + 8 - v_1/1 = 0 \longrightarrow v_1 = 20\,\text{V}$
- Then $v_{oc} = v_1 = 20\,\text{V}$

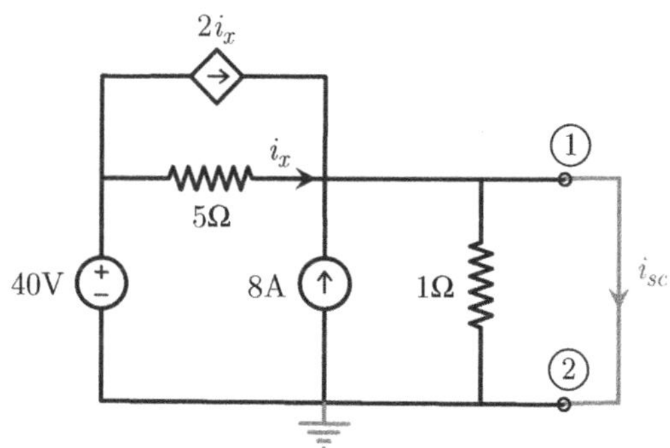

For i_{sc}:

- $v_1 = v_2 = 0\,\text{V}$ and $i_x = 40/5 = 8\,\text{A}$
- KCL (1): $8 + 16 + 8 - i_{sc} = 0 \longrightarrow i_{sc} = 32\,\text{A}$
- $R_{th} = 20/32 = 5/8\,\Omega$

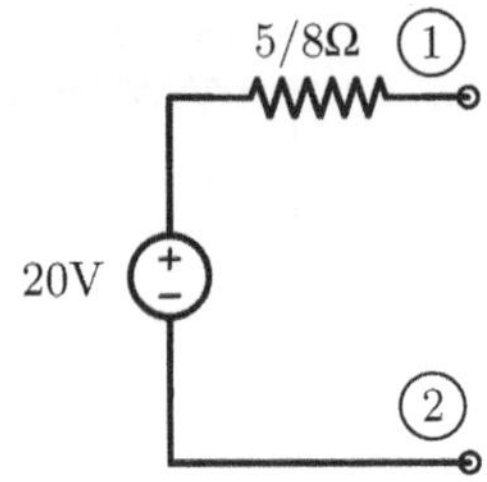

Solution to Exercise 73:

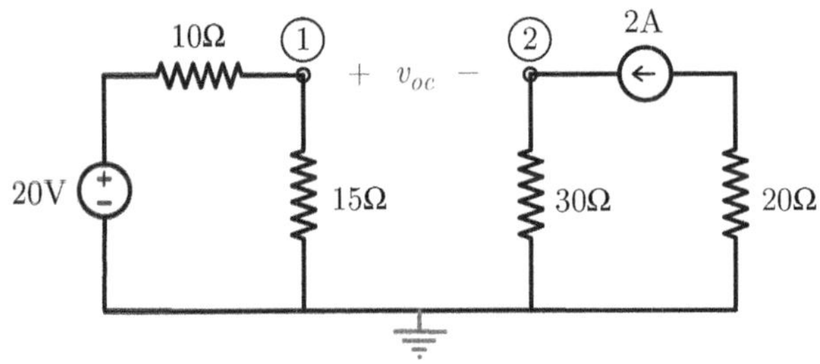

For v_{oc}:

- $v_1 = 20 \times 15/25 = 12\,\text{V}$
- $v_2 = 2 \times 30 = 60\,\text{V}$
- Then $v_{oc} = 12 - 60 = -48\,\text{V}$

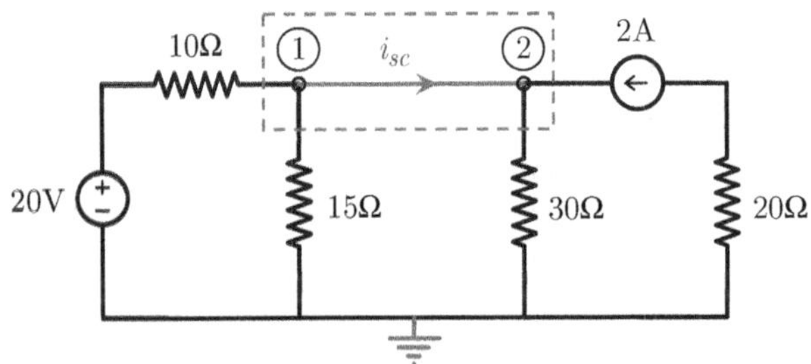

For i_{sc}:

- $v_1 = v_2$
- KCL(1&2): $(20 - v_1)/10 - v_1/15 - v_2/30 + 2 = 0 \longrightarrow v_1 = v_2 = 20\,\text{V}$
- KCL(1): $(20 - 20)/10 - 20/15 - i_{sc} = 0 \longrightarrow i_{sc} = -4/3\,\text{A}$
- $R_{th} = 36\,\Omega$

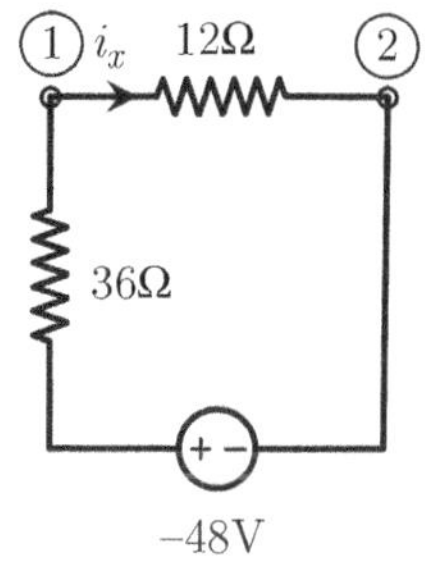

Solution to Exercise 74:

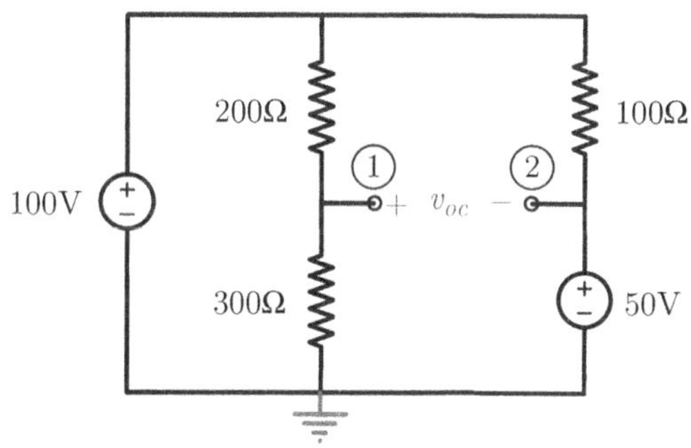

For v_{oc}:

- $v_1 = 100 \times 300/500 = 60\,\text{V}$
- $v_2 = 50\,\text{V}$
- Then $v_{oc} = 60 - 50 = 10\,\text{V}$

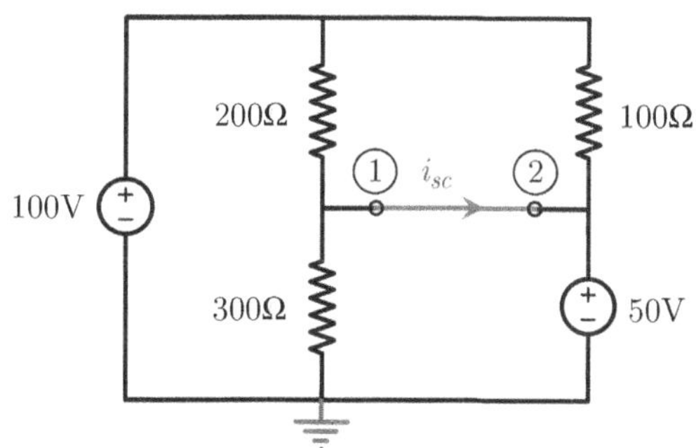

For i_{sc}:

- $v_1 = v_2 = 50\,\text{V}$
- KCL(1): $(100 - 50)/200 - 50/300 - i_{sc} = 0 \longrightarrow i_{sc} = 1/12\,\text{A}$
- $R_{th} = 120\,\Omega$

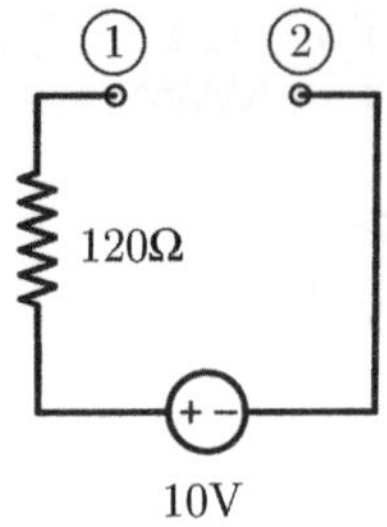

Solution to Exercise 75:

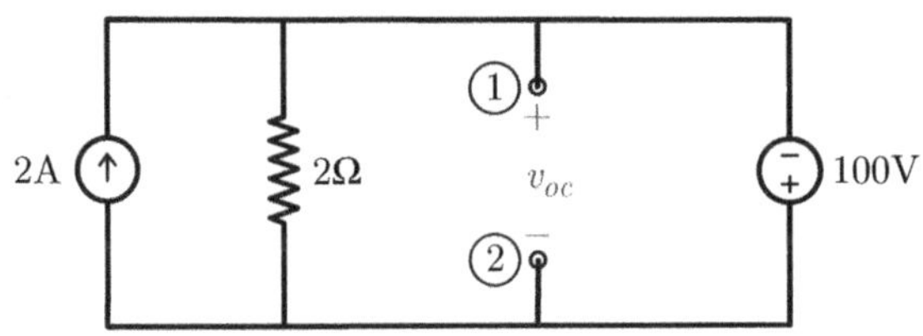

For v_{oc}:

- $v_{\text{oc}} = -100\,\text{V}$

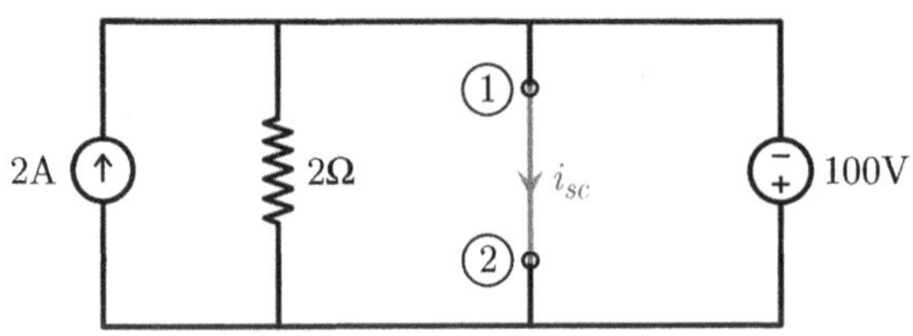

For i_{sc}:

- $i_{\text{sc}} = -\infty \longrightarrow R_{\text{th}} = 0\,\Omega$

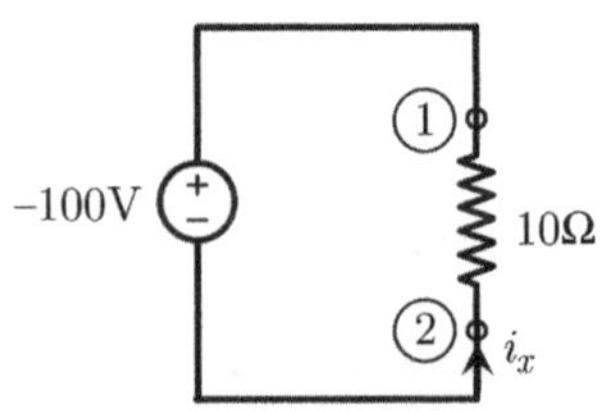

We normally consider a parallel connection of a voltage source and short circuit as an impossible scenario. However, in this case, since the aim is to find the Thévenin seen by a component, we mathematically carry out the process to reach $R_{\text{th}} = 0$. For confused readers, an alternative process is to make the voltage/current source short/open circuit to derive the equivalent resistance seen by the 10 Ω resistor.

Solution to Exercise 76:

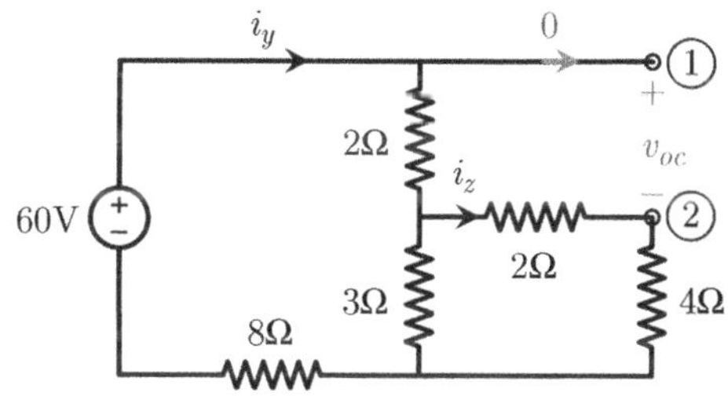

For v_{oc}:

- $i_y = 60/(2+8+3 \parallel 6) = 60/12 = 5\,\text{A} \longrightarrow i_z = 5 \times 3/9 = 5/3\,\text{A}$
- Then $v_{oc} = v_1 - v_2 = 5 \times 2 + (5/3) \times 2 = 40/3\,\text{V}$

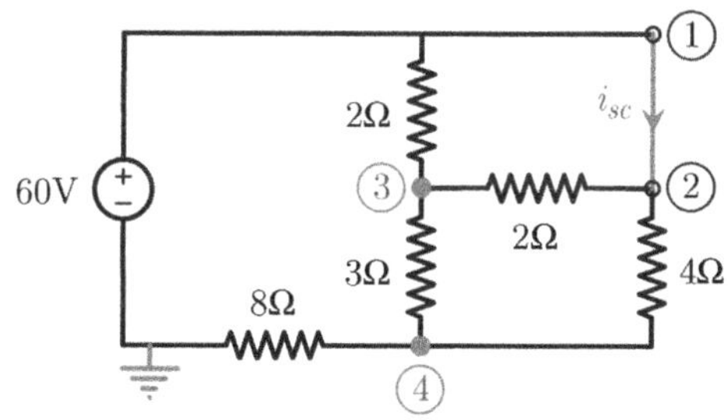

For i_{sc}:

- $v_1 = v_2 = 60\,\text{V}$
- KCL(3): $(60 - v_3)/2 - (v_3 - 60)/2 - (v_3 - v_4)/3 = 0 \longrightarrow 4v_3 - v_4 = 180$
- KCL(4): $(v_3 - v_4)/3 + (60 - v_4)/4 - v_4/8 = 0 \longrightarrow 8v_3 - 17v_4 = -360$
- Then $v_3 = 57\,\text{V}$, $v_4 = 48\,\text{V}$
- KCL(2): $(v_3 - 60)/2 + i_{sc} - (v_2 - v_4)/4 = 0 \longrightarrow i_{sc} = 9/2\,\text{A}$
- And $R_{th} = 80/27\,\Omega$

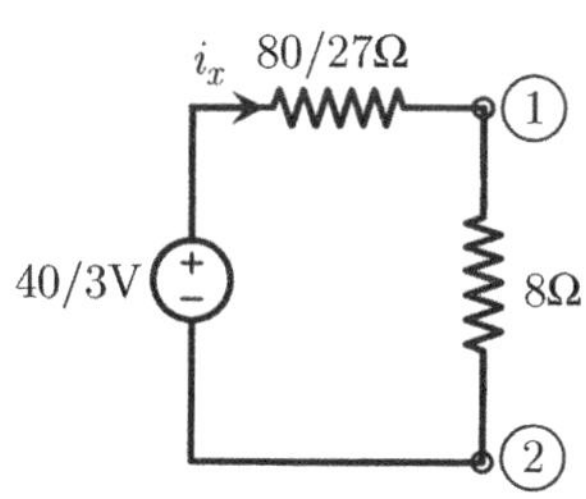

Solution to Exercise 77:

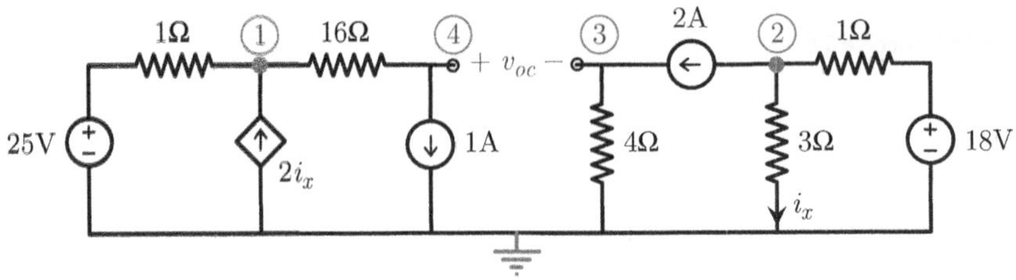

For v_{oc}:

- $v_3 = 8\,\text{V}$
- KCL(2): $-2 - v_2/3 - (v_2 - 18)/1 = 0 \longrightarrow v_2 = 12\,\text{V} \longrightarrow i_x = 4\,\text{A}$
- KCL(1): $(25 - v_1)1/ + 2i_x - 1 = 0 \longrightarrow v_1 = 32\,\text{V}$
- Then $v_4 = 32 - 16 = 16\,\text{V}$ and $v_{oc} = 16 - 8 = 8\,\text{V}$

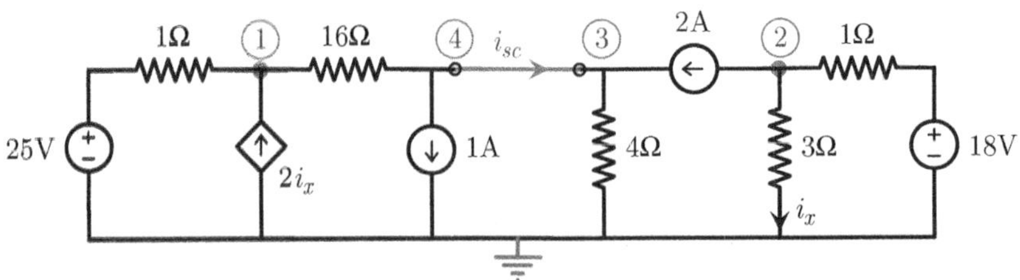

For i_{sc}:

- $v_2 = 12\,\text{V}$ and $i_x = 4\,\text{A}$ (same as before)
- KCL(1): $(25 - v_1)/1 + 2i_x - (v_1 - v_4)/16 = 0 \longrightarrow -17v_1 + v_4 = -16 \times 33$
- KCL(3&4): $(v_1 - v_4)/16 - 1 - v_3/4 + 2 = 0 \longrightarrow v_1 - 5v_4 = -16$
- Then $v_4 = 200/21\,\text{V}$ and $i_{sc} = 8/21\,\text{A}$
- And $R_{th} = 21\,\Omega$

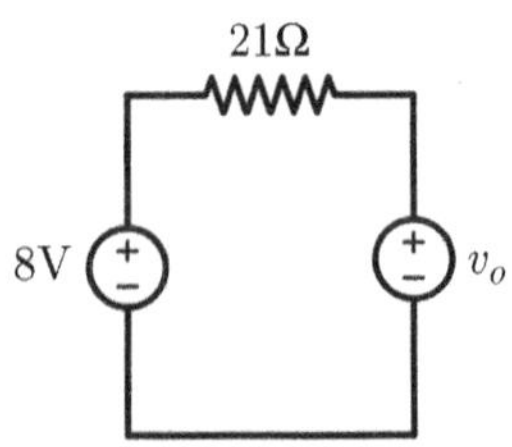

Solution to Exercise 78:

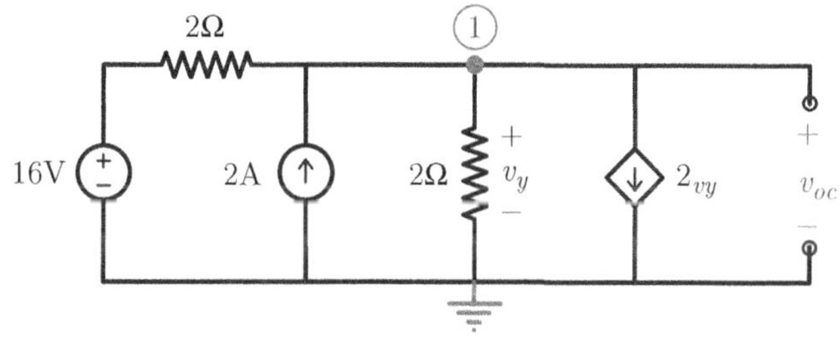

For v_{oc}:

- $v_y = v_1$
- KCL(1): $(16 - v_1)/2 + 2 - v_1/2 - 2v_y = 0 \longrightarrow v_1 = 10/3\,\text{V}$
- $v_{oc} = v_1 = 10/3\,\text{A}$

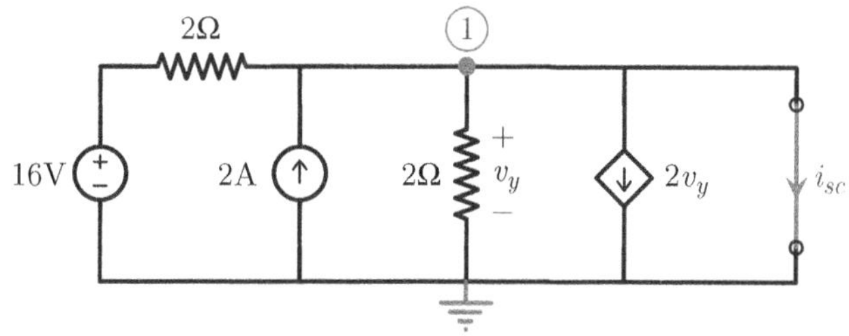

For i_{sc}:

- $v_1 = 0\,\text{V}$ and $v_y = 0\,\text{V}$
- KCL(1): $(16 - v_1)/2 + 2 - v_1/2 - 2v_y - i_{sc} = 0 \longrightarrow i_{sc} = 10\,\text{A}$
- Then $R_{th} = 1/3\,\Omega$

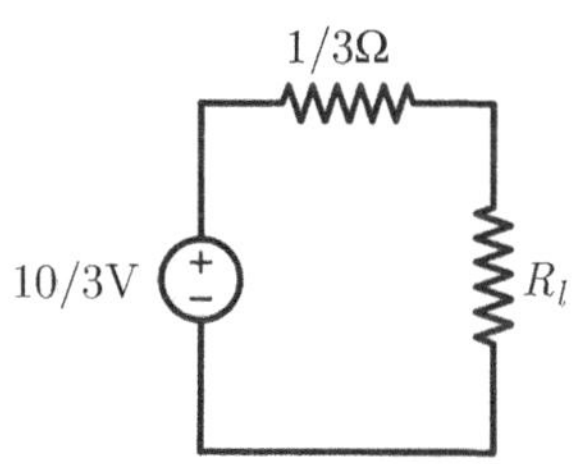

Solution to Exercise 79:

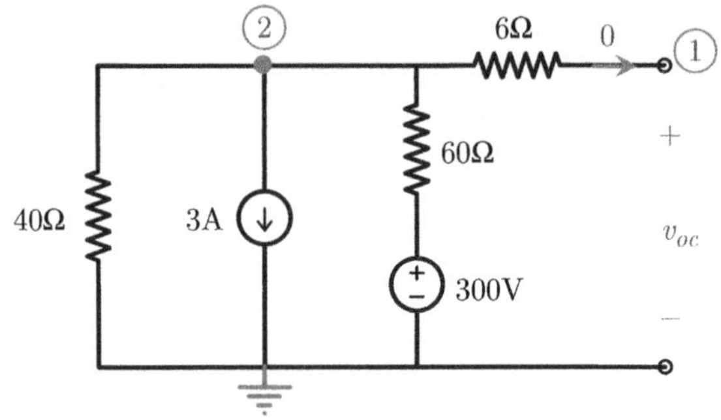

For v_{oc}:

- KCL(2): $-v_2/40 - 3 - (v_2 - 300)/60 = 0 \longrightarrow v_2 = 48\,\text{V}$
- Then $v_{oc} = v_2 = 48\,\text{V}$

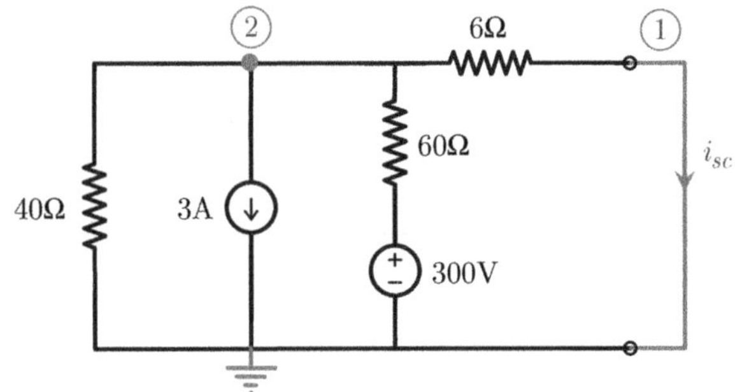

For i_{sc}:

- $v_1 = 0\,\text{V}$
- KCL(2): $-v_2/40 - 3 - (v_2 - 300)/60 - v_2/6 = 0 \longrightarrow v_2 = 48/5\,\text{V}$
- Then $i_{sc} = v_2/6 = 8/5\,\text{A}$
- And $R_{th} = 30\,\Omega$

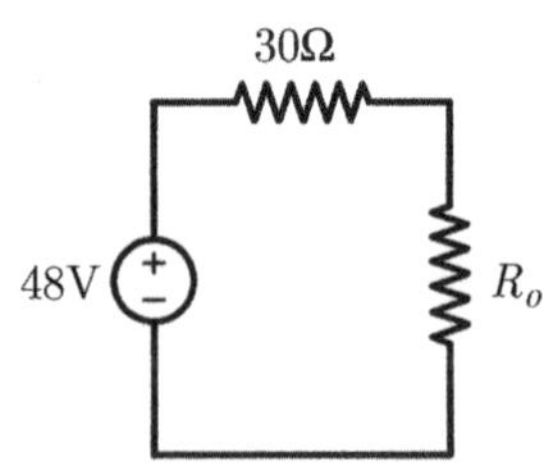

Solution to Exercise 80:

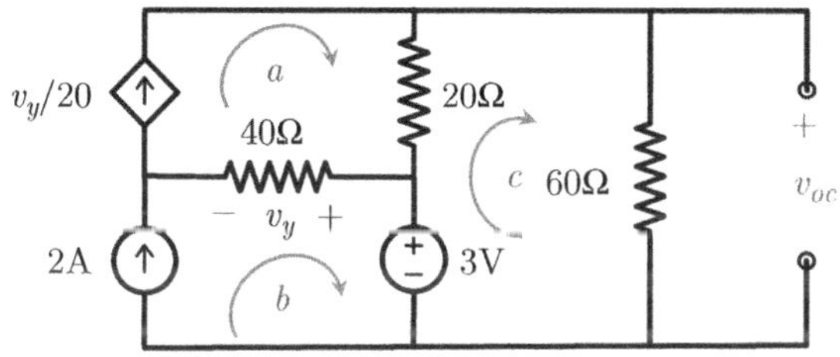

For v_{oc}:

- $i_b = 2\,\text{A}$
- $v_y = 40(i_a - i_b) = 40i_a - 80$
- $i_a = v_y/20 \longrightarrow v_y = 2v_y - 80 \longrightarrow v_y = 80\,\text{V}$ and $i_a = 4\,\text{A}$
- KVL(c): $-3 + 20(i_c - i_a) + 60i_c = 0 \longrightarrow i_c = 83/80\,\text{A}$
- Then $v_{oc} = 60i_c = 249/4\,\text{V}$

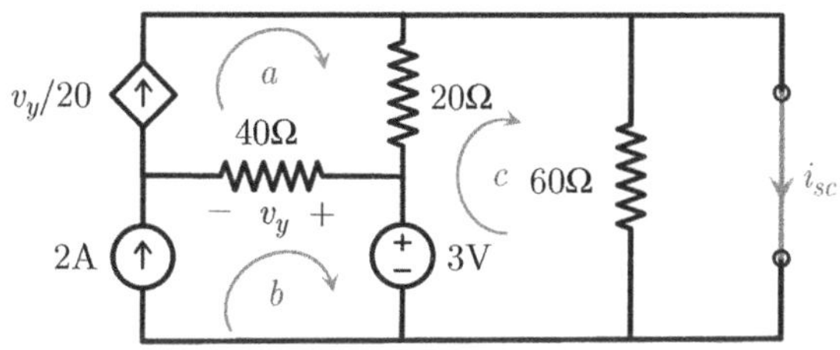

For i_{sc}:

- $i_b = 2\,\text{A}$
- $v_y = 40(i_a - i_b) = 40i_a - 80$
- $i_a = v_y/20 \longrightarrow v_y = 2v_y - 80 \longrightarrow v_y = 80\,\text{V}$ and $i_a = 4\,\text{A}$ (same as before)
- KVL(c): $-3 + 20(i_c - i_a) = 0 \longrightarrow i_c = 83/20\,\text{A}$
- Then $i_{sc} = i_c = 83/20\,\text{A}$
- And $R_{th} = (249/4)/(83/20) = 15\,\Omega$

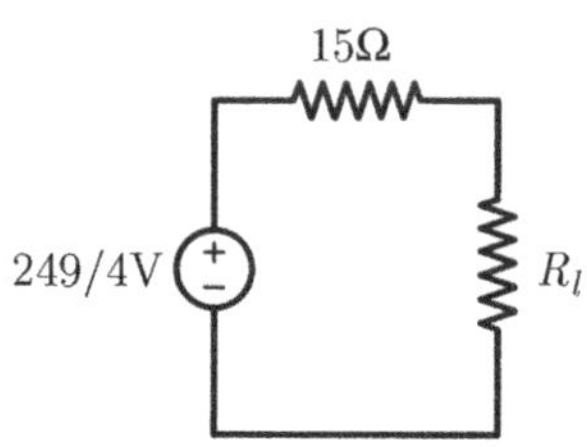

Solution to Exercise 81:

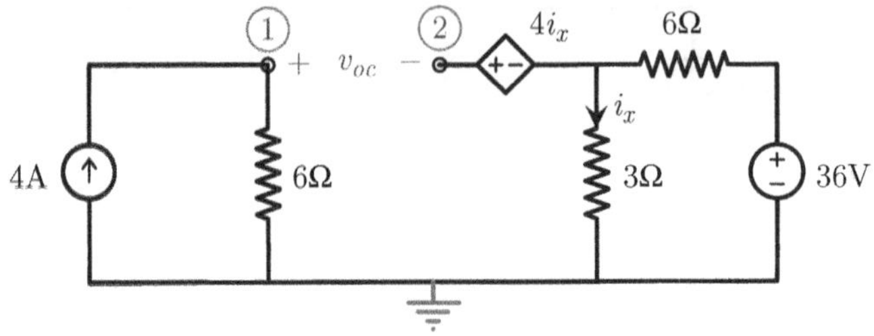

For v_{oc}:

- $v_1 = 24\,\text{V}$
- $i_x = 36/9 = 4\,\text{A}$ and $v_2 = 7i_x = 28\,\text{V}$
- Then $v_{oc} = v_1 - v_2 = -4\,\text{V}$

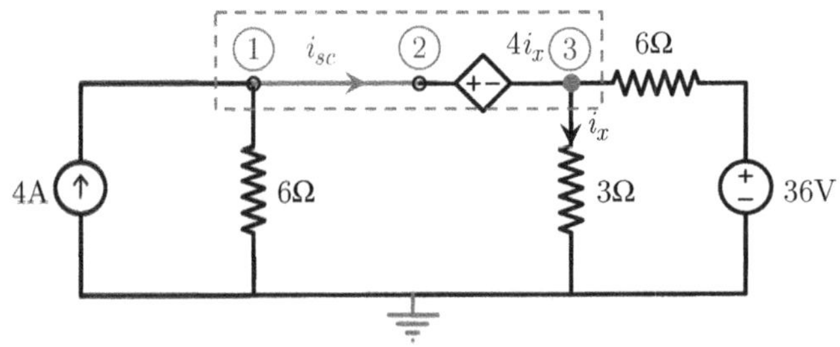

For i_{sc}:

- $v_1 = v_2$ and $i_x = v_3/3$
- KCL(1&2): $4 - v_1/6 - v_3/3 - (v_3 - 36)/6 = 0 \longrightarrow v_1 + 3v_3 = 60$
- Supernode: $v_1 = v_3 + 4i_x = (7/3)v_3$
- Then $v_3 = 45/4\,\text{V}$ and $v_1 = 105/4\,\text{V}$
- $i_{sc} = 4 - v_1/6 = -3/8\,\text{A}$ and $R_{th} = 32/3\,\Omega$
- And choose $R_o = 32/3\,\Omega$ and $p_o = 4/(32/3) = 3/8\,\text{W}$

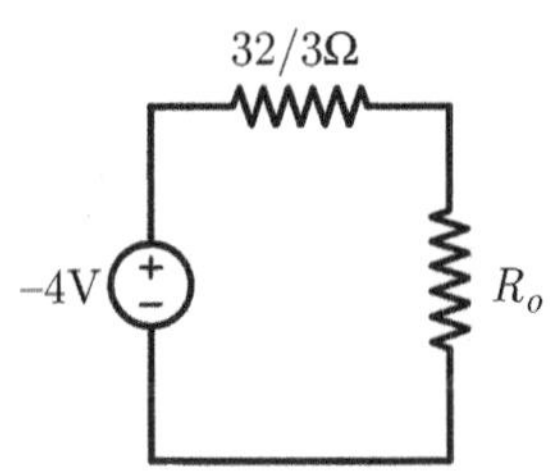

Solution to Exercise 82:

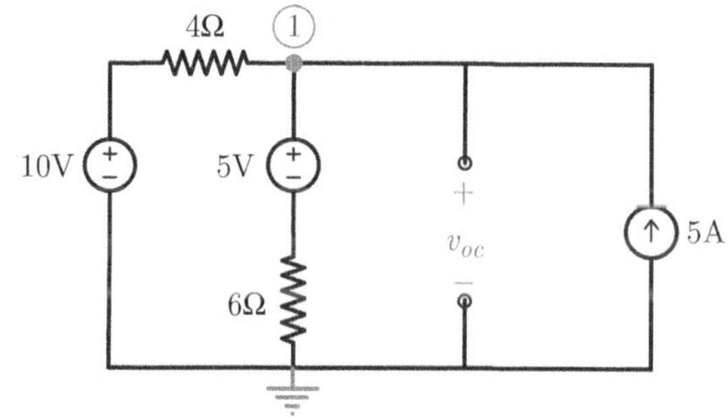

For v_{oc}:

- KCL(1): $(10 - v_1)/4 - (v_1 - 5)/6 + 5 = 0 \longrightarrow v_1 = 20\,\text{V}$
- Then $v_{oc} = v_1 = 20\,\text{V}$

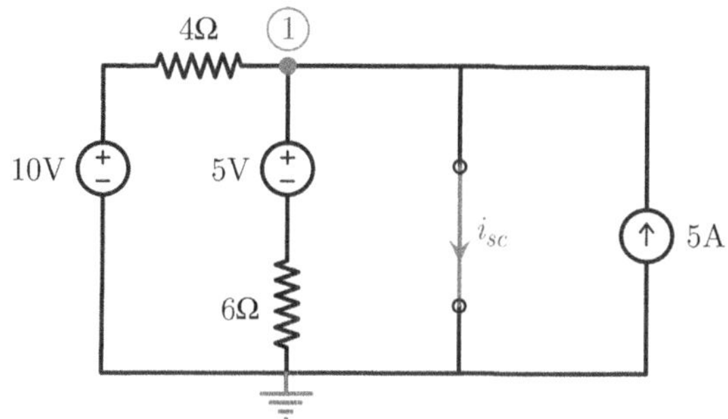

For i_{sc}:

- $v_1 = 0$
- KCL(1): $10/4 + 5/6 + 5 - i_{sc} = 0 \longrightarrow i_{sc} = 25/3\,\text{A}$ and $R_{th} = 12/5\,\Omega$
- Then choose $R_o = 12/5\,\Omega$ and $p_o = 100/(12/5) = 125/3\,\text{W}$

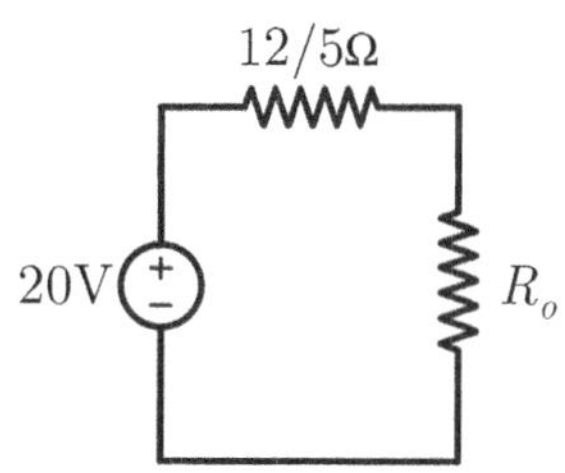

Solution to Exercise 83:

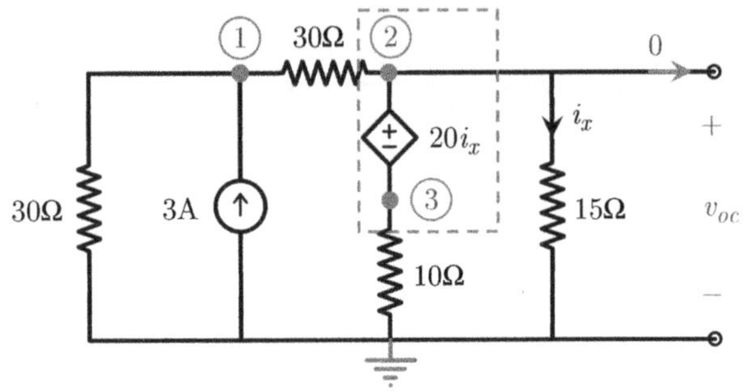

For v_{oc}:

- $v_2 = 15i_x$
- Supernode: $v_2 = v_3 + 20i_x \longrightarrow v_3 = -v_2/3$
- KCL(1): $-v_1/30 + 3 + (v_2 - v_1)/30 = 0 \longrightarrow 2v_1 - v_2 = 90$
- KCL(2&3): $(v_1 - v_2)/30 - v_3/10 - v_2/15 = 0 \longrightarrow v_1 - 2v_2 = 0$
- Then $v_{oc} = v_2 = 30\,\text{V}$

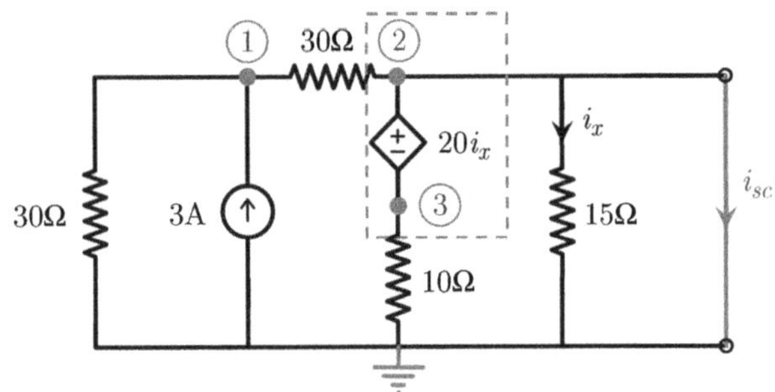

For i_{sc}:

- $i_x = 0$ and $v_2 = v_3 = 0$
- KCL(1): $-v_1/30 + 3 - v_1/30 = 0 \longrightarrow v_1 = 45\,\text{V}$
- $i_{sc} = 45/30 = 3/2\,\text{A}$ and $R_{th} = 20\,\Omega$
- Then choose $R_o = 20\,\Omega$ and $p_o = 225/20 = 45/4\,\text{W}$

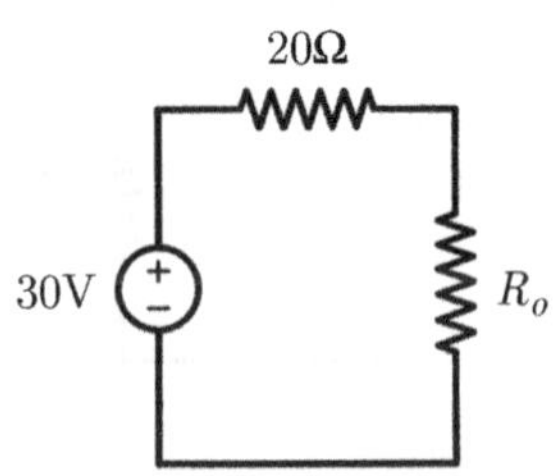

Solution to Exercise 84:

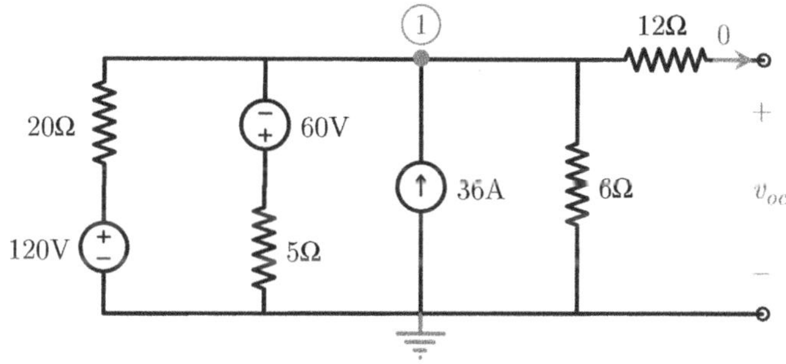

For v_{oc}:

- $(120 - v_1)/20 - (v_1 + 60)/5 + 36 - v_1/6 = 0 \longrightarrow v_1 = 72\,\text{V}$
- Then $v_{\text{oc}} = v_1 = 72\,\text{V}$

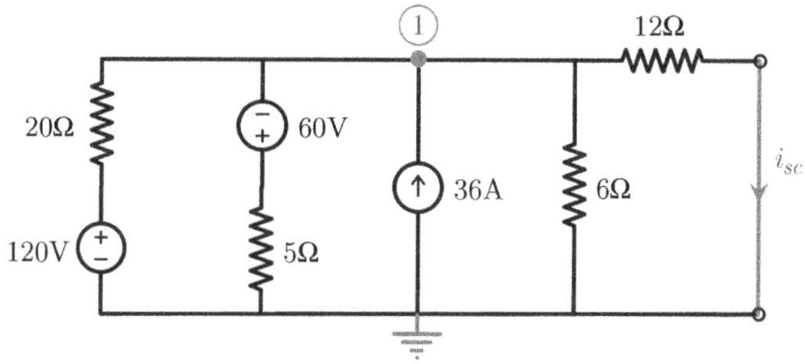

For i_{sc}:

- $(120 - v_1)/20 - (v_1 + 60)/5 + 36 - v_1/6 - v_1/12 = 0 \longrightarrow v_1 = 60\,\text{V}$
- Then $i_{\text{sc}} = v_1/12 = 5\,\text{A}$ and $R_{\text{th}} = 72/5\,\Omega$
- Therefore, choose $R_o = 72/5\,\Omega$ and $p_o = 36^2/(72/5) = 90\,\text{W}$

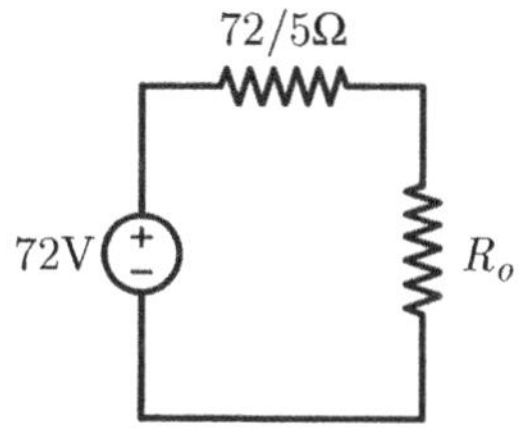

Solution to Exercise 85:

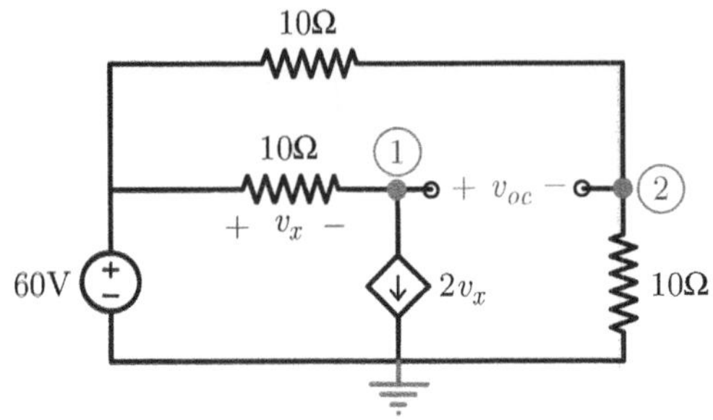

For v_{oc}:

- $v_x = 10 \times 2v_x = 20v_x \longrightarrow v_x = 0 \longrightarrow v_1 = 60\,\text{V}$
- $v_2 = 30\,\text{V}$ and $v_{oc} = v_1 - v_2 = 30\,\text{V}$

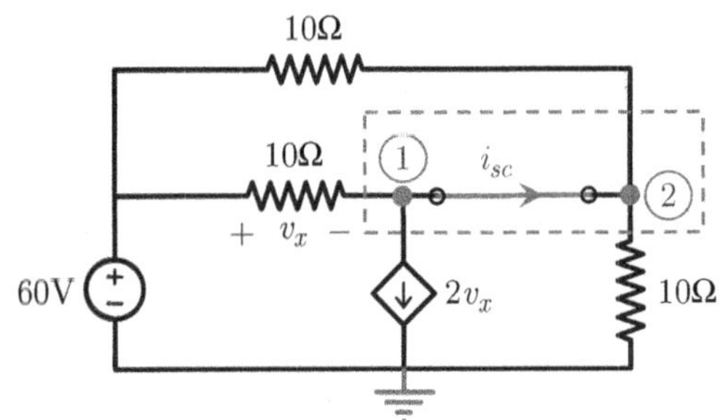

For i_{sc}:

- $v_1 = v_2$ and $v_x = 60 - v_1$
- KCL(1&2): $(60 - v_1)/10 - 2v_x - v_2/10 + (60 - v_2)/10 = 0$
  $\longrightarrow v_1 = 1080/17\,\text{V}$
- Then $i_{sc} = (60 - v_1)/10 - 2(60 - v_1) = 114/17\,\text{A}$ and $R_{th} = 85/19\,\Omega$
- Therefore, choose $R_o = 85/19\,\Omega$ and $p_o = (15)^2/(85/19) = 855/17\,\text{W}$

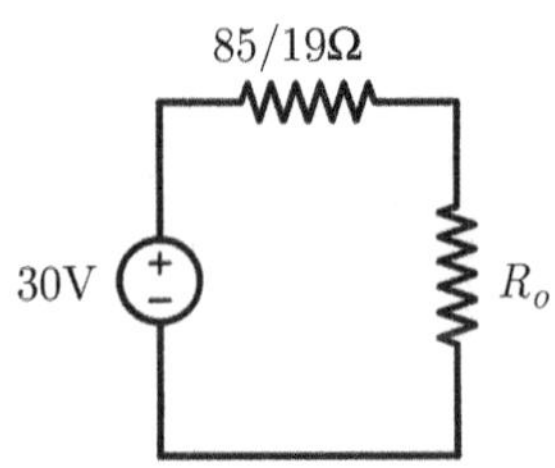

Solution to Exercise 86:

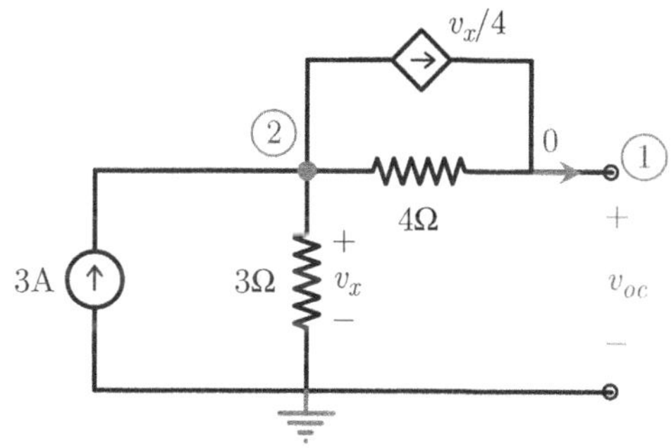

For v_{oc}:

- $v_x = v_2$
- KCL(1): $v_x/4 + (v_2 - v_1)/4 = 0 \longrightarrow v_1 = 2v_2$
- KCL(2): $3 - v_2/3 - (v_2 - v_1)/4 - v_2/4 = 0 \longrightarrow 3v_1 - 10v_2 = -36$
- Then $v_2 = 9\,\text{V}$, $v_1 = 18\,\text{V}$, and $v_{oc} = v_1 = 18\,\text{V}$

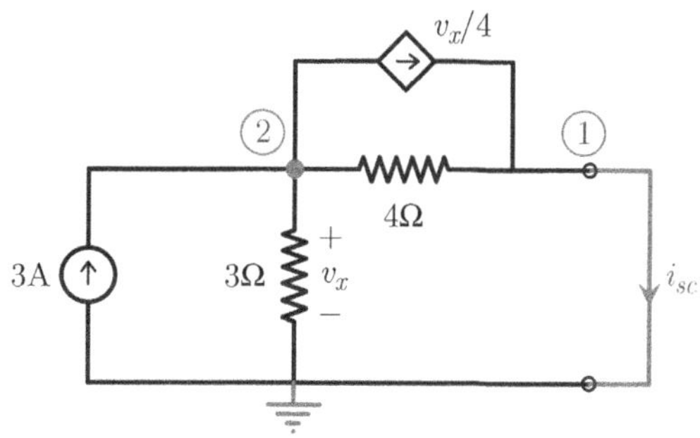

For i_{sc}:

- $v_x = v_2$ and $v_1 = 0\,\text{V}$
- KCL(2): $3 - v_2/3 - v_2/4 - v_2/4 = 0 \longrightarrow v_2 = 18/5\,\text{V}$
- Then $i_{sc} = v_2/4 + v_2/4 = 9/5\,\text{A}$ and $R_{th} = 10\,\Omega$
- Therefore, choose $R_o = 10\,\Omega$ and $p_o = 9^2/10 = 81/10\,\text{W}$

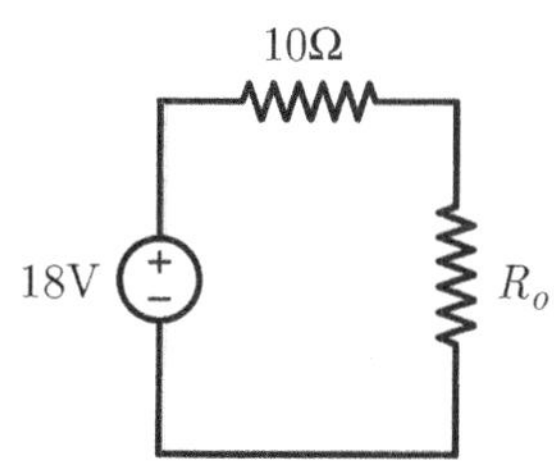

Solution to Exercise 87:

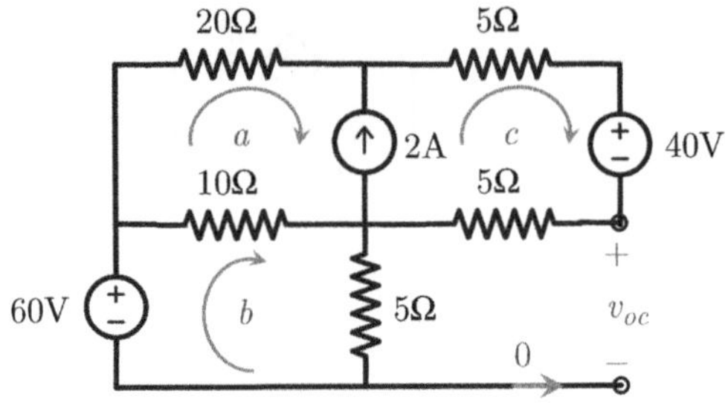

For v_{oc}:

- $i_c - i_a = 2$
- KVL($a\&c$): $20i_a + 5i_c + 40 + 5i_c + 10(i_a - i_b) = 0 \longrightarrow 3i_a - i_b + i_c = -4$
- KVL(b): $-60 + 10(i_b - i_a) + 5i_b = 0 \longrightarrow -2i_a + 3i_b = 12$
- Then $4i_a - i_b = -6 \longrightarrow i_a = -3/5\,\text{A},\ i_b = 18/5\,\text{A},\ i_c = 7/5\,\text{A}$
- And $v_{oc} = 5(i_b + i_c) = 25\,\text{V}$

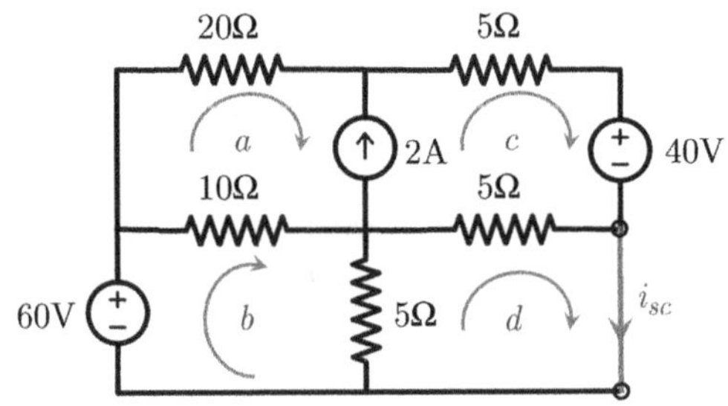

For i_{sc}:

- $i_c - i_a = 2$
- KVL($a\&c$): $20i_a + 5(i_c - i_d) + 40 + 5i_c + 10(i_a - i_b) = 0$
  $\longrightarrow 6i_a - 2i_b + 2i_c - i_d = -8$
- KVL(b): $-60 + 10(i_b - i_a) + 5(i_b - i_d) = 0 \longrightarrow -2i_a + 3i_b - i_d = 12$
- KVL(d): $5(i_d - i_b) + 5(i_d - i_c) = 0 \longrightarrow -i_b - i_c + 2i_d = 0$
- Then, $i_a = 2/5\,\text{A},\ i_b = 28/5\,\text{A},\ i_c = 12/5\,\text{A}$, and $i_d = 4\,\text{A}$
- And $i_{sc} = i_d = 4\,\text{A}$ and $R_{th} = 25/4\,\Omega$
- Therefore, choose $R_o = 25/4\,\Omega$ and $p_o = [(25/2)^2]/(25/4) = 25\,\text{W}$

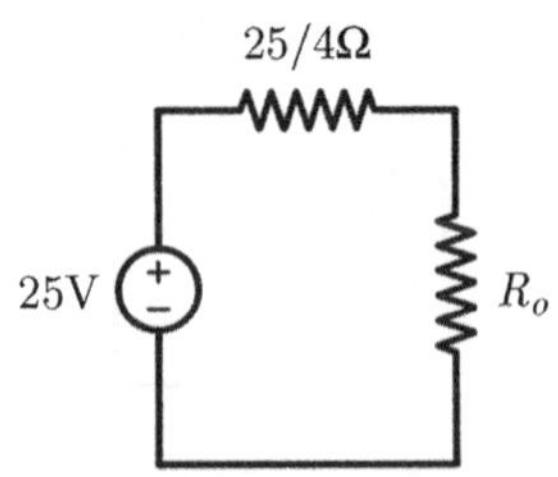

Solution to Exercise 88:

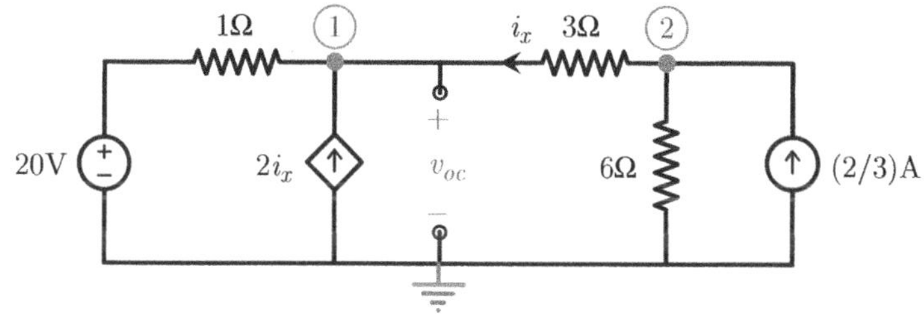

For v_{oc}:

- $i_x = (v_2 - v_1)/3$
- KCL(1): $(20 - v_1)/1 + 2i_x + (v_2 - v_1)/3 = 0 \longrightarrow 2v_1 - v_2 = 20$
- KCL(2): $2/3 - v_2/6 - (v_2 - v_1)/3 = 0 \longrightarrow 2v_1 - 3v_2 = -4$
- Then $v_1 = 16\,\text{V}$, $v_2 = 12\,\text{V}$, and $v_{oc} = 16\,\text{V}$

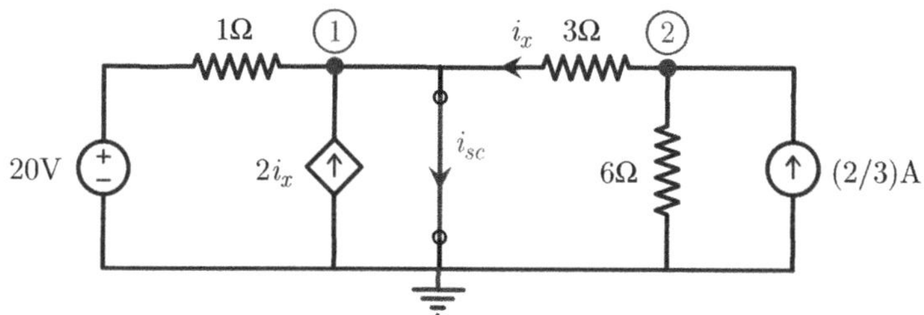

For i_{sc}:

- $v_1 = 0$
- $i_x = (2/3) \times (2/3) = 4/9\,\text{A}$
- KCL(1): $(20 - v_1)/1 + 2i_x + i_x - i_{sc} = 0 \longrightarrow i_{sc} = 20 + 4/3 = 64/3\,\text{A}$
- Then $R_{th} = 3/4\,\Omega$
- Therefore, choose $R_o = 3/4\,\Omega$ and $p_o = 8^2/(3/4) = 256/3\,\text{W}$

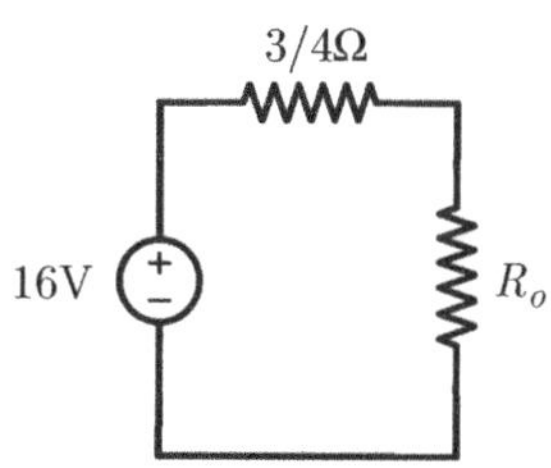

Solution to Exercise 89:

The Thévenin equivalent circuit seen by R_o can be found by recognizing the rest of the circuit as a Norton equivalent with $i_{sc} = i_o$ and $R_{th} = R_1 \parallel R_2$. Then we have

$$v_{oc} = R_{th} i_{sc} = \frac{R_1 R_2}{R_1 + R_2} i_o,$$

and one must select $R_o = R_{th} = R_1 \parallel R_2$ for maximum power transfer. For this selection, the voltage across R_o can be found to be

$$v_o = \frac{v_{oc}}{2} = \frac{R_1 R_2}{R_1 + R_2} \frac{i_o}{2}.$$

Finally, the transferred power can be obtained as

$$p_o^{max} = \frac{v_o^2}{R_o} = \frac{R_1^2 R_2^2}{(R_1 + R_2)^2} \frac{i_o^2}{4} \frac{1}{R_o} = \frac{R_1 R_2}{4(R_1 + R_2)} i_o^2.$$

Chapter 6

Solution to Exercise 90:

- $0 \le t \le 3$ s: $v_C(t) = 0 + \int_0^t (-t'^2 + 4t')dt' = \left(-\frac{t^3}{3} + 2t^2\right)$ V
  and $v_C(3) = 9$ V
- $3 \le t \le 6$ s: $v_C(t) = 9 + \int_3^t (-5t'/6 + 11/2)dt' = \left(-\frac{5t^2}{12} + \frac{11t}{2} - \frac{15}{4}\right)$ V
  and $v_C(3) = \frac{9}{2}$ V
- $3 \le t \le 4$ s: $v_C(t) = 9/2 + \int_3^t (t' - 2)dt' = \left(\frac{t^2}{2} - 2t + 6\right)$ V
  and $v_C(6) = \frac{57}{4}$ V
- $6 \le t \le 8$ s: $v_C(t) = 57/4 + \int_6^t (1/2)dt' = \left(\frac{t}{2} + \frac{45}{4}\right)$ V
  and $v_C(8) = \frac{61}{4}$ V
- $t \ge 8$ s: $v_C(t) = \frac{61}{4}$ V

Solution to Exercise 91:

$$i_C(t) = C\frac{dv_C(t)}{dt} = \begin{cases} 2t, & 0 \le t \le 2 \text{ s} \\ 1, & 2 \le t \le 5 \text{ s} \\ -2t + 12, & 5 \le t \le 6 \text{ s} \\ 0, & t \ge 6 \text{ s} \end{cases} \quad \text{(A)}.$$

Solution to Exercise 92:

- $0 \le t \le 1$ s: $v_C(t) = 1 + \int_0^t 2dt' = (2t+1)$ V and $v_C(1) = 3$ V
- $1 \le t \le 2$ s: $v_C(t) = 3 + \int_1^t (2t'-4)dt' = (t^2 - 4t + 6)$ V and $v_C(2) = 2$ V
- $t \ge 2$ s: $v_C(t) = 2$ V
- $0 \le t \le 1$ s: $p_C(t) = i_C(t)v_C(t) = 2(2t+1)$ W
- $1 \le t \le 2$ s: $p_C(t) = i_C(t)v_C(t) = (2t-4)(t^2 - 4t + 6)$ W
- $t \ge 2$ s: $p_C(t) = 0$

Solution to Exercise 93:

- $0 \le t \le 1$ s: $v_C(t) = 0 + \frac{1}{2}\int_0^t 5t'dt' = (5t^2/4)$ V and $v_C(1) = 5/4$ V
- $1 \le t \le 2$ s: $v_C(t) = 5/4 + \frac{1}{2}\int_1^t 5dt' = (5t/2 - 5/4)$ V and $v_C(2) = 15/4$ V
- $2 \le t \le 3$ s: $v_C(t) = 15/4 + \frac{1}{2}\int_2^t (9 - t'^2)dt' = (-t^3/6 + 9t/2 - 47/12)$ V and $v_C(3) = 61/12$ V
- $t \ge 3$ s: $v_C(t) = 61/12$ V
- Then, for $t \ge 3$ s: $w_C(t) = \frac{1}{2}C[v_C]^2 = (61/12)^2$ J

Solution to Exercise 94:

- $0 \le t \le 1$ s: $v_C(t) = 0 + \int_0^t t'dt' = (t^2/2)$ V and $v_C(1) = 1/2$ V
- $1 \le t \le 3$ s: $v_C(t) = 1/2 + \int_1^t 1dt' = (t - 1/2)$ V and $v_C(3) = 5/2$ V
- $3 \le t \le 4$ s: $v_C(t) = 5/2 + \int_3^t (-1)dt' = (-t + 11/2)$ V and $v_C(3) = 3/2$ V
- $t \ge 4$ s: $v_C(t) = 3/2$ V

$$w_C(t) = \frac{1}{2}C[v_C(t)]^2 = \begin{cases} t^4/8, & 0 \le t \le 1 \text{ s} \\ t^2/2 - t/2 + 1/8, & 1 \le t \le 3 \text{ s} \\ t^2/2 - 11t/2 + 121/8, & 3 \le t \le 4 \text{ s} \\ 9/8, & t \ge 4 \text{ s} \end{cases} \quad \text{(W)}.$$

Solution to Exercise 95:

$$v_L(t) = L\frac{di_L(t)}{dt} = \begin{cases} 400, & 0 < t < 4 \text{ s} \\ 50, & 4 < t < 8 \text{ s} \\ 0, & t > 8 \text{ s} \end{cases} \quad \text{(V)}.$$

$$p_L(t) = v_L(t)i_L(t) = \begin{cases} 16\,000t, & 0 < t < 4 \text{ s} \\ 250t + 7000, & 4 < t < 8 \text{ s} \\ 0, & t > 8 \text{ s} \end{cases} \quad \text{(W)}.$$

$$w_L(t) = \frac{1}{2}L[i_L(t)]^2 = \begin{cases} 8t^2, & 0 \le t \le 4 \text{ s} \\ 0.125t^2 + 7t + 98, & 4 \le t \le 8 \text{ s} \\ 162, & t \ge 8 \text{ s} \end{cases} \quad \text{(kJ)}.$$

Solution to Exercise 96:

- Finding the Thévenin circuit seen by the capacitor:

  $v_{oc} = [30/(2 + 2 \parallel 15)] \times 2/17 \times 10 = 75/8\,\text{V}$
  $i_{sc} = [30/(2 + 2 \parallel 5)] \times 2/7 = 5/2\,\text{A} \longrightarrow R_{th} = 15/4\,\Omega$

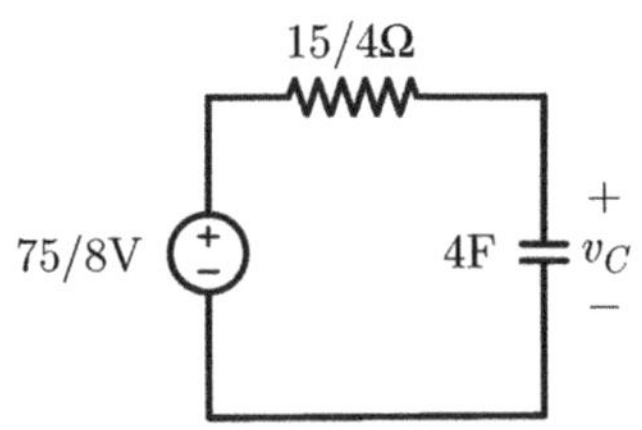

- For $t > 0$, this is an RC circuit with $RC = 15$ s
- Then $v_C(t) = -9 \exp(-t/15) + 75/8\,\text{V}$

Solution to Exercise 97:

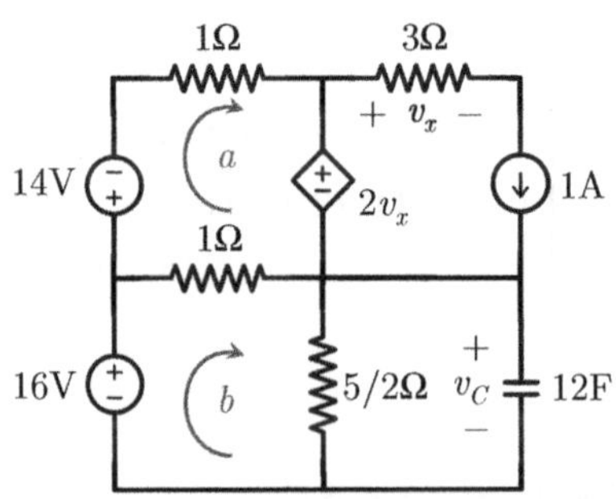

For v_{oc} seen by the capacitor:

- $v_x = 3\,\text{V}$
- $\text{KVL}(a)$: $14 + i_a + 2v_x + (i_a - i_b) = 0 \longrightarrow 2i_a - i_b = -20$
- $\text{KVL}(b)$: $-16 + (i_b - i_a) + (5/2)i_b = 0 \longrightarrow -2i_a + 7i_b = 32$
- Then $i_a = -9\,\text{A}$, $i_b = 2\,\text{A}$, and $v_{oc} = (5/2)i_b = 5\,\text{V}$

For i_{sc} seen by the capacitor: v_x and KVL(a) are the same as above.

- $\text{KVL}(b)$: $-16 + (i_b - i_a) = 0 \longrightarrow -2i_a + 2i_b = 32$
- Then $i_a = -4\,\text{A}$, $i_b = 12\,\text{A}$, and $i_{sc} = i_b = 12\,\text{A}$
- And $R_{th} = v_{oc}/i_{sc} = 5/12\,\Omega$

Equivalent circuit:

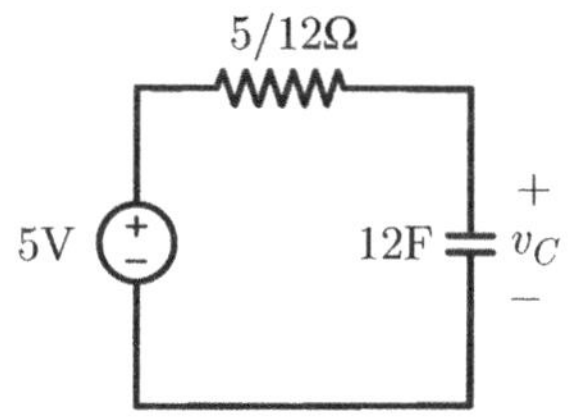

- $v_C(t) = 5\exp(-t/5) + 5\,\text{V}$
- $i_C(t) = C\dfrac{dv_C(t)}{dt} = -12\exp(-t/5)\,\text{A}$
- $t = 5\ln(2)\,\text{s} \longrightarrow v_C(t) = 5\exp(-\ln(2)) + 5 = 15/2\,\text{V}$
  $\longrightarrow w_C(t) = 675/2\,\text{W}$

Solution to Exercise 98:

The equivalent circuit was found in the previous exercise (the only difference is the capacitor instead of the inductor):

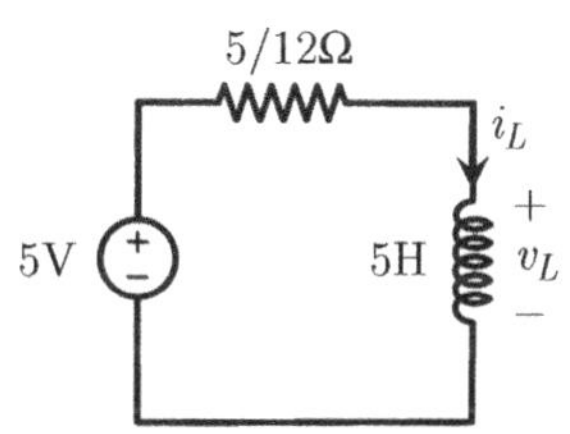

- $i_L(t) = -4\exp(-t/12) + 12\,\text{A}$
- $v_L(t) = L\dfrac{di_L(t)}{dt} = (5/3)\exp(-t/12)\,\text{V}$
- $p_L(t) = (-20/3)\exp(-t/12)[\exp(-t/12) - 3]\,\text{W}$
- Then $p_L(12) = (20/3)\exp(-1)[3 - \exp(-1)] = \dfrac{20(3e-1)}{3e^2} \approx 6.4554\,\text{W}$

Solution to Exercise 99:

Before the switch is opened:

- $i_{4\Omega} = 6\,\text{A}$
- $i_{12\Omega} = 2\,\text{A}$
- $i_{3\Omega} = 4\,\text{A}$
- $v_{1\text{F}} = 3 \times 4 = 12\,\text{V}$

After the switch is opened:

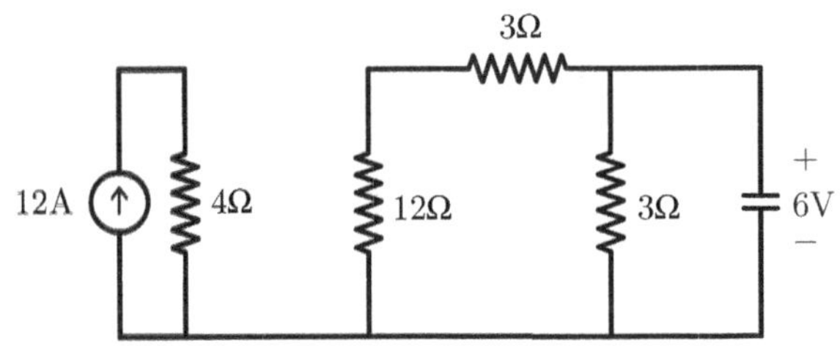

- $v_{12\Omega} = (12/15) \times 12 = 48/5\,\text{V}$

Solution to Exercise 100:

After the switch is closed, we have the following circuit.

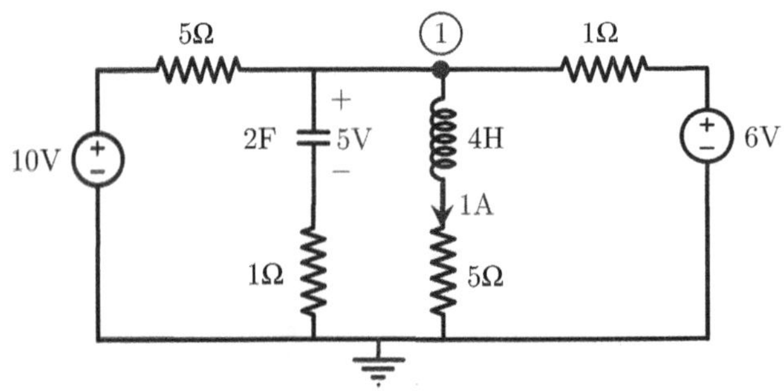

- KCL(1): $(10 - v_1)/5 - (v_1 - 5)/1 - 1 - (v_1 - 6)/1 = 0 \longrightarrow v_1(0^+) = 60/11\,\text{V}$
- Then $i_{6\text{V}}(0^+) = v_1(0^+) - 6 = -6/11\,\text{A}$
- Therefore, $p_{6\text{V}} = 6 \times (-6/11) = -36/11\,\text{W}$

Solution to Exercise 101:

Before the switch is opened ($t = 0^-$):

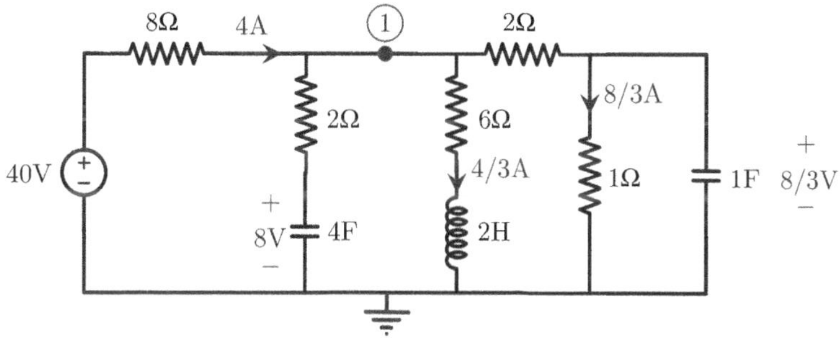

- $v_1(0^-) = 40 - 4 \times 8 = 8\,\text{V}$
- $v_{4\text{F}}(0^-) = v_1(0^-) = 8\,\text{V}$
- $v_{1\text{F}}(0^-) = 8/3\,\text{V}$
- $i_{2\text{H}}(0^-) = 4 \times (3/9) = 4/3\,\text{A}$

After the switch is opened ($t = 0^+$):

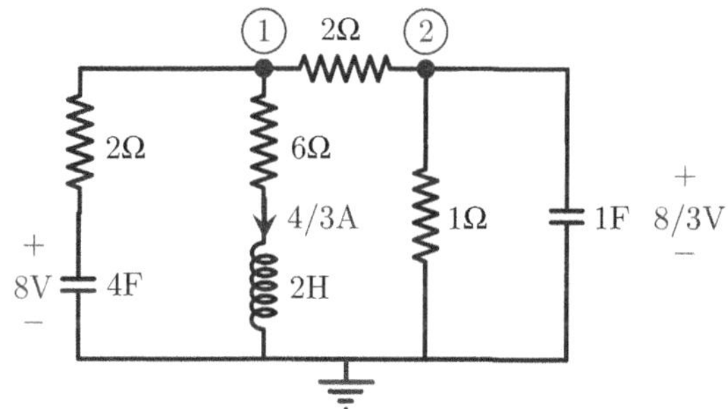

- $v_2(0^+) = 8/3\,\text{V}$
- KCL(1): $-(v_1 - 8)/2 - 4/3 - (v_1 - 8/3)/2 = 0 \longrightarrow v_1(0^+) = 4\,\text{V}$
- KCL(2): $(v_1 - v_2)/2 - v_2/1 - i_{1\text{F}} = 0 \longrightarrow i_{1\text{F}}(0^+) = 2/3 - 8/3 = -2\,\text{A}$
- $v_{2\text{H}}(0^+) = 4 - 6 \times 4/3 = -4\,\text{V} \longrightarrow p_{2\text{H}}(0^+) = (-4) \times 4/3 = -16/3\,\text{W}$
- $p_{1\text{F}}(0^+) = (8/3) \times (-2) = -16/3\,\text{W}$
- $i_{4\text{F}}(0^+) = [v_1(0^+) - 8)]/2 = -2\,\text{A} \longrightarrow \dfrac{dv_{4\text{F}}}{dt}\Big|_{t=0^+} = \dfrac{i_{4\text{F}}(0^+)}{4} = -\dfrac{1}{2}\,\text{V/s}$

Solution to Exercise 102:

Before the switch is opened ($t = 0^-$):

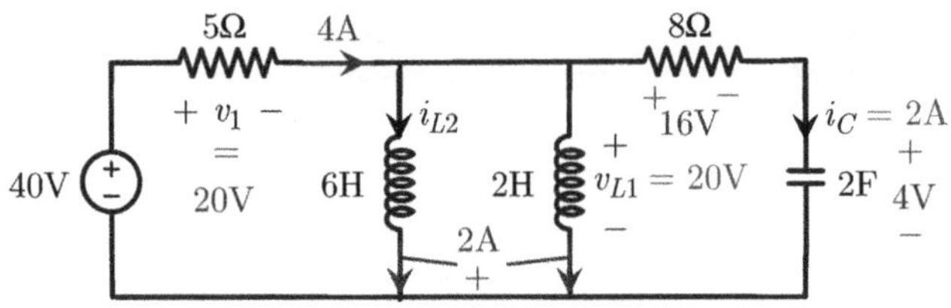

After the switch is opened ($t = 0^+$):

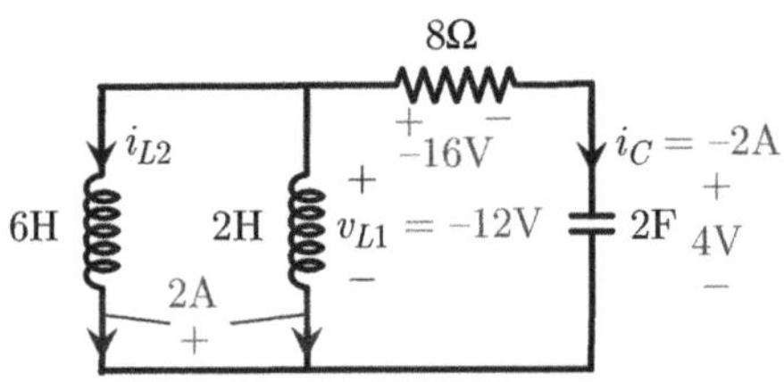

- $i_C(0^+) = -i_{L1}(0^+) - i_{L2}(0^+) = -2\,\text{A}$
- $v_L(0^+) = -16 + 4 = -12\,\text{V}$

Solution to Exercise 103:

After the switch is changed ($t = 0^+$):

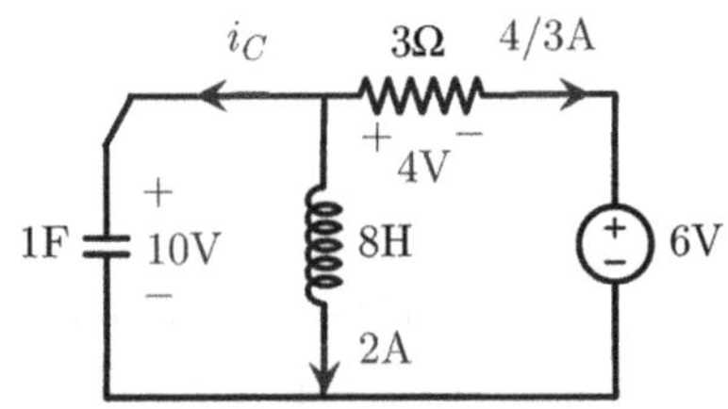

- $i_C(0^+) = -2 - 4/3 = -10/3\,\text{A}$
- $p_C(0^+) = (-10/3)10 = -100/3\,\text{W}$
- $p_L(0^+) = 20\,\text{W}$

Chapter 7

Solution to Exercise 104:

- $3\cos(1000t + 45°)\text{V} \longrightarrow \frac{3}{\sqrt{2}}\underline{/(\pi/4)}\,\text{V}$
- $0.2\ \text{H} \longrightarrow j \times 1000 \times 0.2 = j200\ \Omega$
- $0.4\ \text{H} \longrightarrow j \times 1000 \times 0.4 = j400\ \Omega$
- $0.6\ \text{H} \longrightarrow j \times 1000 \times 0.6 = j600\ \Omega$
- $2\ \mu\text{F} \longrightarrow 1/(j \times 1000 \times 2 \times 10^{-6}) = -j500\ \Omega$
- $4\ \mu\text{F} \longrightarrow 1/(j \times 1000 \times 4 \times 10^{-6}) = -j250\ \Omega$

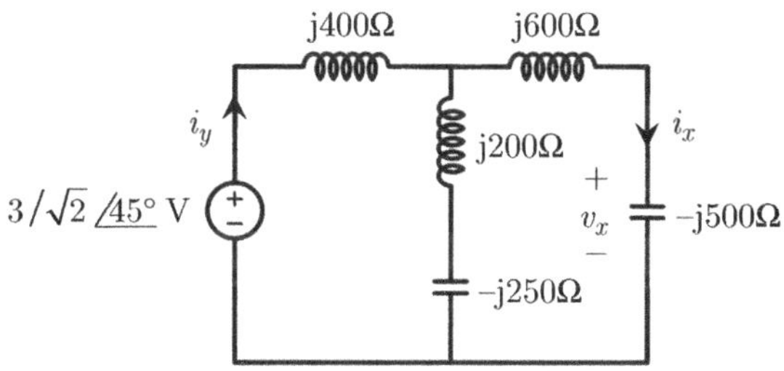

- $Z_{\text{in}} = j400 + (j200 - j250) \parallel (j600 - j500) = j400 - j100 = j300\ \Omega$
- $\underline{i}_y = \frac{3/\sqrt{2}\underline{/45°}}{Z_{\text{in}}} = \frac{3(1+j)}{j300} = \frac{-1+j}{100}\ \text{A}$
- $\underline{i}_x = -j50/(-j50 + j100)\underline{i}_y = -\underline{i}_y = \frac{1-j}{100}\ \text{A}$
- $\underline{v}_x = -j500\underline{i}_y = 5(-1-j) = 5\sqrt{2}\underline{/5\pi/4}\,\text{V}$
- Then $v_x(t) = 10\cos(1000t + 5\pi/4)\ \text{V}$

Solution to Exercise 105:

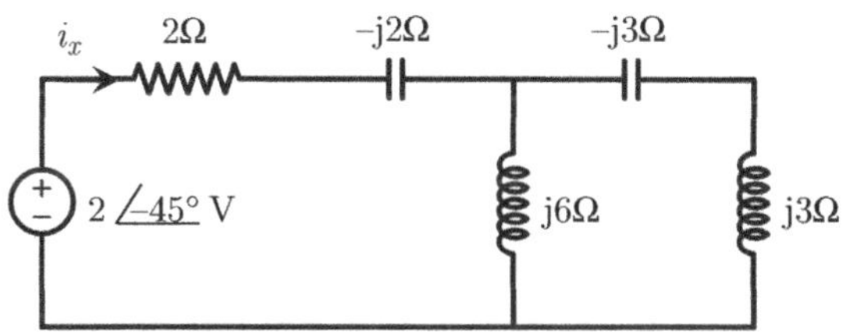

- $Z_{in} = 2 - j2 + 6 \parallel (-j3 + j3) = (2 - 2j) = 2\sqrt{2}\underline{/(-\pi/4)}\,\Omega$
- $\underline{i}_x = 2\underline{/(-\pi/4)}/Z_{in} = \frac{1}{\sqrt{2}}\underline{/0}\,\text{A}$
- Then $i_x(t) = \frac{\sqrt{2}}{\sqrt{2}}\cos(3t) = \cos(3t)\,\text{A}$

Solution to Exercise 106:

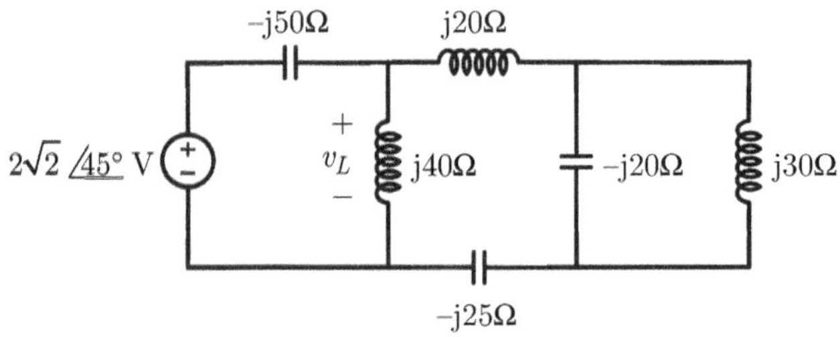

- $Z_{in} = -j50 + j40 \parallel [-j5 + (-j20 \parallel j30)] = -j50 + j40 \parallel (-j65) = j54\,\Omega$
- $i_s = \frac{2+2j}{j54}\,\text{A}$ and $i_L = \frac{-j65}{j40 - j65} \times i_s = \frac{13}{5} \times \frac{(2+j2)}{j54}\,\text{A}$
- Then $v_L = j40 \times i_L = \frac{104}{27}(1+j)\,\text{V}$
- And $v_L(t) = \frac{208}{27}\cos(1000t + \pi/4)\,\text{V}$

Solution to Exercise 107:

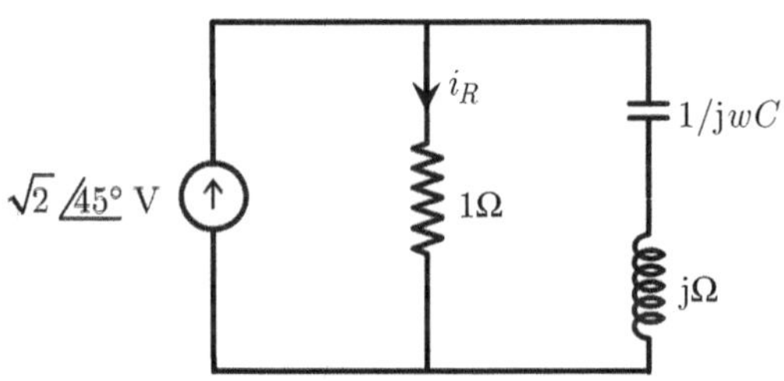

For $C = 20\,\text{mF}$:

- $Z_{CL} = -j + j = 0$
- Then $\underline{i}_R = 0\,\text{A}$ and $i_R(t) = 0\,\text{A}$

For $C = 10\,\text{mF}$:

- $Z_{CL} = -2j + j = -j\,\Omega$

- Then $\underline{i}_R = -j/(1-j)\underline{i}_s = -j(-1+j)/(1-j) = j$ A
- Then $i_R(t) = \sqrt{2}\cos(50t + \pi/2)$ A

Solution to Exercise 108:

- $10\cos(1000t)$ A $\longrightarrow 5\sqrt{2}\underline{/0} - 5\sqrt{2}$ A
- 0.5 mF $\longrightarrow 1/(j \times 1000 \times 0.5 \times 10^{-3}) = -j2\ \Omega$
- 4 mH $\longrightarrow j \times 1000 \times 4 \times 10^{-3} = j4\ \Omega$

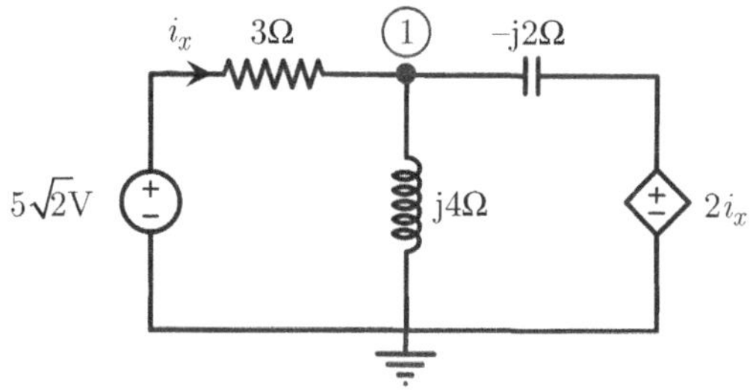

- $\underline{i}_x = \dfrac{5\sqrt{2} - \underline{v}_1}{3}$
- KCL(1): $\dfrac{5\sqrt{2} - \underline{v}_1}{3} - \dfrac{\underline{v}_1}{j4} - \dfrac{\underline{v}_1 - 2\underline{i}_x}{-j2} = 0 \longrightarrow \underline{v}_1 = \dfrac{4\sqrt{2}}{13}(11 - j3)$ V
- Then $\underline{i}_x = \dfrac{7\sqrt{2} + j4\sqrt{2}}{13} = \dfrac{\sqrt{2}}{13}(7 + j4)$ A
- And $i_x(t) = \dfrac{2\sqrt{5}}{\sqrt{13}}\cos(1000t + \tan^{-1}(4/7))$ A

Solution to Exercise 109:

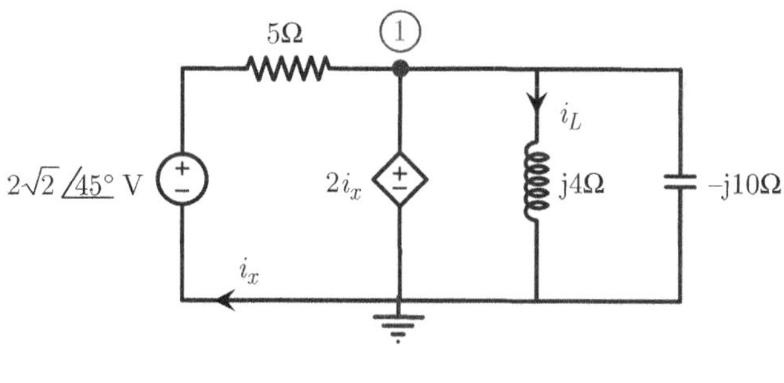

- $\underline{v}_1 = 2\underline{i}_x$ and $2(1+j) - \underline{v}_1 = 5\underline{i}_x \longrightarrow \underline{i}_x = \dfrac{2}{7}(1+j)$ A

- $\underline{i}_L = \frac{2\underline{i}_x}{j4} = \frac{1-j}{7}$ A
- Then $i_L(t) = \frac{2}{7}\cos(2t - \pi/4)$ A

Solution to Exercise 110:

- $10\sqrt{2}\sin(4t)$ V $\longrightarrow 10\underline{/(-\pi/2)}$ V
- 1/16 F $\longrightarrow 1/(j \times 4 \times 1/16) = -j4\,\Omega$
- 0.5 H $\longrightarrow j \times 4 \times 0.5 = j2\,\Omega$

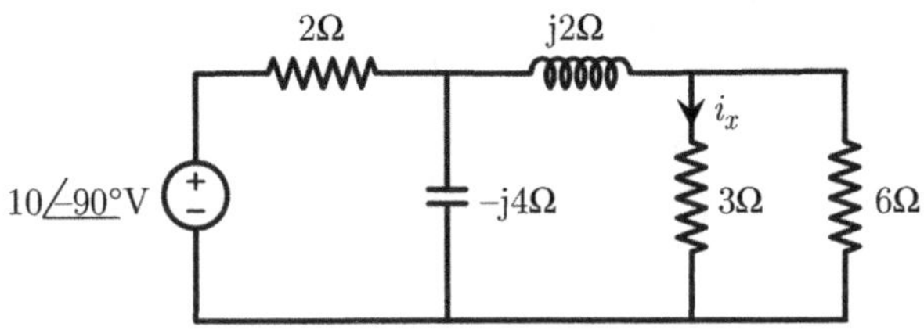

- $Z_{\text{in}} = 2 + [(-4j) \parallel (2j+2)] = 2 - \frac{j4(1+j)}{1-j} = 6\,\Omega$
- $\underline{i}_s = -10j/6 = -5j/3$ A and $\underline{i}_{j2\Omega} = \frac{-j4}{-j4+j2+2} \times \underline{i}_s = \frac{-j5(1-j)}{3}$ A
- Then $\underline{i}_x = \frac{2}{3} \times \underline{i}_{j2\Omega} = \frac{-j10(1-j)}{9} = -\frac{10}{9}(1+j)$ A
- And $i_x(t) = \frac{20}{9}\cos(4t + 5\pi/4)$ A

Solution to Exercise 111:

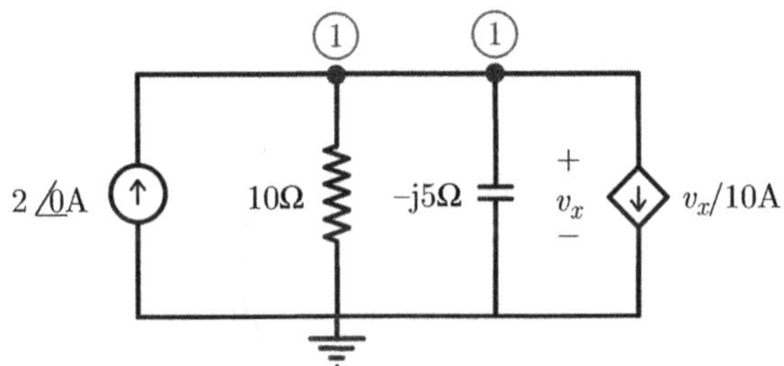

- $\underline{v}_x = \underline{v}_1$
- KCL(1): $2 - \dfrac{\underline{v}_1}{10} - \dfrac{\underline{v}_1}{-5j} - \dfrac{\underline{v}_1}{10} = 0 \longrightarrow \underline{v}_1 = 5(1-j) = 5\sqrt{2}\underline{/(-\pi/4)}\,\text{V}$
- Then $v_x(t) = 10\cos(20t - \pi/4)\,\text{V}$

Solution to Exercise 112:

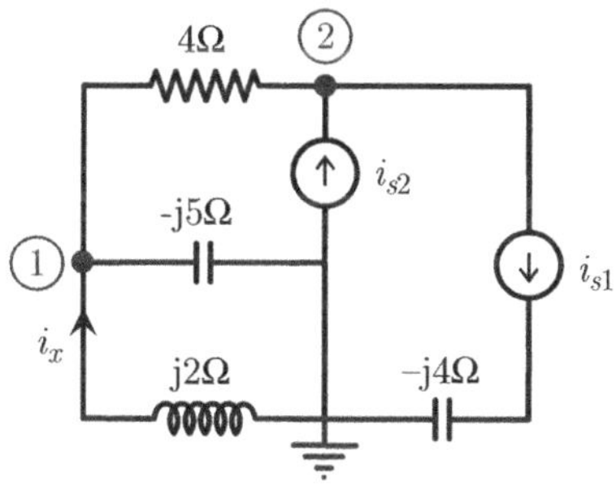

- $i_{s1}(t) = 2\cos(t - 45^\circ)\ \text{A} \longrightarrow \underline{i}_{s1} = \sqrt{2}\underline{/(-\pi/4)} = (1-j)\,\text{A}$
- $i_{s2}(t) = \sqrt{2}\cos(t)\ \text{A} \longrightarrow \underline{i}_{s2} = 1\underline{/0} = 1\,\text{A}$
- KCL(1): $\dfrac{-\underline{v}_1}{2j} + \dfrac{-\underline{v}_1}{-5j} - \dfrac{\underline{v}_1 - \underline{v}_2}{4} = 0 \longrightarrow \underline{v}_2 = \underline{v}_1(1 - j6/5)$
- KCL(2): $\dfrac{\underline{v}_1 - \underline{v}_2}{4} + 1 - (1-j) = 0 \longrightarrow \underline{v}_1 - \underline{v}_2 = -j4$
- Then $\underline{v}_1 = -10/3\,\text{V}$, $\underline{i}_x = -10/(j6) = 5j/3 = 5/3\underline{/\pi/2}\,\text{A}$
- And $i_x(t) = \dfrac{5\sqrt{2}}{3}\cos(t + \pi/2) = -\dfrac{5\sqrt{2}}{3}\sin(t)\,\text{A}$

Solution to Exercise 113:

- $i_{s1}(t) = 12\cos(4t)\ \text{A} \longrightarrow \underline{i}_{s1} = 6\sqrt{2}\underline{/0} = 6\sqrt{2}\,\text{A}$
- $i_{s2}(t) = 8\cos(4t)\ \text{A} \longrightarrow \underline{i}_{s1} = 4\sqrt{2}\underline{/0} = 4\sqrt{2}\,\text{A}$
- $1/16\ \text{F} \longrightarrow 1/(j \times 4 \times 1/16) = -j4\,\Omega$
- $2\ \text{H} \longrightarrow j \times 4 \times 2 = j8\,\Omega$

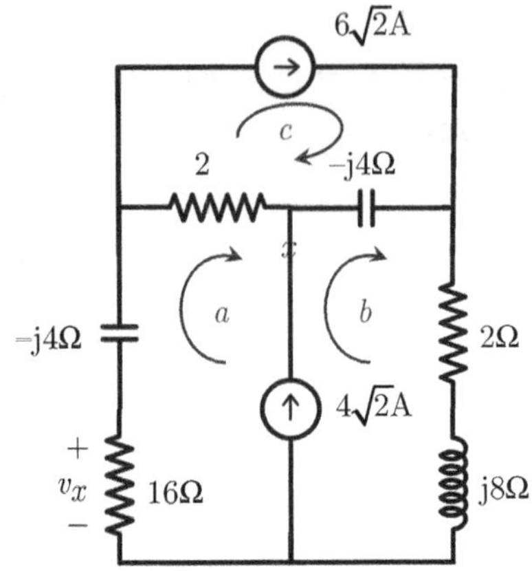

- $i_c = 6\sqrt{2}\,\text{A}$
- KVL(a&b): $2(i_a - i_c) - j4(i_b - i_c) + (2 + j8)i_b + (16 - j4)i_a = 0$
- Supermesh: $i_b = i_a + 4\sqrt{2}$
- Then $i_a = (\sqrt{2}/5 - j2\sqrt{2})\,\text{A}$ and $v_x = (-16\sqrt{2}/5 + j32\sqrt{2})\,\text{V}$
- Then $v_x(t) = \dfrac{32\sqrt{101}}{5}\cos(4t + \pi - \tan^{-1}10)\,\text{V}$

Solution to Exercise 114:

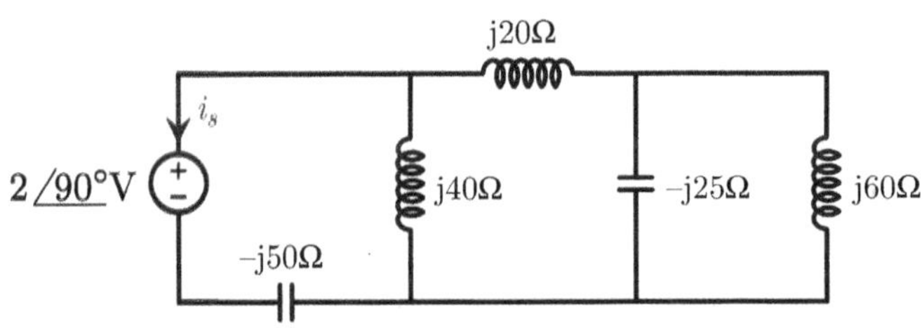

- $Z_{\text{in}} = -j50 + j40 \parallel [j20 + (-j25) \parallel j60] = -j50 + j40 \parallel (-j160/7)$
  $\longrightarrow Z_{\text{in}} = -j310/3\,\Omega$
- $\underline{i}_s = \dfrac{-j2}{-j310/3} = \dfrac{3}{155}\,\text{A}$
- Then $s_s = j2 \times \dfrac{3}{155} = \dfrac{j6}{155}\,\text{VA}$

Solution to Exercise 115:

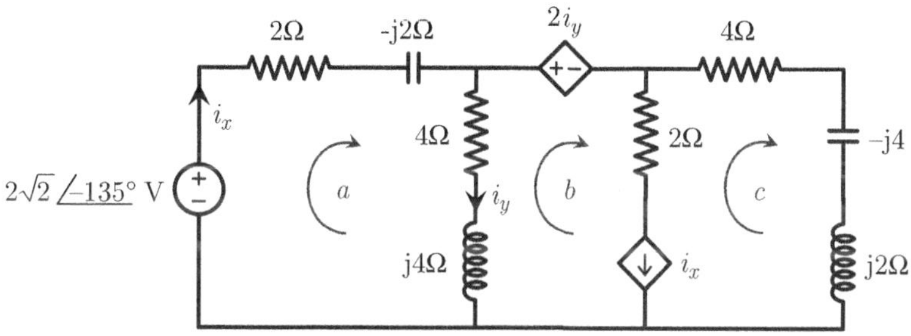

- $\underline{i}_y = \underline{i}_a - \underline{i}_b$
- $\underline{i}_b - \underline{i}_c = \underline{i}_x = \underline{i}_a$
- KVL(a): $2(1+j) + (2-j2)\underline{i}_a + (4+j4)(\underline{i}_a - \underline{i}_b) = 0$
  $\longrightarrow (3+j)\underline{i}_a - (2+j2)\underline{i}_b = -1-j$
- KVL(b&c): $(4+j4)(\underline{i}_b - \underline{i}_a) + 2(\underline{i}_a - \underline{i}_b) + (4-j2)\underline{i}_c = 0$
  $\longrightarrow (-2-j4)\underline{i}_a + (2+j4)\underline{i}_b + (4-j2)\underline{i}_c = 0$
- Then $\underline{i}_a = \underline{i}_b = -j$ A and $\underline{i}_c = 0$
- And $s_{2i_y} = 2\underline{i}_y \times \underline{i}_b^* = 0$
- KVL(b): $(4+j4)(\underline{i}_b - \underline{i}_a) + 2(\underline{i}_a - \underline{i}_b) + 2(\underline{i}_b - \underline{i}_c) + \underline{v}_x = 0 \longrightarrow \underline{v}_x = j2\,\text{V}$
- And $s_{i_x} = \underline{v}_x \times \underline{i}_x^* = j2 \times j = -2\,\text{V A}$
- And $p_{\text{avg},i_x} = -2\,\text{W}$

Solution to Exercise 116:

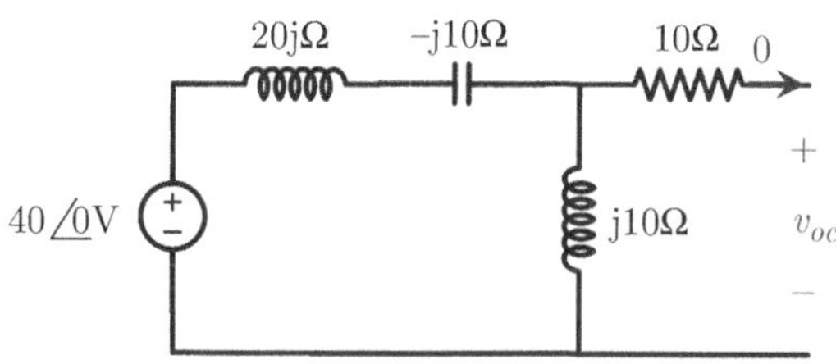

For $\underline{v}_{oc}$:

- $\underline{v}_{oc} = 40(j10/j20) = 20\,\text{V}$

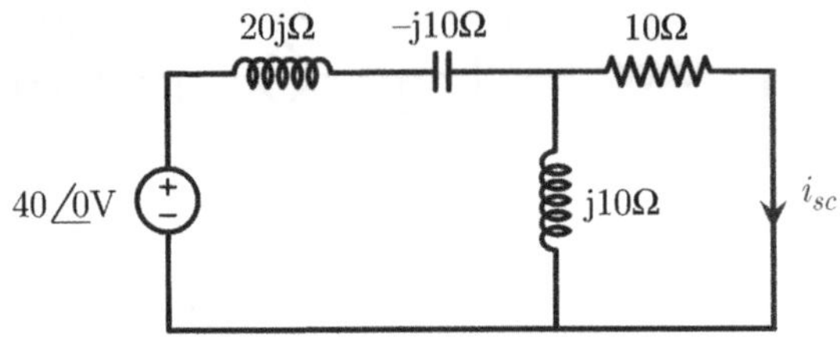

For $\underline{i}_{sc}$:

- $Z_{\text{in}} = j10 + j10 \parallel 10 = j10 + j100/(10 + j10) = j10 + j10/(1 + j)$ $\longrightarrow Z_{\text{in}} = 5(1 + j3)\,\Omega$
- $\underline{i}_{sc} = \dfrac{40}{Z_{\text{in}}}\dfrac{j10}{10 + j10} = \dfrac{j20(1 - j)}{Z_{\text{in}}} = \dfrac{8 - j4}{5}\,\text{A}$
- Then $Z_{\text{th}} = 100/(8 - 4j) = (10 + j5)\,\Omega$
- And choose $Z_l = Z_{\text{th}}^* = (10 - j5)\,\Omega$ for maximum power transfer

Solution to Exercise 117:

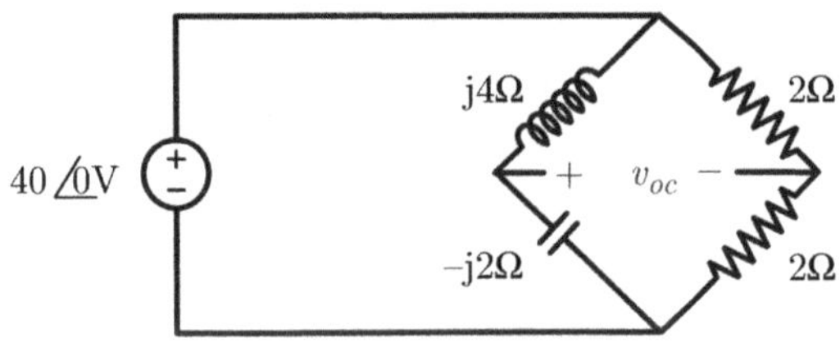

For $\underline{v}_{oc}$:

- $\underline{v}_{oc} = 40[-j2/(-j2 + j4) - 2/(2 + 2)] = 40(-1 - 1/2) = -60\,\text{V}$

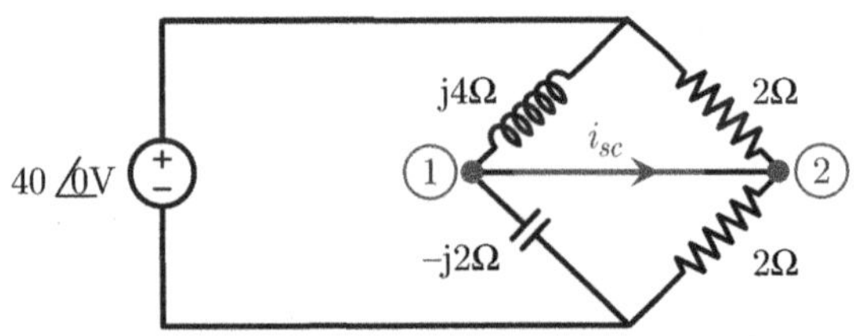

For i_{sc}:

- $Z_{\text{in}} = (2 \parallel j4) + (-j2 \parallel 2) = j8/(2 + j4) - j4/(2 - j2) = (13 - j)/5\,\Omega$
- $\underline{i}_s = 40/Z_{\text{in}} = 200/(13 - j)\,\text{A}$

- $\underline{i}_{j4} = \dfrac{2}{2+4j} \times \underline{i}_s = \dfrac{200}{(1+j2)(13-j)}$ A
- $\underline{i}_{-2j} = \dfrac{2}{2-2j} \times \underline{i}_s = \dfrac{200}{(1-j)(13-j)}$ A
- KCL(1): $\underline{i}_{sc} = \underline{i}_{4j} - \underline{i}_{-2j} = \dfrac{200}{(13-j)} \times \dfrac{3(-1-j3)}{10} = -60\dfrac{(1+j4)}{17}$ A
- Then $Z_{th} = 17/(1+4j) = (1-4j)\,\Omega$
- And choose $Z_l = Z_{th}^* = (1+4j)\,\Omega$ for maximum power transfer
- Considering that $\omega = 2$ rad/s, use a 2 H inductor in series with a 1 Ω resistor

Solution to Exercise 118:

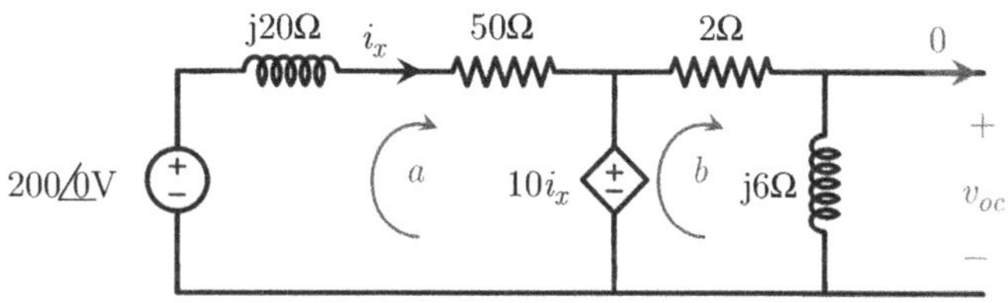

For $\underline{v}_{oc}$:

- $\underline{i}_x = \underline{i}_a$
- KVL(a): $-200 + (50+j20)\underline{i}_a + 10\underline{i}_a = 0 \longrightarrow \underline{i}_a = 10/(3+j) = (3-j)$ A
- KVL(b): $-10\underline{i}_x + (2+j6)\underline{i}_b = 0 \longrightarrow \underline{i}_b = 5\underline{i}_a/(1+j3) = -j5$ A
- Then $\underline{v}_{oc} = j6 \times \underline{i}_b = 30$ V

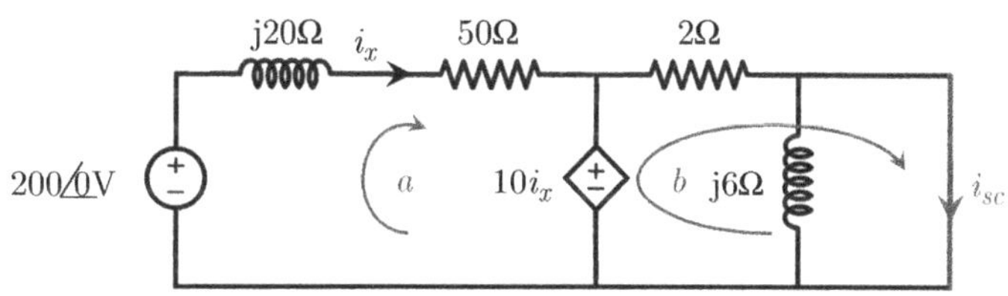

For $\underline{i}_{sc}$:

- $\underline{i}_a = 10/(3+j) = (3-j)$ A (as before)
- KVL(b): $-10\underline{i}_x + 2\underline{i}_b = 0 \longrightarrow \underline{i}_b = 5\underline{i}_a = 5(3-j)$ A
- Then $\underline{i}_{sc} = 5(3-j)$ A and $Z_{th} = 6/(3-j) = 6(3+j)/10\,\Omega$
- And choose $Z_l = Z_{th}^* = \dfrac{9-j3}{5}\,\Omega$
- Considering that $\omega = 50$ rad/s, use a 1/30 F capacitor in series with a 9/5 Ω resistor

Solution to Exercise 119:

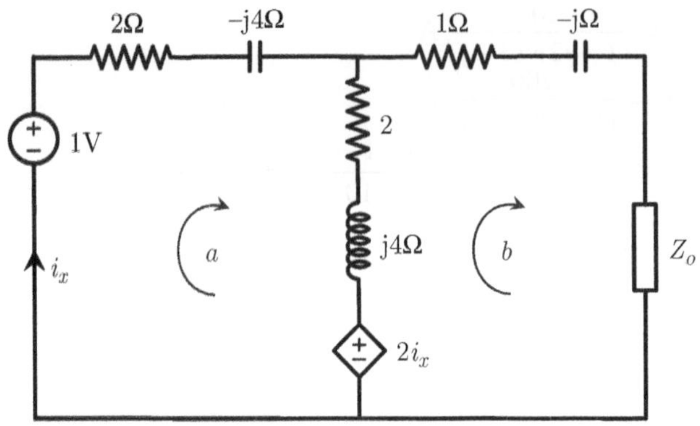

For $\underline{v}_{oc}$:

- KVL(a): $-1 + (2 - j4)\underline{i}_a + (2 + j4)\underline{i}_a + 2\underline{i}_x = 0 \longrightarrow \underline{i}_a = 1/6\,\text{A}$
- $\underline{v}_{oc} = 2\underline{i}_x + (2 + j4)\underline{i}_a = (2 + j2)/3\,\text{V}$

For i_{sc}:

- KVL(a): $-1 + (2 - j4)\underline{i}_a + (2 + j4)(\underline{i}_a - \underline{i}_b) + 2\underline{i}_x = 0$
- KVL(b): $-2\underline{i}_x + (2 + j4)(\underline{i}_b - \underline{i}_a) + (1 - j)\underline{i}_b = 0$
- Then $\underline{i}_{sc} = \underline{i}_b = 2/(5 - j8)\,\text{A}$ and $Z_{\text{th}} = (13/3 - j)\,\Omega$
- And choose $Z_l = Z^*_{\text{th}} = (13/3 + j)\,\Omega$
- Considering that $\omega = 4\,\text{rad/s}$, use a $1/4\,\text{H}$ inductor in series with a $13/3\,\Omega$ resistor

Chapter 8

Solution to Exercise 120:

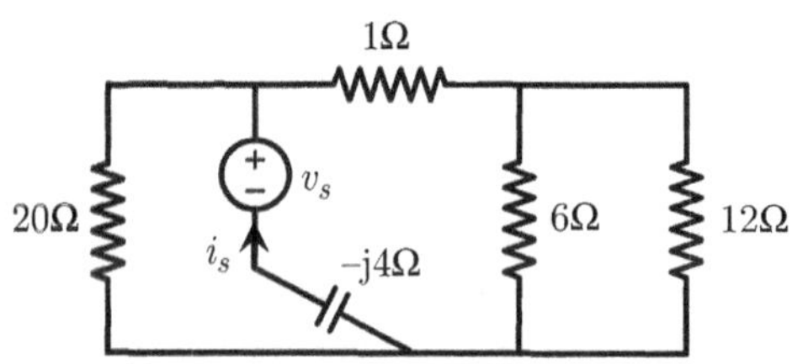

- $\underline{v}_s = 4\sqrt{2}\angle 3\pi/4\,\text{V}$
- $\underline{i}_s = \underline{v}_s/[-j4 + 20 \parallel (1 + 6 \parallel 12)] = \underline{v}_s/(4 - j4) = \dfrac{4\sqrt{2}\angle 3\pi/4}{4\sqrt{2}\angle(-\pi/4)}$
- Then $\underline{i}_s = 1\angle\pi\,\text{A}$
- And $i_s(t) = \sqrt{2}\cos(4t + 180^\circ)\,\text{A}$

Solution to Exercise 121:

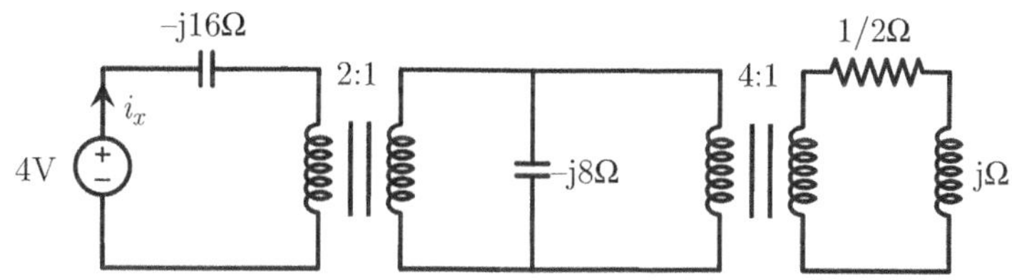

- $(1/2 + j)\,\Omega$ & Transformer$(4 : 1) \longrightarrow (1/2 + j)/(1/16) = 8 + j16\,\Omega$
- $-j8 \parallel (8 + j16) = (4 - j12)\,\Omega$

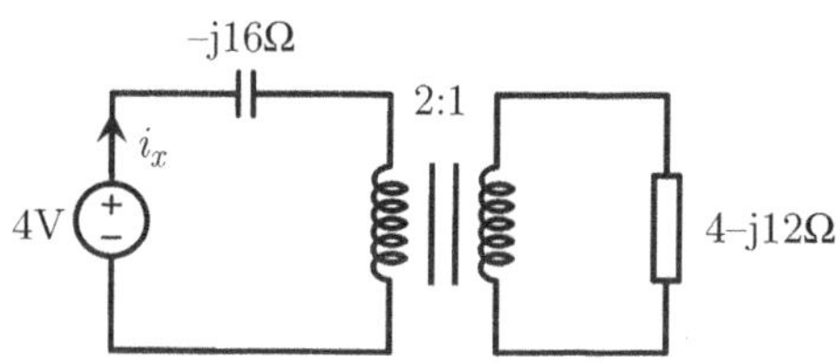

- $(4 - j12)\,\Omega$ & Transformer$(2 : 1) \longrightarrow (4 - j12)/(1/4) = 16 - j48\,\Omega$
- $\underline{i}_x = \underline{v}_x/(-j16 + 16 - j48) = \underline{v}_x/(16 - j64) = 1/(4 - j16) = (1 + j4)/68\,\text{V}$
- Then $i_x(t) = \dfrac{\sqrt{2}}{4\sqrt{17}}\cos(2t + \tan^{-1}4)\,\text{A}$

Solution to Exercise 122:

When $v_{\text{in}}(t) > v_{\text{cl}}$, the diode is on.

- $v_D(t) = 0$
- $i_D(t) = [v_{\text{in}}(t) - v_{\text{cl}}]/R > 0$ (satisfied)
- $v_o(t) = v_{\text{cl}}$

When $v_{in}(t) < v_{cl}$, the diode is off.

- $i_D(t) = 0$
- $v_D(t) = v_{in}(t) - v_{cl} < 0$ (satisfied)
- $v_o(t) = v_{in}$

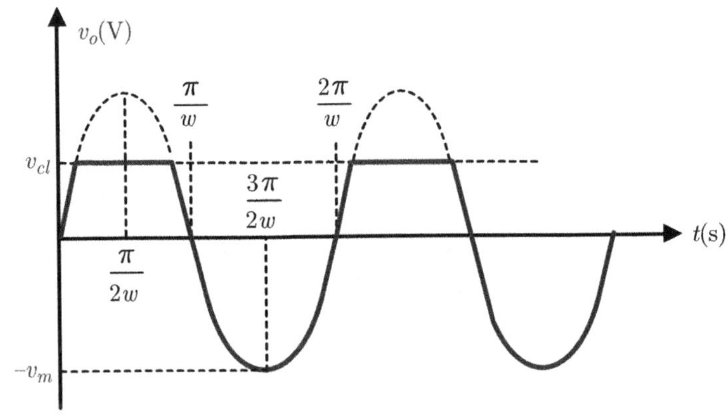

Solution to Exercise 123:

When $v_{in}(t) > v_{c1}$, D_1 is on and D_2 is off.

- $v_{D1}(t) = 0$ and $i_{D2} = 0$
- $i_{D1}(t) = [v_{in}(t) - v_{c1}]/R > 0$ (satisfied)
- $v_{D2}(t) = -v_{c2} - v_{c1} < 0$ (satisfied)
- $v_o(t) = v_{c1}$

When $v_{in}(t) < -v_{c2}$, D_1 is off and D_2 is on.

- $i_{D1} = 0$ and $v_{D2}(t) = 0$
- $v_{D1}(t) = v_{in} - v_{c1} < 0$ (satisfied)
- $i_{D2}(t) = [-v_{c2} - v_{in}(t)]/R > 0$ (satisfied)
- $v_o(t) = -v_{c2}$

When $-v_{c2} < v_{in}(t) < v_{c1}$, both D_1 and D_2 are off.

- $i_{D1}(t) = 0$ and $i_{D2} = 0$
- $v_{D1}(t) = v_{in} - v_{c1} < 0$ (satisfied)

- $v_{D2}(t) = -v_{c2} - v_{c1} < 0$ (satisfied)
- $v_o(t) = v_{in}(t)$

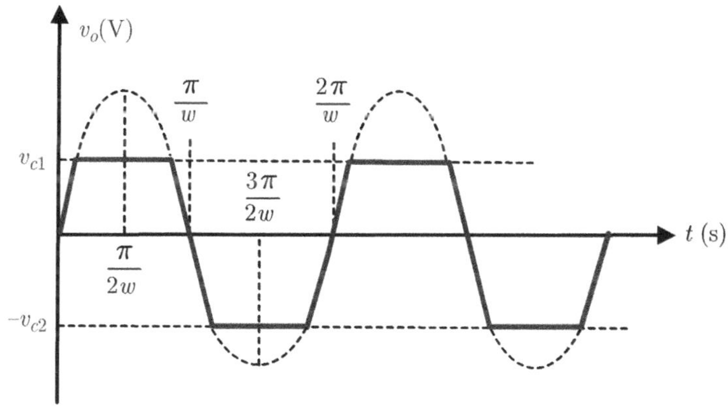

Solution to Exercise 124:

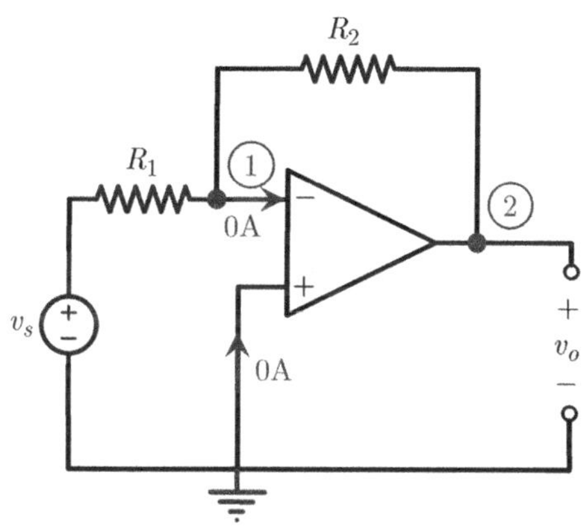

- Ideal OP-AMP: $v_1 = 0$
- KCL(1): $(v_s - v_1)/R_1 - (v_1 - v_2)/R_2 = 0 \longrightarrow v_2 = -(R_2/R_1)v_s$
- Then $v_o = -(R_2/R_1)v_s$

Solution to Exercise 125:

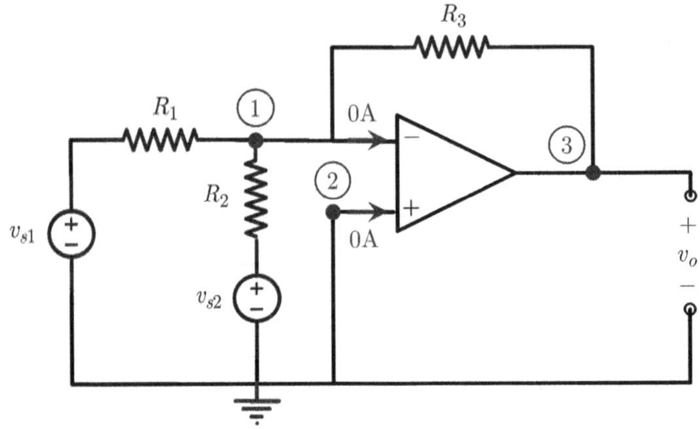

- Ideal OP-AMP: $v_1 = v_2 = 0$
- KCL(1): $(v_{s1} - v_1)/R_1 + (v_{s2} - v_1)/R_2 - (v_1 - v_3)/R_3 = 0$
- Then $v_3/R_3 = -v_{s1}/R_1 - v_{s2}/R_2$
- And $v_o = -(R_3/R_1)v_{s1} - (R_3/R_2)v_{s2}$

Solution to Exercise 126:

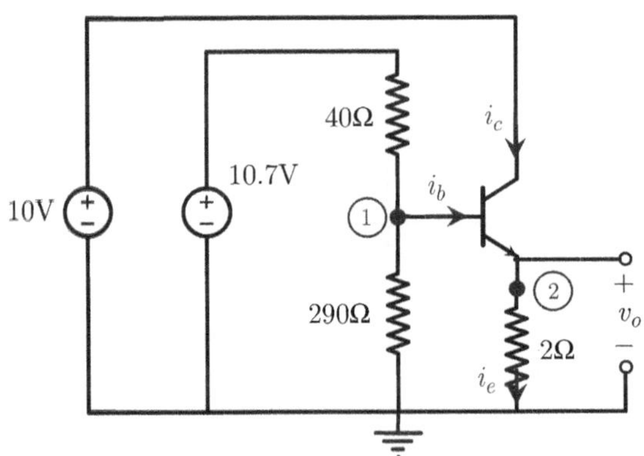

Assuming active mode:

- $i_e = (\beta + 1)i_b = 200i_b$
- $v_2 = 2i_e = 400i_b$
- $v_1 = 0.7 + v_2 = 0.7 + 400i_b$
- KCL(1): $(10.7 - v_1)/40 - v_1/290 - i_b = 0 \longrightarrow v_1 = 8.7\,\text{V}$
- Then $i_b = 0.02\,\text{A}$, $i_e = 4\,\text{A}$, and $v_2 = 8\,\text{V}$
- Check $v_{cb} = 10 - 8.7 = 1.3 > 0$
- And $v_o = 8\,\text{V}$

Index

Introduction to Electrical Circuit Analysis, First Edition. Özgür Ergül.

Companion Website: www.wiley.com/go/ergul4412

Printed and bound by CPI Group (UK) Ltd, Croydon, CR0 4YY

17/09/2026

14964300-0003